**Biological Control of Weeds and Plant Diseases**

# Biological Control of Weeds and Plant Diseases

## *Advances in Applied Allelopathy*

Elroy L. Rice



UNIVERSITY OF OKLAHOMA PRESS : NORMAN AND LONDON

**By Elroy L. Rice**

*Allelopathy* (New York, 1974, 1984)

*Pest Control with Nature's Chemicals* (Norman, 1983)

*Biological Control of Weeds and Plant Diseases: Advances in Applied Allelopathy* (Norman, 1995)

**Library of Congress Cataloging-in-Publication Data**

Rice, Elroy L. (Elroy Leon), 1917–
Biological control of weeds and plant diseases : advances in applied allelopathy / Elroy L. Rice.
p. cm.
Includes bibliographical references (p. ) and index.
ISBN 0-8061-2698-1
1. Weeds—Biological control. 2. Phytopathogenic microorganisms—Biological control. 3. Allelopathy. I. Title.
SB611.5.R53 1994 94-23242
632'.96—dc20 CIP

*Book design by Bill Cason*

The paper in this book meets the guidelines for permanence and durability of the Committee on Production Guidelines for Book Longevity of the Council on Library Resources, Inc. ♾



1 2 3 4 5 6 7 8 9 10

# CONTENTS

# PREFACE

I PREVIOUSLY WROTE two monographs on *Allelopathy* (Rice 1974, 1984), and I deliberated at length whether I should write a new edition or an update covering selected topics that have been researched most actively. A careful survey of some 1500 papers I had collected since my second edition of *Allelopathy* indicated that the most activity has occurred in agriculture and forestry, with considerable research in those areas directed at the biological control of weeds and plant diseases.

Biological control of diseases and other pests is of paramount concern in agriculture and forestry. The development of satisfactory techniques would enable us to lessen or eliminate many of the uses now made of synthetic chemicals, many of which have caused serious pollution and other undesired side-effects. Unfortunately, the United States Department of Agriculture and other funding agencies have not supplied sufficient leadership and funds to push biological control, although it is clear that various allelopathic techniques offer real promise in solving many of the problems involved in the biological control of weeds and plant diseases. Thus, it was decided that an update of the more active areas of allelopathy is needed more than a new edition on the broad subject of allelopathy. This book is based chiefly on references not used in Rice (1984), although a few were published prior to that book.

I am grateful to Susan Barber for her excellent work in word processing the entire manuscript for this book. I am thankful also to Beverly Richey, who did the word processing for all the permissions requests and for the publicity packet requested by the University of Oklahoma Press. I extend my thanks as well to the scientists and copyright holders who gave me permission to use tables and figures from their publications. Finally, I thank the staff at the University of Oklahoma Press who performed the numerous tasks required to publish this book.

ELROY L. RICE

**Biological Control of Weeds and Plant Diseases**

CHAPTER 1

# INTRODUCTION AND ALLELOPATHIC EFFECTS OF CROP PLANTS ON CROP PLANTS

## I. INTRODUCTION

ALLELOPATHY IS relatively new as a science, even though statements were made about the phenomenon over 2000 years ago. The term was coined by Molisch (1937) to refer to biochemical interactions between plants of all kinds, including the microorganisms typically placed in the plant kingdom. These microorganisms include, of course, many microscopic algae, fungi, and bacteria. Molisch indicated in his discussion that he meant the term to include stimulatory as well as inhibitory biochemical interactions. Fortunately, most scientists involved in research in allelopathy around the world have followed Molisch's definition of the term. Only a very few scientists in the United States have deviated from his terminology.

It is important to keep in mind that *allelopathy* involves the addition of a chemical compound or several chemical compounds to the environment, while *competition* involves the removal or reduction of some factor in the environment, factors like water, minerals, food, and light. Muller (1969) recommended use of the term *interference* to refer to the overall influence of one plant on another. Thus interference would include both allelopathy and competition. Since it is still impossible to eliminate allelopathic effects in so-called competition experiments, all such studies should be termed interference experiments. Those readers desiring to learn more about the historical development of allelopathy should read at least the first chapter of Rice (1983).

The rate of allelopathic research has accelerated rapidly in the past two or three decades, particularly in relation to agroecosystems. One of the most exciting areas is in phytopathology, where much progress has been made concerning the chemical signals required for the completion of life cycles of pathogens; the use of microorganisms in biological disease control; and the possible use of pathogens, or phytotoxins produced by pathogens, in weed control. More is being learned about the chemical signals necessary from plant residues to stimulate germination of dormant spores of nonpathogenic fungi, which are important in decomposition of residues and in mineral cycling, in biological disease control, and in other phenomena. The horizons for future research in allelopathy are unlimited, and applications of the knowledge gained should be phenomenal.

A brief note here: Common names have been used in this text for the better-known crop plants, with the scientific name given in parentheses when the crop is first mentioned. On the other hand, for the less commonly known crops, only the scientific names have been used. All the allelopathic crops have been discussed in alphabetical order, except in instances where several allelopathic crop plants have been investigated and reported in the same paper.

## II. ALLELOPATHIC EFFECTS OF CROP PLANTS

Research on the allelopathic potentials of crop plants has been very active in recent years, and I have found several important papers that I missed in my earlier books on allelopathy. Those papers are included in the discussion here.

Mishustin and Naumova reported as early as 1955 that alfalfa (*Medicago sativa*) secretes toxic substances that influence growth of cotton (*Gossyplum hirsutum*) and soil microflora. Guenzi et al. (1964) examined the phytotoxic effects of water extracts of forage samples from Buffalo and Ranger alfalfa at three cuttings, at six stages of growth, and for two years on germination and growth of corn seedlings. Stages of growth and cuttings differed significantly in their effects on root growth, whereas years and cuttings differed significantly in effects on shoot growth. There was no significant effect at any time on percent germination of corn (*Zea mays*). Pedersen (1965) found that alfalfa meal added to an agar medium reduced the percentage germination and the weight and length of the radicle of Acala 4-42 cotton in proportion to the amount of alfalfa meal added. Alfalfa saponin also reduced radicle length in proportion to the amount of saponin used, but the effects on germination were not conclusive.

Webster et al. (1967) reported that alfalfa was growing poorly on some soils in central Alberta, Canada where it previously grew well. Plants were usually short, spindly, yellowish-green, and poorly nodulated. Macro- and micronutrients applied in the field did not markedly improve the situation. Aqueous extracts of "poor" soils depressed growth and nodulation of alfalfa in greenhouse experiments. Liming and nutrient solution increased yields somewhat for all soils, but failed to promote nodulation or produce healthy plants in soil taken from areas of poor growth. They concluded, therefore, that the soils from the areas with poor growth contained a factor that was toxic to alfalfa. McElgunn and Heinrichs (1970) supported the conclusions of Webster et al. in experiments on the "alfalfa-poor soils" of Alberta. It was significant that McElgunn and Heinrichs used Rambler and Alfa cultivars of alfalfa, whereas Webster et al. used Grimm. In a greenhouse experiment Lahontain and Moapa 69 alfalfa cultivars were grown in soil in which alfalfa had grown the previous two years and in a fallow soil of the same origin (Jensen et al., 1981). Dried alfalfa shoots or roots at 0.5% of the weight of the soil were added either before or after steam pasteurization or fumigation. Seedlings of both varieties were smaller when grown in soil in which alfalfa had been grown the previous years when compared with those grown in fallow soil, even though both soils were either steam-pasteurized or fumigated. Addition of ground foliage or roots to both soils after steam pasteurization or fumigation also reduced the size of the plants compared with soils where alfalfa residues were not added. Again it was inferred that alfalfa was autotoxic (allelopathic to itself).

Several thorough and complex experiments were conducted at the University of Illinois to determine the role of allelopathic effects of alfalfa on establishment of alfalfa, whether there are genetic differences among alfalfa cultivars for resistance to autotoxicity, and whether saponin appear to be phytotoxic to alfalfa (Miller, 1983). Great care was taken to keep fertility adequate in all tests, to control weeds, to control other pests, and to control pathogens. It was concluded that there is a release of phytotoxic factors from the previous crop on reestablishing alfalfa, that the best preceding crop at Urbana for this purpose is corn, followed by various small grains and soybeans (*Gly-*

*cine max*), and that the worst preceding crop is alfalfa. There were no apparent differences among seven cultivars tested for resistance to autotoxicity, and there was no evidence supporting saponins as being phytotoxic to alfalfa.

The results of experiments on alfalfa establishment and production with continuous alfalfa at Mead, Nebraska (Kehr et al., 1983), were not as clear-cut as those discussed above. Two alfalfa cultivar experiments were spring-seeded in 1970 on a Sharpsburg silty-clay loam with no known history of prior alfalfa cropping. The alfalfa was allowed to grow until late May 1975, at which time it was plowed again and fallowed until the spring of 1976. Soybeans were seeded in 1976 on subareas reserved for seeding alfalfa in 1977 and 1978, and in 1977 on the subarea reserved for seeding alfalfa in 1978. Two alfalfa experiments were spring-seeded per year in 1976, 1977, and 1978—a foliar fungicide and a fungicide seed treatment. Forage yields of alfalfa did not differ among foliar fungicides or between Arc and Ranger cultivars in the year of seeding or in subsequent years. Moreover, stands did not differ among fungicide seed treatments or between Agate and Dawson cultivars in the year of seeding. Stands in the year of seeding were excellent on the same land area where alfalfa was seeded in 1970 with no known previous alfalfa cropping, after plowing the five-year-old stand of alfalfa in 1975 and seeding in 1976, and after one or two years of soybeans, and seeding alfalfa in 1977 and 1978. Average forage yield of the experiments seeded in 1970 were, however, 26% higher in the year of seeding and 40 and 35% higher, respectively, in the first and second year after seeding than in the experiments seeded in 1976 to 1978. These scientists concluded that there were no problems with stand establishment, and forage yields were at an acceptable level in the year of seeding following plowing and fallowing of a previous five-year old alfalfa stand.

Stand and yield did decline more rapidly, however, under continuous alfalfa than when alfalfa is grown in rotation with annual crops in the silty-clay loam soils at Mead, Nebraska. Tesar (1984) conducted five experiments to determine if alfalfa could be reestablished, without an intervening crop or fallow, by sod seeding or conventional seeding on a plowed alfalfa stand. The experiments were carried out on fertilized loam soils of pH 6.5 or above, apparently in Michigan. The first four experiments indicated that no autotoxicity occurred if an adequate period existed between killing an old stand with glyphosate or plowing an old stand and seeding. In the fifth experiment, alfalfa sod-seeded two days after spraying a one-year-old stand with glyphosate had a relatively poor stand and a lower yield in the first cutting in the next year when compared with alfalfa that was sod-seeded on alfalfa killed with glyphosate the previous fall.

Hegde and Miller (1990) found that growth of alfalfa was significantly reduced on soil previously cropped to alfalfa compared with that cropped to sorghum (*Sorghum bicolor*). Alfalfa germination, however, was not affected by the previous crop. In the case of sorghum, fresh weight per plant was significantly lower when the preceding crop was alfalfa than when it was sorghum. On the other hand, differences in germination and plant height due to preceding crops were not significant. Germination percentage, plant height, and dry weight per plant of alfalfa were all significantly lower with incorporation of alfalfa fresh residue than in the control. Preceding crops did not make a difference in this experiment, except with plant height. The germination percentage decreased with incorporation of roots alone and was further decreased by incorporation of shoots along with roots. These results support other evidence indicating that alfalfa is also allelopathic to plants other than itself.

Oleszek and Jurzysta (1987) found that water and alcohol extracts of alfalfa and red clover (*Trifolium pratense*) roots and crude saponins of alfalfa roots reduced seed germination and seedling growth of wheat (*Triticum aestivum*) and growth of *Trichoderma viride*. The effects of the saponins indicated that these glycosides were the important allelochemicals involved. Powdered alfalfa and red clover roots inhibited wheat growth when incorporated in sand, with the red clover roots being less allelopathic. The wheat root system was completely destroyed by concentrations of alfalfa roots as small as 0.25% (w/w). Read and Jensen (1989) reported that water-soluble substances present in methanol extracts of soil cropped to alfalfa or barley (*Hordeum vulgare*) decreased seedling root length of L-220 alfalfa, Nugaines winter wheat, and Crimson Giant radish. Extracts of crop residue screened from soil cropped to alfalfa or barley significantly reduced seedling root length also, with extracts of the alfalfa residue causing greater inhibition than those of barley residue.

Miller et al. (1988) extracted dried Moapa alfalfa foliage sequentially via percolation with pentane/hexane and 95% ethanol. The ethanol solids were dissolved in methanol-water (1:1) and extracted with hexane. The aqueous phase was adjusted to $H_2O$-methanol (3:1) and extracted with ethyl acetate (EtOAc). The EtOAc solubles produced 14 fractions, 6 of which completely inhibited lettuce (*Lactuca sativa*) seed germination. The first of the active fractions was purified by preparative TLC to yield 30 mg of what was thought to be a single component. GC-MS revealed at least eight components: several unidentified compounds, medicarpin, sativan, and the methoxy analogues of the last two named compounds. Miller et al. concluded that medicarpin and 4-methoxymedicarpin are contributors to the autotoxicity of alfalfa, as medicarpin inhibits germination of alfalfa seeds (59% after six days at $10^{-1}$ *M*) and 4-methoxymedicarpin inhibits growth of alfalfa seedlings 55% after five days at 1.7 x $10^{-3}$ *M*.

In greenhouse studies, Vanguard alfalfa plant material decreased alfalfa emergence by 87 and 62%, respectively, in a Kokomo silty loam previously cropped in alfalfa and corn (Hall and Henderlong, 1989). Autoclaving the soil and plant material eliminated the autoallelopathic effect. The autotoxic compound was water-soluble and was not the direct result of microbial action. It had an $R_f$ in paper chromatography similar to phenolic acid, but the phenolic-absorbent polyvinylpolypyrrolidone did not affect the autotoxic activity. Use of scanning electron microscopy of WL-316 alfalfa roots inhibited by aqueous extracts of alfalfa shoots revealed a 46% reduction in density and a 54% reduction in length of root hairs (Hegde and Miller, 1992). Extract-treated roots also had larger cortical cells, resulting in a greater bulging of the primary root than in controls. The shoot extract did not, however, cause clogging of xylem vessels.

Ells and McSay (1991) air-dried a Nunn clay loam and mixed it with air-dried sand to give a 75% sand and 25% soil medium. This medium was mixed with oven-dried alfalfa roots that had passed through a 1-mm screen to make alfalfa root concentrations from 0 to 0.7% (w/w). Triple-Mech cucumber seeds were planted in these mixtures, and there was a linear decrease in both percentage germination and seedling weight, with both reaching zero when the concentration of alfalfa root reached 0.7% in the sand-soil medium. The alfalfa residue was as inhibitory to newly planted seeds as it was to five-day-old seedlings that were transplanted into the medium. Thus the allelopathic effect of alfalfa root residue on cucumber was clearly demonstrated and calls into question the planting of a cucumber crop in soil previously cropped to alfalfa.

Saponins and canavanine have been shown to be biologically active allelochemicals produced by alfalfa; thus Gorski et al. (1991) studied the time course of accumulation of these materials in alfalfa seedlings. During the first eight days, saponin concentrations in Verko alfalfa rose from zero in seeds to 8.7% in roots and 1.8% in shoots on the eighth day and then slowly decreased to 7.6% in roots and 0.8% in shoots on the twenty-fourth day. Canavanine was found in seeds at a concentration of 1%, then increased to 3.2% in seedlings on the sixth day and rapidly decreased to 0.2% per dry mass in roots and shoots on the twenty-fourth day. No saponins were found in the culture solution during these tests. In contrast, saponin efflux was detected from cut and dead sections of both roots and shoots. Canavanine did move out of the alfalfa roots, however. The saponin, medicagenic-acid sodium salt, was very inhibitory to radicle growth of both *Lepidium* and *Amaranthus.* Canavanine sulfate was very bioactive, inhibiting radicle growth of *Lepidium* by 21% and of *Amaranthus* by 13% at a concentration of 10 $\mu M$. Medicagenic-acid sodium salt was strongly inhibitory to tomato (*Lycopersicon esculentum*) cell growth in tissue culture, even at $10^{-5}$ *M* where all cells were lysed on the fourth day of the experiment. Medicagenic-acid glycosides (also saponins) caused only slight inhibition of tomato cell growth. It appears therefore that saponins and canavanine may be active allelochemicals in at least some allelopathic actions of alfalfa.

Wyman-Simpson et al. (1991) analyzed six cultivars of alfalfa that differed in dormancy rating for numbers and kinds of saponins in the roots during a period from January to August 1988 at the Oklahoma State University agronomy farm in Payne County. They also tested water extracts of the roots for activity against cheat (*Bromus secalinus*), wheat, and *Trichoderma viride.* Wheat and cheat bioassays indicated no significant differences among cultivars, whereas *T. viride* was inhibited most during months of high rainfall and rapid growth. Cheat seedling roots were inhibited 8 to 10% more than roots of wheat, suggesting that extracted alfalfa root saponins were more effective as allelopathic compounds in preventing growth of cheat than wheat plants alone. An average of 14 different saponins per cultivar were separated by TLC. Saponins and the aglycones produced by acid hydrolysis of the May samples were separated by TLC. The presence of medicagenic acid, soyasapogenol B, and hederagenin was shown in all the cultivars.

Miersch et al. (1992) reported that L-canavanine, the structural analog of L-arginine, is found in 26 cultivars of alfalfa, and its concentration ranges from 6 to 16 mg/g of dry seeds. Practically all of the canavanine was stored in the cotyledons. During sprouting, 29% of the canavanine in the cv. Verko was translocated into the hypocotyl and radicle in 24 hours. In the early stage of seedling development, canavanine increased 3-fold, whereas the protein amino acid arginine increased 11-fold and asparagine increased 35-fold. Canavanine and arginine were metabolized rapidly, however, in seedlings grown for 24 days. Canavanine entered the environment of swelled seeds and into the rhizosphere of young seedlings and increased in the growing medium to 3 to 57 $\mu M$. This concentration range inhibited growth of cabbage (*Brassica oleracea*) radicles and tomato cell cultures. These results definitely prove the allelopathic effects of very young alfalfa seedlings and of canavanine, at least under the conditions of the experiments.

Failures in reestablishing commercially profitable asparagus (*Asparagus officinalis*) stands in old asparagus fields in Europe, Japan, and North America have often been reported (Yang, 1982). The major cause of this decline has generally been diagnosed as an accumulation of pathogenic organisms. Yang speculated that at least part of the decline might be due to autoallelopathy, and his experimental results substantiated this hypothesis. Root extract inhibited root and shoot growth of asparagus seedlings; stem and crown extracts reduced root growth but not shoot growth. The highest concentration of extract from crown plus root tissues (5 g tissue/100 mL $H_2O$) inhibited radicle growth and killed the seedlings. Crude aqueous extracts from microorganism-free tissue of cultured plants were as toxic to asparagus seedlings as were extracts of field-grown plants. Young (1984) tested the effects of root exudates of three cultivars of asparagus on the same three cultivars, using a simple donor-receptor pot system. The height and number of stalks of asparagus seedlings of the three cultivars were significantly reduced compared with control seedlings. No significant differences occurred in the growth of seedlings of the three cultivars, and the N, P, and K contents in the tops of asparagus seedlings were not affected by the root exudates. Root exudates were also collected in an XAD-4 resin column and similarly inhibited radicle and shoot growth. These results definitely indicated that asparagus is an autoinhibited species. Subsequent experiments by Young and Chou (1985) demonstrated that asparagus residues also inhibited growth of asparagus seedlings in both vermiculite and sand cultures. Root and stem extracts strongly inhibited growth of asparagus seedlings, and the quantities of total phenolics and catechol-type phenolics from root, stem, and old-root litter extracts corresponded to the autotoxicity in the seed bioassays.

Greenhouse studies at Michigan State University showed that the severity of crown or root rot of asparagus seedlings increased in direct proportion to increased amounts of dried root tissue added to soil along with either *Fusarium oxysporum* f.sp. *asparagi, F. moniliforme,* or a combination of these two pathogens (Hartung et al., 1989). When excised asparagus roots were treated with increasing concentrations of a water extract of dried asparagus root tissues, electrolyte efflux increased, peroxidase activity decreased, and respiration decreased. They concluded that allelochemicals of asparagus may have direct physiological and biochemical effects on asparagus plants that predispose them to *Fusarium* infection. Hazebroek et al. (1989) reported that crude aqueous extracts of *A. officinalis* inhibited germination of tomato and lettuce, with several cultivars of lettuce having varying sensitivities. Cucumber was not affected in germination. The extracts reduced hypocotyl growth in lettuce and shoot growth in asparagus and inhibited radicle elongation in barley, lettuce, and asparagus. Seedling growth in two cultivars of wheat and in tomato was not affected. Aqueous root extracts of *A. racemosis* also inhibited germination and radicle growth in Grand Rapids lettuce. Solvent partitioning, charcoal adsorption, cation exchange, and TLC gave a band that fluoresced under UV light, reacted with a phenolic reagent, and inhibited growth of lettuce and asparagus radicles. Thus these scientists agreed with Young and Chou (1985) that the active allelochemicals in asparagus are probably phenolic compounds.

Hartung et al. (1990) separated several active fractions in aqueous extracts of asparagus roots by HPLC. These included ferulic, isoferulic, malic, citric, and fumaric acids. Soxhlet extraction of the residues also indicated the presence of caffeic acid. An extract from lyophilized fresh root tissue also contained methylenedioxycinnamic acid that was one order of magnitude more active than any compound obtained from dried roots. Miller et al. (1991) identified caffeic acid in methanol extracts of fresh asparagus roots and stated it was the most bioactive compound identified by them.

Wacker et al. (1990) tested effects of ferulic acid on hyphal elongation and colonization of asparagus by the mycorrhiza *Glomus fasciculatum.* Spore germination was not affected, but hyphal elongation decreased significantly with increasing ferulic acid concentration. In greenhouse studies, mycorrhizal colonization of asparagus roots and growth of mycorrhizal asparagus also decreased with increasing ferulic acid concentration. Growth of nonmycorrhizal plants was not affected by ferulic acid. The production of ferulic acid by asparagus apparently reduces the effectiveness of the mycorrhizal fungus and consequently reduces growth of the asparagus.

Pedersen et al. (1991) reported that ferulic, caffeic, and methylenedioxycinnamic (MDC) acids and soil extracts from an asparagus field soil all inhibited colonization of asparagus roots by *G. fasciculatum.* Extracts of nonasparagus soil did not affect colonization of asparagus roots by *Glomus.* The inhibitory effects increased linearly with increasing concentrations of the added substances. Reduction of mycorrhizal plant growth was significantly correlated with increasing concentration of caffeic and MDC acids. In the absence of mycorrhizae, asparagus plant growth was inhibited by ferulic and MDC acids, but not by caffeic acid. The fact that ferulic acid significantly inhibited growth of nonmycorrhizal plants but not mycorrhizal plants indicates that the mycorrhizal fungi may alleviate the phytotoxic effects of ferulic acid and may act as detoxifiers.

As I reported above, Read and Jensen (1989) found that extracts of soil cropped to barley inhibited root growth of alfalfa, wheat, and radish (*Raphanus sativus*). Considerably before that date, Müller-Wilmes and Zoschke (1980) studied the effect of continuous growing of barley on grain, straw, and total yield. They found on a five-year average in a field trial that there was a grain yield reduction of 18% if no N was applied and a 5% decrease with 100 kg/ha of N compared with the first winter barley after potatoes. The minimum grain yields occurred in the second and third years of continuous barley. Examination proved that foot-rot diseases and weeds could not be regarded as causes of the yield reduction; a decrease in root formation at heading time in the early years was instead suggested as the cause of the yield decrease. Moreover, certain phenolic compounds were found to increase in soil cropped to barley during the shooting phase of the second and third crops of continuous winter barley. Evidence indicated that these compounds probably caused the different root formation of barley at heading time, which probably decreased nutrient uptake at this critical time.

Tsuzuki et al. (1977) pointed out that Tsuzuki et al. (1975) had demonstrated that aqueous extracts of shoots and roots of perennial buckwheat (*Fagopyrum cymosum*) inhibited or delayed seed germination of several crop plants and weeds and inhibited growth of rice seedlings. Tsuzuki et al. (1977) found that the inhibitors in buckwheat were readily soluble in water, and separation by paper chromatography and subsequent bioassays indicated there were three inhibitors in water extracts. One of these was identified as abscisic acid, but the others were not identified. Tsuzuki et al. (1987) isolated and identified four long-chain fatty acids in extracts of perennial buckwheat—palmitic, stearic, arachidic, and behenic acids. They were identified by GC-MS and bioassays with rice seedlings indicated that all except for behenic acid were inhibitory to seedling growth at 250 ppm.

A rather limited amount of research has been done recently on allelopathic effects of clovers and hairy vetch. MacFarlane et al. (1982) found that aqueous extracts of white clover (*Trifolium repens* cv. Grasslands Huia) shoots collected near Lincoln, Canterbury, New Zealand, in mid-September 1978 caused inhibition of germination and seedling abnormalities in several of the five legumes and five grasses tested. A rate equivalent to 800 kg/ha of white clover dry matter almost totally inhibited germination of white clover, alsike clover (*T. hybridum*), red clover, lotus (*Lotus pedunculatus*), cocksfoot (*Dactylis glomerata*), and perennial ryegrass (*Lolium perenne*). Addition of polyvinylpolypyrrolidone removed up to 75% of the observed inhibition, and chromatographic analyses implicated phenolics as the active allelochemicals.

There is growing interest in the use of annual winter legume cover crops in no-till farming, and research indicates that planting into legume cover crops may result in inadequate stands of no-till corn, grain sorghum (*Sorghum bicolor*), and cotton and reduced crop yields. White et al. (1989) decided to determine, therefore, if crimson clover (*T. incarnatum*) and hairy vetch (*Vicia villosa*) have allelopathic effects on some crops and weeds. Germination and seedling growth of corn, Italian ryegrass (*Lolium multiflorum*), cotton, pitted morning glory (*Ipomoea lacunosa*), and wild mustard (*Sinapis arvensis*) decreased progressively in increasing concentrations of aqueous extracts of crimson clover and hairy vetch (8.3 to 33.3 g debris/*L*). Mustard and ryegrass germination and growth were almost completely inhibited by full-strength extracts of both legumes. Equivalent osmotic concentrations had no effect on any species, except for cotton, where part of the inhibition may have been due to increased osmotic pressures in the more concentrated extracts. Cotton and pitted morning glory germination and dry weight decreased 60 to 80% when these plants were grown in greenhouse conditions with increasing amounts (0.8 to 6.7 mg debris/g) of field-grown crimson clover or hairy vetch incorporated into the soil medium. Corn dry weight increased 20 to 75% when clover or vetch debris was placed on the surface, but germination and dry weight were changed very little when the debris was incorporated into the soil. Emergence and growth of corn and cotton were not affected when those plants were seeded in soil that contained root biomass and possible leaf and root exudates and that had been collected from fields where crimson clover or hairy vetch was growing. Morning glory dry weight, however, increased 35% in both the clover and the hairy vetch soils. Overall, allelopathic effects of crimson clover and hairy vetch were definitely demonstrated, and these scientists concluded that incorporating legume cover crops into weed-control strategies or no-till crops appears to be practical and of potential benefit when coordinated with herbicide applications.

Bradow and Connick (1990) found that volatile emissions of residues of winter-cover berseem clover (*T. alexandrinum*), hairy vetch, and crimson clover inhibited germination and seedling development of onion (*Allium cepa* cv. Texas Early Grano 502), carrot (*Daucus carota* cv. Danvers Half-Long), and tomato (cv. Homestead). Employing GC-MS, the researchers identified 31 $C_2$-$C_{10}$ hydrocarbons, alcohols, aldehydes, ketones, esters, furans, and monoterpenes. Time- and concentration-dependent studies with these compounds, plus a few from shoot residue of *Cyperus rotundus* and shoot and root residues of cotton, demonstrated that (E)-2-hexenal was most inhibitory, followed by nonanal, 3-methylbutanal, and ethyl-2-methyl-butyrate. Many of the volatiles in all mixtures tested, however, were quite inhibitory to seed germination of the three test species. Obviously, the volatiles can be very important allelochemicals in allelopathy and yet are generally ignored.

Yakle and Cruse (1983, 1984) pointed out that the increased use of no-till farming has resulted in large amounts of corn residue on the soil surface in newly planted corn fields. This results in reduction in early corn growth. They postulated that this growth reduction might be due, at least in part, to allelopathic effects of the residue. In initial experiments, they collected both fresh and partially decomposed corn residues from the field, oven-dried the residues, and ground them. Corn seeds were planted in sand or soil with one residue band placed either above, below, or at seed depth. A no-residue control was included. Root and shoot weights of corn were reduced by both fresh and partially decomposed residues, with fresh residue having the greatest impact. The effects were most notable when the roots came in contact with the residues. In subsequent experiments corn residues were incubated for different time periods, with or without soil, under wet, poorly aerated conditions. Water extracts were made of the various residues; some were filtered through columns of sterilized soil before being used to grow corn seedlings. All types of extracts inhibited growth of corn seedlings compared with water controls, but incubation of the residue with soil reduced the inhibitory effects of the extracts, as did filtration of the residue extracts through sterilized soil columns. Apparently, phytotoxins extracted from the corn residues incubated with soil were inactivated in the soil by microbial decomposition. However, if they were not adsorbed on the soil micelles, the allelochemicals were apparently able to move through sterile soil without significant change in their inhibitory properties.

Aqueous extracts of residues of corn, soybean (*Glycine max*), oats (*Avena sativa*), and mixed-grass hay were all found to reduce germination of corn (Martin et al., 1990). Nonsterile extracts of soybean and oat straw reduced germination by 74%, and extracts of corn and hay residues reduced germination by 27%. Nonsterile extracts of soybean and oat hay did not reduce coleoptile length but did reduce radicle and secondary root length by 34% compared with the water control. Nonsterile corn and mixed-hay extracts reduced coleoptile lengths by 42% and radicle and secondary root lengths by 81%. Sterile extracts of soybean and oat residues decreased radicle and secondary root lengths by 81%. Overall, oat and soybean residue extracts were only mildly inhibitory, whereas corn and hay residue extracts were extremely detrimental to seed germination and seedling growth of corn.

Turco et al. (1990) stated that average yields in continuous corn at the Purdue agronomy farm over an 11-year field study were decreased considerably compared with corn in rotation with soybeans in either a plowed or a no-till system. They stated also that their studies have shown that chemicals originating directly from corn residue decomposing under field or laboratory conditions are of limited consequence to corn growth. They suggested that the decline in yield in continuous corn cropping might be due, at least in part, to a change in population of rhizosphere microorganisms, which might then influence growth and yield of the corn. They selected plots in continuous corn and in rotated corn-soybean fields and treated some plots in each with methyl bromide three days prior to planting, leaving some plots in each untreated. In the first year, corn yields were similar in fumigated continuous corn plots to nontreated corn-soybean plots. Rotated-corn yields were less affected, however, by fumigation. In the second year, the effects were similar but not as significant. Many bacteria were isolated from the soil and residues and tested against growth of corn seedlings. About 22% were able to inhibit plant growth, and 72% of these were from the continuous-corn system. The bacteria isolated were assumed to be *Pseudomonas* sp. because of the techniques used. It should also be pointed out that 33% of the isolates enhanced either shoot or root growth of corn. It is also significant that the effect of fumigation was transient since the number of bacteria in soil samples in fumigated plots returned to normal in one month. Of course, fumigation may enhance early-season root vigor. The overall effect in continuous corn cropping is still allelopathic even if the initial effect is simply an increase in populations of allelopathic microorganisms. It should be recalled that Molisch clearly included production of bioactive allelochemicals by microorganisms in his definition of allelopathy.

Crookston et al. (1991) conducted a nine-year field study at two locations in Minnesota to determine the impact of various corn and soybean cropping patterns on the yields of both crops. They studied continuous monoculture with the same cultivar, monoculture with two cultivars alternated annually, annual rotation of the two crops, and one, two, three, four, and five years of monoculture following five years of the other crop. Annually rotated soybean yielded 10% better and first-year corn 15% better than corn under monoculture. Annually rotated soybean yielded 8% better, and first-year soybean 17% better than soybean under monoculture. For both corn and soybeans, alternating two cultivars annually gave the same yields as continuous cultivation with one cultivar. The lowest corn yield was from the second year of continuous corn production, whereas the lowest yield of soybeans was from continuous monoculture.

It is noteworthy that, despite all the evidence indicating yield benefits in crop rotations of many kinds of crops and in many areas, the practice of monoculture farming in the United States became widespread in the 1950s and continues to the present day.

Agroecosystems in Tlaxcala, Mexico, are good examples of traditional polycultures. Anaya et al. (1987) identified all plants in typical examples of these polycultures and determined allelopathic potentials of the various crop plants and weeds. Corn, beans (*Phaseolus vulgaris*), and squash (*Cucurbita moschàta*) showed a clear allelopathic effect, as did several weeds, including *Chenopodium murale, Tradescantia crassifolia, Melilotus indicus,* and *Amaranthus hybridus.* The researchers concluded that allelopathy can be used in the biological control of weeds and other pests, which is a necessity in any sustainable system of agriculture.

Jiménez et al. (1983) confirmed that corn pollen has allelochemicals that inhibit growth of other plants. In fact, to study the claim of peasants from Tabasco, Mexico, that fruiting of watermelon is reduced when corn pollen falls on the plant, Cruz-Ortega et al. (1988) sprinkled corn pollen on watermelon (*Citrullus lanatus*) seeds and found that as little as 50 mg of pollen in 5 mL of water reduced radicle growth by more than 50%. They also observed swelling and a grayish coloration at the base of the radicle. Ethanolic extracts of corn pollen strongly inhibited stem growth and inhibited root growth slightly less. The extract added to watermelon hypocotyl mitochondria respiring with malateglutamate caused a decrease in the state III respiratory rate, which was not relieved by DNP. The extract acted like an electron transport inhibitor, preventing stimulation of oxygen consumption by ADP and phosphorylation of ADP to ATP. Ethanol alone in equal concentration did not produce any effect on respiration by the mitochondria. Pollen, and extract of the pollen, also strongly reduced the mitotic index in meristematic tissue of watermelon seedlings.

Anaya et al. (1992) isolated and identified phenylacetic acid (PAA) in corn pollen by GC-MS. Subsequently, whole pollen, methylene chloride extract of corn pollen, and pure PAA were tested against germination and radicle growth of *Amaranthus leucocarpus* and *Echinochloa crusgalli.* Whole pollen inhibited germination and radicle growth of both weedy species, as did methylene chloride extracts of the pollen and some TLC fractions of the extracts. PAA was also very inhibitory to both weeds, but more so to *Echinochloa* than to *Amaranthus.*

The genus *Brassica,* which includes many crop plants and weeds, is notorious for its allelopathic potential (Rice, 1984), and much research continues on this genus and some other genera of the Cruciferae. Mason-Sedun et al. (1986) and Mason-Sedun and Jessop (1988) extended such studies to a comparison of the relative allelopathic effects of several species and cultivars of *Brassica* on wheat in Australia. Aqueous extracts of thirteen *Brassica* residues from six species inhibited wheat coleoptile growth by 56.7 to 91.4% compared with water controls. Root growth was inhibited from 59.1 to 97.8%. Field tests indicated that all *Brassica* residues reduced grain yield, plant dry weight, plant height, and tiller production. The order of increasing allelopathy was as follows: *B. campestris* cv. Torch, *B. juncea* cv. Zem 1, *B. napus* cv. Wesroona, *B. nigra* cv. Vince, *B. juncea* cvs. Lethbridge 22A and Zem 2. Field residues that were stored dry became less toxic with time, and increasing humidity increased the rate of loss of activity.

Akram and Hussain (1987) reported that aqueous extracts of fresh and dried roots and root exudates of Chinese cabbage (*Brassica oleracea* cv. Chinensis) reduced its own radicle growth, fresh- and dry-weight, and also inhibited growth of mustard (*B. campestris*). Extracts of dried roots were more inhibitory than extracts of fresh roots, and mustard was inhibited in growth more than was Chinese cabbage. The researchers concluded that the allelopathic effects of the root exudates might play a significant role in reducing productivity in hydroponics. Oleszek (1987) grew *B. juncea* (undetermined cv.), *B. juncea* cv. 603, *B. nigra, B. napus* cv. Bronowski, *B. napus* cv. Górczański, *B. rapa* var. *rapifera, B. oleracea* var. *acephala,* and *Sinapis alba* in the field. He tested the youngest fully expanded leaves from 35-day-old plants for bioactive volatiles against germination and growth of lettuce, *Echinochloa crusgalli,* and wheat. All species and cultivars slightly inhibited or retarded germination of barnyard grass and some severely reduced germination of lettuce, but none affected germination of wheat. Volatiles from three species or cultivars markedly reduced coleoptile length of lettuce and slightly reduced coleoptile length of *Echinochloa* and wheat. The same three—*B. juncea* cv. 603, *B. juncea* (no cv. determined), and *B. nigra*—produced volatiles that completely inhibited root growth of lettuce and lightly inhibited root growth of wheat. All produced volatiles that strongly reduced root growth of *Echinochloa.*

Gubbels and Kenaschuk (1989) conducted field experiments in

Manitoba, Canada, to determine the possible allelopathic effects of crop residues of canola (*Brassica napus*), barley, and flax (*Linum usitatissimum*) on the growth and yield of subsequent flax crops. They also did additional studies to include barley as a test crop on canola and barley stubble and to compare conventional tillage and no tillage. Flax yields were generally lower on canola and flax stubble than on barley stubble when conventional tillage was used. With no tillage, flax yielded as well on canola as on barley stubble. Spring volunteer seedlings of canola and flax often lowered flax yields, but fall volunteer seedlings had little effect. Barley yielded better on canola stubble than on its own stubble, and better with tillage than without. It was concluded that some of he reduced yields could have been caused by toxic compounds released from the stubble.

Politycka et al. (1984) used two substrates to determine what allelochemicals are responsible for the allelopathic effects in substrates used repeatedly in cucumber growing in greenhouses. Some plots had been used for six years in early cucumber (*Cucumis sativus*) and autumn tomato growing before this two-year experiment was begun, but fresh plots using new substrates of the same types were also included. Marked inhibition of cucumber growth occurred in all plots during the second year; there were also very large increases in the total phenolic content in all the plots. Addition of Polyclar AT to extracts of the allelopathic substrates eliminated the toxic effects of the extracts, indicating that the allelopathic effects were due to increases in the concentrations of phenolics in the substrates. Seven phenolic compounds were identified in the substrates: ferulic, p-hydroxybenzoic, p-coumaric, protocatechuic, salicyclic, syringic, and vanillic acids.

Extracts of leaves of Jerusalem artichoke (*Helianthus tuberosus*) inhibited seed germination of five species of plants tested: *Daucus carota, Inula helenium, Acer negundo,* radish, and pea (*Pisum sativum*). The percentage inhibition of germination also increased with increases in concentrations of the extracts. These results suggest allelopathic potential for Jerusalem artichoke.

Other plants discussed here as crop plants are grasses used for grazing by domestic animals and many cultivated and several range grasses that do not fall under the category of weeds. *Cenchrus ciliaris* and *Chrysopogon aucheri* are important range grasses in Pakistan, and Akhtar et al. (1978) reported that root exudates and water extracts of both species inhibited their own growth plus that of other species used in bioassays. The inhibition in both cases increased with increasing concentration and soaking time. Comparative tests indicated that *Cenchrus* was considerably more allelopathic than *Chrysopogon*. All species of grasses being considered for introduction into the rangelands of Pakistan are tested for allelopathic potential at the Pakistan Forest Institute at Peshawar (Khanum et al., 1979). Pot and laboratory experiments demonstrated that *Chloris gayana* and *Panicum antidotale* are potentially allelopathic. Root exudates of each inhibited the plant's own growth and that of the other species. Shoot extracts of each inhibited radicle growth of both of these species plus *Pennisetum americanum* and *C. aucheri*. The allelopathic effect of each depended on the amount of material soaked, the soaking duration, freshness of the residue used, and the test species.

Begum and Hussain (1980) subsequently tested *Panicum antidotale* against several other range grasses. Aqueous extracts of inflorescences, shoots and roots, and root exudates in which *P. antidotale* grew inhibited germination and seedling growth of *Brassica campestris, Cenchrus ciliaris, Lolium multiflorum, Pennisetum americanum, Setaria italica,* and *Sorghum almum*. These scientists concluded that *P. antidotale* would not exhibit its benefits as a range grass because of its allelopathic effects.

Hussain et al. (1982) extended the studies on allelopathic effects of *Cenchrus ciliaris* to other range grasses and also examined the allelopathic potential of *Bothriochloa pertusa*. They first determined the frequency of every species in three stands with and without *C. ciliaris* or *B. pertusa*. All associated species exhibited low importance values in association with either of these species. It appeared therefore that both species suppressed growth of associated species. Very complex field and greenhouse experiments were established with many different species combinations—some with roots allowed to mix, some not—and with numerous other kinds of controls. *C. ciliaris* and *B. pertusa* exhibited autointerference, and all combinations of species exhibited mutual suppression in root-mixed tests. The growth was definitely inhibited through roots, since shoot-separated treatments had no effects on associated species. The researchers concluded that the suppressed growth and density of the species under similar physical environments suggested the presence of a biochemical inhibition, which is probably assisted by competition for the physical factors of the environment.

When new grass is sown into old swards, establishment can sometimes be poor. To help determine the cause, Gussin and Lynch (1981) collected fresh leaves of *Agrostis stolonifera, Alopecurus pratensis, Anthoxanthum odoratum, Festuca rubra, Holcus lanatus, Lolium perenne,* and *Poa trivialis;* cut them into 3.5-cm lengths; and fermented them aerobically or anaerobically in flasks containing a soil inoculum and distilled water. Extracts from the flasks were added to petri dishes containing sand and seeds of one of the test species: *Alopecurus myosuroides, A. stolonifera, F. rubra, H. lanatus, L. perenne, Poa annua, P. trivialis,* and *Trifolium repens.* After 10 days of anaerobic fermentation, solutions from most of the grass residues were allelopathic to most of the test species. After 20 days of fermentation, however, only *A. stolonifera, A. pratensis,* and *F. rubra* produced extracts that significantly inhibited root growth of all species.

After aerobic fermentation for 10 days, most of the residues stimulated growth of test seedlings, but *A. pratensis* and *L. perenne* were toxic to all the test species. After 20 days, all residues were stimulatory to test species. After 20 days of anaerobic fermentation, extracts from *A. stolonifera, A. pratensis,* and *F. rubra* contained the greatest concentrations of volatile fatty acids. On the other hand, concentrations of volatile fatty acids produced during aerobic fermentation were at least an order of magnitude smaller. Aromatic acids were detected in both anaerobic and aerobic extracts, and phenylacetic and hydrocinnamic acids were common to most extracts. Phenolics were not found in any of the extracts. After four days' incubation in water, the concentration of acetic acid could be large. Ryegrass was more susceptible than clover to volatile fatty acids. Propionic and butyric acids were much more toxic than acetic and lactic acids to both these species.

Exudates from frost-killed cells of frost-damaged grasses reduced regrowth of those grasses (Habeshaw, 1980). Cold (4°C) leachates of dead leaves were also inhibitory to seed germination of these grasses, and chromatographic separation showed the allelochemical to be the same in both cases. Experiments with *Sorghum almum* (Columbus grass) in Pakistan demonstrated that it is strongly autotoxic, and root exudates in mixed culture and soil taken from underneath this grass were allelopathic to *Brassica campestris, Cenchrus ciliaris, Lolium multiflorum, Pennisetum americanum, Setaria italica,* and *Panicum antidotale* (Qureshi and Hussain, 1980). Shoot extract also reduced seed germination, fresh and dry weight, water content, and survival of seedlings of the test species. It was concluded again that introduction of this species to rangelands in Pakistan would be counterproductive because of its strong allelopathic potential against several important grass species in those grasslands.

Stephenson and Posler (1988) studied the effects of tall fescue (*Festuca arundinacea*) on germination and seedling growth of birdsfoot trefoil (*Lotus corniculatus*) from 1983 to 1985 at Manhattan, Kansas. Tall fescue produced allelopathic compounds, particularly during the spring and autumn months. In a sand medium under greenhouse conditions, fescue extracts reduced trefoil growth by 50 and 56% in spring and autumn, respectively. Under field conditions, full-strength (15 g fescue/100 mL water) fescue extracts reduced trefoil plant populations by 14 and 57% with spring and autumn extracts, respectively. Full-strength fescue extracts reduced trefoil seedling growth by an average of 37%, and trefoil dry weight by 53%. The many reports concerning the allelopathic effects of tall fescue against numerous species caused Luu et al. (1989) to identify some of the allelochemicals possibly involved. Water extracts of the herbage were fractionated into anion, cation, and neutral fractions. Birdsfoot trefoil germination and seedling development were reduced primarily by the anion fraction, which suppressed root growth by two-thirds and hypocotyl growth by one-half. Lactic, succinic, malic, citric, shikimic, glyceric, fumaric, and quinic acids plus several unknowns were found to be possible inhibitors, with lactic and succinic acids as the apparent major inhibitors.

Buta and Spaulding (1989) found that extracts of excised leaves of Rebel and Kentucky 31 tall fescue grown in the greenhouse did not give any activity in the lettuce seedling bioassay; however, when the excised leaves were dried in air in the greenhouse for 30 days, pronounced activity occurred against seedling growth of *Lolium multiflorum, Festuca rubra* cv. Ensylva, *Poa pratensis* cv. Kenblue, *Lolium perenne,* and Rebel tall fescue. Subsequent partitioning and analysis by GC-MS identified three principal inhibitory compounds: abscisic (ABA), caffeic, and p-coumaric acids. Quantitative analysis indicated that ABA was the predominant inhibitor. A 10-fold increase in ABA concentration in leachates occurred after one day of desiccation of the tall fescue leaves (in both Rebel and Kentucky 31). The ABA concentration in Kentucky 31 leachate was, however, 40% higher in Kentucky 31 than in Rebel tall fescue. This difference was also found in leachates of these grasses that had been air-dried for 30 days, although the ABA concentrations in both were reduced by about 60% from the high concentrations after one day of desiccation. On the basis of the quantities of ABA found on leaching after the extended, 30-day desiccation, Buta and Spaulding concluded that the likelihood that bioactive quantities of ABA would be leached out of senescent or dehydrated plant parts of tall fescue was reasonable.

Young and Bartholomew (1981) found that residues of limpograss (*Hemarthria altissima*) in soil inhibited growth of *Desmodium intortum* cv. Greenleaf. The scientists subsequently studied the effects of exudates of two cultivars of limpograss, Bigalta and Greenalta, on growth, nodulation, and nitrogen fixation in Greenleaf *Desmodium* (Young and Bartholomew, 1987). They grew the plants in vermiculite in a greenhouse and watered them with Hoagland's and N-free nutrient solutions. Competition between limpograss and desmodium for nutrients and light was decreased by growing the plants in divided pots, which kept the plants separate but permitted transfer of root exudates between the halves of the pots. Exudates from Bigalta limpograss and Greenalta inhibited top growth of Greenleaf desmodium in pots with dividers irrigated with Hoagland's solution, but not in the N-free solution. Fresh-shoot yields of Greenleaf desmodium were less in the presence of root exudates of Bigalta limpograss as compared with those in the presence of exudates of Greenalta limpograss. Average root weight, total nodule weight, and nodule size were very low in the presence of root exudates from both grasses in the Hoagland's culture solution. Under N-free regimes, nitrogenase activity (acetylene reduction) was lower in nodules of Greenleaf desmodium plants whose roots were intermingled with limpograss roots. These scientists concluded that retardation of Greenleaf desmodium growth in vermiculite cultures was due to allelochemicals in root exudates produced by the limpograsses and that, when N was limiting, some of the effect may have been due to interference with nodulation and nitrogen fixation.

Murphy and Aarssen (1989) tested extracts of pollen from five test species in a grassland community in Ontario, Canada, against pollen germination of 40 sympatric species. The five test species were *Agrostis stolonifera, Erigeron annuus, Melilotus alba, Phleum pratense,* and *Vicia cracca*. The 40 target species were selected from 39 genera. Seventeen of the target species were used in tests in 1987 and 23 were used in tests in 1988. In 1987 *P. pratense* inhibited germination of pollen of all 17 target species except *Linaria vulgaris,* and it exhibited significant phenological divergence with seven of the target species. Pollen extracts of *A. stolonifera, M. alba* and *V. cracca* inhibited pollen germination in some of the target species, but the scientists involved suggested that these effects were pH-mediated. The pollen extract of *E. annuus* did not inhibit pollen germination of any of the target species. In 1988 the pollen extract of *P. pratense* significantly inhibited pollen germination of 21 of 23 target species. It did not inhibit pollen germination of *A. stolonifera* and *P. pratense,* indicating it is not pollen-autoallelopathic. Murphy and Aarssen concluded that the breeding system, relatively tall growth, and relatively large amount of pollen produced all support the in vitro evidence of *P. pratense* as a pollen-allelopathic species.

Chou and Lee (1991) studied an alpine grassland community in Nantou County, Taiwan, where *Miscanthus transmorrisonensis* and *Yushinia nütakayamensis* were the principal species, with coverages of 25 and 19.5%, respectively. To elucidate the mechanism of dominance of *M. transmorrisonensis,* several tests were conducted concerning possible allelopathic effects of that species. Water extracts of two ecotypes of that species significantly inhibited seed germination and radicle growth in four test species: *Lolium multiflorum* (ryegrass), lettuce, and two varieties of Chinese cabbage. Moreover, rhizosphere soils under *Miscanthus* significantly inhibited test species. Leaf extracts and leachates contained caffeic, gallic, p-hydroxybenzoic, ferulic, and m-hydroxybenzoic acids; root extracts and soil extracts contained the first four of these acids. The concentrations in soil extracts were similar to those in leaf extracts. It was concluded therefore that allelopathy was an important contributing factor to the dominance of *M. transmorrisonensis* in the alpine grassland community.

During screening for allelopathic substances in roots and root exudates of higher plants, Schenk and Werner (1991) noted a strong growth-inhibiting activity of hot-water extracts of pea (*Pisum sativum*) seedling roots against baker's yeast and seedlings of several crop species. Baker's yeast is an ideal screening organism for the presence of the nonprotein heterocyclic amino acid, β-(3-isoxazolin-5-on-2yl)-alanine (βIA). Thus the researchers decided to determine if this compound was an effective allelochemical in pea roots. They found that root extracts of members of the tribe Vicieae to which pea belongs and βIA inhibited growth of baker's yeast, whereas root extracts of members of the Phaseoleae and βIA did not. Moreover, pea root exudates and βIA inhibited seedling growth of several crop plants, but not of pea. The principal bioactive allelochemical in pea root extracts was subsequently identified as βIA by paper electrophoresis, two-dimensional TLC, mass and NMR spectra. The minimum inhibitory concentration against baker's yeast was found to be 0.5 ppm (3 μ*M*). βIA is a component of root exudates of pea and is thus released into the soil; it is definitely an allelopathic compound. It is notable that Roundup, a commercial herbicide, is also a nonprotein amino acid, but a different one.

Pigeon pea (*Cajanus cajan*) litter caused root-tip necrosis, stunting, and reduced germination in seeds and seedlings of soybeans, lablab beans (*Lablab purpureus*), pigeon peas, and several local weeds in Puerto Rico (Hepperly and Diaz, 1983). Weed species varied in their sensitivity to pigeon pea litter, with grasses being inhibited more than dicotyledons. Minimum thresholds for detecting plant inhibition by leaf litter in the laboratory was 5–10 mg of litter per square meter of surface area. At harvest time, leaf litter in the field was greater than 150 $g/m^2$, which was close to the optimum amount for plant inhibition in laboratory tests. According to Hepperly and Diaz, it appears that in pigeon pea leaves numerous glands secreting volatile terpenoids may be responsible for allelopathy, whereas seed phytotoxicity appears to be related to polyphenols, such as tannins concentrated in the endoderm of the seed coat.

Much research (Rice, 1984) has been done on the allelopathic potential of rice (*Oryza sativa*), and this interest continues. Sadhu and Das (1971a) reported that root and shoot growth of test seedlings of two cultivars of rice, CB-1 and Rupsail, were equally inhibited by root exudates from CB-1. On the other hand, CB-1 test seedlings were severely inhibited, and Rupsail less so, in the root exudates of Rupsail donor plants. Preliminary tests indicated the presence of phenolic compounds in the root exudate, most likely cinnamic and salicylic acids, according to Sadhu and Das. Chandrasekaran and Yoshida (1973) found that addition of fresh green *Sesbania* sp. leaves to three different Philippine soils at the rate of 2.5% (w/w) and submersion in water caused large increases in concentrations of acetic, propionic, and butyric acids within two weeks, with the highest concentrations generally occurring between the fifth and seventh days. Acetic acid reached a peak in five days in all three soils, at about 2.5 m*M*/100 g soil. Tests with the same three acids added separately to the three soils indicated that each acid was toxic to rice seedlings at 0.5 mmol/100 g soil. The effects were accentuated at 1.0 mmol/100 g soil. Propionic and butyric acids were produced in smaller amounts when green material was added to soil, but they were more phytotoxic at the peaks than acetic acid and were more persistent in soil.

Sadhu (1975) came to a different conclusion concerning the important inhibitors in root exudates of the same two cultivars of rice, CB-1 and Rupsail, examined by Sadhu and Das (1971b). He concluded that abscisic acid was the most important inhibitor in both cultivars, but that a second significant inhibitor in root exudates of Rupsail was coumarin.

Rice callus (cv. Sekiguchi-asahi) markedly inhibited growth of soybean (*G. max* cv. Kihoshu) when grown together in the same culture bottle (Yang and Futsuhara, 1991). Subsequent experiments demonstrated that the supernatant fluid of rice culture cells also inhibited soybean callus growth just as strongly; in a similar fashion, volatile materials from the rice callus inhibited soybean callus growth. Calli of two cultivars of rice (Sekiguchi-asahi and Nipponbare) strongly inhibited growth of the calli of four species of *Glycine* and one species of *Vigna,* but did not affect callus growth of *Atropa belladonna* and *Datura stramonium* (two species of the Solanaceae). The volatile inhibitor was not identified.

Tapaswi et al. (1991) examined the effects on yield in two rice cultivars (Subarna and Pankaj) of different spacings in field plantings of a single cultivar and mixed cultivars. They found pronounced yield differences in both cultivars, depending on spacing and on whether cultivars were single or mixed. They speculated that the variations were due to allelochemicals in root exudates that varied in concentration as they moved through the soil, with stimulatory concentrations occurring in some zones and inhibitory concentrations occurring in others. Experiments were being continued to test their hypothesis.

Rye (*Secale cereale*) is a very useful plant in several cropping systems. Overwintered rye produces large amounts of biomass early in the growing season. It is thus a good green manure crop and as such has created much interest concerning its possible allelopathic effects (Rice, 1984). This interest continues. Barnes et al. (1986) reported that 40-day-old greenhouse-grown rye reduced total weed growth by 80% compared with a nontoxic control mulch that reduced weed biomass by

44% in similar tests. Fall-planted, spring-killed rye residue also reduced total weed biomass by 68 to 95% when compared with controls with no residue. In other tests, rye residue had no effect on seed germination of *Echinochloa crusgalli* (barnyard grass) and *Panicum miliaceum* (proso millet) at any distance, but it reduced seed germination of cress (*Lepidium sativum*) and lettuce when placed next to the seed. The primary effect in all cases was inhibition of root growth, and rye shoot tissue was more inhibitory than root tissue. The researchers' evidence demonstrated that microorganisms were not responsible for the toxicity, but instead decreased it. Barnes et al. identified 2,4-dihydroxy-1,4(2H)-benzoxazin-3-one (DIBOA) and 2(3H)benzoxazolinone (BOA) as two potent inhibitors in water extracts of rye residue. Both were strongly allelopathic to seedling growth of barnyard grass, cress, and lettuce. BOA is a breakdown product of DIBOA and probably contributes more to the allelopathic effect of rye residue after a period of decomposition. Additional details concerning these allelochemicals were given in a later paper (Barnes et al., 1987).

Shilling et al. (1986) harvested field-grown rye (cv. Abruzzi) at early flowering stage at Clayton, North Carolina, air-dried it for seven days, extracted it with water, acidified it to pH 2.5, and partitioned it with diethyl ether and ethyl acetate. The organic phase was separated by preparative TLC and biologically active bands were removed, extracted, and further separated by reversed-phase, high-performance liquid chromatography. The collected fractions were evaporated, and silyl and deuterated silyl derivatives were prepared and analyzed by GC-MS. The researchers identified β-phenyllactic acid (β-PLA), β-hydroxybutyric acid (β-HBA), and succinic acid and separated three other unidentified acids. Additional testing indicated that β-PLA and β-HBA were both inhibitory to hypocotyl and root growth of *Chenopodium album* and *Amaranthus retroflexus,* with β-PLA being most active. The isomeric α-PLA was 17 times more active than β-PLA to hypocotyl growth of *A. retroflexus.* It appears therefore that several potent bioactive allelochemicals occur in rye residue, depending on cultivar, growing conditions, and methods of identification. All the allelopathic compounds identified by the various researchers are soluble in water and could thus be leached from rye residue under field conditions to inhibit weed and crop growth.

Wójcik-Wojtkowiak et al. (1990) allowed rye seedlings, tillering plants, and crop residues to decompose for 14 days. Water extracts were made of each and tested against growth of rye seedlings; the extracts of seedlings and tillering plants were very allelopathic, whereas extracts of mature plants were not. Seven phenolic acids were identified in the investigated materials, but the concentrations of the separate phenols and total content of phenolics did not compare with the level of phytotoxicity determined in the bioassays of the water extracts. The researchers concluded therefore that other, unidentified water-soluble compounds were also responsible in part for the toxicity of rye residues. This would certainly agree with results described previously. Raimbault et al. (1991) pointed out that studies in Ontario, Canada, have shown that corn yields are reduced when corn is seeded immediately after rye harvest or chemical rye kill. In order to determine what could be done to eliminate the detrimental effect of rye residue on corn yields, they designed several treatments, among them conventional moldboard plowing and five variations of no-till with supplementary treatments. They also seeded plots in which the rye was killed two weeks before corn planting and plots in which the rye was killed just before planting. Corn yields were about 10% higher for the early-kill than for the late-kill plots handled with the same tillage practices. Corn yield in the moldboard-plow treatment was higher than in strip tillage and than the average of the no-till treatments. Use of disc furrowers along with no-till gave yields similar to the moldboard treatment. It was concluded that the furrowers removed enough residue out of the row to prevent the occurrence of an allelopathic effect.

Nair et al. (1990) isolated a microbially produced 2,2'-oxo-1,1'-azobenzone (AZOB) from soil enriched with BOA and showed that DIBOA, BOA, and AZOB were inhibitory to barnyard grass and garden cress (*Lepidium sativum*) at concentrations of 67 to 250 ppm. AZOB was more inhibitory than the others tested. Chase et al. (1991a) tested activity of the same three compounds against two weeds, garden cress and barnyard grass, and against two crop plants, cucumber and snap bean (*Phaseolus vulgaris*). All three compounds were applied alone at 50, 100, and 200 ppm and in two- and three-way combinations, with each applied at 50 and 100 ppm. AZOB was most active against barnyard grass, reducing both root and shoot growth at all concentrations. Combinations of the compounds containing AZOB showed 41 to 67% more activity on roots and 20 to 33% more activity on shoots when compared with combinations of DIBOA and BOA at the same concentrations. With garden cress, AZOB inhibited all parameters measured with all applications. The order of activity on cress was AZOB > BOA > DIBOA. With cucumber, DIBOA was the most active, followed by BOA and AZOB. Applications of BOA of 100 and 200 ppm and combinations of BOA and AZOB had the greatest activity against snap beans. In general, the plant-produced inhibitors, DIBOA and BOA, were more inhibitory to crops than to weeds. Thus, according to these scientists, improved herbicidal activity would be expected with rapid transformation of the DIBOA to BOA to the microbially produced AZOB. Chase et al. (1991b) reported that *Acinetobacter calcoaceticus,* a gram negative bacterium isolated from field soil, was responsible for the biotransformation of BOA to AZOB.

Gagliardo and Chilton (1992) found that some BOA was transformed into 2-amino-3H-phenoxozin-3-one in nonsterile soil also. This compound was an order of magnitude more inhibitory to barnyard grass than was AZOB. Thus rye either produces or gives rise to several very potent allelopathic compounds in soil.

Kalantari (1981) reported that, in some tests in Iowa, corn following soybeans produced 7 to 10% more grain yield than corn after corn, even under circumstances where there were no nitrogen fertilizer or moisture advantages following soybeans. Kalantari was able to isolate a fraction from a chloroform extract of the soil-soybean residue that increased radicle weight of corn by 20% and coleoptile weight by 19%. A similar fraction from ground soybean residue increased radicle fresh weight and dry weight of corn by 24.7 and 9.2%, respectively. This fraction increased coleoptile dry weight by 11.7%. Nelson (1985) made a much more thorough study of the allelopathic potential of soybeans using *Lemna minor* plants in the bioassays. Soil was collected in late summer from a soybean field (cv. Corsoy) near Ames, Iowa. Some soil was collected underneath plastic covers so that the chief source of allelochemicals would be from root exudates or decaying roots; some was collected from uncovered areas. Some soil samples were collected from both open and covered areas between soybean rows, and some were collected from covered soil beneath soybean plants. When soil extracts were fractionated, some fractions in all plots were inhibitory to *Lemna* growth and a few were stimulatory. The most bioactive soil samples came from uncovered plots between soybean rows. Thus it appears likely that at least part of the increased corn yield following soybeans is due to allelopathic stimulation by allelochemicals deposited in the soil by soybean plants or soybean residues.

Much previous research has been done on the allelopathic effects of sorghum (*Sorghum bicolor* = *Sorghum vulgare*) (Rice, 1984). The striking results have caused considerable research to continue. Hussain and Gadoon (1981) grew sorghum in combination with sorghum, *Setaria italica, Pennisetum americanum,* and corn having their roots separated or intermingled with each other in 20 × 20-cm plots, with equal volumes of similar litter-free soil. In controls, the roots of interacting species were restricted to their respective halves of the plot by a polythene sheet. Water and nutrient deficiency were prevented by watering regularly with a nutrient solution. Plots were frequently weeded by hand, and all plants were harvested at 10 weeks. Sorghum reduced its own growth and significantly reduced height and fresh and dry weight of all test species in the root-mixed treatment. Aqueous extracts of sorghum significantly reduced germination of *S. italica, P. americanum,* and *Brassica campestris,* but root extracts retarded germination of *Brassica* only. Radicle growth was significantly inhibited by all extracts, except for sorghum in root extract of sorghum. Water extract of soil in which sorghum previously grew significantly reduced germination of sorghum and radicle growth of all test species. Hussain and Gadoon's results indicate that sorghum must be cultivated in rotation with some suitable crop and that the field should be well plowed and irrigated before sowing to remove or decrease amounts of allelochemicals remaining in the soil.

Ruiz-Sifre and Ries (1983) reported that seedling growth of corn in the greenhouse was increased by residues of sorghum shoots but not by residues of sorghum roots. On the other hand, growth of seed corn in the field were always decreased by residues, of Bird-a-Boo sorghum roots and whole plants. The growth and yield of snap beans (*Phaseolus vulgaris*) in the field were increased or decreased by sorghum residues, depending on the sorghum plant part, quantity, cultivar, and soil condition. In field tests in 1980, the dry weight of cucumber plants was increased 37% by Milkmaker sorghum compared with controls.

Lehle and Putnam (1983) obtained 12 bioactive fractions from an aqueous extract of field-grown sorghum cv. Bird-a-Boo as determined by indexing three aspects of cumulative cress cv. Curlycress seed germination. They reported that some of the inhibitory components included chemical classes not previously associated with herbage allelopathy. The combined inhibitory potentials and individual specific inhibitory activities of both the low- and high-molecular-weight fractions were greater than that expressed by the crude aqueous extract. The researchers suggested that both fractions were antagonistic to each other in the overall activity.

Weston et al. (1989) tested the allelopathic effects of sudex (*Sorghum bicolor* x *S. sudanense*) on tomato, cress, *Setaria italica,* and barnyard grass. As sudex age increased, allelopathic potential of the tissue decreased against radicle elongation of test species. Sudex tissues were generally more inhibitory under sterile conditions, indicating that soil microorganisms may decompose some of the allelochemicals. Two major inhibitors were isolated from aqueous extracts of sudex shoot material by partitioning with diethyl ether, followed by TLC and column chromatography. The inhibitors were p-hydroxybenzoic acid and p-hydroxybenzaldehyde, the enzymatic breakdown products of dhurrin. The $I_{50}$ values of these compounds using a cress seed bioassay were 140 and 113 μg/mL for the acid and aldehyde, respectively. As sudex tissue age increased, the percentage of p-hydroxybenzaldehyde increased, whereas the percentage of p-hydroxybenzoic acid decreased.

Einhellig and Souza (1992) stated that root exudates of sorghum consist primarily of a dihydroquinone that is quickly oxidized to a p-benzoquinone named sorgoleone. They also pointed out that Dr. Larry Butler and his colleagues at Purdue University had recovered sorgoleone from soil in which sorghum had grown the previous year. Tests were therefore run by Einhellig and Souza to determine the herbicidal activity of sorgoleone against *Eragrostis tef, Lemna minor,* and six weedy species. Over a 10-day treatment period, 10 μ*M* of sorgoleone in the nutrient medium reduced growth of all six weeds. Bioassays demonstrated that 125 μ*M* reduced radicle growth of *E. tef* and 50 μ*M* stunted growth of *L. minor.* Einhellig et al. (1993) reported that concentrations as low as 10 μ*M* of sorgoleone inhibited oxygen evolution from soybean leaf discs by more than 50%. Thus sorgoleone apparently contributes greatly to sorghum allelopathy.

Pardales and Dingal (1988) pointed out that taro (*Colocasia esculenta*) decreases in vigor and yield when that crop is grown year after year in the same fields. They collected vegetative parts of mature taro plants from the Research and Training Centre experimental field at Pangasugan, Leyte, dried them, and ground them for tests of allelopathic effects. Water extracts of the residue were quite inhibitory to germination of mungbean (*Vigna radiata*), corn, and okra (*Hibiscus esculentus*) seeds but only slightly so to germination of eggplant (*Solanum melongena*) seeds. The taro extract activities were dependent on concentration. Aqueous extracts of cassava (*Manihot esculenta*) and sweet potato (*Ipomoea batatas*) residues did not affect germination of mungbean seeds.

ul Haq and Hussain (1979) reported that root exudate of tobacco (*Nicotiana rustica*) collected on filter paper, in seedbeds, and in culture solution significantly inhibited germination and radicle growth of *Brassica rapa,* lettuce, *Setaria italica* and corn. Inhibition in germination varied from 64% in *B. rapa* to 90% in corn, and radicle growth was inhibited from 52% in *S. italica* to 80% in corn. It was apparent that the root exudates of tobacco were thus allelopathic to several crop plants and a common weed.

Kim and Kil (1987) found that aqueous extracts and leachates of tomato plants inhibited germination, elongation, and dry-weight increases in several crop species. Root exudates of tomato inhibited growth of all test species, which included *Brassica campestris,* sorghum, *Perilla frutescens,* and corn. Volatiles from tomato plants suppressed dry-weight increases of lettuce and elongation and dry-weight increases of grapevines planted near the tomato plants in a greenhouse.

Germination inhibitors in the chaff of spring triticale (X *Triticosecale*) could provide a significant means of reducing sprouting caused by prolonged periods of rain during harvest (Salmon et al., 1986). Therefore, from 1982 to 1984, eight spring triticales were grown in the field at Lacombe, Alberta, and evaluated for germination inhibitors against their own seed in their aqueous chaff extracts. Extracts evaluated on seed that had been stored at 20°C for three months briefly reduced germination in all lines except one. Regardless of the year, extract concentration, or storage temperature of the seed, the chaff extracts provided the greatest protection against germination when combined with high levels of seed dormancy. The researchers concluded that selection for high concentrations of germination inhibitors in the chaff will be a valuable means of prolonging sprouting resistance in spring triticale.

Many scientists have demonstrated that wheat straw residues and residues of other grasses often markedly inhibit growth of wheat seedlings as well as seedlings of other crop plants (Kimber, 1967, 1973; Cochran et al., 1977; Srivastava et al., 1986; Lynch, 1977, 1978; Lynch et al., 1980a; Lovett and Jessop, 1982; Hamilton-Kemp and Andersen,

1986; Lodhi, et al., 1987; Waller et al., 1987; Thorne et al., 1990). Kimber (1967) reported that aqueous extracts of wheat straw rotted for periods up to six weeks inhibited growth of wheat and oats, with maximum effects after two to six days of rotting. Roots were more sensitive than shoots to the inhibitors produced, and wheat shoots were inhibited more than oat shoots, but oat roots were inhibited more than wheat roots. Cochran et al. (1977) found that decaying residues of lentil (*Lens culinaris*), pea, wheat, barley, and bluegrass (*Poa pratensis*) inhibited root growth of winter wheat seedlings, but only after conditions became favorable for microbial growth. Lentil and pea extracts were most toxic in the fall and early winter, causing up to 90% root growth inhibition. Wheat and barley residue extracts were less inhibitory, but they intermittently produced seedling inhibitors in the fall, late winter, and until April. Bluegrass extracts were toxic primarily in late winter and early spring. Srivastava et al. (1986) reported that water extract of fresh wheat straw inhibited wheat seed germination by about 18% but did not affect water absorption by germinating wheat seeds. The maximum germination inhibition was caused by straw rotted for 31 days.

Lynch (1977, 1978a) found that acetic acid was the chief phytotoxin produced by aerobic and anaerobic decomposition of wheat straw. Aerobic decomposition stimulated root extension of barley seedlings, whereas anaerobic decomposition yielded products that inhibited growth. Straw from wheat, barley, oats, and rape *(Brassica napus)*, and decaying rhizomes of couch grass (*Agropyron repens*) had the same effect. The breakdown of acetic acid was slow in flooded soil, and the maximum accumulation took place under these conditions. Freshly harvested wheat straw contained 180 m*M* acetic acid and fell to about 10 m*M* within six hours of incorporation of the straw in the soil; it then remained relatively constant for a period of 12 days, irrespective of soil moisture content (Lynch et al., 1980b). The researchers concluded that acetic acid is the major cause of poor establishment and growth when seeds and seedling roots come into contact with straw.

Prasad and Rao (1981) found that binary mixtures of three spring wheat cultivars caused allelopathic effects on root number, root growth, and fresh weight of the seedlings, compared with results of pure stands of each cultivar.

Lovett and Jessop (1982) tested effects of residues (dry, fresh, partly decayed) of 12 crop species [pea, broad bean (*Vicia faba*), soybean, lupin (*Lupinus angustifolius*), chickpea, safflower (*Carthamus tinctorius*), sunflower (*Helianthus annuus*), rape, sorghum, oats, barley, and wheat] against germination and root length of wheat (cv. Songlen) in petri dishes, greenhouse, and field. At least some residues of all crops significantly inhibited growth in length of the longest seminal root of wheat, and all residues except oats significantly inhibited growth of the coleoptile under laboratory conditions. Surface residues of sorghum, rape, lupin, and pea stimulated coleoptile and root growth of wheat under glasshouse conditions, whereas incorporated residues of the same species inhibited growth of the same plant parts. The same crop residues, under field conditions, inhibited growth of both the coleoptile and seminal roots of wheat, although the differences failed to attain statistical significance.

Chapman and Lynch (1983) found that anaerobically decomposed wheat straw inhibited seedling root growth of barley, whereas aerobic decay of *Anthroxanthum odoratum* residue stimulated root growth of barley. Separation of the associated microorganisms from their metabolites demonstrated the effect to be largely chemical. Lynch and Elliott (1983) reported that Daws wheat straw degraded for five days by microorganisms decreased growth of seedling roots of the same wheat cultivar less than nondegraded straw in laboratory tests. Subsequent tests of straw decomposed for 10 days on two types of soil demonstrated that the degraded straw was much less inhibitory to wheat root elongation in two cultivars than was fresh straw. These results suggest that the responsible allelochemicals could be degraded sufficiently by native populations of microorganisms under field conditions if proper conditions and time are allowed.

Hall et al. (1986) found that wheat straw collected from the soil surface nine weeks after the 1983 harvest in Pakistan contained abscisic acid. A 10-$cm^3$ leachate from 1 g of the straw contained sufficient ABA to retard wheat root extension. In contrast, straw incorporated into the soil shortly after harvest contained no detectable ABA nine weeks after harvest. In the wet autumn of 1984, ABA in surface straw decreased steadily to below the detection level seven weeks after grain harvest. Fractions of aqueous extracts of fresh straw inhibited root extension of barley. Lodhi et al. (1987) found that germination rates of cotton and wheat were significantly inhibited by extracts of wheat mulch and soils collected from the wheat fields in Pakistan. The toxicity was even greater against seedling growth. Five allelochemicals—ferulic, p-coumaric, p-hydroxybenzoic, syringic, and vanillic acids—were identified from the wheat residue and its associated soil. Ferulic acid was found at higher concentrations than p-coumaric acid. Several tested concentrations of ferulic and p-coumaric acids were inhibitory to growth of radish seedlings. Organic solvent and water extracts from monoculture wheat soils under conventional-tillage (CT) and no-tillage (NT) conditions indicated that both soils contained some inhibitory compounds. The total ion current chromatograms of extracts obtained by steam distillation of CT and NT soils were very complex, containing many peaks, but no specific allelopathic compounds were identified.

Thorne et al. (1990) followed up the previous work using somewhat similar techniques for bioassays and identification of allelochemicals. They found inhibitory biological activity in the lyophilized aqueous extract, methanol, methylene chloride, chloroform, and final water fractions of new straw, whereas the inhibitory activity of old wheat straw was present only in the lyophilized aqueous extract and the methanol fraction. The new straw was collected on the day of combining; the old straw represented the straw left on the surface of the soil and accumulated over four years. Some of the allelopathic compounds identified were dimethyl malonate; dimethyl fumarate; dimethyl succinate; dimethyl malate; dimethyl nonanoate; 1,2:3, 4:5,6-tris-0-(1-methylethylidene)-D-mannitol; methyl palmitate; methyl stearate; 11,14-eicosadienoate; methyl oleate; dimethyl terephthalate; and disopropyl phthalate. The mannitol derivative and the phthalates were not shown to be allelopathic compounds, but the quantities of the phthalates were high.

Cast et al. (1990) obtained allelochemicals from NT and CT soil by a mild alkaline aqueous extraction procedure, bioassayed them with Pioneer brand 2157 wheat, and analyzed them with GC-MS. The most significant inhibition occurred in extracts from soil collected immediately after harvest in June, July, and August. No-tillage soils also produced significant inhibition during the rest of the year. Fatty acids were the most abundant compounds identified, but the five free fatty acids showed no significant activity against wheat. The scientists concluded that there are one or more organic allelochemicals present in wheat soils, especially NT soils, but they had not identified them. They suggested that the allelochemicals responsible for the inhibition of wheat growth are microbial products that accumulate more in NT than in CT soils because that soil is not disturbed and dispersed by cultivation.

Pérez and Ormeño-Nuñez (1991a) found that wheat and rye (*Secale cereale*) cultivars with high hydroxamic concentrations in the leaves had these compounds in the roots of all cultivars analyzed and in root exudates of rye. Moreover, bioassays of root exudates from wheat and rye demonstrated that only rye exudates inhibited root growth of wild oats (*Avena fatua*), a weed whose root growth is inhibited by hydroxamic acids. Blum et al. (1991) obtained soil samples from NT wheat, CT wheat, and fallow CT soybean cropping systems from July to October of 1989 in North Carolina, extracted samples with water in an autoclave, and analyzed them for p-coumaric, vanillic, p-hydroxybenzoic, syringic, caffeic, ferulic, and sinapic acids by HPLC. The highest concentration determined was 4 $\mu g/g^{-1}$ soil for p-coumaric acid. Total phenolic acid content of the 0–2.5-cm core samples from wheat NT systems was significantly higher than those from all other cropping systems. The individual phenolic acid contents and total phenolic acid content were highly correlated. The researchers concluded that the primary source of available phenolic acids would be from the action of microorganisms on plant litter. Thus the most likely place for allelopathic interactions in the soils tested would be in the soil surface of NT plots.

Blum et al. (1992) used a water-autoclave extraction procedure in CT and NT wheat straw and residues over an immediate water or a five-hour EDTA extraction method because the autoclave procedure was effective in extracting solution and reversibly bound ferulic acid as well as phenolic acids from wheat debris. A mixture of phenolic acids similar to that obtained from NT wheat soils did not affect germination of crimson clover (*Trifolium incarnatum*) or ivy-leaved morning glory (*Ipomoea hederacea*), nor the radicle and hypocotyl length of morning glory. The mixture *did* reduce radicle and hypocotyl length of the clover. Individual phenolic acids did not inhibit germination of the clover and morning glory, but they reduced radicle and hypocotyl growth of both species. 6-MBOA, a conversion product of 2–0-glucosyl-7-methoxy-1,4-benzoxazin-3-one, a hydroxamic acid in living wheat plants, inhibited germination and radicle and hypocotyl growth of clover and morning glory. Unfortunately, 6-MBOA was not detected in wheat debris, stubble, or soil extracts. Total phenolic acids in both CT and NT soil extracts were not related to germination of clover or morning glory, but extracts from NT soils were inversely related to radicle and hypocotyl length of both species. According to these authors, data derived from the water-autoclave extraction procedure, total phenolic acid analysis, and slope analysis for extract activity in conjunction with data on extract pH and solute potential can be used to estimate allelopathic activity of NT wheat soils.

Many producers are planting cotton (*Gossypium hirsutum*) into wheat straw in the southern high plains of Texas, so Hicks et al. (1989) conducted tests on the allelopathic potential of wheat residues on cotton germination, emergence, seedling growth, and lint yield. Laboratory tests indicated that cotton seedling development was inhibited by aqueous extracts of wheat straw. Thus cotton cultivars were screened for the ability to tolerate inhibitive effects of wheat straw. In field tests, emergence was reduced by an average of 9% for Paymaster 404 and 21% for Acala A246 when wheat stubble residues were present in the seedbed. The allelopathic effect of wheat straw indirectly affected lint yield by affecting population densities. It was concluded that the negative effects of wheat residue could be overcome by limiting the amount of aboveground residue at the time of cotton planting, increasing the seeding rate, and planting tolerant cultivars.

Young et al. (1989), using a wheat seed bioassay, found that wheat rhizosphere soil extract was phytotoxic to wheat seedlings, and more allelopathic materials were found in the wheat-rice rotation soils than in the rice soils. Moreover, extracts obtained under basic conditions (pH 8) were more inhibitory than those obtained by acid extraction (pH 5). Concentration of the aqueous extracts by rotary evaporation gave greater phytotoxicity than concentration by lyophilization.

**Conclusion** It is obvious that many crop species have allelopathic effects on themselves and other crop plants. It is equally obvious that this allelopathic potential makes it essential to return to crop rotation systems in farming. Moreover, it is notable that practices are being developed in agriculture to help eliminate some of the problems caused by crop allelopathy, and continued research will result in more and better practices to prevent the harmful effects.

CHAPTER 2

# ALLELOPATHIC EFFECTS OF WEEDS ON CROP PLANTS

HOLM (1972) FOUND that freshly harvested *Abutilon theophrasti* (velvetleaf), *Ipomoea purpurea,* and *Brassica kaber* seeds, which normally germinated well in light or dark, failed to germinate in soil if buried too deeply. Flushing the soil with air for two minutes per day overcame much of the inhibition due to burial, and nitrogen flushing provided similar relief. Experiments using KOH to trap $CO_2$ gave no relief from the inhibition. Thus the remaining factor was possible volatiles produced by the seeds. Ethanol, acetone, and acetaldehyde were identified, with ethanol being present in the highest concentration in *B. kaber* and *I. purpurea* (0.1 μL/L). Acetaldehyde and acetone were present in the concentration of 0.01 μL/L in both species. When these volatile components were added to the air flush of soil containing seeds, germination of the seeds was reduced by about 80%. This inhibition was largely overcome by a two minute per day soil flush. Addition of the volatiles to the air flush again reimposed most of the inhibition caused by burial. The same results were obtained in sterile sand with surface-sterilized seeds, indicating that the gases came from inside the seeds and not from external microbial action.

Dekker and Meggitt (1983) reported that, in field tests, 2.4 to 4.7 plants per $m^{-2}$ of *Abutilon theophrasti* reduced soybean dry matter, flowering node, and seed production. These effects were due to the presence of *A. theophrasti* and not to changes in plant population of soybeans.

Bhowmik and Doll (1984) evaluated the effects of the aboveground residues of *A. theophrasti,* common lambsquarters (*Chenopodium album*), redroot pigweed (*Amaranthus retroflexus*), common ragweed (*Ambrosia artemisiifolia*), and yellow foxtail (*Setaria glauca*) for their allelopathic effects on the growth and uptake of N, P, and K in corn and soybean under controlled environments. The growth inhibition was more (8 to 14%) when the residues were incorporated in the soils than when they were surface-applied. All except ragweed residue reduced dry-matter production by 6 to 20% in corn and 2 to 20% in soybeans. The inhibition or stimulation of N, P, and K uptake in both crops was not consistent, depending on the residue source, residue placement, and soil texture. Nevertheless, addition of N and P to various residues did not alleviate growth effects on either crop. It was inferred that the weed residues caused an allelopathic inhibition of growth that was not related to nutrient uptake.

Sterling and Putnam (1987) experimented with the liquid globules that are exuded by glandular trichomes on velvetleaf stems and petioles. They attempted to separate the influence of these glandular exudates from other mechanisms of interference by removing exudates from some velvetleaf plants and not from others. Removing exudates from some plants and not from others did not affect interference by velvetleaf in field or greenhouse studies. A velvetleaf accession from Rosemount, Minnesota, produced 1.4 times more exudate per stem dry weight than an accession from Stoneville, Mississippi, but the exudate from the latter accession was about 1.4 times as toxic as the Minnesota exudate in petri plate assays. Greenhouse-grown plants had 3.3 times as much exudate per unit stem dry weight as field-grown plants, but the field exudate was 1.6 times as toxic. The exudate of the trichomes was active in autoclaved soil. Thus microorganisms apparently detoxify the exudate rapidly. Subsequent research by Sterling et al. (1987) reported that the quantity of trichome exudate produced did not vary at 16, 24, or 36°C, but the exudates collected from velvetleaf plants grown at 24 and 36°C were about twice as toxic as the exudates from plants grown at lower temperatures.

Aqueous extracts of seed coats of velvetleaf were slightly inhibitory to germination of cress, radish, and soybean seed and were very inhibitory to radicle growth of the same species (Paszkowski and Kremer, 1988). Fungal growth was either inhibited or not affected by the extracts, depending on the test species (five species were tested). Six flavonoids were isolated and identified from the extracts: delphinidin, cyanidin, quercetin, myricetin, (+)-catechin, and (–)-epicatechin. All of these compounds significantly inhibited germination and radicle growth of all test plants at 1.0 $\mu M$. They also inhibited growth and sporulation of potential seed-decomposing fungi, thus appearing to function in defensive roles against such fungi and competing seedlings.

Murthy and Zakharia (1980) tested water extracts of 10 weeds from bajra (*Pennisetum typhoides*) fields against seed germination, radicle elongation, and dry weight of bajra. The weed species investigated were *Acalypha indica, Commelina benghalensis, Cassia pumila, Cleome simplicifolia, Corchorus antichorus, Tribulus terrestris, Pulicaria wightiana, Ipomoea biloba, Leucas cephalotes,* and *Amaranthus spinosus.* Inhibition values based on dry weight of bajra indicated that *A. indica, L. cephalotes,* and *P. wightiana* had maximum allelopathic potential. Sugha (1979) reported that water extracts of roots, stems, and leaves of *Ageratum conyzoides, Euphorbia geniculata,* and *Plantago lanceolatus* and seed extracts of *E. geniculata* caused statistically significant reductions in wheat seed germination.

*Agropyron repens* (couch grass) has been implicated in allelopathy for many years (Rice, 1984). Lynch et al. (1980a,b) therefore investigated the causative allelochemicals resulting from the anaerobic decomposition of couch grass rhizomes. They reported that the short chain aliphatic acids (acetic, propionic, and butyric) appear to be chiefly responsible for the allelopathic effects, but hexanoic, succinic, phenylacetic, cinnamic, p-coumaric, 4-hydroxyphenylpropionic, and 3,4-dihydroxyphenylpropionic acids are also present in the allelopathic solutions formed during the decomposition.

Poor establishment and growth of a succeeding crop has sometimes been reported after glyphosate was used to kill couch grass (Lynch and Penn, 1980). These scientists concluded that herbicide residue in the rhizomes or soil did not appear to be the cause because the damage symptoms to new crop plants were similar whether the rhizomes of the couch grass were killed by heat or glyphosate. This conclusion was consistent with reports that glyphosate is rapidly degraded by microorganisms. As previously discussed, decomposition of the rhizomes by plant enzymes or microorganisms is known to produce allelopathic compounds, and these compounds could have allelopathic activity against the new crop plants. Lynch and Penn also found that the most common microorganism in decomposed rhizomes of couch grass is *Fusarium culmorum,* an organism that causes foot rot and severe stunting of barley seedlings. Allelopathic compounds also make plants more susceptible to infection by pathogens, so these scientists concluded that the inhibition and death of succeeding crop plants is probably due to a synergistic effect between allelopathy and *F. culmorum.* In a subsequent project, Penn and Lynch (1981) reported that growth of barley seedlings was significantly inhibited by the presence of live couch grass rhizomes near their roots; the addition of N did not decrease this effect. Decaying rhizomes of couch grass caused severe damage and death at times. The scientists reported that the damage was greater after five-week decay than after four-week. However, if the decay period was extended to seven weeks, the inhibitory effect disappeared and growth was stimulated, indicating that microorganisms were either producing the allelopathic compounds or were changing them to more active forms and subsequently degrading them.

Weston and Putnam (1985) planted inoculated soybeans, navy beans (*Phaseolus vulgaris*), and snap beans (*P. vulgaris*) in (1) living quackgrass (couch grass) sod that was regularly mowed, (2) glyphosate-treated quackgrass sod, (3) soil from quackgrass sod from which plant material was removed, and (4) a control soil of similar type and structure but free from quackgrass invasion. Legumes grown in mowed quackgrass sod in the greenhouse and field had decreased nodule numbers, nodule fresh weight, and $N_2$ fixation compared with legumes grown under similar conditions in screened quackgrass soil or control soil. Shoot and root weights were also significantly decreased in legumes grown in living quackgrass sod. In many cases, legume nodulation and growth were decreased in glyphosate-treated quackgrass sod as compared with screened quackgrass soil or control soil. Thus quackgrass was obviously allelopathic to growth and $N_2$ fixation of the three legumes.

In subsequent research, Weston and Putnam (1986) found that aqueous extracts of quackgrass shoots and rhizomes inhibited seed germination and root growth of alfalfa, soybeans, navy bean, and curly cress (*Lepidium sativum*) at concentrations less than 2.5 mg dried extract per milliter. The extracts of shoots were generally more inhibitory than those of rhizomes. Root and shoot dry weights of snap beans grown under sterile conditions were reduced by aqueous extracts of shoots, and root systems were stunted and necrotic and lacked root hairs (Fig. 1). Growth of *Bradyrhizobium japonicum, Rhizobium meliloti,*

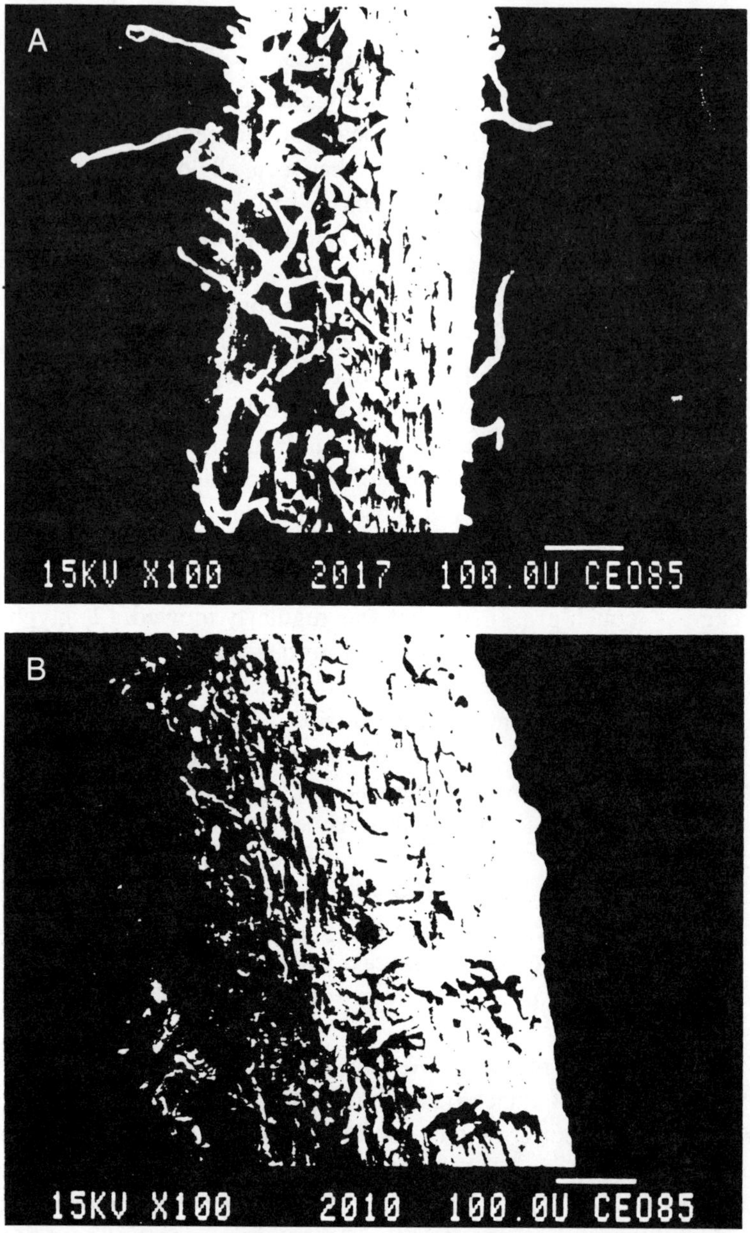


FIGURE 1 (*A*) Scanning electron micrograph (×100 magnification) of a typical untreated, aseptically grown snap bean root (control). Root segment was taken 1.5 cm from root tip. Note the presence of numerous root hairs. (*B*) Scanning electron micrograph (×100 magnification) of a typical aseptically grown snap bean root cultivated in extract of quackgrass shoots. Root segment was taken 1.5 cm from root tip. Note the absence of root hairs and flaking of epidermal tissue. *From Weston and Putnam (1986), with permission of Weed Sci. Soc. of America.*

*R. phaseoli,* and *R. leguminosarum* was not at all affected by either 40 or 80 mg/mL concentrations of extracts of shoots or rhizomes. Thus the researchers concluded that quackgrass may inhibit the legume-*Rhizobium* or legume-*Bradyrhizobium* symbiosis indirectly by inhibiting root hair formation. They also inferred that soil microorganisms were not necessary for the development of quackgrass toxicity in soil or agar and that soil microorganisms reduced toxicity of quackgrass residues in soil.

Hagin (1989) identified the major allelopathic compounds in quackgrass as 5-hydroxyindole-3-acetic acid (5-HIAA) and 5-hydroxytryptophan (5-HTP) using alfalfa as the bioassay plant and TLC, HPLC, MS, UV, IR, and C,H,N analysis. He found that these compounds accumulate to high concentrations throughout the plant as glucosides attached to the 5-O-indolyl moiety in ß linkage. Molecular weights of the glucosides ranged from 353 to 4159. Roots accumulated glucosides rich in 5-HIAA, whereas tops were rich in phenolic glucosides containing about twice as much 5-HTP as 5-HIAA. Tests demonstrated that 5-HTP inhibited root growth and shoot growth of corn at some concentrations and stimulated growth at other concentrations. 5-HIAA mostly stimulated root growth of corn, except at the higher concentrations of $10^{-6}$ and $10^{-5}$ *M*. There was no significant effect on corn shoot growth of 5-HIAA. All concentrations of HIAA ($10^{-3}$ to $10^{-11}$ *M*) tested inhibited root growth of red kidney beans but only the two highest concentrations inhibited shoot growth. None of the similar 5-HTP concentrations affected root or shoot growth of red kidney beans.

Bhowmik and Doll (1983) reported that redroot pigweed (*Amaranthus retroflexus*) reduced corn height, leaf area, and dry weights at all growth stages. Yellow foxtail residue reduced total dry weights of corn 10, 20, and 30 days after planting (DAP) but inhibited corn height and leaf area only at 30 DAP. Larger shoot-root weight ratios occurred with the pigweed residue, indicating a greater inhibitory effect on root growth. Redroot pigweed residue inhibited height, dry weight, and leaf area of soybeans at all growth stages. Yellow foxtail residue did not affect soybean growth at 10 DAP but reduced dry weights and leaf areas 20 and 30 DAP. Temperature and photosynthetic photon flux densities were found to affect the actions of the weed residues to some extent against both corn and soybeans.

Menges (1987) examined the effects of residues of several crop and weed species on the subsequent growth of the same or other species over several years in field plots. Crops involved were cabbage (*Brassica oleracea*), carrot, onion, cotton, and sorghum (grain); the weeds were *Amaranthus palmeri, Sisymbrium irio* (London rocket), *Helianthus annuus* (common sunflower), and *Sorghum halepense* (Johnsongrass). Palmer amaranth biomasses of 8.5 and 5.1 kg/m$^2$ decreased the fresh weight of onions and carrots in 1982 and 1983, respectively, but increased the weight of Palmer amaranthus in 1983. Cabbage, *S. irio*, and common sunflower were not affected. A Johnsongrass residue of 6.6 kg/m$^2$ decreased the growth of onion 71% in 1982, and a residue of 5.4 kg/m$^2$ decreased the growth of cabbage 26%, onion 62%, and sunflower 10% in 1983. Palmer amaranth grown at population densities of 300 plants/m$^2$ in 1984 and 1985 attained biomasses of 2 to 3 kg/m$^2$ in 10 weeks, and phytotoxicity persisted 11 weeks in cabbage and carrot plantings. In 1985 phytotoxicity disappeared 16 weeks after incorporation of Palmer amaranth in soil. Residues of this weed were obviously very allelopathic to several crop plants and other weeds, at least under some conditions.

Bradow and Connick (1987) reported that Palmer amaranth residues in soil reduced fresh-weight accumulation in onions and carrots and strongly decreased establishment of carrots in field tests. Water-soluble organic compounds were extracted from Palmer amaranth residues and soils, and most of these compounds were inactive to seed germination of onion, carrot, and tomato seeds at concentrations of 20 to 100 mg/L. The aqueous extracts from roots of Palmer amaranth did increase 72-hour germination percentages in carrots and tomatoes. A time study of residue decomposition demonstrated that there was an increase in extractable inhibitors of onion germination after 62 days, but no other significant changes occurred. The most active allelochemicals from Palmer amaranth proved to be volatile compounds. Volatiles from soil containing residues and from partially rehydrated leaf and flower residues reduced carrot and tomato seed germination to less than 7%; onion seeds were also inhibited by volatiles from Palmer amaranth residues. The researchers concluded that the volatiles emitted from Palmer amaranth residues could definitely explain the decreased stands of carrots in fields amended with such residues, and the residual effects of exposure to the volatiles could be an important factor in reducing fresh weight and in delayed maturity of crops.

In subsequent research, Bradow and Connick (1988a,b) identified nine methyl ketones, 3-pentanone, and eight alcohols and aldehydes in the mixtures of volatiles released from Palmer amaranth residues. A three-day exposure of onion, carrot, and tomato seeds to each of the compounds at a concentration of 34.4 $\mu M$ significantly inhibited germination of each crop plant in every test. Using time- and concentration-dependent assays, the scientists found that the activity series for the methyl ketones was 2-octanone, 2-nonanone > 2-undecanone > 2-heptanone > 2-hexanone, 3-methyl-2-butanone, 2-pentanone, 3-hydroxy-2-butanone > 2-butanone. Using similar assays, they found that the series for the 3-pentanone and alcohols and aldehydes was 2-heptanol > 3-methyl-1-butanol, 1-hexanol > hexanal, 1-pentanol, 3-pentanone, acetaldehyde > ethanol, 2-methyl-1-propanol. 2-Heptanol and 2-heptanone also significantly inhibited germination of *Capsella bursa-pastoris,* soybeans, lettuce, alfalfa, *Portulaca oleracea,* oats, and weeping lovegrass (*Eragrostis curvula.*) The researchers concluded that these volatiles released from various plant residues must be important factors in the various food webs encountered in both conventional and no-tillage ecosystems.

Menges (1988) reported that seedling growth of grain sorghum, cabbage, carrot, and onion were adversely affected by residues of Palmer amaranth. Root and shoot growth were equally affected. Cultivar differences in sensitivity of some crop plants were pronounced; Grand slam cabbage was 17 to 30% more sensitive than Sanibel cabbage. Growth of onion and carrot seedlings was less inhibited than either cabbage or grain sorghum. Grain sorghum root growth was severely inhibited by 8000 and 16,000 ppm of Palmer amaranth residue in soil; activity of the residue was not affected by oven drying.

Hegazy et al. (1990) investigated the allelopathic potential of *Anastatica hierochuntica,* a plant of exposed depressions in Egypt which generally grows in pure stands. Shoot-water extracts inhibited seed germination, seedling growth, and cell division of *Rumex cyprius, Trigonella stellata, Diplotaxis harra, Cleome droserifolia,* and *Farsetia aegyptia,* all of which are plants of the Egyptian deserts. An extract concentration of 8% reduced the mitotic index from the control by 55% in *C. droserifolia,* 54% in *T. stellata,* 45% in *F. aegyptia,* 43% in *A. hierochuntica,* and 35% in *R. cyprius.* The researchers concluded that the inhibitory compounds are released onto soil by repeated washing of the standing plants by rain and dew.

Linear furanocoumarins (psoralens) are known to be growth and germination inhibitors to several plants, and so Zobel and Brown (1991) decided to determine concentrations of these compounds on the outsides and insides of the fruits and seeds of various plants in the Rutaceae, Umbelliferae, and Leguminosae. Surface concentrations varied between traces and 40 μg/g for the mature fruits or seeds. Concentrations were very high in the whole fruits of *Angelica archangelica, Heracleum lanatum, Pastinaca sativa,* and *Psoralea bituminosa,* reaching concentrations of mg/g fresh weight of fruits. Seeds of the Rutaceae had much smaller concentrations of furanocoumarins, both on the surface and in the seed.

Manners and Galitz (1986) found that an ether extract of *Antennaria microphylla* (small everlasting) was inhibitory to the growth of the radicle of lettuce and to root elongation and cell culture growth of *Euphorbia esula* (leafy spurge). They isolated and identified hydroquinone, arbutin, and caffeic acid from the ether extract and found that hydroquinone was very inhibitory to root growth of leafy spurge at 50 ppm (w/v) and that arbutin and caffeic acid were moderately inhibitory to root growth at 300 ppm. Arbutin is a water-soluble, easily hydrolyzed, monoglucoside of hydroquinone.

Sharma and Nathawat (1987) reported that *Argemone mexicana* was strongly allelopathic to the growth of four crop plants cultivated in the semiarid area of Jaipur, Rajasthan: wheat, mustard, radish, and bajra. *A. mexicana* did not, however, affect their germination.

Ito et al. (1981a) reported that *Artemisia princeps* (wormwood) and *Digitaria sanguinalis* (crabgrass) strongly depressed the number of roots on apple trees after three years of growth in the orchard; the distribution of apple rootlets to the upper layer of soil was also suppressed. Ito et al. (1981b) found also that wormwood, crabgrass, *Rumex japonicus, Polygonum longisetum, Cyperus rotundus*, and *Lolium multiflorum* markedly increased the shoot-root ratio of peach seedlings when used as mulches in peach orchards. Aqueous extracts of these species inhibited root growth in lettuce seedlings but promoted shoot growth, again increasing the shoot-root ratio. Yun and Kil (1989, 1992), Kil et al. (1991), and Kil and Yun (1992) reported *Artemisia princeps* var. *orientalis* to be allelopathic to some eight weed species and three crop species. Yun and Kil (1989) isolated and identified some 18 allelochemicals in leaf extracts, most of which were phenolic compounds. They did not bioassay these compounds to determine which were allelopathic. In other tests, however, they found that aqueous extracts of leaves, stems, and roots of wormwood were inhibitory to seed germination and to seedling growth of most test species. Radicle growth of all test species except two was inhibited by volatile compounds from leaves. Kil et al. (1991) identified 10 volatile terpenoids from wormwood leaves: α-pinene, ß-myrcene, α-terpinine, cineole, γ-terpinine, (-)-thujone, camphor, bornyl acetate, (-)-transcarophyllene, and α-humulene. These were not bioassayed against test species, but most have been shown by other scientists to be allelopathic against many plant species.

Schumacher et al. (1983) reported that *Avena fatua* (wild oats) root exudates had no allelopathic effects on spring wheat until the wild oats reached the four-leaf stage of development. At this stage, the root exudate caused a 34% decrease in root dry weight of wheat compared with untreated controls. Leaf dry weight of wheat was significantly reduced by exudates from wild oats, starting at the two-leaf stage. Tests of isolated compounds from wild oats root exudates indicated that two allelopathic compounds were present, scopoletin and vanillic acid. Porwal and Gupta (1986) found that root exudates of wild oats, *Chenopodium murale,* and *Phalaris minor* reduced the shoot and head lengths and the dry matter of wheat. Aqueous extracts of fresh roots of *Chenopodium album* also reduced germination, plumule, and radicle lengths of wheat. Qureshi et al. (1987) reported that aqueous extracts of shoots and roots of wild oats, shoot mulch, soil collected under the mulch, and litter reduced germination, radicle and plumule growth, water content, and chlorophyll content of wheat, corn, mustard, and *Pennisetum americanum* in both field and laboratory studies. Paper and gas chromatography were used to identify the following compounds in water extracts of roots and shoots: chlorogenic, ellagic, caffeic, ferulic, *p*-coumaric and *p*-OH-benzoic acids. All are proven allelopathic compounds, and all are water-soluble. Pérez and Ormeño-Nuñez (1991b) also found that root exudates of wild oats inhibited root and coleoptile growth of spring wheat. They identified scopoletin, coumarin, *p*-OH-benzoic, and vanillic acids in the exudates by HPLC, all of which are known allelopathic compounds.

Kuti et al. (1990) reported that the trichothecene roridins and baccharinoids occur naturally in the Brazilian plants *Baccharis coridifolia* and *B. megapotamica* and that biosynthesis of these trichothecenes appears to be related to intraspecific pollination in the species. Removal of seed coats from the trichothecene-producing Brazilian species and from the nontrichothecene-producing American species, *B. halimifolia* and *B. glutinosa*, resulted in improved seed germination of the American species but complete inhibition of seed germination of the Brazilian species. Addition of seed-coat extracts of the Brazilian *Baccharis* species in dilute solutions ($10^{-6}$ μg/mL) to decoated seeds of Brazilian species resulted in their germination, but addition of the extract to decoated American seeds killed the seeds.

Root exudates of *Bidens pilosa* were found to significantly inhibit root growth of lettuce, beans, corn, and grain sorghum. In these experiments, Stevens and Tang (1985) used a root-exudate recirculating system that allows continuous exposure of plants to allelopathic chemicals and eliminates physical contact and competition. When hydrophobic compounds were eliminated in the system by XAD-4 resin, considerably less root inhibition occurred in all species except corn, indicating that the hydrophobic exudates played an important role in growth inhibition in all plants except corn. Meissner et al. (1986) collected soil from a plot of pure *Bidens bipinnata,* a plot of pure *Tagetes minuta*, and a bare plot having similar soil. Water extracts of the soils where the weeds previously grew and from the bare plot were tested against germination of seven crop species. The *Tagetes* soil extract reduced germination of two crop species; the *B. bipinnata* soil extract also reduced germination of the same crop species. Soil from the *Tagetes* plot reduced the dry weight of the tops and plant height of all nine crop plants tested. Soil from the *Bidens* plot significantly affected growth of all nine crop plants also, but the effects were not as marked as with the *Tagetes* soil. It is notable that roots of *T. minuta* produce alphaterthienyl and leaves of *B. bipinnata* produce phenylheptatriyne, naturally occurring polyacetylenes.

Biswas and Chakraborti (1984) found that water extracts of leaves of *Borreria articularis* (*Spermacoce hispida*) inhibited either the root or hypocotyl growth, or both, of jute (*Corchorus olitorius*), lettuce, mustard, rice, and wheat and five weed species.

*Bothriochloa pertusa* is a weed of cultivated fields on relatively mesic habitats in Pakistan, it replaces or reduces growth of associated species within the habitat. This weed significantly reduced growth of tomato, lettuce, corn, German millet (*Setaria italica*), and *Capsicum annuum* in root-mixed cultures by releasing water-soluble chemicals from the roots in the field. Aqueous extracts from shoots, roots, seeds, and root exudates; shoot mulch; soil underneath the grass; and soil extracts also reduced germination, radicle growth, and dry matter of the crop plants listed above. The results indicated marked allelopathy of *B. pertusa* against the tested species.

*Brassica napus* is a common and widespread weed and crop crucifer. The allelopathic potential of various *Brassica* species has been attributed to the release of the mustard oil glycosides, which on hydrolysis yield isothiocyanates with strong allelopathic properties (Choesin and Boerner, 1991). Choesin and Boerner were interested in determining if, and how much, allyl isothiocyanate (AI) is released into the substrate from the wild type and from the low-glucosinolate mutant *B. napus* grown in soil low in organic matter. The wild type generally has a concentration range of 186–524 $\mu$mol/g$^{-1}$, whereas the mutant averages about 10.4 $\mu$mol/g$^{-1}$. Wild-type plants released more AI into the substrate than did low mutant types. Alfalfa was not affected, however, by addition of 20 $\mu$g/kg$^{-1}$ or 195 $\mu$g/kg$^{-1}$ of AI to the substrate. It was concluded that under test conditions, *B. napus* showed no indication of being allelopathic.

Srivastava and Das (1974) found that water extracts of *Cannabis sativa* and *Argemone mexicana* inhibited germination of tubers of *Cyperus rotundus* (purple nutsedge), but water extracts of *Eichornia crassipes, Sanseveria* sp., and *Holarrhena antidysentrica* promoted germination of purple nutsedge tubers. No follow-up experiments were done to determine what active compounds were produced or whether they moved from the active plants into the substrate.

Bhatia and Chawan (1976) found that *Cassia tora* and *C. auriculata* did not germinate when all conditions were favorable. When seeds of these species were mechanically scarified, they germinated but the seedlings were stunted. When the seeds were chemically scarified with concentrated $H_2SO_4$, followed by washing for several hours in running water, germination and seedling growth were normal. When the seeds were extracted with a series of organic solvents—petroleum ether, benzene, acetone, ethanol, methanol, and water—and the fractions were tested against til (*Sesamum indicum*), the ethanol, methanol, and water fractions strongly inhibited radicle and hypocotyl growth, with methanol being the most active. All three of these fractions gave positive tests for phenols, and the methanol fraction of *C. auriculata* was found to contain proanthocyanidin. The other active fractions were not identified.

Datta and Dasmahapatra (1984) found that water leachates and extracts of different organs of *Cassia sophera* var. *purpurea* and *Crotolaria pallida* var. *pallida* were usually inhibitory to seedling growth and sometimes to seed germination of wheat, lettuce, mustard, and black gram (*Phaseolus mungo*). The extracts were generally more active than were leachates of the same organs. When seeds of test plants were exposed to leaf extracts for more than 48 hours, seeds of all test plants failed to germinate. Upon treatment with Norit, extracts of *Cassia* and *Crotolaria* became colorless and lost their activity.

Interactions between 12 cultivars of soybean and *Cassia obtusifolia* (sicklepod) were studied in greenhouse and field tests by James et al. (1988). In field tests, sicklepod increased the plant height of three cultivars of soybean significantly at harvest, decreased the canopy width of half the cultivars, decreased the dry weight of ten cultivars, decreased the number of leaves of half the cultivars, and significantly decreased the leaf area of five cultivars of soybean. The nature of the experiments was such that one could not tell whether the interference was due to allelopathy or competition. Nevertheless, according to James et al., the potential exists to develop soybean cultivars having improved interference characteristics with sicklepod.

*Celosia argentea* is a herbaceous weed infesting monsoon fields in India; it has been reported by several scientists to be allelopathic. Pandya et al. (1984) decided therefore to determine what effects interference by this weed would have on growth and yield of pearl millet (*Pennisetum typhoides*), grain sorghum, and corn. They found that increase in density of the weed decreased dry-weight production and yield of all the crops tested. The weed was more detrimental within the crop row and when associated for a longer time period. No doubt, at least some of the pronounced interference was due to the reported allelopathic potential of the weed.

Lahiri and Kharabanda (1962) reported that germination inhibition of spikelet-enclosed grains of *Cenchrus ciliaris, C. setigerus*, and *Lasiurus sindicus* was caused by naturally occurring water-soluble inhibitors located in the enveloping structures of the spikelets and was not due to restricted moisture or gaseous diffusion, or to immaturity of the embryos. Roder et al. (1988) found that establishment of *Panicum virgatum* from seed was limited by current year's growth of *Cenchrus longispinus* (sandbur). *P. virgatum* germination was not influenced by root, shoot, or whole-plant leachate from sandbur plants composited over several phenological stages. Leachate reduced the length of the primary root, however, and increased shoot length at 11 days in *P. virgatum*. It was concluded that sandbur may affect *P. virgatum* germination and early root growth through allelopathy.

Glandular trichomes on the epidermal surfaces of *Centaurea maculosa* contain the sesquiterpene lactone cnicin (Locken and Kelsey, 1987). Highest concentrations occur in leaf tissue, with low concentrations in inflorescence branches, stems, and heads. Nocnicin occurs in the roots, and only trace levels occurred in soils at the study site in Montana. Cnicin was allelopathic to all test species under laboratory conditions, including three range grasses, western larch, and lodgepole pine. According to Locken and Kelsey, the potential for cnicin to function as an allelopathic compound is enhanced when live or dead *C. maculosa* plants are trampled or mowed, bringing large quantities of cnicin-rich tissues in direct contact with the soil, seeds, and seedlings. *Centaurea solstitialis* (yellow starthistle) is an aggressive weed in the western United States and other parts of the world. Merrill (1989) isolated and identified two chromenes, eupatoriochromene and encecalin, from yellow starthistle and found both to retard seed germination and decrease radicle and hypocotyl growth of lettuce, red millet (*Panicum miliaceum*), alfalfa, cucumber, *Echinochloa crusgalli* (barnyard grass), *Lolium perenne* (perennial ryegrass), and yellow starthistle. She concluded that these chromenes, plus the sesquiterpene lactones found in yellow starthistle, may play a key role in the allelopathic potential of this species.

Several species of *Chenopodium* have been reported to be allelopathic (Datta and Ghosh, 1982, 1987; Biswas and Ghosh, 1984; Qasem and Hill, 1989). Datta and Ghosh (1982) reported that treating mustard (*Brassica juncea*) seeds with leaf extracts of *C. murale* in concentrations of 1:2.5, 1:5, and 1:10 (w/v) lowered germination values at all concentrations in Petri dishes, but only at the highest concentrations in soil. On the other hand, the two highest concentrations of inflorescence extract stimulated the seed output and seed weight of mustard. It was inferred that an oily residue having carboxyl and hydroxyl functions and a solid residue containing oxalic acid are probably involved in the allelopathy of the *C. murale* extracts. Later, Datta and Ghosh (1987) reported that in numerous instances, leaf and inflorescence extracts and leachates, decaying leaves and inflorescences, and field soil collected beneath *C. ambrosioides* and *C. murale* were active against germination and/or seedling growth of five test weeds: *Abutilon indi-*

*cum, Cassia sophera* var. *purpurea, C. tora, Evolvulus numularius,* and *Tephrosia hamiltonii.* Three active terpenes (*p*-cymene, ascaridole, aritazone) were isolated and identified from *C. ambrosioides,* and oxalic acid was identified again from *C. murale.* Qasem and Hill (1989) developed a new drip technique to determine possible allelopathic effects of *Chenopodium album* and *Senecio vulgaris* on tomatoes. *C. album* leachates significantly reduced shoot fresh and dry weight and accumulation of N, P, K, Ca, and Mg in tomato shoots. No effect was found, however, on quantities of these elements in tomato roots. Leachate of five plants of *C. album* per pot was sufficient to reduce tomato plant growth. Addition of 10 or 20 g $kg^{-1}$ of *C. album* dried shoots to the soil mixture significantly decreased the fresh and dry weights of tomato plants also. On the other hand, *S. vulgaris* leachates did not show significant effects on growth or nutrient accumulation by tomato plants. The researchers concluded that *C. album* had effects on tomato plants through allelopathy.

*Cirsium arvense* is the most common weed of wheat fields in the plains of Pakistan, and Hussain et al. (1987) found that *Cirsium*-wheat mixtures in fields for three to seven weeks reduced plant height; number of tillers; nodes per tiller; and number, length, and breadth of leaves. In controlled laboratory tests, litter leachate, litter-soil mixture, field soil from under *Cirsium,* and rain leachate markedly reduced radicle growth, plumule growth, dry weight, and fresh weight of barley, as well as radicle and plumule growth of wheat, in most tests. Germination of both species was reduced in a few tests. Ferulic, caffeic, chlorogenic, *p*-coumaric, *p*-hydroxybenzoic, benzoic, and vanillic acids were identified as allelochemicals in the rain leachate of dried *Cirsium.* Although most of these acids have been shown by other scientists to be allelopathic compounds, they were not tested against growth of wheat and barley by Hussain et al. Ballegaard and Warncke (1985) observed that no, or very sparse, germination of *C. palustre* achenes occurred under and in the nearest surroundings of the older rosettes in a spring area at Jutland, Denmark. Ethanolic extracts of foliage of *C. palustre* significantly reduced germination of *C. palustre* achenes. In fact, germination was almost completely prevented in the first 10 to 16 days, after which some germination occurred, perhaps because of partial decomposition of the allelopathic compound(s). The inhibition was not mediated by pH or osmotic effects. Incorporation of *Cirsium* top residue in soil (0.25%) reduced growth of *C. palustre* by 52% compared with the control (no organic matter added) or the peat moss–soil mixture. Roots did not affect growth of *Cirsium*. It appears that the autoallelopathic effects of *C. palustre* may explain the failure of its achenes to germinate in natural areas, at least under some conditions.

Datta and Chakrabarti (1982a) found that all plant parts of *Clerodendrum viscosum,* an undershrub and serious weed in some parts of India, bear inhibitors to mustard (*Brassica nigra*). The three-day leachate of plant parts is more toxic than seven-day leachates, and all leachates cause greater germination and growth inhibition during the monsoon than during other seasons. Chlorosis develops in black gram and mustard plants receiving a spray or soil dressing of the leachates. Moreover, nodule formation is reduced in black gram plants by soil-applied leachates, and no nodulation occurs when these plants are sprayed with leaf leachates. An active terpene was isolated from the leaves and identified as clerodin. Datta and Chakrabarti (1982b) later reported that decaying plant parts and field soils collected beneath *Clerodendrum* plants reduced germination and seedling growth of five weed species—*Abutilon indicum, Amaranthus spinosus, Cassia sophera, C. tora*, and *Tephrosia hamiltonii*—in most tests. *C. tora* was generally stimulated in growth by decaying leaves, but was inhibited by decaying roots. Clerodin was again suggested as the chief allelopathic compound.

In previous work, Datta and Sinha-Roy had demonstrated that *Croton bonplandianum* was strongly allelopathic to other plants with which it was associated. Thus they determined to try to identify the allelopathic compounds in the leaves of that species (Datta and Sinha-Roy, 1983). They used pea (*Pisum arvense*), mustard (*Brassica juncea*), lettuce, and rice as assay plants and water extracts of the leaves of *Croton* in their experiments. The extracts had virtually no effect on germination of peas, but germination of mustard, lettuce, and rice was strongly reduced. A leachate made by soaking *Croton* leaves for three days was also inhibitory, but the inhibitory effect was much greater after soaking for seven days. The inhibitory effect of the water extract decreased with the age of the *Croton* leaves. Moreover, activity occurred in darkness but was accentuated in light. Soil to which the leaf extract was added was inhibitory to growth of some assay plants, but spraying test plants with the extract usually had greater effects. The use of TLC, UV absorption spectra, and IR spectra led these scientists to conclude that the active compounds were probably abscisic and phaseic acids in the extract and leachate of leaves, in the leaf litter, and in the soil under *Croton* plants.

Meissner et al. (1989) reported that shoot growth was reduced when nine crop species—carrot, cucumber, lettuce, corn, onion, radish, squash, sunflower, and tomato—were grown in soil previously infested with *Cynodon dactylon* (Bermuda grass). Height growth was reduced also in all test plants except sunflower. The scientists concluded that there were probably biologically active substances in the soil because of the previous presence of *Cynodon* and that Bermuda grass was probably allelopathic.

Many scientists have reported allelopathic effects of purple nutsedge (*Cyperus rotundus*) (Rice, 1984). Thus Komai and Ueki (1975, 1977, 1980) and Komai et al. (1977, 1978, 1991) isolated and identified numerous allelochemicals from purple nutsedge that were biologically active against several test organisms. Komai and Ueki (1975) identified catechol tannin consisting of leucocyanidin and leucocyanidin glucoside. Catechol tannin was high in mature tubers, seed heads, and rhizomes. Subsequently, Komai and Ueki (1977) reported a low concentration of polyphenolic substances in mature tubers and a high concentration in young tubers. Moreover, there was a marked negative correlation between the polyphenol content of tubers and the percentage germination. Komai et al. (1977) isolated and identified four sesquiterpenes from purple nutsedge tubers: cyperene, ß-selinene, cyperenone, and α-cyperone. α-cyperone was most inhibitory against elongation of wheat coleoptiles in the presence of indole acetic acid and against growth of the second leaf sheath of rice in the presence of gibberellin $A_3$ ($GA_3$). In a ($10^{-3}$ *M*) solution, α-cyperone strongly suppressed growth of hypocotyls and roots of lettuce in the presence of $GA_3$, but hardly affected germination. Activity of several compounds derived from α-cyperone was much lower than that of α-cyperone.

Komai et al. (1978, 1991) reported four genetic types of purple nutsedge based on the types of sesquiterpenes produced and determined the geographical locations of those types. Tests of the essential oils of the four types against lettuce and oats suggested that the different chemotypes of purple nutsedge may have different allelopathic activity in crop-weed activity. Komai and Ueki (1980) identified cyperene, β-elemene, caryophyllene, α-humulene, β-selinene, cyperenone, and α-cyperone in the steam distillate of soil in which purple nutsedge was growing. This definitely proved the allelopathic potential of that weed. Meissner et al. (1982) reported that the growth and water economy of barley, grain sorghum, cucumber, radish, onion, squash, and tomato were impaired when grown in soil previously infested with *C. rotundus.* Komai and Tang (1989) reported that *C. brevifolius* and *C. kyllingia* are common weeds in Hawaii and that the rhizomes and roots contain essential oils that may contribute to the aggressive invasion of grasslands and lawns by these weeds. *C. kyllingia* contains more essential oils than does *C. brevifolius*, and the former is high in terpenes, including α-cyperone, β-selinene, and α-humulene, whereas the latter is high in $C_{17}$ to $C_{25}$ n-paraffins. These differences agree with the weeds' inhibitory activities and with the fact that *C. kyllingia* is more prevalent, according to Komai and Tang.

Mubarak and Hussain (1978) tested the aqueous extracts of the whole seed, the seed coat, and the cotyledon (including the embryo of *Datura innoxia*) against germination and seedling growth of *D. innoxia, Brassica campestris,* and *Setaria italica.* All extracts were inhibitory to germination and early seedling growth of all test species. Whole-seed and seed-coat extracts were more inhibitory, however, than was the cotyledon extract. *Datura* itself was most strongly inhibited, which probably helps explain the dormancy of its seeds in early stages of germination. If the inhibitors are allowed to accumulate to physiologically significant concentrations in the soil, they can inhibit the germination and early growth of other species in the vicinity of *D. innoxia*. Lovett et al. (1981), Levitt and Lovett (1984a), and Levitt et al. (1984) reported that *Datura stramonium* (jimsonweed or thorn-apple) was allelopathic to *Linum usitatissimum* (flax) and *Helianthus annuus* (sunflower). Lovett et al. (1981) found that aqueous leachates of seeds and leaves of *D. stramonium* inhibited both seed germination and radicle elongation of flax at relatively high leachate concentrations; the effect was greater on radicle growth than on germination. Two allelopathic compounds, scopolamine and hyoscyamine, were isolated and confirmed by GC-MS. The scientists demonstrated that various concentrations of pure scopolamine, or scopolamine and hyoscyamine, had effects on flax germination and seedling growth similar to seed and leaf leachates of jimsonweed.

According to Levitt and Lovett (1984b), alkaloids washed from jimsonweed seeds inhibited early growth of sunflower seedlings in Black Earth and Lateritic Podzolic soils. The effects were greater in the Lateritic Podzols, but the effects persisted for eight months in a jimsonweed-infested Black Earth soil under field conditions, whereas they lasted for 20 weeks in Lateritic Podzols under controlled conditions. Thus under some conditions, severe summer invasions of jimsonweed may inhibit crop seedling growth the following spring because of the persistence in the soil of alkaloids from the *D. stramonium* seeds and the release into the soil of newly synthesized alkaloids by newly emerging jimsonweed seedlings. Levitt et al. (1984) found that the allelopathic effect of *D. stramonium* seeds persisted in Lateritic Podzolic soil kept under field conditions for 15 weeks. The amount of rainfall was average for the locality and time of year, but its distribution was sporadic. The rate of phytotoxic-alkaloid release from the jimsonweed seeds in soil was rapid following the initial wetting, with microbial decomposition apparently continuing after alkaloid release ceased. Sunflower cells from treated seedlings showed no obvious structural damage, having distinct nuclei and other organelles and intact membranes. A large proportion of the cells, however, contained aggregations of electron-dense structures bounded by double membranes and identified as amyloplasts. Treated cells also contained greater quantities of lipids than did control cells. It was concluded that the alkaloids retarded metabolism of food reserves, particularly starch, in germinating sunflower seeds.

*Dicanthium annulatum* is a range grass in Pakistan and also a weed in the moist lowland and plains areas of Pakistan. Dirvi and Hussain (1979) found that aqueous extracts of shoots, root exudates, and soil taken from areas infested with *D. annulatum* inhibited germination and growth of *Pennisetum typhoideum, Setaria italica,* lettuce, and mustard (*Brassica campestris*). Moreover, the extracts retarded water absorption and water content of test plants. It was recommended that the residue of *D. annulatum* be removed from the field during weeding to prevent at least part of its allelopathic effects.

*Digera alternifolia* is a common weed in parts of India; the earliest report found concerning its allelopathic effects was one by Dubey (1973). He found that a leaf leachate of this weed (12 g/100 mL $H_2O$) and dilutions of the leachate markedly decreased germination of cotton, tomatoes, pigeon peas (*Cajanus cajan*), *Solanum melongena,* chilies (*Capsicum annuum*), cluster beans (*Cyamopsis psoralioides*), pearl millet (*Pennisetum typhoideum*), onions, and grain sorghum. A 50-fold diluted leachate decreased germination of all test crops except cotton. Later, Ashraf and Sen (1980) reported that water extracts of leaves, stems, and roots of *D. alternifolia* strongly decreased the growth of radicles and hypocotyls of pearl millet and til (*Sesamum indicum*). In field experiments, maximum effects were found where the weed residues were buried in the soil. Even the burned weed residue was inhibitory to the growth of til. Suseelamma and Raju (1992) demonstrated that extracts of stems, leaves, inflorescences, and roots of *Digera muricata* were allelopathic to the germination and seedling growth of the legume horsegram. The relative effects on seed germination were in the order listed above—stems, leaves, inflorescences, and roots—as were the effects on reduction in length and dry weight of roots and shoots and decreases in protein concentrations in roots and shoots.

*Echinops echinatus* is a winter weed of north India and a weed of the wet season in the arid zone. Jha and Sen (1981), found that this weed reduced growth and yield of pearl millet and til in the field. Aqueous extracts of leaves, stems, and roots inhibited germination and seedling growth of pearl millet and til in laboratory tests as well. In field tests, maximum inhibition occurred in pearl millet when the weed residues were buried, whereas the burned residue of the weed increased yield of this crop. The effects of the residues under field conditions definitely confirmed the allelopathic effects of that weed. Soni and Mohnot (1988) reported that *E. echinatus* showed no sign of germination under laboratory conditions until the fruit carpels were removed from the seeds, after which 100% germination occurred. Aqueous extracts of the fruit carpels (0.05 to 10%) caused marked decreases in germination of the naked seeds, even in a 4% concentration. The extracts also moderately inhibited root and shoot growth of the *E. echinatus* seedlings. Phenolics and terpenoids were identified in the extracts.

Sircar and Chakravarty (1961) reported that aqueous root extracts of *Eichhornia speciosa* sprayed on jute plants (*Corchorus capsularis*) once a week, starting at the age of 10 days, markedly increased shoot height, basal circumference, number of leaves, and number of nodes. These effects first appeared at 21 days after the first application and continued until harvest. The quantity of fiber was also increased by about 50% at harvest time. This phenomenon has not yet been tested in an economic way, but it definitely should be. Unfortunately, few of the stimulatory effects of allelopathy have been reported and virtually no uses of those effects have been made to this time. Ahmed et al. (1982) found that water extracts of leaves, rhizomes, roots, and whole plants of *Eichhornia crassipes* (water hyacinth) strongly reduced radish germination, hypocotyl growth, radicle growth, and seedling dry weight. Extracts of decaying water hyacinth had slight inhibitory effects on seedling dry weight and hypocotyl growth in about the first two weeks of decay, but later than that these extracts usually had no or only slight, stimulatory effects. Root growth was definitely stimulated in the later stages of decay of the residue.

*Eragrostis poaeoides* is a common weed in cultivated fields from the low plains up to 3660 m in elevation in Pakistan. Hussain et al., (1984) found that water extracts of shoots, inflorescences, and roots were inhibitory to germination of mustard (*Brassica campestris*) and that the extracts of the inflorescences were inhibitory to germination of *Setaria italica* (German millet) and *Pennisetum americanum,* with the dried inflorescence extracts being most toxic. Extracts of dried inflorescences and dried shoots were also significantly inhibitory to radicle growth of *P. americanum* and German millet, and dried root extracts were also inhibitory to radicle growth of *P. americanum.* All extracts were inhibitory to radicle growth of mustard. Dried straw residues of *E. poaeoides* significantly inhibited germination of all three test species, and fresh straw beds significantly inhibited germination of mustard. Both fresh and dried straw residues significantly inhibited radicle growth of all three test species. Aqueous extracts of soil in which *E. poaeoides* previously grew significantly inhibited radicle growth of all three test species and germination of mustard seeds. Washing the soil for 30 minutes prior to the tests eliminated all inhibitory activities, thus indicating that water-soluble inhibitors were present in the residual soils prior to the washing process and clearly demonstrating that *E. poaeoides* has allelopathic potential.

Carballeira and Cuervo (1980) previously demonstrated that *Erica australis* has pronounced allelopathic potential. They decided, therefore, to determine if there is seasonal variation in phytotoxicity of soils dominated by *E. australis.* They collected soil at two depths, 0 to 10 and 10 to 20 cm, in an area dominated by that species, in the months of April, October, and January. Aqueous extracts of the soil were tested against germination of white clover (*Trifolium repens*), timothy (*Phleum pratense*), red clover (*Trifolium pratense),* and *Lolium perenne* (perennial ryegrass). Allelopathic effects of soil extracts were present throughout the year but generally were slightly greater in the January and April collections. The effects were usually stronger against the legumes than against the grass and generally more pronounced in the 0 to 10 cm than in the 10 to 20 cm zone of the soil. Protocatechuic, vanillic, *p*-hydroxybenzoic, and *p*-coumaric acids were identified from the aqueous soil extracts and were suggested to be the chief allelopathic compounds in the heath soil under investigation.

*Eupatorium adenophorum* is a native perennial herb of Mexico but it has become a noxious weed of many parts of the world, including parts of India (Dhyani, 1978). At Dehra Dun, India, it is gradually replacing *Lantana camara* var. *aculeata*. Dhyani found that a 5% aqueous extract of the shoots of *E. adenophorum* virtually prevented germination of the seeds of *Lantana camara*. Dhyani suggested that the very water-soluble inhibitors may be leached from the shoots by precipitation, thereby inhibiting seed germination of that vicious weed. Tripathi et al. (1981) reported that water extracts of leaves, roots, and litter of *E. adenophorum, E. riparium, Anaphalis araneosa, L. camara*, and *Galinsoga ciliata* strongly inhibited wheat seed germination. In subsequent tests, leaf, root, and litter extracts of *E. adenophorum* inhibited seed germination and plumule and radicle growth of seedlings of *Trifolium repens, Rumex nepalensis, E. riparium*, and *Paspalum dilatatum*. The populations of fungi, bacteria, and actinomycetes in soil were considerably reduced because of the presence of *E. adenophorum*, with the populations of bacteria being affected more than those of fungi and actinomycetes. Ambika and Jayachandra (1980) found that *E. odoratum* has also spread widely in many parts of India, where it is very detrimental to legume cover crops; it has caused the failure of several new plantations of soft wood and teak in forest areas of Karnataka. Aqueous leachates of leaves, roots, and cypsella of *E. odoratum* inhibited seed germination and root and shoot growth of wheat and fenugreek (*Trigonella foenum-graecum*). Ambika and Jayachandra inferred therefore that at least part of the detrimental effect of *E. odoratum* on legumes and on plantation crops is due to allelopathy.

Hussain (1980) found that the aqueous leachate of the shoots of *Euphorbia granulata* prepared in Hoagland's solution and soil from underneath this weed significantly inhibited seed germination and radicle growth of *Dicanthium annulatum*, Bermuda grass, German millet, *Pennisetum americanum, Euphorbia pilulifera, Oxalis corniculata*, and lettuce. *E. granulata* litter artificially decomposed in a nutrient medium significantly reduced germination and fresh and dry weight of the same test species. The strong allelopathic effects of this weed caused Hussain to infer that allelopathy was primarily responsible for the local dominance of this weed over its counterparts in areas investigated in Pakistan. Alsaadawi et al. (1990) observed that *E. prostrata* interferes strongly with Bermuda grass in fields in Iraq. Analyses of physical and chemical soil factors indicated that competition was not the dominant factor in that interference. The researchers found that soil collected from underneath *E. prostrata* was very suppressive to seed germination and seedling growth of Bermuda grass, pigweed (*Amaranthus retroflexus*), cotton, and alfalfa. Aqueous extracts of *E. prostrata,* decaying residues of that weed, and root exudates inhibited germination of the test species and markedly inhibited root and top growth of their seedlings also.

*Galium aparine* (catchweed or cleaver) is one of the most widespread biennial weeds in arable lands in Japan, and it is often injurious to crop plants. The seeds appeared to be most toxic, so Komai et al. (1983), decided to isolate and identify the causative allelochemicals. Using Sephadex LH-20 chromatography, TLC, UV, IR, and NMR, the researchers identified the inhibitor as the asperuloside of a monoterpene lactone glucoside. Asperuloside was inhibitory to lettuce seed germination and to growth of the hypocotyl, radicle, and cotyledon of lettuce seedlings. That allelochemical was also inhibitory to germination and seedling growth of crabgrass and alfalfa, but was inhibitory only to seed germination in white clover and catchweed. Thus it appears that at least part of the injurious action of the catchweed seeds on certain crop plants and weeds is due to allelopathy.

*Gomphrena decumbens* is a common weed in fields of sorghum and sesamum (til) in parts of India. It was found that water leachates of fresh and dry roots, stems, and leaves inhibited seedling growth of sorghum and sesamum in concentrations of 1, 5, and 10% (Solomon and Bhandarai, 1981). The root leachates of the weed were more inhibitory to seedling growth of the crop plants than were leachates of leaves and stems. It was concluded that this weed not only competes with sorghum and sesamum, but it also interferes with their germination and growth through its allelopathic effects. It was recommended that control of the weed at a very early stage of crop plant growth, including uprooting and removal of the weed from the fields, is essential.

Alán and Barrantes (1988) were interested in finding woody plants that might be used in controlling common weeds in Costa Rican fields. They collected leaves, branches (stems), and sections of roots of

*Gliricidia sepium* (madero negro) and tested water extracts of these parts against seed germination and seedling growth of four weed species—*Hyptis capitata, Asclepias curassavica, Ipomoea* sp., and *Momordica charantia.* The most effective extracts against the weeds were those from leaves of madero negro, followed by those from the roots. The resulting allelopathic effects were stronger in *H. capitata* and *Ipomoea* sp. than in the other test species. The stem extracts did not significantly affect seed germination or seedling growth of *A. curassavica.* A significant response was obtained in seed germination of *M. charantia* with leaf extracts and in the germination and growth of *A. curassavica* with leaf and root extracts in soil-sample trials.

Sugha (1978) was intrigued by a report that tubers of *Gloriosa superba* (superb lily) cause abortion in women, and he decided to determine the effects of water extracts of the tubers on the growth of pea plants. He found that soaking of pea seeds in the water extract of superb lily for four hours stimulated plant height at flowering time, number of pods per plant at 67 days after planting, pod length at 67 days, number of seeds per pod at 67 days, total number of seeds per plant, and weight of 100 seeds. However, soaking the pea seeds for eight hours decreased germination, soaking for eight hours also decreased height at flowering time and—after 84 days from planting—number of pods per plant, pod length, number of seeds per pod, total number of seeds per plant, and weight of seeds. The bioactive chemicals were water-soluble and thus the results certainly suggested a strong allelopathic potential for superb lily.

*Helenium amarum* (bitter sneezeweed) is a warm-season annual widely distributed in the southern United States. It is a weed of economic importance to the dairy industry because it is toxic to livestock and causes bitterness in milk. Smith (1989), who made these findings, also found that, even when he eliminated potential osmotic effects, water extracts of bitter sneezeweed markedly inhibited seedling growth of alfalfa and Italian ryegrass—as much as 50% at concentrations of 0.5% (w/v). Leaf extracts were more toxic than stem or root extracts. Bitter sneezeweed tissue mixed in potting soil at concentrations as low as 0.3% (w/w) reduced alfalfa seedling numbers by 43%, plant height by 26%, and foliage dry matter production by 54% compared with plants grown in soil without sneezeweed residue. Volatile chemicals from sneezeweed reduced alfalfa seedling growth by 70%. It was concluded, therefore, that a strong potential exists for bitter sneezeweed allelopathy against alfalfa and Italian ryegrass where those plants are interseeded in the fall in pastures infested with bitter sneezeweed.

Umber (1978) reported that *Helianthus rigidus* clones in grassland areas of western Oklahoma had fewer nonclimax species than the surrounding area. The species excluded from the clones were *Sorghum halepense, Chaerophyllum tainturieri, Gutierrezia dracunculoides, Andropogon saccharoides,* and two unidentified species. Species partially suppressed by *H. rigidus* were *Achillea lanulosa, Erigeron philadelphicus, Eragrostis trichodes, Sisyrinchrum angustifolium,* and *Sporobolus cryptandrus.* All of these are basically nonclimax species. The climax species were unaffected by the increasing frequency of *H. rigidus.* Umber found that the exclusion of weed species was not the result of competition for the physical factors of light, soil moisture, and soil nutrients. Biosassays of leachates of various parts of *H. rigidus* did have marked effects against germination and/or seedling growth of subclimax species in many instances. The season of the year and the part of *H. rigidus* tested, as well as the concentration of the extract and the test species, markedly affected the reaction to the leachate. Climax species were generally not affected, or were affected less strongly. Allelopathy did not appear to be entirely responsible for the succession toward the climax, but it did appear to have a strong influence.

Spring and Hager (1982) isolated two sesquiterpene lactones from the leaves and stems of *Helianthus annuus,* both of which strongly inhibited IAA-induced elongation growth of oat coleoptile segments and *H. annuus* hypocotyl segments. Neighter allelochemical inhibited acid-induced growth or growth triggered by fusicoccin.

Srivastava (1969) reported that water extracts of *Heliotropium eichwaldi* seeds caused gradual decreases in germination of *Phaseolus aureus* seeds with increasing concentrations of the extracts in continuous light, whereas germination was increased under continuous darkness. It was therefore concluded that *H. eichwaldi* contains a chemical that is activated in light and acts as an inhibitor, while under darkness it acts as a growth promoter for the seed germination of *P. aureus.*

Seed germination and seedling growth of *Brassica rapa,* tomato, radish, and *Paspalum notatum* were remarkably inhibited when leaves and stems of *Hydrocotyl sibthorpioides* were buried in the soil of pots and when a solution of those plant parts extracted with water was sprayed on the seedlings (Tsuzuki et al., 1978). Seedling height, length of second-leaf sheath, and roots in rice plants were also strongly repressed by the aqueous extracts of that weed. Paper chromatography of methanol extracts of *H. sibthorpioides* indicated that several active allelochemicals were present, but they were not identified. The biological activity of the residue demonstrated that the weed was definitely allelopathic.

Inderjit and Dakshini (1991a) carried out a thorough set of experiments to determine the interference mechanisms of cogongrass (*Imperata cylindrica*). They found that leachates of the leaves and root-rhizomes of cogongrass reduced seed germination, root growth, shoot growth, fresh weight, and dry weight of radish, mustard (*B. juncea*), fenugreek, and tomato. The leachates and soils from three sampling sites (with cogongrass and at 1.5 and 3 m away from cogongrass) were analyzed with HPLC on a C18 column. No significant differences in nutrient availability were found, but there were qualitative and quantitative differences in phenolic fractions in the three sampling sites. About five phenolics that did not occur in control soil were found in noncontrol soil, and most of the phenolics found in both noncontrol and control soil were in considerably higher concentrations in the test soils than in the control soil. This clearly demonstrated that several phenolics came from the cogongrass. *Melilotus parviflora* growing with cogongrass in field plots had significantly lower root growth, lower root-shoot ratio, lower nodule number, lower nodule weight, and lower $N_2$ fixation than *M. parviflora* growing in a cogongrass-free area (Table 1). Of the 19 fungi recorded in the various soils, a decrease in the numbers of *Aspergillus fumigatus, A. niger, and A. candidus* and an increase in *A. flavus* occurred in soils with cogongrass. All the data accumulated clearly indicate that allelopathy plays a very important role in the interference of cogongrass against other plants.

Soybean fields with a significant morning glory (*Ipomoea lacunosa*) infestation often have a marked reduction in yield (LaBonte and Darding, 1988). Although these yield reductions are generally attributed to competition for minerals, water, and light, these scientists experimented to determine if allelopathy might also be involved in the interference. They found that water extracts (5%) of the tops of six-week-old morning glory plants significantly reduced germination of soybean seeds by 11% and radicle weight by 17%, whereas 5% root extracts significantly reduced germination by 11% and radicle growth by 13% compared with the water control. In contrast, no significant effects were noted when the extracts were allowed to stand for one

**Table 1** Root length, root-shoot ratio, and nodule characteristics of *Melilotus parviflora* growing with (treated) and without (control) cogongrass

| Characteristics | Control | | | Treated | | |
|---|---|---|---|---|---|---|
| Nodule number | 58.50 | ± | 19.01 | 26.00 | ± | 7.88* |
| Nodule weight (g) | 0.0073 | ± | 0.0026 | 0.0041 | ± | 0.0017† |
| $N_2$ fixation (μmol $C_2H_4$/g fresh wt) | 10.9588 | ± | 3.4078 | 2.1814 | ± | 0.8494‡ |
| Root length (cm) | 10.540 | ± | 1.390 | 7.460 | ± | 2.48† |
| Root-shoot ratio | 0.367 | ± | 0.101 | 0.228 | ± | 0.055* |

*$0.01 \leq P < 0.05$.
†$0.05 \leq P < 0.1$.
‡$0.001 < P < 0.01$.

*Source.* From Inderjit and Dakshini (1991a), with permission of Plenum Pub. Corp.

week prior to the bioassays. This indicated that the responsible allelochemicals were either chemically decomposed or decomposed by microorganisms present in the extracts. There were no pH changes; all solutions were found to be neutral. When soybean plants were watered with aqueous leaf leachates (shoots submerged in water for one hour) every three days for five weeks, the soybean root weight was decreased 40% and nodule weight was reduced 23%, and both values were highly significant. When soybean plants were watered with root exudates of morning glory for five weeks, the root weight of soybeans was decreased 22%, shoot weight 14%, and nodule weight 31%, compared with the water control. It was concluded that morning glory leaf leachate (rain drip) and root exudate are likely to contain phytotoxins that have an inhibitory effect on the growth and nitrogen-fixation of soybean plants.

Kochia (*Kochia scoparia*) is an early pioneering annual plant in denuded areas of the southwestern United States and on mine spoils elsewhere. Karachi and Pieper (1987) decided to determine the phytotoxic effects of aqueous extracts of different plant parts and different phenological stages of that weed on germination and seedling growth of blue grama (*Bouteloua gracilis*). Seed germination was not inhibited but radicle and shoot growth were significantly inhibited. Inhibition declined significantly with declining concentrations of the extracts and with advancing phenological stages. Hot-water extracts inhibited growth more than cold-water extracts.

Volatile allelochemicals that escaped from the leaves of *Lantana camara, Ocimum sanctum, Tridax procumbens,* and *Cyperus rotundus* and from the rhizomes of *C. rotundus* significantly reduced the growth in length and increase in dry matter in the roots of rice seedlings (Das and Pal, 1970). The maximum effect was caused by *O. sanctum* leaves in which the volatiles reduced the root length by 32% and dry-matter content by 28%, as compared with the controls. The growth of rice coleoptiles was reduced significantly (20 to 30%) by all treatments except the leaves of *C. rotundus.* Thus all four aromatic weeds were definitely shown to be allelopathic, because the active allelochemicals were definitely escaping from the producing plants and causing significant growth effects. Unfortunately, volatile allelochemicals have often been overlooked in allelopathic research.

The earliest subsequent report concerning the potential allelopathic effect of *Lantana camara* of which I am aware was that of Wadhwani and Bhardwaja (1981) in which they demonstrated that aqueous extracts of roots, stems, leaves, and inflorescences of lantana inhibited exine bursting, rhizoid initiation, and protonemal initiation of spores of the fern *Cyclosorus dentatus.* The leaf extract was most potent. This research was done because it was noted that the fern component in the vegetation of Mt. Abu in Rajasthan, India, was decreasing in importance as the spread of *L. camara* increased. The scientists concluded from their study that further spread of lantana on Mt. Abu may lead to the complete destruction of the sizeable pteridophyte component there.

Many subsequent papers have emphasized the allelopathic potential of *L. camara* in agriculture (Achhireddy and Singh, 1984; Achhireddy et al., 1985; Mersie and Singh, 1987a; Jain et al., 1989; Singh et al., 1989). Achhireddy and Singh (1984) reported that foliar leachates and soil in which lantana had previously grown did not affect the final germination percentage or the seedling growth of milkweed vine (*Morrenia odorata*). Incorporation of dried lantana shoot or root material into soil, however, significantly reduced milkweed vine growth over a 30-day test period, and roots were more inhibitory than shoots. In fact, with 1% (w/w) dried lantana root incorporation in soil, 50% of milkweed vine seedlings died within 15 days of germination. Incorporation of lantana roots into the soil produced foliar symptoms such as wilting and desiccation, whereas lantana shoot incorporation in soil caused yellowing of the milkweed vine foliage. The allelopathic activity of the lantana residue was still strong after four weeks' decomposition. Achhireddy et al. (1985) reported that the active allelochemicals in field-grown lantana were both polar and slightly acidic. TLC separation on silica gel of the acidic butanol fraction by a solvent mixture of butanol–acetic acid–water (4:1:5) yielded an active fraction at an $R_f$ value of 0.89–1.0 that inhibited both root and shoot growth of Italian ryegrass. Two other fractions with $R_f$ values of 0.04–0.23 and 0.41–0.57 were inhibitory only to root growth of Italian ryegrass. Jain et al. (1989) separated out several active fractions from leaf extracts of lantana. They found the acidic and neutral fractions to be most inhibitory to duckweed growth, and they subsequently isolated and separated out the phenolic compounds by HPLC. The acidic extract contained p-hydroxybenzoic acid (0.09 m*M*) as the most abundant phenolic compound, whereas the neutral and basic extracts contained p-coumaric acid (0.11 and 0.26 m*M*, respectively). All phenolic compounds (at least 14 found), except p-hydroxybenzoic acid, were toxic to duckweed growth at a concentration of 1 m*M* or less. Salicylic acid was the most toxic to duckweed growth of the phenolics detected. The researches suggested that the phytotoxicity of the lantana-leaf extracts is probably due to a complex interaction between all the phenolic compounds present. Using Italian ryegrass as the bioassay plant, Singh et al. (1989) separated 13 phenolic compounds from lantana-leaf extracts. Most of the phenolics were inhibitory to growth of ryegrass seedlings. It appears that more research needs to be done on the active allelochemicals released by lantana, with emphasis on possible nonphenolics.

*Leersia hexandra* is a companion plant of *Oryza perennis* in about 40% of the habitats observed in India and Thailand, and an experiment to introduce *O. perennis* in different lowland habitats after denudation in Taiwan suggested that *L. hexandra* was a key species determining the regenerating success of wild rice (Chou et al., 1984). Their interaction mechanisms were examined by looking at their allelopathic interrelationships. Both species showed phytotoxic effects on radicle growth of rice and lettuce, but *Leersia* generally showed higher phytotoxicity than did the wild rice. A study of plants arising from a buried seed pool in soil to which powdered plant material of one or the other of the two grasses was added also showed both to exert a toxic action, but with *Leersia* being the more toxic. These scientists concluded that allelopathy probably plays an appreciable role in the successional replacement of the two species.

*Lycoris radiata* is a dominant species along roadsides and on slopes in Japan; it appears to inhibit the growth of other weeds and occasionally to eliminate them. Ueki and Takahashi (1984) made water extracts of the bulbs of this plant and of the soil adjacent to it.

Subsequent fractionation with ether and separation by TLC on silica gel indicated there were two potent water-soluble inhibitors present. Each of the inhibitors was diluted in a series starting with 1000 ppm, up to 128-fold, and tested against root elongation of 11 weed species. One of the allelochemicals inhibited radicle growth of all test species, and the other inhibited root growth of nine of the test species. In undiluted solutions, root growth of all species stopped after five days. The pronounced toxicity of the bulbs of *L. radiata* causes these plants to be used to eliminate weeds on the footpaths between crops in fields and to prevent paddy fields from leaking. (In the proximity of *L. radiata,* rats and moles will not make holes.) The bulbs are ground and rubbed into the plaster in barns to keep rats away and are used as a paste for industrial arts.

Chou et al. (1991) continued the research of Chou et al. (1984) on the allelopathic activity of wild rice (*Oryza perennis*) by comparing the effects of 24 strains of wild rice and two cultivars of *O. sativa* on Chinese cabbage (*Brassica oleracea*) and cultivated rice. They reported a wide range of biological activities of leaf-stem aqueous extracts of the 24 strains of wild rice against root growth of Chinese cabbage, from a slight stimulation to 68% reduction. Tests of the phenolic acids produced by each strain also gave strong differences. The distribution of nine phenolic acids differed markedly among the strains. Only *o*-coumaric acid showed a correlation with the bioassays with Chinese cabbage. Stepwise-regression analysis showed that a combination of four phenolic acids had a significant multiple correlation with the root-suppression rate, and when interactions were taken into account, a high multiple correlation ($r = 0.84$) was obtained with six phenolic acids. It was suggested that certain phenolic acids interact to suppress plant growth when present at the proper concentrations, but some others at different concentrations are ineffective or may even promote growth. A strain from the Malay Peninsula was highest in both allelopathic potential and aggressiveness, and this strain is now under observation in an experimental garden in southern Taiwan. As *suspected*, the two rice cultivars tested showed a low allelopathic potential.

*Parthenium hysterophorus,* a common weed of tropical America, was reported for the first time in India in 1956, and in two decades it had spread to several parts of the country. Kanchan and Jayachandra (1976) reported that different parts of that weed contain water-soluble plant-growth toxins. In field trials, crop plants grown in soil containing dried roots and leaves were reduced in growth and yield. The size, number, and dry weight of root nodules of bean and cowpea were also reduced by the same treatment. Pollen of parthenium dusted on stigmatic surfaces of crop plants inhibited fruit development by those plants. Kanchan and Jayachandra published the details of their experiments on the allelopathic potential of parthenium in the years following 1976 (Rice, 1984). In 1980 they described their research concerning identification of the inhibitors produced by *P. hysterophorus.* They collected root exudates; leachates of leaf trichomes, roots, stems, leaves, inflorescences, and fruits; and extracts of soil adhering to parthenium. The acidic and nonacidic fractions of each leachate and root exudate and the ethyl acetate fraction of the rhizosphere soil extract were chromatographed on silica gel plates. The acidic and nonacidic fractions were also analyzed by column chromatography, UV absorption, melting-point determinations, and NMR. All isolated spots from the TLC plates and other isolated compounds were tested against coleoptile growth of *Eleusine coracana.* Total phenols were determined in the aqueous leachates, the soil extracts from the rhizosphere, and the soil extracts taken 10, 20, and 30 cm from parthenium plants. Sesquiterpene lactones and phenolics were the most important groups of active allelochemicals identified in leachates and soil extracts. Parthenin was the major sesquiterpene lactone involved, although dampsin was found in trace amounts. Caffeic, vanillic, ferulic, chlorogenic, and anisic acids among the phenolics, and fumaric acid among the organic acids, were the important constituents of the air-dried parts of parthenium. Moreover, most of them were traced in the root exudates and in the leachates of leaves, pollen, and trichomes. It was concluded that the addition of large amounts of sesquiterpene lactones and phenolics to the soil by various routes from parthenium renders the soil highly unsuitable to the growth of other species. Identification of the sesquiterpene lactone parthenin in leaves of *P. hysterophorus* was substantiated by Patil and Hegde (1988). They also found that this allelochemical significantly inhibited germination and seedling growth of wheat in a concentration of 500 ppm or above and of *Aspergillus* spp. at a concentration of 1000 ppm.

Several scientists have demonstrated that *Parthenium hysterophorus* is allelopathic to numerous other crop plants and also to itself (Kohli et al., 1985; Srivastava et al., 1985; Kumari et al., 1985; Kumari and Kohli, 1987; Mersie and Singh, 1987b, 1988).

Megharaj et al. (1987 a,b) demonstrated that soil amended with 1, 2, or 5% (w/w) concentrations of powder of air-dried leaves, inflorescences, or roots of *P. hysterophorus* markedly changed the types of algae present in the soil. This was particularly true of the 2 and 5% concentrations. The numbers were also greatly changed in many cases where the algal species under study remained in the treated soil. The inhibition of algae was greatest in soil amended with leaf material followed by treatment with inflorescence powder; it was least inhibited by the root-powder treatment.

Sharma et al. (1982) reported that aqueous extracts of air-dried roots, stems, leaves, and seeds of fresh plants of *Peganum harmala,* as well as aqueous extracts of similar parts of plants kept for one year after collecting, markedly repressed germination and seedling growth of *Pennisetum typhoideum* (bajra). Chromatographic studies revealed that the types of active allelochemicals changed with time, and chemical analysis revealed that the active compound(s) may be alkaloid in nature.

Edwards et al. (1988) reported that water extracts of most plant parts of *Phytolacca americana* (pokeweed) reduced or prevented germination of pokeweed seeds. Juice of pokeweed fruits, whether diluted fivefold by water or undiluted, completely prevented seed germination of that species. Root extracts appeared to be less inhibitory to seed germination than extracts of other plant parts. Even there, the highest concentration tested completely inhibited seed germination, whereas the most dilute (10 and 20% v/v) stimulated germination. Edwards et al. stated that fruit extracts also inhibited seed germination of two other species in preliminary tests. Obviously, this work needs to be expanded.

Kleiman et al. (1988) identified a total of eight phenylpropenyl esters in extracts of seeds from six *Pimpinella* species: *P. anisum, P. diversifolia, P. major, P. peregrina, P. saxifraga,* and *P. tragium.* Six of the compounds were isolated and tested against seed germination and seedling growth of *Abutilon theophrasti,* Italian ryegrass, carrot, lettuce, tomato, radish, and cucumber. Those with epoxypropenyl groups were active against several different species, whereas compounds with olefinic groups instead of epoxy groups were either inactive or had very low activity.

*Pluchea lanceolata* has become a serious weed in cultivated fields in the semiarid regions of India, and Inderjit and Dakshini (1990) decided to determine the nature of its interference potential against crop plants. They tested water leachates of leaves of *P. lanceolata* and filtrates of leaf-soil suspensions against seed germination and seedling growth of radish, okra (*Abelmoschus esculentus*), mustard (*Brassica campestris*), tomato, bajra, wheat, gram (*Cicer arietinum*), and carrot. The leaf leachate markedly reduced the germination rate of okra, mustard, tomato, and carrot but did not affect the germination of the others. However, all leaf leachates and soil filtrates suppressed growth rates of the seedlings of all the test plants. Ten spots were found on chromatograms of leaf extracts of *P. lanceolata,* but only four were found on those of the soil filtrates. The four in the soil filtrates were matched with four from the leaf filtrates. Of the four, three markedly reduced the seed germination rate of mustard, but the fourth had no effect. It was concluded that *P. lanceolata* attains at least part of its success in interference against crop plants through allelopathy. Subsequently, Inderjit and Dakshini (1991b, 1992a) isolated and identified three important allelochemicals from *P. lanceolata* soils that did not occur in control soil. These were a flavonone (hesperetin-7-rutinoside, or hesperidin), a dihydroflavanol (taxifolin-3-arabinoside), and the isoflavonoid (formononetin-7-O-glucoside, or ononin). A $10^{-4}$-*M* solution of hesperidin or taxifolin significantly inhibited the seedling growth of mustard, radish, and tomato and the germination of tomato. Aqueous solutions of ononin significantly inhibited root and shoot growth of mustard (the only test plant used) at $10^{-4}$, $5 \times 10^{-4}$, and $10^{-3}$ *M* concentrations. A pool of these three compounds is maintained in the soil by *P. lanceolata,* and thus these compounds are implicated in the allelopathic effects of this weed.

Inderjit and Dakshini (1992b) did a thorough study of the allelopathic effects of *P. lanceolata* on the asparagus beans (*Vigna unguiculata* var. *sesquipedalis*). They found that leaves of the weed significantly reduced seed germination, number of nodes, internode length, shoot and root lengths, nodule number and weight, chlorophyll a and b content, and chlorophyll a-b ratio of asparagus beans growing in soil to which the leaves were added. The concentrations of $Mg^{2+}$, $Zn^{2+}$, $PO_4^{-3}$, $Mg^{2+}$, and $NO_3^-$ were higher in plants growing in the treated soil. These data provided strong evidence that the interference effects of *P. lanceolata* were primarily due to allelopathy and less so to competition.

Several species of *Polygonum* were previously reported to be allelopathic (Rice, 1984). Inoue et al. (1992) found that *P. sachalinense* root exudates in a recirculating system significantly inhibited lettuce seedling growth. Bioassay of the neutral-acidic fraction of an 80% acetone extract of roots and rhizomes of *P. sachalinense* on a TLC agar plate showed that the inhibitory activity corresponded to two yellow pigment bands. Two compounds with orange needles were isolated and identified as the anthraquinone compounds: emodin and physcion. Both compounds inhibited seedling growth of *Amaranthus viridis,* lettuce, and timothy grass (*Phleum pratense*). Contents of emodin and physcion in fresh rhizomes were about 158 and 32 mg/kg fresh weight and in the aerial parts about 72 and 22 mg/kg fresh weight, respectively. Large amounts of these anthraquinones were still present in dry fallen leaves more than four months after defoliation: emodin, 213 mg/kg dry weight and physcion, 180 mg/kg dry weight. The concentrations of emodin and physcion in the soil were 55 and 30 mg/kg dry weight, respectively; these levels were sufficiently high to inhibit plant seedlings. It also indicated that the anthraquinones are very stable in the ecosystem. The results indicate that these anthraquinones are responsible for the observed interference and are potent allelopathic substances.

Mercer et al. (1987) reported that densities of two, four, and eight unicorn (devils claw) plants (*Proboscidea louisianica*) per 10 m of cotton row reduced height of the cotton plants by 20, 28, and 43% in the three densities. As unicorn plant number increased from 1 to 32 plants per 10 m of row, lint yield reductions ranged from 84 to 146 kg/ha. Thus unicorn plants had a very bad interference effect on cotton. Riffle (1988) decided to determine whether allelopathy was involved in the marked interference of the unicorn plant on lint yield of cotton. He collected almost-mature unicorn plants in August in field plots in central Oklahoma. He isolated the essential oils primarily from the roots and pods by steam distillation and separated and identified some of the major components of the essential oils by CGC/MS/DS analysis. Cotton and wheat were used for bioassays of the suspected allelopathic compounds. The compounds from the essential oils that were inhibitory to radicle growth of cotton and/or wheat at 1-m*M* concentration are listed below, from most inhibitory, to least inhibitory, and the plant parts from which they were obtained are also noted: piperitenone (roots), *p*-cymen-9-ol (pods), vanillin (roots, pods), α-bisabolol (roots), phenethyl alcohol (pods), and δ-cadinene (roots). It was concluded that the essential oils collected from the upper portions of unicorn plants in late August were inhibitory to cotton; this is a very sensitive growth stage for cotton in Oklahoma, since flowers and bolls are initiated at that time. Thus the allelopathic activity of the volatile compounds from unicorn plants must definitely be an important factor in the interference of unicorn plants on cotton yield. Riffle et al. (1988) demonstrated that aqueous extracts of unicornplant testa, leaves, stems, roots, and exocarps of the fruits were inhibitory to the radicle growth of cotton seedlings. They found also that aqueous extracts of stems, roots, and exocarps were inhibitory to wheat radicle growth and that extracts of endocarps, leaves, and exocarps were inhibitory to unicorn plant radicle growth. Thus both water-soluble allelochemicals and volatiles from unicorn plants are allelopathic to certain crop plants, and at least some water-soluble allelochemicals are allelopathic to the unicorn plant itself.

*Rorippa sylvestris* (yellow fieldcress) is one of the worst weeds in wet fields and pastures in Hokkaido, Japan. Both the neutral and acidic fractions of the acetone extract of the roots inhibited lettuce seed germination (Yamane et al., 1992a). Salicylic, *p*-hydroxybenzoic, vanillic, and syringic acids were identified in the acidic fraction. Hirsutin (8-methylsulfinyloctyl isothiocyanate), 4-methoxyindole-3-acetonitrile, and pyrocatechol were identified in the neutral fraction. A root-exudate recirculating system demonstrated that *R. sylvestris* inhibited lettuce seedling growth during flowering. Hirsutin (13μg/plant/day) and pyrocatechol (9.3 μg/plant/day) were the major compounds released into the rhizosphere, and several combinations of pyrocatechol, *p*-hydroxybenzoic acid, vanillic acid, and hirsutin reduced lettuce seedling growth. Thus allelopathy was definitely demonstrated. Yamane et al. (1992b) subsequently found that the ethyl acetate extract of the roots of *R. indica* contained hirsutin, arabin, camelinin, and three ω-methylsulfonylalkyl isothiocyanates. These compounds strongly inhibited lettuce root and hypocotyl growth at 0.1 m*M* or above. The precursor gloucosinolates of all the above were isolated and identified. The researchers used a continuous root-exudate trapping system and GC-MS to isolate and identify hirsutin and the three ω-methylaulfonylalkyl isothiocyanates in the root exudates of *R indica.* They concluded that those allelopathic compounds contribute to the aggressiveness of that weed.

*Saccharum spontaneum* is a perennial rhizomatous weed infesting various crop fields near Ujjain, India. Amritphale and Mall (1978) tested aqueous leachates of rhizomes and roots of *S. spontaneum* against root and shoot growth of three cultivars of wheat: Kalyan Sona HD-1593, Narmada-4, and NP-404. Germination of the three wheat cultivars was not affected, but both root and shoot leachates of *Saccharum* reduced both root and shoot dry weights of all three wheat cultivars and root and shoot lengths of the first two cultivars listed above. Root and shoot lengths of NP-404 were hardly affected. It was concluded that *S. spontaneum* is a noxious weed, not only for its competitive effects against wheat but also because its allelopathic effects may make the soil unsuitable for wheat.

Qasem and Abu-Irmaileh (1985) reported that water extracts of shoots and rhizomes of *Salvia syriaca* (Syrian sage) slowed germination and inhibited growth in length of roots and shoots of wheat (cv. Stork). The effects were more pronounced at 20 than at 10 or 15°C. Moreover, fresh and dried shoots added to soil drastically decreased germination, growth, and yield of wheat. This definitely demonstrated that allelochemicals were getting out of the sage residue and into the soil and affecting wheat growth and development. Thus the allelopathic potential of *S. syriaca* was definitely established. Abu-Irmaileh and Qasem (1986) extended their investigation of *S. syriaca* to effects on barley (nine cultivars), wheat (ten other cultivars), chickpea (eight cultivars) and lentil (*Lens culinaris,* -nine cultivars). Aqueous shoot extracts of Syrian sage drastically reduced seed germination and seedling development in certain cultivars of each crop. Barley and wheat cultivars were more sensitive to the sage extracts than were cultivars of chickpea and lentil. Root development was more sensitive to the extracts than shoot development in most lines of each crop. Relatively dilute extracts stimulated seedling growth of some chickpea and lentil cultivars. This type of research with known allelopathic weeds is extremely important to help eliminate or lessen the allelopathic effects of the weeds on crops.

*Sasa cernua*(sasa) is a serious weed in Japan and other regions of Asia and Europe. Li et al. (1992) decided to study its possible allelopathic effects on lettuce, wheat, timothy, and *Amaranthus viridis* (green amaranth). Seedlings of these plants cultured in rhizosphere soil of sasa were inhibited by 42 to 80% compared with controls cultured in soil free of sasa to which vermiculite was added. The phenolic fraction extracted from the rhizosphere soil of sasa significantly inhibited seed germination and seedling growth of the species listed above, plus barnyard grass (*Echinochloa crusgalli*). HPLC and NMR were used to isolate and identify *p*-coumaric, ferulic, vanillic, and *p*-hydroxybenzoic acids and *p*-hydroxybenzaldehyde as the main allelochemicals in sasa soil. The concentrations for those phenolics in sasa soil were 5640, 1060, 860, 810, and 630 μg/100 g of soil, respectively. The neutral fraction of the soil extract also inhibited seed germination and seedling growth of lettuce. Volatile compounds released from sasa leaves inhibited growth of lettuce, wheat, timothy, and green amaranth grown under light as well as growth of etiolated seedlings of barley and wheat. Li et al. thus confirmed that *Sasa cernua* produces allelopathic effects through its soil and air space.

Lovett (1982) reported that *Stevia eupatoria* (Kempton's weed) interferes with white clover, in part by allelopathy. That weed has many trichomes and is strongly aromatic. When air-dried material was placed in a closed system, sufficient chemicals were released in the air to significantly decrease the radicle length of white clover seedlings. Leaves, stems, or flowers of *S. eupatoria* placed between moist filter papers upon which white clover seed was placed strongly inhibited seed germination and radicle growth of the seedlings.

Zuberi et al. (1989) found that water leachates or extracts of fresh or dried *Striga densiflora* whole plants or underground stems, collected in sugarcane and rice fields, strongly inhibited seed germination and radicle and plumule growth of rape (*Brassica napus*). Boiling the leachates greatly increased the inhibitory effects. Sugarcane and rice fields badly infested with *Striga* are generally plowed in the summer and sown with rape or a related crop. Thus the water-soluble inhibitors may suppress seed germination and seedling establishment of the crops. *Tagetes patula* (marigold) contains secondary metabolites such as thiophenes and benzofurans, so Tang et al. (1987) used a continuous root-exudate trapping system, along with GC/MS/DS, to determine whether those compounds were exuded into the rhizosphere. They found that four thiophenes and two benzofurans were exuded into the undisturbed rhizospheres of marigolds. It is clear, therefore, that these compounds are normal constituents in the root-soil interface of marigolds. Moreover, these compounds have been shown to be toxic to plants, bacteria, and numerous other organisms.

*Tephrosia purpurea* is a serious weed in some of the dryland farming areas of India, and Sundaramoorthy and Sen (1990) suspected that it might have strong allelopathic potential along with a marked competitive ability. Soil collected near *T. purpurea* was very inhibitory to germination and growth of pearl millet (bajra), til (sesame), and clusterbean compared with germination and seedling growth of the same test plants in soil collected in an area free of tephrosia. Volatile chemicals from tephrosia and water leachates from leaves, stems, and roots of that weed markedly depressed germination and seedling growth of the same test crop plants. Thus *T. purpurea* was shown to be strongly allelopathic to common crop plants of the study area.

*Trianthema portulacastrum* is also a common weed in arid lands of Rajasthan, India. Sethi and Mohnot (1988) made aqueous extracts of air-dried leaves of *T. portulacastrum,* in concentrations from 0.01 to 10%, and tested them against germination and seedling growth of moth bean (*Vigna aconitifolius*). All concentrations from 0.1% and above inhibited seed germination and seedling growth, and concentrations of 5 and 10% completely inhibited seedling growth. The 10% concentration completely prevented seed germination.

El-Ghareeb (1991) observed that, in an abandoned field in the sandy desert of Kuwait, annual plants were less numerous in stands dominated by *Tribulus terrestris* than in adjacent stands dominated by other species. Moreover, in the *Tribulus* stands, annuals were smaller in stature. Measurements of physical conditions, such as moisture, light, and soil characteristics, did not appear to be limiting factors. Next a fine mist was sprayed on living plants of *T. terrestris;* the drip was collected and tested against seed germination and seedling growth of common species in the neighboring area: *Amaranthus lividus, Bromus tectorum, Chenopodium murale, Hordeum murinum, Conyza bonariensis, Solanum nigrum, Senecio glaucus, Malva parviflora, Brassica tournefortii,* and *Eruca sativa.* Seed germination and radicle growth of all species was significantly reduced, except for *M. parviflora* and *S. glaucus.* It was concluded that *T. terrestris* is able to exert effective allelopathic interference on certain annuals in the surrounding flora and thus gain an advantage in dominating some sites in abandoned fields in the desert of Kuwait.

*Trichodesma sedgwickianum* is another serious weed in arid lands in India. Bansal and Sen (1982) found that interference by that weed decreased growth of bajra and til significantly and reduced yields by 29 and 42%, respectively. Buried residue of the weed was also toxic to bajra and til and reduced crop yields by 8 and 46%, respectively. The residue experiment definitely indicated that allelopathy is an important part of the interference effect of the weed.

*Urgenia indica* is a perennial weed in cultivated fields with sandy soils in some parts of India; it is the source of interference that causes decreased yields in various crops (Khare, 1980). Khare found that aqueous leachates of its bulbs, seeds, and leaves reduced germination and seedling growth of corn, grain sorghum, black gram (*Phaseolus mungo*), and pigeon pea (*Cajanus cajan*). Leachates of bulbs were must inhibitory, followed by seeds and leaves. In so far as chemical inhibitors were concerned, the vegetative phase of the bulbs was most inhibitory to crop growth.

*Xanthium strumarium* is a common weed in Pakistan and its interference against crop plants is pronounced. Inam et al. (1987) found that aqueous extracts of leaves, stems, fruits, and inflorescences strongly reduced radicle growth in lettuce, mustard (*B. campestris*), *Pennisetum americanum,* and corn and occasionally reduced germination of lettuce. The extracts of leaves and inflorescences were usually most inhibitory. The germination of lettuce and mustard was reduced in shoot litter, and radicle growth of all test species was reduced by both shoot and root litter. Litter incorporated into a nutrient-rich medium significantly decreased germination of lettuce and *P. americanum* and fresh and dry weights of all test species. Nonconcentrated rain leachate of the tops reduced radicle growth only of *P. americanum.* In one case, radicle growth was stimulated. Concentrated rain leachates reduced germination of lettuce and *P. americanum* and radicle growth of all test species. Caffeic, p-coumaric, p-hydroxybenzoic, chlorogenic, and ellagic acids were identified as possible inhibitors in shoot extracts and rain leachates. The evidence was thus strong that *Xanthium strumarium* has considerable allelopathic potential in fields in Pakistan.

**Conclusion** Even a quick perusal of the first two chapters will demonstrate that many important crop plants have been clearly shown to have strong allelopathic effects against other crop plants and often to weeds. The same can be noted concerning allelopathic effects of many weed species against crops and often against other weeds. Modern methods of chemical identification have enabled investigators to identify potent allelopathic compounds in the plants and in the substrate—and, in the case of volatile compounds, also in the air. These compounds have also been quantified in many cases in the plant extracts, in the substrate, and in the air and have been shown to be present in concentrations sufficient to affect the growth of suspected target plants.

It is now time for practitioners in the various biological sciences to accept the fact of allelopathic effects and to make use of the scientific data in solving agroeconomic problems. Those persons working in agriculture should pursue the problems of determining how the negative allelopathic effects can be lessened or eliminated and how the positive effects can be used in the field. Most of the rest of this book will be concerned with suggestions as to how allelopathy can be useful to farmers, foresters and others.

CHAPTER 3

# OTHER ROLES OF ALLELOPATHY IN AGRICULTURE

## I. ALLELOPATHIC EFFECTS OF MICROORGANISMS ON PLANTS

AS EARLY AS 1951, Brian et al. found that lettuce plants grown with their roots in nutrient solution containing the antibiotic griseofulvin not only had this compound in their leaves after a few days but also gave evidence of stunted growth. Oat seedlings placed with their roots in a solution containing this same antibiotic had the compound in guttation drops from their leaves after seven days. In both experiments no antibiotic was found in leaves of plants growing in similar nutrient solutions that lacked griseofulvin. In addition, lettuce plants growing in the solution with griseofulvin were sprayed with a suspension of spores of *Botrytis cinerea* isolated from diseased lettuce; two weeks later most control plants were dead, but at least 60% of the treated plants were free of *Botrytis* infection. In another experiment, oat plants were grown in soil inoculated with *Penicillium griseofulvum* and in untreated soil. Guttation drops from plants grown in the treated soil were found to contain griseofulvin, while plants grown in the control soil showed no trace of the substance. This indicated that *P. griseofulvum* did produce the antibiotic in soil and that it was taken up through the roots of the oat plants. Moreover, the oat plants were severely retarded in growth in the soil in which griseofulvin was produced.

Brian et al. (1961) subsequently reported that several strains of *Fusarium equiseti* were highly phytotoxic to growth of pea stems. They tested 45 species of *Fusarium,* based on a broad classification of the genus, testing up to six strains of some species. Slight scorching of pea foliage was occasionally noted for culture filtrates of *F. avenaceum, F. culmorum, F. dimerum,* and *F. lateritium,* but severe scorching of the foliage and marked depression of stem growth were consistently caused by four strains of *F. equiseti* (in the broad sense). Several phytotoxic substances were isolated, but the most important was diacetoxyscirpenol, which was highly toxic to pea plants but stimulated elongation of cress roots at 0.01 to 0.5 μg/ml. Two less-abundant phytotoxins were isolated that had physical and chemical properties similar to diacetoxyscirpenol. Scirpentriol, a hydrolysis product of diacetoxyscirpenol, was much less phytotoxic. These reports of Brian et al. (1951, 1961) have been discussed out of alphabetical sequence of the genera investigated because of their early dates.

During a study of growth-promoting rhizobacteria, Suslow and Schroth (1982) isolated other root-colonizing bacteria that were deleterious to seed germination and seedling growth of sugar beets. The researchers designated these bacteria as deleterious rhizobacteria (DRB). When these bacteria were pelleted on sugar beet seeds, they reduced seed germination, caused root distortions and root lesions, reduced root elongation, increased infection by root-colonizing fungi, and significantly reduced plant growth. The genera of DRB tentatively identified were *Enterobacter, Klebsiella, Citrobacter, Flavobacterium, Achromobacter, Arthrobacter,* and *Pseudomonas.* The DRB caused significant weight reductions of seedling sugar beets and up to 48% reduction in weight of sugar beet tops. The scientists concluded that the group of bacteria closely related to *Pseudomonas syringae* may produce toxins while colonizing roots.

Mishra et al. (1987) tested culture filtrates of 796 isolates of aerobic actinomycetes for plant-growth activity using *Chlamydomonas reinhardtii* as an indicator. There were 266 isolates of *Streptomyces,* 28 unidentified actinomycetes, and 502 isolates of novel actinomycetes represented by 18 genera. Sixty isolates caused growth inhibition greater than 30%; 37 of these isolates belonged to the genus *Streptomyces.* Other inhibitors included eight isolates of *Actinomadura;* six of *Actinoplanes;* two each of *Thermomonospora, Streptoverticillium,* and *Promicrononospora;* and three unidentified. Filtrates of 70 isolates stimulated growth of the alga by greater than 20%. Six isolates caused dry-weight increases in one or more of the following plants: corn, soybeans, cucumbers, tomatoes, and sorghum. Nine other isolates decreased the dry weight of one or more of the following plants: flax, soybeans, corn, and cucumbers. Only the filtrates of some of the more promising isolates were tested against the crop plants.

Friedman et al. (1989) isolated actinomycetes from the upper 3 cm of the soil layer in a well-developed forest and in an adjacent area where Douglas fir (*Pseudotsuga menziesii*) regeneration had been prevented for two decades. They found that the population density of the actinomycetes in the clear-cut area was two times as high as in the forested area. Moreover, the percentage of actinomycetes that inhibited seed germination of test plants was significantly higher among isolates obtained from the clear-cut area than among those obtained in the forest, and isolates from the clear-cut area had five times the phytotoxic effect of those from the forest. Four percent of the isolates from the clear-cut area and 2.6% of the isolates from the forest significantly reduced growth of two common ectomycorrhizal fungi of Douglas fir. It was concluded that phytotoxic and antifungal actinomycetes may suppress natural regeneration or seedling establishment, either directly or indirectly through inhibition of seed germination or of mycorrhizal fungi.

Odamitten and Clerk (1988) reported that undiluted filtrate and dilutions of 1:1 and 1:2 of *Aspergillus niger* and *Trichoderma viride* prevented radicles of cocoa (*Theobroma cacao*) from developing lateral roots; eventually, the radicles died. Radicles of cocoa seedlings in the same volumes (20 and 30 mL) of higher dilutions (1:5 and 1:10) developed lateral roots and survived. Filtrates of *A. niger* caused hypocotyls of cocoa seedlings to have greater cambial activity and to form xylem vessels and tracheids with larger lumina.

The microbial activity in the rhizosphere has been rather extensively studied, but the microbial activity in the corresponding area around seeds, the spermosphere, has generally had much less study (Lynch, 1978b). Lynch found that microbial colonization of barley seeds usually starts at the tip, which is closest to the embryo. The tip contains a plug of fibrous material that is a continuation of the inner layer of the husk; this material is the site of oxygen and water entry into the seed. This is also the site of exudation of soluble organic and inorganic materials. Inoculation of barley seeds with *Gliocladium roseum* reduced germination. *Azotobacter chroococcum* also suppressed germination of barley when the *Azotobacter* had grown on a medium containing nitrate. In contrast, when this bacterium had grown on a nitrogen-free medium, it either promoted barley seed germination or had no effect. When *A. chroococcum* grown on an N-free medium was added to barley seeds along with *G. roseum,* the inhibitory effect of *Glocladium* was overcome.

El-Shanshoury et al. (1989) reported that when tomato plants grown in sterilized sandy soil, low in available N and P, were inoculated with both *Azotobacter chroococcum* and *Glomus fasciculatum* (mycorrhizal fungus), *Azotobacter* enhanced root infection by *Glomus* and stimulated plant growth and increased shoot N, Ca, Mg, P, and K compared with inoculation with *Glomus* only. The increased plant growth due to the dual inoculation with *Azotobacter* and *Glomus* could have been due solely to the increased availability of minerals. However, the authors pointed out that *A. chroococcum* is known to improve plant growth through production of plant-growth regulators; this is, of course, a form of allelopathy, which may have been operating in this situation.

*Bacillus subtilis* (A13) and *Streptomyces griseus* (2-A24), antagonists to the phytopathogen *Rhizoctonia solani,* were applied individually to barley, oats, and wheat planted at three sites with a known incidence of *R. solani* (Merriman et al., 1974). Root disease was not effectively controlled, yet the seed treatments increased grain yield and dry-matter production at one site, advanced time of heading at another site, and increased tiller number at two sites. Moreover, the organisms persisted on pericarps in soil for five weeks after sowing. In other trials, application of *B. subtilis* and *S. griseus* increased marketable yields of carrots by 48 and 15%, respectively, over controls. The results of all these trials suggested that the plant-growth responses may have been due to factors other than biological control of root pathogens.

Turner and Backman (1991) reported that *Bacillus subtilis* added as a seed treatment consistently colonized the roots of peanut plants at rates exceeding $10^4$ colony-forming units (cfu) per gram of root tissue 120 days after planting. Yield increases of peanuts from 1982 to 1985 at Auburn, Alabama, ranged from –3.5 to 37%, with only two incidences of negative responses in 24 tests. The inoculated peanut plants responded most favorably when subjected to stresses such as limited water availability, poor rotational practices, or cool soils. Treatment of peanut seeds was associated with improved germination and emergence, increased nodulation by *Rhizobium* spp., enhanced plant nutrition, reduced levels of root cankers caused by *Rhizoctonia solani,* and increased root growth. All these effects did not necessarily happen at the same time. Unfortunately, no specific causes of the observed changes were investigated.

Chanway et al. (1988a) investigated the effects of *Bacillus polymyxa* on plant growth, using genotypically defined mixtures of white clover and perennial ryegrass growing in flats. Addition of *B. polymyxa* to a mixture of these plants did not induce significant growth effects in perennial ryegrass but caused a 23% yield increase in white clover. The clover yield advantage increased further when clones of the legume were inoculated with *B. polymyxa* genotypes with which they had previously coexisted in the field where collections were made. The highest white clover yield was attained when clones of all three organisms that had previously coexisted in the field were grown together. Again, no experiments were run to determine if *B. polymyxa* produced allelochemicals that caused the growth stimulation of the white clover.

Chanway and Holl (1991) found that inoculation of one-month-old *Pinus contorta* seedlings with the mycorrhizal fungus *Wilcoxina mikolae* (isolate R947) resulted in an increase of ectendomycorrhizae in the seedlings but decreased shoot biomass. Inoculation with the nitrogen-fixing bacterium (*Bacillus polymyxa* strain L6) alone did not affect seedling biomass or foliar nitrogen content. Coinoculation of seedlings with *Wilcoxina* and *Bacillus,* increased shoot biomass; however, it resulted in ectendomycorrhizal infection similar to fungal inoculation alone. Total foliar nitrogen was lower in seedlings inoculated with *Wilcoxina,* and coinoculation with *Bacillus* did not reverse that trend. Associative nitrogen fixation by *Bacillus* contributed 4% of seedling foliar nitrogen, whether or not the seedlings were mycorrhizal. Coinoculated seedlings, however, had lower foliar nitrogen than did uninoculated controls. Apparently, growth of mycorrhizal pine can be stimulated by inoculating with beneficial bacteria, but growth promotion does not result from increased mycorrhizae and may result only in part from associative nitrogen fixation. Chanway et al. (1991) reported that root growth of lodgepole pine (*Pinus contorta*) was promoted by inoculation of the seeds with *Bacillus* strain L5 in a sterilized, but not in a nonsterile, growth medium. *Bacillus* strain L6 promoted pine root growth in a sterilized medium and also caused significant increases in seedling emergence, shoot weight and height, root weight and surface area, and root diameter in nonsterile peat-vermiculite medium. Shoot growth promotion occurred also when one-year-old lodgepole pine seedlings were planted in pots and inoculated with strain L6. Seed inoculation with *Bacillus* L5 and L6 significantly increased the rate of seedling emergence of white spruce (*Picea glauca*) but did not affect subsequent seedling growth. Douglas fir (*Pseudotsuga menziesii*) seedlings grown from seed inoculated with strain L6 had increased root surface area, whereas seeds inoculated with strain L5 produced seedlings with increased collar diameters. Chanway and his co-workers stated that the consistent stimulation of root dry weight, branching, or surface area in experiments with pine—and the ability of strain L6 to produce compounds with IAA-like activity—suggest that bacterially mediated production of IAA or related compounds may be involved in conifer growth promotion by *Bacillus.* It appears that inoculation with *Bacillus* strains L5 and L6 or other emergence-promoting bacteria may be a useful operational technique to employ in the nursery at sowing time.

Inoculation of lodgepole pine seeds with *Bacillus polymyxa* isolate L6–16R, a strain marked with antibiotic resistance to rifamycin, resulted in statistically significant seedling biomass increases eight weeks after planting (Holl and Chanway, 1992). The L6–16R rhizosphere population declined by an order of magnitude per month, from approximately $10^6$ cfu/g dry weight of pine root tissue four weeks after inoculation to about $10^4$ cfu/g twelve weeks after inoculation. The size of the L6–16R population was not correlated with the magnitude of shoot growth at four weeks, but there was a strong positive correlation between shoot growth eight weeks after inoculation and the rhizosphere population at four weeks. It appeared from these data that promotion of pine seedling growth may be a function of the rhizosphere population size four weeks after inoculation. Tests of lodgepole pine provenances from four different locations demonstrated that inoculation of the seeds with L6–16R stimulated shoot and root dry weight increases up to 35% in three of the four provenances, but growth of the seedlings from the fourth provenance was inhibited.

Chanway et al. (1988b) pointed out that there is apparently selective pressure for growth-promoting microbes to adapt to host plants; therefore the probability of observing a positive effect on plant performance because of inoculation with coexistent bacterial strains should be greater than if plants are inoculated with strains to which they have not been previously exposed. The researchers tested this hypothesis by isolating seven strains of *Bacillus* from rhizosphere soil of spring wheat cv. Katepwa. The field involved had been continuously cultivated in spring wheat for 27 years, with cv. Katepwa having been grown for the last 5 years. A closely related cultivar, Neepawa, had been cultivated there for the 12 years previous to the last 5. When the seven isolates were tested separately against three spring wheat cultivars—Katepwa, Neepawa, and HY320—six of the seven promoted root growth of cv. Katepwa, but none promoted growth of the other cultivars. Shoot height was also increased in cv. Katepwa. These results substantiated the hypothesis and indicate that beneficial bacteria may be isolated by using a natural plant cultivar enrichment technique.

Chanway et al. (1990) reported that inoculation of plants in the greenhouse with *Bacillus* isolates that had coexisted in a permanent pasture with the perennial ryegrass component of a ryegrass–white clover mixture increased ryegrass root and shoot weight. Root and nodule weight of white clover in mixture, regardless of its genotype, were greater when coexistent *Bacillus*-ryegrass combinations were present. The presence of other coexistent *Bacillus*-plant or plant-plant combinations did not enhance performance of either pasture species.

Nakajima et al. (1990) found that a neutral ethyl acetate extract from *Cochliobolus spicifer* (D-5), a strain isolated from diseased wheat in Japan, yielded spiciferone A after several purification steps. This compound was previously reported to be a new plant inhibitor. Activated-charcoal and silica gel partition-column chromatography of the acidic ethyl acetate extract and recrystallization gave two new compounds, spicifernin 1 and 2. It was subsequently found that spicifernin exists as an equilibrium mixture in solvents. A bioassay with rice (cv. Akinishiki) indicated that spicifernin at concentrations of $1 \times 10^{-5}$ and $3 \times 10^{-5}$ *M* stimulated root growth to 119 and 129%, respectively, and shoot growth to 119 and 124%, respectively, of the controls. The length of second-leaf sheath treated with $3 \times 10^{-5}$ *M* spiciferinin was increased by 38%. A concentration exceeding $1 \times 10^{-4}$, however, caused both root and shoot growth to be reduced.

Seven of twelve South African isolates of an undescribed *Fusarium* sp. caused stunting of alfalfa and wheat plants as well as discoloration, necrosis, and dieback of the taproot of alfalfa and the primary roots of wheat (Lamprecht et al., 1989). The fungus could not, however, be isolated from the necrotic roots as could the pathogenic forms. Cultures of the 12 isolates were tested for neosolaniol monoacetate (NMA), and this compound was detected in 10 isolates of the *Fusarium* sp. at concentrations ranging from 310 to 2060 mg/$g^{-1}$. The mortality of alfalfa plants and NMA yields of *Fusarium* isolates in corn cultures were significantly correlated ($r = 0.84$, $p < 0.05$). Exposure of alfalfa and wheat seedlings to a solution containing 10 mg/$L^{-1}$ of pure NMA caused marked mortality of alfalfa and reduction of shoot length in wheat. A concentration of 5000 ng NMA/$g^{-1}$ soil killed alfalfa and wheat plants. These strains of *Fusarium* were apparently phytotoxic but not pathogenic to these crop plants and were readily isolated from plant debris in wheat soil.

The yeast *Metschnikowia reukaufii* is a natural contaminant of nectar and is transferred to the flowers of the milkweed *Asclepias syriaca* by insects, some of which are pollinators of the plants (Eisikowitch et al., 1990). Pollen grains of *A. syriaca* are encapsulated in paired packages, termed *pollinia*, which in the process of pollination are moved from one flower to the stigmatic chamber of another. The nectaries are part of the wall of the stigmatic chamber, and the nectar flows from there to the nectar reservoirs. It has been proved that the nectar is the natural medium for pollen germination. In this natural habitat, the yeast *M. reukaufii* inhibits germination of the milkweed pollen; this inhibition is irreversible after about eight hours' exposure to the yeast. Two strains of the yeast were isolated and investigated for their effects on pollen germination. The strains affected pollen germination adversely by reducing its amount and vigor and by causing any pollen tubes that were produced to burst. One strain was more virulent than the other, and a mixture seemed to have an additive effect. These results suggest that *M. reukaufii* has potential as a biocontrol agent to reduce the fecundity and spread of field milkweed in agricultural systems.

Probably more research has been done on the plant-growth-modifying effects of *Pseudomonas* than on any other genera of bacteria, aside from the possible effects of some of the nitrogen-fixing and nitrifying bacteria. The studies on *Pseudomonas* have been particularly prolific since about 1984. Lynch and Clark (1984) isolated many microorganisms from seed, rhizosphere soil, and straw and found most to be pseudomonads. The colonization potentials of the different isolates varied between 0.04 and $47 \times 10^6$ viable cells/mg dry root. Some of the microorganisms stimulated barley root and shoot growth and dry matter production, whereas others were inhibitory. The effects were not directly related to the colonization potentials. Elliott and Lynch (1984) examined the effects of pseudomonads isolated from the roots of winter wheat on growth of this crop. They found that plants produced from direct drilling of seeds into plots where crop residues had been burned were larger and appeared healthier than those from plots where crop residues remained. Roots from plants drilled directly into plots where the residues were burned were colonized by fewer pseudomonads that inhibited wheat plant growth than were roots from plots where crop residues remained. When several wheat cultivars were bioassayed against each of two of the inhibitory pseudomonads, they differed greatly in susceptibility to the adverse effects of the bacteria. Elliott and Lynch (1985), Fredrickson and Elliott (1985 a,b), and Cherrington and Elliott (1987) considerably extended the findings of Elliott and Lynch (1984) concerning the growth effects of pseudomonads on winter wheat. Fredrickson and Elliott (1985b) found that root-colonizing pseudomonads that inhibited winter wheat root growth in an agar assay also significantly inhibited wheat root growth in vermiculite. They also found that the same pseudomonads inhibited growth of *Escherichia coli* (C-la) and *Bacillus subtilis.* It is noteworthy that the inhibitory chemical produced was relatively labile in laboratory culturing. Toxin production by the pseudomonads appeared to be necessary for root-growth inhibition; toxin-negative mutants no longer inhibited bacterial or root growth. Antibiosis toward *E. coli* as well as wheat root inhibition was reversed by L-methionine, which, according to the authors, provided further evidence that a toxin produced by the pseudomonads is involved in growth inhibition.

Cherrington and Elliott (1987) found inhibitory pseudomonads on winter wheat, pea, lentil (*Lens culinaris*), and barley and on the weed downy brome (*Bromus tectorum*). Neither tillage management nor site affected colonizing numbers. Both inhibitory and stimulatory isolates were found, but there were more inhibitory than stimulatory isolates. Several isolates severely reduced root growth of downy brome, which suggests they might be effective in biological weed control under some conditions.

Alström (1987) studied the effects of *Pseudomonas fluorescens* (S596 and S628) on seedling growth of the common bean (*Phaseolus vulgaris* cv. Bonita) and lettuce cv. Montana. S596 was originally isolated from roots of winter rye and S628 from roots of winter wheat. Both isolates affected growth of both plant species. The isolate S596, however, induced necrosis and wilting of leaves, whereas S628 induced epinasty in the first expanding leaves and interveinal chlorosis in subsequent leaves. Inhibition of shoot development was observed under axenic conditions. The two strains differed in nutritional requirements for affecting plant growth. S596 required sucrose, whereas S628 needed sucrose plus peptone or yeast extract. Alström stated that S628 liberates high amounts of potentially toxic volatiles such as hydrogen cyanide and ethylene when grown in vitro, while S596 does not. This suggests different modes of action. Bakker and Schippers (1987) reported that about 50% of the pseudomonads isolated from the potato rhizosphere produced HCN in vitro. These results support the suggestion of Alström.

Bolton et al. (1989) isolated a toxin from the growth medium of *Pseudomonas* sp. strain RC1 after removing the cells. This toxin inhibited both winter wheat and *E. coli* growth. The toxin was stable for at least 24 hours at 40°. It was extremely polar and could not be extracted from culture filtrates with organic solvents. The scientists were not able to purify the toxin completely, but the partially purified toxin inhibited several different microorganisms, whereas the producing strains were resistant.

Åström and Gerhardson (1989) examined effects of *Pseudomonas* sp. isolates Å 112 (fluorescent) and Å 313 (nonfluorescent) on the growth of several wheat cultivars. They found that both induced leaf symptoms and shoot and root growth inhibition and that effects similar to those obtained with living inoculum could be induced by treating plants with sterile culture filtrates of Å 313 or with volatile bacterial metabolites from Å 112.

Vrany et al. (1990) did massive field tests in Czechoslovakia on effects of inoculation of numerous cultivars of potato with isolates of *Pseudomonas putida* and *Trichoderma* sp. in suspension and in preparations with kaolin and peat carriers. The tests were carried out on seven locations of the potato-rye and sugar beet regions. Increased potato yields occurred in 57.5% of the experiments. Physiological properties and the quality of the harvested tubers were comparable to those of the controls. De Freitas and Germida (1990) tested 75 isolates from the rhizosphere of winter wheat for their effects on plant growth and development of winter wheat in two soils. Twelve isolates stimulated growth of the wheat. The most effective of these were nine isolates that significantly increased plant height, root and shoot biomass, and number of tillers. The plant-growth-promoting effects of the isolates were different in the two soils. Three of the effective isolates were identified as *P. aeruginosa,* and two each as *P. cepacia, P. fluorescens,* and *P. putida.* Results of these two experiments demonstrate the potential use of plant-growth-promoting rhizobacteria as inoculants for winter wheat.

Höfte et al. (1991) reported that the plant-growth-stimulating strains, *Pseudomonas aeruginosa* (7NSK2) and *P. fluorescens* (ANP15), protected corn seeds from cold-shock damage. They significantly increased (by 30 to 60%) the germination of corn seeds subjected to temperatures of 2 to 8°C for 10 days in soil in a greenhouse. Moreover, plants originating from *Pseudomonas*-inoculated corn seeds had a 7 to 10% higher dry-matter content than control plants. The ANP 15 strain increased the germination of two-year-old seeds to the same percentage germination as one-year old seeds.

Åström (1991) tested lettuce and wheat cultivars that differed in reaction to root inoculation with plant-growth inhibitory bacteria; the cultivars were tested for sensitivity to pure cyanide and to gaseous metabolites produced by deleterious cyanogenic isolates of *Pseudomonas fluorescens.* Salad Bowl lettuce was significantly less sensitive, both to bacterial volatiles and to pure cyanide, than was Montana lettuce. A similar difference between those cultivars was also obtained where bacteria were inoculated directly on the roots. These differences did not occur, however, when the bacteria were grown on a medium that did not support cyanide synthesis. In wheat, a difference in sensitivity to bacterial-produced volatiles occurred between cultivars Drabant and Besso, but these cultivars did not react differently to pure cyanide. Thus differential cultivar responses to these bacteria did not appear to be related to cyanide.

Mitchell and Coddington (1991) reported still another compound with phytotoxic properties that was produced by another species of *Pseudomonas.* They found that *P. andropogonis* rapidly converted ($^{14}C$) aspartic acid into rhizobitoxine and hydroxythreonine; hydroxythreonine was later incorporated into rhizobitoxine.

It is obvious that there is massive evidence that many strains of rhizosphere pseudomonads produce compounds that affect seedling growth of many crop plants. It is evident that the commercial use of several of these strains is already economically feasible, and many more could become so with additional research.

Dewan and Sivasithamparam (1988a, 1989, 1991) isolated a sterile red fungus from the roots of wheat and ryegrass (*Lolium rigidum*) in Western Australia. The fungus produced clamp connections, hyphal strands, and swellings and was assumed to be a Basidiomycete. Inoculation with the fungus increased fresh shoot and root weight of wheat in nonsterilized soil and of ryegrass in sterilized and nonsterilized soil. It increased root lengths of both plants in both sterilized and nonsterilized soil. The sterile fungus provided significant protection to the plants from infection by the take-all fungus (*Gaeumannomyces graminis* var. *tritici*), so part of the stimulatory activity in nonsterilized soil could have been due to this effect, but there was obviously another factor (or other factors) operating in the sterilized soil. It is noteworthy that the take-all fungus had no effect on the development of the sterile red fungus on the roots of wheat or rye when they were coinfested in sterilized and nonsterilized soil. It is also significant that enhancement of shoot and root growth of wheat seedlings by the sterile red fungus was evident even when extra nutrients were supplied and the density of plants per pot was doubled. The researchers concluded that the stimulatory effect of the fungus was probably due to production of a hormonallike substance or substances.

Sneh et al. (1986) examined the effects in field plots in Israel of inoculation of several crop species with a nonpathogenic *Rhizoctonia solani*. The increases for treated compared with untreated controls were as follows: Giant Butler radish fresh weight, 13.4 to 19.8%; dry weight, 28.4 to 36%; Chantane carrot fresh weight, 30.0 to 97.6%; dry weight 55.0 to 150.5%; Arab lettuce fresh weight, 58.4%; dry weight, 61.8%; Pima cotton fiber weight, 28.7%; Lachish, Miriam, and Beth Lehem wheat weight per grain, 10.6 to 23.3%; grain yield, 15.4 to 36.5%. In the case of Desire potato, there was an increase in leaf, shoot and tuber weight 63 to 70 days after planting, but there was no increased yield at harvest. Unfortunately, the mechanisms for the increased yields due to inoculation with the nonpathogenic *R. solani* were not determined. In some instances, there may have been a decrease in plant disease, but in several of the crops, that was not likely. In such cases, yield increases may have been due to the production of growth stimulators.

Underwood and Baker (1991) inoculated axenic duckweed (*Lemna minor*) plants separately with five different species of freshwater bacteria, *Pseudomonas* sp., *Vibrio* sp., *Klebsiella* sp., *Enterobacter* sp., and *Serratia* sp. *Lemna* inoculated with *Vibrio* sp. had a population density significantly greater after 52 days than that of the axenic controls. Inoculation with *Pseudomonas* sp. caused the final population of *Lemna* to be significantly higher than that of *Lemna* inoculated with a mixed natural population of bacteria. Inoculation with *Klebsiella* sp., *Enterobacter* sp., or *Serratia* sp. increased plant populations of *Lemna* compared with the controls but the differences were not statistically significant. None of the five single bacterial taxa appeared to significantly affect the senescence of duckweed. The growth-stimulating factors were not determined.

## II. ALLELOPATHIC EFFECTS OF PLANTS ON MICROORGANISMS

Levin et al. (1988) extracted *Aloe* leaves with ethanol, filtered the extract, and carried out a double TLC separation. The fraction that was active against *Bacillus subtilis* was extracted, dried, dissolved in water and used in various tests with *B. subtilis.* The active principle was originally a brownish color that turned to a dark red if kept in water, and it precipitated in high concentrations of ethanol. These characteristics plus its $R_f$ value on TLC plates indicated that the active compound was an anthraquinone-type substance. The compound primarily inhibited nucleic acid synthesis in *B. subtilis,* after which protein synthesis was inhibited. *Aloe vera, A. candelabrum, A. dichotoma, A. bohri, A. pelicatelis, A. spectalalis, A. arborescens,* and *A. ferox* were all tested and found to produce the inhibitor, but in different concentrations. On a dry-weight basis, the inhibitor was equally distributed between the skin and gel fraction.

Rafiq et al. (1984) tested water extracts of *Anagallis arvensis, Asphodelus tenuifolius, Chenopodium album, C. murale, Convolvulus arvensis, Coronopus didymus, Cuscuta reflexa,* and *Fumaria parviflora* growing wild near Faisalabad, Pakistan, against growth of some eight species of fungi. Water extracts of only three of the species inhibited any of the fungi. Extracts of *Anagallis arvensis* inhibited growth of *Helminthosporium oryzae, H. carbonum,* and *H. turcicum;* extracts of *Chenopodium album* inhibited growth of *H. carbonum;* and *Colletotrichum falcatum* and *Cuscuta reflexa* inhibited *H. turcicum* only.

The acetylenic allelochemical phenylheptatriyne (PHT), from species of the Asteraceae, was examined by Arnason et al. (1986) for its mechanism of action against *Fusarium culmorum.* They reported that it disrupts membrane function by both phototoxic and nonphototoxic mechanisms. Mild treatment with only 10 ppm PHT and near-UV light inhibited respiration, $^{14}C$-phenylalanine uptake, and enhanced $K^1$ leakage. Total growth of *F. culmorum* was also reduced, but a decline in respiration rate appeared to be the primary inhibitory effect since it occurred at lower concentrations of PHT than did reduction in growth.

Vierheilig and Ocampo (1990) hypothesized that the lack of vescicular-arbuscular mycorrhizal (VAM) colonization in Cruciferae may be due to compounds released in root exudates instead of to intrinsic characteristics of the cortex, as usually attributed. Glucosinolates are a large group of sulfur-containing glycosides found in the Cruciferae, but they have no antifungal activity. They can be easily hydrolyzed, however, by the endogenous enzyme myrosinase after mechanical injury to a variety of products, including isothiocyanates, which are volatile and thermolabile and have marked antifungal activity. To test their hypothesis, the scientists first flooded spores of the VAM fungus, *Glomus mosseae,* with a *tris*-HCl buffer, aqueous cabbage, or aqueous tomato root extracts. Compared with the controls, tomato extracts did not affect germination of the spores, whereas cabbage extracts markedly reduced spore germination. Heating the cabbage root extract to 48°C or autoclaving the roots prior to extraction eliminated the inhibitory effect of the extract. These results suggested that isothiocyanates may act as the inhibitors of spore germination. Other experiments were performed to test this idea. Vierheilig and Ocampo tested the effect of sinigrin alone (a glucosinolate), myrosinate alone, and the two together on spore germination of *G. mosseae.* The compounds applied separately to the spores had no effect on germination, but the combination completely prevented spore germination. Thus considerable evidence supported their hypothesis.

Sokolov et al. (1972) extracted juglone (5-hydroxy-1,4-napththoquinone) from unripe fruits of pecan (*Carya pecan*) and English walnut (*Juglans regia*) and plumbagin (2-methyl-5-hydroxy-1,4-naphthoquinone from *Ceratostigma plumbaginoides* and tested these and 1,4-napyhthoquinone against twelve species of microorganisms, including six phytopathogenic ones. They found that 1,4-naphthoquinone was least active and that the OH group in position 5 greatly increased the antimicrobial action. Addition of the $CH_3$ group in position 2, however, reduced the activity in some cases and increased it in others.

Lethal concentrations of 1,4-naphthoquinone against the 12 test species ranged from 5 to 50 μg/mL, of juglone from 2 to 20 μg/mL, and of plumbagin from 1 to 20 μg/mL. Didry et al. (1985) continued this line of research on plumbagin and juglone and also experimented with lawsone (2-hydroxy-1,4-naphthoquinone) from *Lawsonia* spp. Plumbagin was very inhibitory to seven of the eight aerobic bacteria tested and to a yeast, juglone inhibited two of the aerobic bacteria and a yeast, and lawsone inhibited three of the aerobic bacteria. The minimum inhibitory concentration (MIC) of plumbagin against *Candida albicans, Staphylococcus aureus, Acinetobacter calcoaceticus,* and *Streptococcus faecalis* was 5, 10, 15, and 25 μg/mL, respectively. The MIC of juglone against *Candida albicans* was 5 μg/mL. Plumbagin was inhibitory to all of the 13 anaerobic bacteria tested, juglone was inhibitory to 6 of the 13, and lawsone was inhibitory to 9 of the 13.

Tada and Sakurai (1991) isolated two antimicrobial compounds from the methanol extract of leaves of the Japanese plant *Cercidiphyllum japonicum.* They identified one of these as 3,3',4-O-trimethylellagic acid and found it to be active against *Bacillus subtilis.* The second compound, which they named cercidin, was active against both *B. subtilis* and *Escherichia coli.* With high-resolution MS and NMR, they were able to identify cercidin. It has a benzodioxane ring, which the authors suggested could be biosynthesized from maltol and gallic acid. They gave the likely structure but gave no chemical name for the compound.

Mallik and Tesfai (1987, 1990) reported that small amounts of filter-sterilized-water extracts of shoots of foxtail (*Setaria viridis*), lambsquarters (*Chenopodium album*), Johnsongrass (*Sorghum halepense*), and Pennsylvania smartweed (*Polygonum pennsylvanicum*) stimulated growth of *Bradyrhizobium japonicum* 3I1b110 in yeast extract–mannitol broth by three- or four-fold after a 48-hour incubation. Similar extracts of several other weeds were marginally stimulatory. Autoclaved water extracts and organic extracts were less stimulatory than filter-sterilized-water extracts (Fig. 2). Filter-sterilized-water extracts from shoots of foxtail and lambsquarters increased growth of the bacterium linearly up to 6 mg/mL$^{-1}$. Comparable growth stimulation by these extracts was found in four other strains of *B. japonicum.* Dialysis of the aqueous extract of lambsquarters shoots through a membrane (MWCO 1000) indicated that the molecular weight of the growth factor was less than 1000. The active fraction had an $R_f$ of 0.91 when separated by paper chromatography using 6% acetic acid as

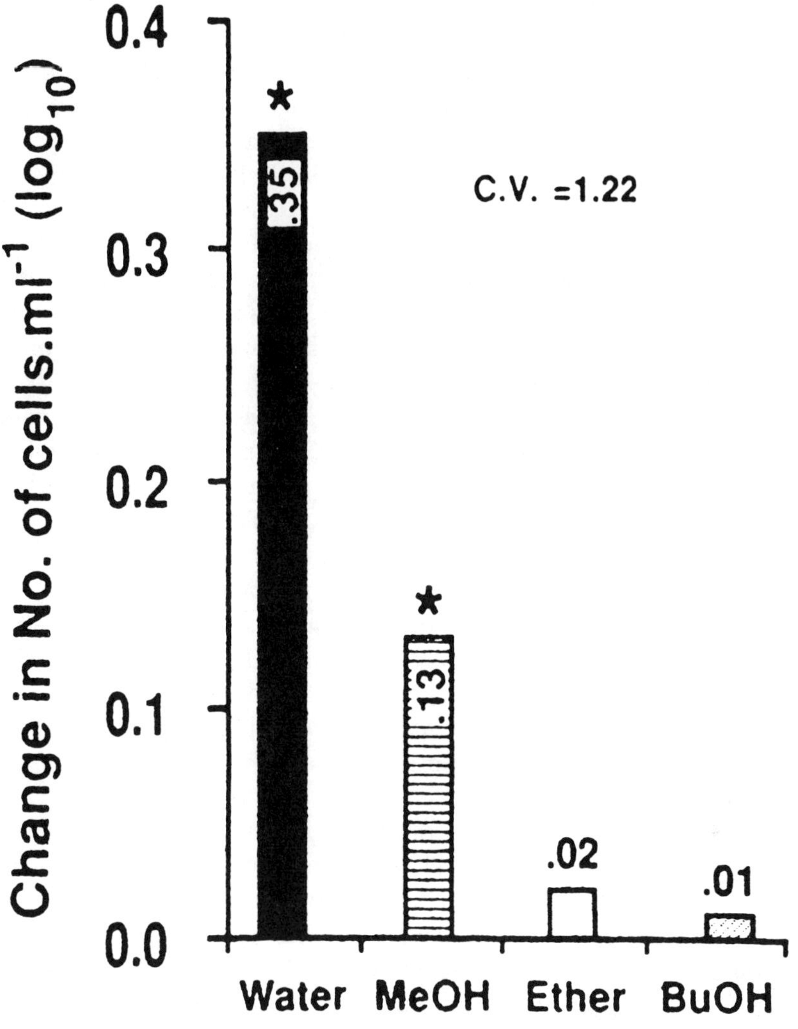


FIGURE 2 Net stimulation of *B. japonicum* in yeast-mannitol broth (YMB) supplemented with extracts of lambsquarters shoots made in different solvents. Growth of control flask was subtracted from that of treated flasks. Cell number was determined by plate count; incubation period was 48 hours. *Significantly different from control by t-test at 5%. *Adapted from Mallik and Tesfai (1990), with permission of Kluwer Academic Pub.*

developer. The stimulatory factor is highly biodegradable; partially thermolabile; highly water-soluble; only partially soluble in methanol; and not soluble in ether, ethyl acetate, hexane, or butanol. Mallik and Tesfai stated that the factor has the potential of being useful in the commercial production of legume inoculants and in the fermentation industry.

Toyota et al. (1990) isolated an antifungal triterpenoid saponin from a methanolic extract of the roots of *Clerodendrum wildii* (an African medicinal plant). They used MS, NMR, TLC, and HPLC to identify the compound as Mi-saponin A. Less than 3.3 μg of this compound inhibited spore formation in *Cladosporium cucumerinum.*

*Croton sonderianus* is a common shrub of the northeast region of Brazil; its durability suggests that it plays an important role in the equilibrium of the ecological system. McChesney et al. (1991) found that the organic resin from the roots showed significant antimicrobial activity. Fractionation of the resin yielded two acidic diterpenes, hardwickic acid and 3,4-secotrachylobanoic acid. The MIC of the former against *Bacillus subtilis* was 0.78 μg/mL in a 24-hour test and 1.56 in a 48-hour test. It was 25 μg/mL against *Trichophyton mentagrophytes.* As a comparison, 3,4-secotrachylobanoic acid had a MIC of 25 μg/mL against *Bacillus subtilis,* but it also inhibited *Saccharomyces cerevisiae* at a MIC of 100 μg/mL. Streptomycin had a MIC of 3.12 μg/mL against *B. subtilis*; amphotericin B had a MIC of 0.78 μg/mL against *T. mentagrophytes* and 6.25 μg/mL (24 hours) and 12.5 μg/mL (48 hours) against *S. cerevisiae.*

Several phenolic acids, commonly produced by plants, were tested by Shafer and Blum (1991) to determine their effects on the rhizosphere microorganisms of cucumber. Cucumber seedlings watered five times (every other day) with aqueous solutions of ferulic, p-coumaric, or vanillic acid (each at 0, 0.25, or 0.5 μmol/g of soil) suppressed leaf growth by up to 42%. Over 600% increases in populations of fast-growing bacteria were detected in the rhizosphere after two but not after five treatments; increases over 400% in numbers of fungal propagules were found after five treatments. Subsequently, chronic treatments with 0.025 or 0.1 μmol/g soil of ferulic acid were found to have no effects on numbers of bacteria and fungi. Furthermore, chronic treatments did not alter responses of plants or microorganisms to a subsequent acute treatment of 0.5 or 1.0 μmol/g soil. Thus microbially mediated acclimation of plants to relatively high concentrations of ferulic acid was not found.

Komai et al. (1982) tested essential oil from underground parts of *Cyperus rotundus,* purple nutsedge, against six species of fungi, including two yeasts, and two species of bacteria. The two bacteria, *Bacillus subtilis* and *Escherichia coli,* were considerably more sensitive to the essential oil than were the fungi. *Aspergillus orizae, A. niger, Saccharomyces cerevisiae,* and *Candida didensii* were inhibited somewhat in growth by the highest concentrations tested, 1000 ppm. *A. niger* was also inhibited slightly by 750 ppm of the oil. Seven sesquiterpenes were isolated from the essential oil; cyperene, ß-selinene, ß-elemene, caryophyllene, α-humulene, cyperenone, and α-cyperone. All of these inhibited both test bacteria, but the last two listed were by far the most active. Only those two and ß-selinene had activity against growth of *Fusarium oxysporum* and *F. lini.* Concentrations of cyperene, ß-selinene, cyperenone, and α-cyperone of $10^{-3}$ *M* or higher inhibited spore germination of *Cochliobolus mibabeanus.* Again, the last two were most inhibitory.

Marston et al. (1988) observed that the methanol extract of the roots of the African medicinal plant, *Dolichos kilimandscharicus,* was active against spore formation in *Cladosporium cucumerinum.* TLC, LPLC, and gel filtration gave three saponins, which were identified by TLC, MS, and NMR to be the 3-O-ß-D-glucopyranosides of hederagenin, bayogenin, and medicagenic acid. The minimum amount required to inhibit spore formation in *C. cucumerinum* in a TLC bioassay was 2.5 to 5 μg of any one of the three saponins.

The increasing resistance of several pathogenic fungi in India to synthetic fungicides and the residual toxicity of some of them caused Singh et al. (1988) to examine *Echinops echinatus,* a composite, for antifungal compounds. They isolated four phenolic compounds—apigenin, apigenin-7-O-glucoside, echinacin, and echinaticin—from the whole plant. They made two derivatives, echinacin permethyl ether and apigenin-5–4'-dimethyl ether, by methylation of echinacin and apigenin-7-o-glucoside permethylate. All the compounds were assayed against conidial germination of *Alternaria tenuissima,* which causes leaf blight of pigeon pea. All were effective at concentrations ranging from 25 to 150 μg/mL. Echinacin was highly effective and is considered the most promising of these compounds for control of the leaf blight of pigeon pea under field conditions.

Many water bodies in cities are heavily polluted. The abundance of algae growing in these waters causes them to become turbid and dirty. Sun et al. (1988, 1989) found that planting water hyacinth (*Eichhornia crassipes*) in such eutrophic waters suppresses the algae and clarifies the water. In addition to competition with the algae for light and minerals, the chief mechanism of the inhibitory effect is the excretion of organic substances from the root system of the water hyacinth that injure or kill the algae. When algae were treated with water in which water hyacinth had previously grown, the photosynthetic rate of the algae declined markedly, and chlorophyll a was destroyed. Moreover, the water hyacinth increased the population of *Hippeutis,* a mollusk that feeds on algae. Sun et al. (1990) found that an extract from the water in which water hyacinth was cultured under sterile conditions still showed allelopathic effects. This indicated that the allelopathic compound(s) was exuded by water hyacinth and not by microbes.

Five essential oil components—anethole, carvacrol, cinnamaldehyde, eugenol, and safrole—were tested against 20 species of food-spoilage fungi belonging to *Aspergillus, Mucor,* and *Rhizopus* by Thompson (1989). All compounds completely inhibited growth of all 20 species, but most required more than 100 μg to do so. Carvacol was most inhibitory to *Rhizopus* spp., with only 5 μg completely inhibiting seven of nine species. It was also most inhibitory to *Mucor* spp., with 5 μg completely inhibiting growth of three of four species and 10 μg completely inhibiting the fourth. Carvacrol and eugenol were most inhibitory to *Aspergillus* spp.; 10 μg of carvacrol completely inhibited all eight species tested, whereas 10 μg of eugenol completely inhibited six species, 50 μg completely inhibited one species, and 100 μg completely inhibited one species.

Kumar et al. (1990) isolated and identified two fungicidal phenylethanones from the root bark of *Euodia lunu-ankenda,* a southern Asian species used as a medicinal plant in Sri Lanka. Using the TLC bioassay technique, the researchers found that the two compounds were strongly active against the fungus *Cladosporium cladosporioides.*

*Ficus septica* is a small tree that grows in coastal provinces of Papua, New Guinea; the leaves of this tree are used locally as medicinals. Baumgartner et al. (1990) found that 4 μg of a crude methanolic extract of the tree's leaves had strong activity against *Penicillium oxalicum* and weak activity against *Bacillus subtilis* and *Micrococcus luteus.* Using TLC, the workers isolated two indolizidine alkaloids and identified (with MS and NMR) ficuseptine, 4,6-*bis*-(4-methoxyphenyl)-1,2,3-trihydroindolizidinium chloride, and antofine, which was previously known. The MIC of each on TLC against *Bacillus subtilis* and *Escherichia coli* was 5 and 10 μg, respectively. The MIC of ficuseptine against *Micrococcus luteus* and *Penicillium oxalicum* was 5 and 1 μg, respectively; the MIC of antofine against the same organisms was 1 μg and 3 μg, respectively. For comparison, the MIC of tetracycline-HCl against *B. subtilis* and *M. luteus* was 0.4 μg and against *E. coli* was 0.8 μg. The MIC of miconazole-$NO_3$ against *P. oxalicum* was 0.8 μg in the TLC bioassay.

Spring et al. (1992) isolated and identified three new sesquiterpenes from noncapitate glandular hairs of the cultivated *Helianthus annuus*. They gave the names glandulone A, B, and C to the three compounds. All three compounds demonstrated cytostatic effects against *Bacillus brevis* Migula ATCC9999 in agar diffusion tests. The MIC of A and C was 200 μg/mL, and the MIC of B was 180 μg/mL.

Tomás-Barberán et al. (1988) collected *Helichrysum nitens* of the Compositae in Malawi and found that soaking the aerial parts in dichloromethane caused the solvent to have antifungal activity. They isolated and identified 5,7-dimethoxyflavone; 5,6,7-trimethoxyflavone; 3,5,7-trimethoxyflavone; 5,6,7,8-tetramethoxyflavone; 3,5,6,7-tetramethoxyflavone; and 3,5,6,7,8-pentamethoxyflavone from the dichloromethane. The new compounds were externally deposited on the leaf and stem surfaces. Only 1 to 5 μg of each of the identified compounds was required to prevent spore germination and mycelial growth of *Cladosporium cucumerinum* in the TLC bioassay. The researchers pointed out that these compounds had about the same antifungal activity as tangeretin, which is responsible in part for the resistance of *Citrus* spp. to the pathogenic fungus, *Deuterophoma tracheiphila*.

Tomás-Barberán et al. (1990) subsequently isolated and identified several antibacterial phloroglucinol and acetophenone derivatives from aerial parts of *Helichrysum decumbens, H. stoechas,* and *H. italicum*. Three antifungal phloroglucinol derivatives were identified, but only one was present in sufficient amounts to be completely characterized. One acetophenone derivative was identified as 4-hydroxy-3 (isopentent-2yl) acetophenone; it was the main antifungal compound in *H. italicum*. Both the main phloroglucinol derivative and the acetophenone derivative showed antifungal activity against *Cladosporium herbarum,* but the acetophenone derivative was three times more active. Both compounds were active against gram-positive bacteria with an MIC of 12.5 and 100 μg/mL; the phloroglucinol derivative was about eight times as active as the acetophenone derivative. The acetophenone derivative, however, was also active against the gram-negative bacterium, *Escherichia coli,* at a MIC of 25 μg/mL. The scientists concluded that the fact that different *Helichrysum* species produce different secondary metabolites as a biochemical defense mechanism against fungi and bacteria is of considerable significance, since it indicates the use of different metabolic pathways to produce chemical barriers that play a single ecological role in higher plant defense against pathogenic bacteria and fungi.

Welch et al. (1990) reported that the placing of barley straw in an unused canal reduced the amount of filamentous algae downstream, beginning with the second year after introduction of the straw and continuing for three years. Algal growth on microscopic slides suspended in the water downstream from the straw was reduced by 90% compared with growth on slides suspended upstream from the straw. Phosphate, nitrate, and ammonium concentrations were not altered significantly by the presence of straw, but nitrite concentrations were increased during summer months. The researchers stated, however, that the nitrite increase was not considered a likely cause of algal growth inhibition. In a subsequent paper, Gibson et al. (1990) pointed out that the presence of barley straw in water, at least in laboratory studies, inhibited growth of several planktonic and filamentous algae. The inhibitory effect was produced progressively during decomposition of the straw at 20°C and reached a maximum after six months. When the straw was autoclaved, all inhibitory activity was lost. Addition of liquor from rotting straw also inhibited algal growth. The capacity to inhibit growth, moreover, remained in the liquor after passage through a 0.2 μm filter but was lost when the liquor was filtered through activated charcoal. The researchers concluded that the inhibitory effect of straw shows promise as a practical way to limit the growth of a range of algae that cause problems in aquatic habitats.

Hubbell et al. (1983) isolated and identified the terpenoid, caryophyllene epoxide, as the repellent of the leaf-cutting ant, *Atta cephalotes,* in *Hymenaea courbaril,* a leguminous tree in the lowland forests of Guanacaste Province, Costa Rica. These ants cut leaves of several trees to serve as a substrate for culturing a specific fungus for their food supply, but they do not use *Hymenaea.* The researchers tested the caryophyllene epoxide against the attine fungus and found that, within 48 hours, all but two of 10 cultures at 100 μg/mL, and all cultures at 3 and 30 mg/mL, were shriveled and brown. At a concentration of 100 μg/mL, caryophyllene epoxide was also fungistatic or fungicidal to about one-third of 45 species of animal or plant pathogens against which it was tested.

Downum et al. (1989) assayed 115 plant species representing 57 genera and 8 families of plants native to arid and semiarid habitats throughout the southwestern United States, Baja California, and northern Mexico for phototoxic natural products. They employed standard antimicrobial techniques using *Escherichia coli* and *Saccharomyces cerevisiae.* The highest frequency of phototoxins in the Compositae occurred in the subtribe Pectidinae in the tribe Heliantheae. Leaf resin from *Larrea tridentata* (Zygophyllaceae) had strong antimicrobial activity in the absence of light; however, the toxicity was enhanced slightly by UV light. No phototoxic antimicrobials were detected in test species from the Asclepiadaceae, Chenopodiaceae, Hydrophyllaceae, Lamiaceae, Polygonaceae, or Solanaceae.

Weaks (1988) investigated the attached algal community of a freshwater marsh in Stewart County, Tennessee. One transect that was analyzed was in close proximity to dead vegetation and a dense stand of *Leersia lenticularix;* a second transect was located parallel to the first but in open water largely free of flood-stressed and living macrophytes. Mean species diversity, cell density, and chlorophyll a were found to be generally lower along the transect located near dead vegetation and rooted plants. Studies of test water from the zone where attached algal inhibition occurred indicated the presence of heat-labile toxic substances. Moreover, light intensity was higher along the transect where numbers of attached algae were lowest. No differences between the transects occurred in turbidity, pH, and conductivity. These results all pointed to an allelopathic inhibition of the attached algae along the transect nearest the dead or emergent vegetation.

Lane et al. (1987) extracted the roots of three-month-old field-grown *Lupinus angustifolius* cv. Uniharvest with 95% ethanol and isolated and identified nine 5-hydroxyisoflavones that acted as feeding deterrents to larvae of *Costelytra zealandica* and *Heteronychus arator.* The identified compounds were licoisoflavone B, licoisoflavone A, luteone, 2'-hydroxygenistein, angustone A, angustone B, angustone C, genistein, and wighteone. At either 20 or 30 μg/mL, the first three listed and wighteone prevented growth of 95% of the sporelings of the fungi *Cladosporium cladosporoioides* and *Colletotrichum gloeosporioides*. The other compounds had little or no antifungal activity up to 100 μg/mL.

Lindblad et al. (1990) pointed out that azoxyglycosides are compounds unique to cycads; they are divided into cycasin, macrozamin, and neocycasins, depending on the nature of their carbohydrate moiety. The same toxic aglycone, methylazoxymethanol, is found in all azoxyglycosides. The researchers used HPLC to determine the presence of azoxyglycosides in the Australian cycad *Macrozamia riedlei* and found cycasin and macrozamin in all tissues examined. Cycasin was 17 times more abundant than was macrozamin. The coralloid root, which contains *Nostoc* sp., a nitrogen fixing cyanobacterium, had 0.16% g/g (fresh weight) of cycasin and 0.01% macrozamin. Addition of these azoxyglycosides, in the same concentrations, to *Nostoc* PCC73102 inhibited light and dark nitrogenase activity. Concentrations of cycasin of 1.6% and macrozamin of 0.1% decreased phycobiliprotein content by 25 and 45%, respectively, one and four hours after the addition.

Moujir et al. (1990) isolated several netzahualcoyone-skeleton triterpene quinones from *Maytenus horrida* and *Schaefferia cuneifolia* in the Celastraceae and tested them against 18 different bacteria and 1 yeast. Many of the gram-positive bacteria were inhibited in growth but none of the gram-negative bacteria was affected. Antimicrobial activity was associated with the presence of functional groups of the E ring. Netzahualcoyone, with a ketone group at C-22 and a ß-OH group at C-21 in ring E, was 60 times as active as netzahualcoyene, which does not have those groups. The groups in ring A did not appear to be particularly significant.

Orabi et al. (1991) isolated and identified two antimicrobial compounds from the fleshy, netlike skin that covers the seeds of *Myristica fragrans.* These compounds were identified as the resorcinols, malabaricone B and malabaricone C. Both were inhibitory to the following bacteria and a yeast at the MIC indicated for each: *Staphylococcus aureus* (1 to 4 μg/mL), *Bacillus subtilis* (1 to 2 μg/mL), *Streptococcus durans* (1 to 4 μg/mL), *Candida albicans* (three strains, 4 to 32 μg/mL). Structural modification by methylation or reduction caused diminished activity of each.

Asthana et al. (1986) tested water extracts of leaves of 27 genera and species from 18 plant families for their volatile antifungal activity against *Aspergillus flavus.* All of the taxa inhibited mycelial growth by 22 to 100%. *Ocimum adscendens* was the most inhibitory (100%), and its leaves were more inhibitory than its other parts. The essential oil, which was the volatile antifungal fraction, was thermostable with the maximum activity remaining after autoclaving and storing for a year. A 0.04% concentration of the oil completely inhibited growth of *A. flavus.* A 0.04% concentration of the oil gave a 98 to 100% inhibition of mycelial growth in 13 fungal genera, including 29 different species. It is remarkable that the oil was 10 to 100 times as active against *A. flavus* as five different commercial fungicides—Bavistin, Blitox-50, Dithane M-45, Agrosan GN, and Dithane Z-78. In subsequent research, Dube et al. (1989) reported that essential oil from leaves of *O. basilicum* was also very inhibitory to the mycelial growth of 22 species of fungi, including mycotoxin-producing strains of *A. flavus* and *A. parasiticus.* Except for eight species, all the fungi were completely inhibited in mycelial growth by a 1.5 mL/L concentration of the oil. Seven of those eight species were inhibited by that concentration 85% or more, and even the eighth was inhibited by 68.4%. The minimum lethal dose of the oil was one-fourth to slightly over one-sixth of the minimum lethal doses of the commercial fungicides—Celphos, Sulphex, Agrozim, Bavistin, and Emison—against *A. flavus* and *A. parasiticus.*

Megharaj et al. (1987 a,b) collected 250 g each of the inflorescences, leaves, and roots of *Parthenium hysterophorus* from a natural stand in Anantapur, India, and surface-sterilized all these parts with 0.1% mercuric chloride. The samples were subsequently washed thoroughly with sterile distilled water, and each plant part was soaked in 1 L of distilled water at 10°C for 84 hours. Each leachate was filtered and concentrated to one-tenth of its original volume. This concentration was taken as 100%. The leachates were tested against axenic cultures of the green alga *Chlorella vulgaris* and the cyanobacterium *Synechococcus elongatus,* isolated from field soil. Leaf-leachate concentrations of 0.5 and 1.0% and all root concentrations stimulated growth of *C. vulgaris,* but the 2.5% leaf concentration and all inflorescence concentrations significantly reduced growth of that species. All concentrations of the inflorescence and leaf leachates significantly inhibited or killed *S. elongatus,* as did the 2.5% root leachate. The 0.5 and 1.0% root leachates had either no effect or stimulated growth of *S. elongatus.* A dense stand of *P. hysterophorus* produces over 300

$g/m^2$ $yr^{-1}$ of dry leaves; thus it was concluded that the allelopathic effects on soil algae may be considerable.

Misra et al. (1989) stated that mycotoxicoses primarily involving the liver or kidneys are largely caused by the consumption of food contaminated with *Aspergillus flavus* or *A. parasiticus.* Thus it is very important in agriculture to control or eliminate these fungi. The scientists screened leaves of numerous plants and found that those of *Pinus roxburghii* were very toxic to the two species of *Aspergillus.* The essential oil from the *P. roxburghii* needles had 100% toxicity against the two important fungal species at a concentration of 800 ppm. The antifungal activity was not affected by autoclaving or storing the oil, at least up to 370 days. At 800 ppm, the oil killed *A. flavus* in 50 seconds and *A parasiticus* in 25 seconds. Moreover, the oil did not have any phytotoxic effect on seed germination of peanuts; in fact, seeds bathed in the essential oil had almost double the germination percentage of the control seeds. It was concluded that either the *Pinus* needles or the essential oil may be a good, cheap, and easily available nontoxic sterilizer or preservative against toxic strains of *Aspergillus.*

The leaves of *Piper sarmentosum* are used for wrapping food in some parts of Thailand and the Philippines, which suggested to Masuda et al. (1991) the possible existence of antimicrobial compounds in those leaves. Those scientists collected leaves of *P. sarmentosum* in Los Baños, the Philippines, air-dried them and extracted them successively with benzene and methanol. The benzene fraction slowed antimicrobial activity, and the researchers isolated four phenylpropo-noids from it by silica gel column chromatography and TLC. They were identified by MS and NMR as 1-allyl-2,6-dimethoxy-3,4-methylenedioxybenzene; 1-allyl-2,45-trimethoxybenzene; 1-(1-E-propenyl)-2, 4,5-trimethoxybenzene; and 1-allyl-2-methoxy-4,5-methylenedioxybenzene. All the compounds inhibited growth of *Escherichia coli* and *Bacillus subtilis* by an agar dilution method at 25 ppm at 30°C.

According to Bernard and Pesando (1989), the marine phanerogams are an ecologically important group in the coastal zone of the Mediterranean Sea. They tested aqueous and lipid extracts from the rhizomes of the Mediterranean seagrass, *Posidonia oceanica,* for antimicrobial activity against six species of gram-positive terrestrial bacteria, two species of gram-negative bacteria, ten species of marine bacteria, nine yeast species, four mold species, five human dermatophytes, and five taxa of phytopathogenic fungi responsible for diseases in Mediterranean plants. All bacteria except for one gram-negative one were inhibited in most tests, only one yeast was inhibited, as were two molds and all dermatophytes; no phytopathogenic fungi were inhibited.

Species of *Salvia* are extensively used in folk medicine, and the extracts studied have shown a wide variety of activity. González et al. (1989) decided therefore to assay 20 diterpenes collected from *Salvia canariensis, S. texana,* and *S. cardiophylla* for antibacterial activity against two gram-positive bacteria, *Staphylococcus aureus* and *Bacillus subtilis,* and two gram-negative bacteria, *Escherichia coli* and *Salmonella* sp. At least one or both of the gram-positive bacteria were inhibited by all the diterpenes, but neither of the gram-negative bacteria was affected. Almost all of the diterpenes were about as active as streptomycin against the gram-positive bacteria but only one-fourth to one-half as active as ampicillin.

Methanolic extracts of fresh material and litter of *Solidago* sp. were remarkably inhibitory to the soil algae (Akiyama and Nishigami, 1991). The chief allelochemical appeared to be *cis*-dehydromatricaria ester (DME), which was found in the soil as well as in the plant. Aqueous extracts of the soil under *Solidago* inhibited the soil algae, and a 0.5-m*M* solution of DME inhibited growth of *Chlorella* sp., *Monoraphidium* sp., and *Fritzschella* sp. in liquid cultures. Volatile allelochemicals from *Solidago* sp. also inhibited growth of soil algae; the chief effects appeared to be due to menthol and camphene.

Gören et al. (1990) collected *Tanacetum argyrophyllum* var. *argyrophyllum* in Turkey and isolated and identified six known compounds and one new sesquiterpene lactone from the aerial parts of the plants. The compounds identified were α-amyrin acetate and dihydro-1-carvone; the sesquiterpene lactones 8 α-hydroxyanhydroverlotorin, tanachin, tabulin, isospeciformin; a germacranolide with a 1,5-ether linkage; dentatin A; and a new sesquiterpene lactone named tanargyrolide. The ether-petrol extracts of the tops of the plants inhibited six of nine bacteria, yeasts, and fungi. After structure determination, the five sesquiterpene lactones were tested against growth of five bacteria: *Staphylococcus aureus, Bacillus subtilis, B. megaterium* (NRRL-B1368), *Salmonella typhosa* (T-5501), and *Escherischia coli.* Four of the five were inhibited by most of sesquiterpene lactones but *S. typhosa* was not affected.

Hufford et al. (1988) found that ethanol extracts of the rhizomes and tops of *Trillium grandiflorum* (Liliaceae) had significant in vitro activity against the yeast *Candida albicans.* The researchers isolated and identified two antifungal compounds from the extract. One of these was identified as the saponin glycoside, dioscin, which has diosgenin as the aglycone and three sugar fractions, two of rhamnose and one glucose. The second allelochemical was the saponin glycoside, pennogenin rhamnosyl chacotrioside, which contains the aglycone pennogenin and four sugars (three rhamnose and one glucose). Both of these compounds were active against three strains of *C. albicans,* one of *Cryptococcus neoformans,* and one of *Saccharomyces cerevisiae,* with MICs from 1.56 to 25 μg/mL. The two compounds also inhibited growth of the filamentous fungi *Aspergillus flavus* (ATCC 9170), *A. flavus* (ATCC 26934), and *Trichophyton mentagrophytes* (ATCC 9972), with MICs from 3.12 to 100 μg/mL. The standard antibiotic amphotericin B was somewhat more active than the saponin glycosides against *Candida* and *Cryptococcus,* about the same against *Saccharomyces* and *Aspergillus,* but much less active against *Trichophyton.*

Dube et al. (1990) collected fruits of *Zanthoxylum alatum* in Gorakhpur, India. These fruits are used in the Indian system of medicine to make cures for toothaches and stomachaches. The scientists hypothesized that the fruits might contain compounds that inhibit two aflatoxin-producing fungi, *Aspergillus flavus* and *A. parasiticus.* They surface-sterilized the fruits with 70% ethanol, washed them thoroughly with sterile distilled water, and subjected them to hydrodistillation for eight hours. The vapor was collected as a water-immiscible oil; $2 \times 10^3$ μL/L was found to be fungistatic to the species of *Aspergillus* listed above. It remained active after heating at 100°C for one hour and after 300 days' storage. Moreover, it was tested against the growth of 26 other species of fungi, and at $2 \times 10^{-3}$ μL/L, it completely inhibited growth of all the fungi tested except for *A. ochraceus* and *A. ruber,* which were inhibited by about 90%.

## III. ALLELOPATHIC EFFECTS OF MICROORGANISMS ON MICROORGANISMS

There is unquestionably a deficiency of scientific information, acquired under field conditions, on this potentially important topic. There is no doubt in my mind that highly significant information will eventually be obtained and will help answer many questions in basic ecology, agriculture, and forestry. I will therefore discuss the topic rather briefly, if only to call attention to its potential importance. It will be made clear in Chap. 6 that there is already much pertinent information available on this topic in relation to the biological control of plant diseases, and the information is increasing at a very rapid rate.

Lemos et al. (1991) did some highly significant research on the competitive dominance of antibiotic-producing marine bacteria in mixed cultures. The researchers concentrated on the genus *Alteromonas,* which has many strains, some that are antibiotic-producing and many that are not. All strains used in the study were isolated from intertidal seaweeds in the Ría de Arosa and Ría de Pontevedra estuaries of northwest Spain. The scientists did all experiments in unsupplemented seawater collected from the same area where the strains were collected. The water was filtered through 0.65-μm membranes and then autoclaved. Logarithmic MB cultures, adjusted to an equal optical density of each strain, were used as inocula (0.2 mL of each culture). Samples were taken at zero time, and every one to three hours throughout the first twelve hours, and then every day thereafter for three to five days. Cultures of each strain growing alone in seawater served as controls. Appropriate serial dilutions were incubated for three days on marine agar at room temperature, and the number of colony-forming units (cfu) of each strain was recorded. In mixed cultures, strains were distinguished by different pigmentations. There were four types of mixed cultures: (1) two producer strains, (2) one producer and one nonproducer strain, (3) two nonproducer strains, and (4) three strains with different combinations of producers and nonproducers. When an antibiotic-producing strain was cultured with a nonproducer, a rapid inhibition of the nonproducer occurred. In all cases, after an initial increase of the number of viable cells of nonproducer strains, a rapid decrease from $10^5$ or higher to less than 10 cfu/mL occurred in only one to two hours. This corresponded to the middle of the log phase of the producer strain. Moreover, the growth curves of the producer strains were similar to those in control cultures. In mixed cultures with two producer strains, the effect was variable, depending on the strains used. In most cases, one of the strains disappeared after a variable culture period, from a rapid inhibition after three to four hours to a slower decrease in the number of cells in twenty to thirty hours. In all cases, one strain (EP14) had the greatest inhibitory spectrum, since it was able to inhibit all other strains tested and was not inhibited by any of them. Mixed cultures between two nonproducers indicated that both coexisted without any adverse effect on each other. The mixed cultures with three strains gave variable results; but generally one of the producer strains maintained a constant density, whereas the others disappeared after a variable time period. When three nonproducer strains were used, none was inhibited and the three coexisted with similar numbers. It was concluded that antibiotics could play an important role in interference relationships between marine bacterial populations.

Toyota and Hostettmann (1990) reported that the methanolic extract of *Boletinus cavipes* was antifungal to *Cladosporium cucumerinum.* This fungus occurs in coniferous forests in mountainous areas of Europe. Toyota and Hostettmann collected it in Switzerland and isolated and identified five diterpenic esters of 16-hydroxygeranylgeraniol and mesaconic and/or fumaric acid in its methanolic extract. These esters were named cavipetin A, B, C, D, and E. Each ester inhibited spore formation of *C. cucumerinum* in a TLC bioassay.

Yamamoto and Suzuki (1990) identified and quantified the myxobacteria in the surface sediments of eutrophic Lake Suwa in Japan in 1985 and 1986. Four species of *Myxococcus* and 10 unidentified colonies were identified in September 1985. *M. fulvus* made up 51.3% of the myxobacterial colonies. Fruiting myxobacteria isolated from the sediments had a lytic effect on a wide variety of cyanobacteria. The growth liquor of myxobacteria was also able to lyse cells of cyanobacteria and other bacteria.

Spore germination of the ectomycorrhizal hymenomycetes is generally very difficult to achieve. Therefore Ali and Jackson (1989) examined the effects of different isolates of bacteria and fungi for their ability to stimulate such germination. They collected basidiocarps of the mycorrhizal fungi *Paxillus involutus, Laccaria laccata, Amanita fulva, A. rubescens, Lactarius turpis,* and *Russula nitida* beneath birch trees at Surrey, England. Those of *Hebeloma crustuliniforme* were collected beneath willow (*Salix* spp.) at the University of Surrey. Basidiospores of these fungi were collected from clean and mature sporocarps in sterile petri dishes under humid conditions and used within 24 hours. The researchers collected potential stimulating organisms as follows: 42 bacteria from spores or sporophores of mycorrhizal fungi, 23 fungi from the same sources, 36 bacteria from mycorrhizal roots, 24 fungi from the same source, 40 bacteria from the soil, and 30 fungi from the soil. *Pseudomonas* sp. No. 1, *Pseudomonas* sp. No. 2, *Pseudomonas* sp. No. 3, *Pseudomonas* sp. No. 4, and *P. stutzeri* isolated from the sporophores of *Hebeloma crustuliniforme* stimulated spore germination of *H. crustuliniforme;* and *Corynebacterium* sp. No. 1 and a member of the Entobacteriaceae isolated from sporophores of *H. crustuliniforme* stimulated spore germination of *H. crustuliniforme* and *Paxillus involutus. Micrococcus roseus* and *Tritirachium roseum* isolated from spores of *H. crustuliniforme* and *Arthrobacter* sp. and *Corynebacterium* sp. No. 6a isolated from mycorrhizae of *Salix* spp. with *H. crustuliniforme* stimulated germination of spores of *H. crustuliniforme. Pseudomonas* sp. No. S30 and an unidentified bacterium No. S19 from birch forest soil stimulated spore germination of *H. crustuliniforme* and *Laccaria laccata.* These findings indicate that there is some specificity in the stimulation of mycorrhizal spore germination. The substances produced by microorganisms that stimulated were not identified, and Ali and Jackson stated that such substances have not yet been identified by other scientists either. This information is badly needed for use in agriculture as well as in other areas.

Paulitz and Linderman (1989) also studied certain interactions between pseudomonads and VA mycorrhizal fungi. They treated cucumber seeds with rifampin-resistant derivatives of *Pseudomonas putida* (A12, N1R, or R-20) or *P. fluorescens* (2–79 or 3871) and planted them in soil with or without inoculum of the VA mycorrhizal fungus *Glomus intraradices* or *G. etunicatum.* Populations of *Pseudomonas* in the soil were determined, by the dilution-plating method, at one to nine weeks after planting. At one to three weeks, the populations of all strains except R-20 were 1.5 to 7 times lower in the rhizosphere of cucumber roots colonized by *G. intraradices,* when compared with nonmycorrhizal plants. No significant difference was found in populations of *Pseudomonas* strains between roots colonized by *G. etunicatum* and nonmycorrhizal roots. Strains 3871 and 2–79 (antibiotic producers) delayed germination of *G. etunicatum* spores in raw soil, but no significant differences in frequency of germination were noted after seven days. Moreover, no fluorescent *Pseudomonas* strains affected colonization of cucumber roots by *G. etunicatum.*

Anbu and Sullia (1990) examined the chemical interactions between *Rhizobium* sp. (strains CB-1024 and CB-530 from peanut plants in Australia and strains BU-1 and BU-2 from peanut plants in India) and 19 genera and species of rhizosphere fungi. All of the fungi were inhibited in growth by one or more of the rhizobial strains; most of the fungi were inhibited by all the strains. Conversely, six of the fungal species inhibited one or more of the test strains of *Rhizobium* sp. It was concluded that not only are rhizobia inhibited by soil fungi, but apparently most of the soil fungi are also inhibited by antibiotic effects of rhizobia. Obviously, such microbial interactions could cause broad effects wherever plant systems occur and should be investigated much more thoroughly.

Benallaoua et al. (1990) isolated and purified an antifungal metabolite from *Streptomyces spectabilis,* strain BT 352, isolated from soil. The compound was identified as desertomycin, a macrocyclic compound containing five nonconjugated double bonds and a D-mannose residue. Desertomycin inhibited the growth of six yeast species in three genera with a MIC (for 80% growth reduction) of 50 to greater than 100 μg/mL, and five filamentous fungal species (in five genera) with a MIC of 50–100 μg/mL. The compound had a MIC of 50 μg/mL for four of the five species. The antifungal metabolite did not immediately affect yeast respiration, even at high concentrations, but subinhibiting concentrations reduced respiration by up to 40% when tested against yeast cells collected during their chief growth phase. Yeasts liberated potassium in large amounts following a 30-minute exposure to desertomycin, indicating that plasma membranes were affected.

Golubev (1989) pointed out that killer strains of saccharomycetes secreting protein or glycoprotein toxins were discovered about a quarter of a century ago and that such organisms have been discovered in recent years in other yeast genera. He cross-tested 17 strains of *Rhodotorula glutinis* against each other and found two strains—VKM Y-749 from the Central Culture Collection of Fungi (the Netherlands) and 1133 isolated from the river Ili and now deposited in the All-Union Collection of Microorganisms in the Institute of Microbiology, Academy of Sciences of the former Kazakh SSR. This was the first report of killer strains in *Rhodotorula.* The toxins of both strains were heat-labile and were inactivated after maintenance at 100°C for five minutes. These strains inhibited growth of all strains examined, including the type strain of *R. glutinus,* VKM Y-332. Golubev found that all strains of this species, with the exception of the producer strains themselves, and also other species from the genera *Rhodotorula, Rho-*

*dosporidium, Sporobolomyces, and Sporidiobalus* and smut fungi from the genera *Sphacelotheca* and *Ustilago* are sensitive to these toxins. It is noteworthy that taxonomically rcmotc organisms were resistant to the *R. glutinis* toxins. It appears significant also that species of *Sporobolomyces* and *Rhodotorula* are among the most commonly occurring epiphytic yeast flora. Golubev suggested, therefore, that because of their killer activity they may play a key protective role in the plant phyllosphere by preventing the development and dissemination of smut fungi.

Zekhnov et al. (1989) investigated the killer action of 193 strains of several yeast genera. The genera in which activity was found included *Saccharomyces, Hansenula, Pichia, Debaryomyces, Torulopsis, Sporobolomyces,* and *Rhodotorula.* The actual number of killer strains depended on the test strains used to detect the action. *Pachysolen tannophilus* (VKM Y-274) was relatively sensitive to the killer toxins, and 49 of the 193 strains tested showed various degrees of killer activity when it was used as the test strain. This result strongly supports the final conclusion of Golubev (1989).

**Conclusion** The allelopathic effects of microorganisms on plants and on other microorganisms have been little dealt with by scientists working in the field. This is particularly true of the effects of microorganisms on other microorganisms, as discussed in Sec. III of this chapter. Obviously, many tremendously important interrelationships have been ignored, especially the relationships that do not pertain directly to the biological control of weeds and plant diseases.

It is clear that all interrelationships discussed in this chapter fit perfectly into the definition of allelopathy given by Dr. Molisch, who originated the term. (See the introductory paragraphs of Chap. 1.) It is also obvious that in numerous interrelationships discussed in this chapter the causative allelochemicals were not identified, nor was it clearly demonstrated in many other instances that identified allelochemicals escaped from the organisms that produced them. The types of experiments performed in most cases, however, made it clear that the causative allelochemicals had to enter the environment from the producers. In interrelationships between microorganisms, there are fewer problems involved in the escape of allelochemicals from the producers. It has been clearly demonstrated in many experiments involving closely related strains of the same microbial taxon that only those that produce certain antibiotics can cause the observed results. This clearly demonstrates that the antibiotics are escaping from the donors into receptor cells, even though it has not been possible to detect the antibiotic outside the donor cells. I sincerely hope that the experiments discussed in this chapter will stimulate many other scientists to become involved in research on the topic.

CHAPTER 4

# ALLELOPATHY IN THE BIOLOGICAL CONTROL OF WEEDS

## I. SUGGESTED METHODS FOR BIOLOGICAL WEED CONTROL USING ALLELOPATHY

IT IS UNFORTUNATE that to many persons, biological weed control means control of weeds by use of animals such as insects, etcetera. This was due to the historical sequence of events. That type of biological weed control is obviously not concerned with allelopathy and will not be discussed further in this book.

One of the earliest suggestions concerning use of allelopathy in biological weed control related to the selection or breeding of crop plants for the control of major weeds in a given area. Putnam and Duke (1974) tested 526 accessions of cucumber (*Cucumis sativis*) and 12 accessions of eight related *Cucumis* species for allelopathic activity against a forb, *Brassica hirta,* and a grass, *Panicum miliaceum.* One accession inhibited indicator plant growth by 87%, and 25 inhibited growth by 50% or more. The researchers concluded that selection of an appropriate crop cultivar, or incorporation of an allelopathic character into a desirable crop cultivar, could provide the plant with a means of gaining a competitive advantage over certain important weeds. Fay and Duke (1977) examined 3000 accessions of the U.S.D.A. World Collection of *Avena* spp. for root exudation of scopoletin, a naturally occurring root inhibitor. Twenty-five accessions exuded more scopoletin than the standard oat cultivar Garry. Four accessions exuded up to three times as much scopoletin as Garry oats, and one of these was grown with crunchweed (*Brassica hirta* var. *pinnatifida*) for 16 days, at which time the crunchweed was stunted, chlorotic, and twisted, which indicated allelopathic effects rather than simple competition.

Very little research has been done on the genetics of allelopathic agents; much more needs to be done, and many more screening programs need to be carried out before large-scale breeding programs can be established to produce commercial cultivars of important crop plants that are allelopathic to selected weeds. Selections should also be made for accessions or wild types that are resistant to allelopathic compounds produced by certain important weeds and/or by crop cultivars.

Another way in which allelopathy may be used in weed control is to apply residues of allelopathic weeds or crop plants as mulches or to plant an allelopathic crop in rotational sequence and allow the residues to remain on the field. Putnam and DeFrank (1979) tested residues of several cover crops for weed control in Michigan. The plants were desiccated by herbicides or were allowed to freeze. Tecumseh wheat reduced weed weights by 76 and 88% (spring and fall, respectively). Fall-killed Balboa rye was similar in effectiveness, but spring-killed rye or fall-killed Garry oats had no toxic effect on weeds. Mulches of Bird-a-Boo sorghum or Monarch Sudangrass applied to apple orchards in early spring reduced weed biomass by 90 and 85%, respectively. An equivalent mulch of peat moss had virtually no effect. Tree growth in mulched plots was equal to or better than that in unmulched plots. Putnam and DeFrank (1983) expanded this investigation in a three-year series of field trials to determine the influence of allelopathic crop residues on emergence and growth of annual weeds and selected vegetable crops. Sorghum residues reduced populations of purslane (*Portulaca oleracea*) and smooth crabgrass (*Digitaria ischaemum*) by 70 and 98%, respectively. Total weed biomass and weight increases of several species were also consistently reduced with residues of barley, oats, wheat, rye, and sorghum. Use of *Populus excelsior* as a control indicated that the effects of the cover crop residues were chemical, not physical. Generally, the larger-seeded vegetables, particularly the legumes, grew normally or were even stimulated sometimes by the cover crop residues. Several species of smaller seeded vegetables, however, were severely injured.

Putnam and his colleagues found that a major variable in the effectiveness of weed control was the maturity of the cover or mulch crop at the time of killing. Early fall planting appeared to be crucial in allowing the sorghum to achieve enough growth before being winter-killed if it was subsequently to be effective in suppressing weed germination and growth. Evidence was strong that stressed plants produced stronger allelopathic effects than unstressed plants. This agrees, of course, with a large body of research evidence (Rice, 1984, Chap. 11).

Before the widespread use of herbicides, weeds were controlled chiefly by cultivation and crop rotation. The practice of crop rotation declined in the late 1940s with the advent of cheap inorganic fertilizers and synthetic herbicides. Appropriate crop rotations can often suppress weed growth effectively, as I pointed out in Chap. 1. Moreover, they can help alleviate detrimental allelopathic effects of some crops on themselves and on other sensitive crop plants.

Pathogenic bacteria and fungi produce phytotoxins that affect growth and survival of host plants. This is obviously an allelopathic phenomenon according to Molisch's definition of "allelopathy" (Chap. 1), although the term is seldom used in plant pathology literature. There is increasing use of—and much promise in—applications of effective weed pathogens to control weeds. This subject will be discussed further in Sec. III of this chapter. The toxins produced by the pathogens can be used to control weeds, even if the weeds are resistant to infection by the microorganisms that produce them.

This is probably an appropriate time and place to point out some of the problems encountered when microorganisms are applied to soil, plant parts, or plants for weed control—as well as for growth stimulation of the plants or other microorganisms, inhibition of microorganisms, or control of plant disease or weeds. Microbial strains frequently mutate in culture; any mutation may greatly affect performance (Burr and Caesar, 1984). Effects of long-term storage on strains must be determined, and excellent quality control maintained. Effects of all commonly applied agricultural chemicals on microbial strains selected for various uses will have to be determined, and an effective inoculum identified for each microbial strain. Generally, suspension of culture media in water is not satisfactory. Pelleting seeds with a methylcellulose powder formulation makes for ease in storage, transport, and handling.

Hasan and Ayres (1990) described some of the principles, problems, and prospects of selecting appropriate fungi for control of selected weed species. They pointed out that biocontrol of weeds by the introduction of exotic fungi has proved successful in some cases, most notably in the control of the skeleton weed (*Chondrilla juncea*) in Australia by the introduced rust, *Puccinia chondrillina*. The scientists also stated that "The use of integrated control methods in which fungi are used in pairs, or in combination with either reduced rates of chemical herbicides or with arthropods, also offers hope for development."

Another way in which allelopathy may be used indirectly in weed control is to select allelochemicals from various sources, such as plants or microorganisms, and use them as herbicides in place of synthetic chemicals. This procedure can have desirable results, because most natural products are broken down rather rapidly by common microorganisms and thus are not persistent pollutants in the environment, as are many of the synthetic herbicides. A few examples from some earlier research should suffice here, though more will be discussed in Sec. IV of this chapter. Rhizobitoxine is an allelochemical produced by certain strains of *Bradyrhizobium japonicum* in soybean nodules (Owens et al., 1972). Anonymous (1969) pointed out that rhizobitoxine is an effective herbicide in amounts as low as 3 oz/acre. Owens (1973) found that Kentucky bluegrass (*Poa pratensis*) was much less sensitive to rhizobitoxine than was crabgrass, a common weed in bluegrass lawns. On the other hand, amitrole, which is sometimes used to control crabgrass in bluegrass lawns, is almost as toxic to bluegrass as to crabgrass. Thus it appears that rhizobitoxine could be a good selective herbicide in bluegrass lawns.

Because herbicides from plant sources are generally more easily biodegradable than synthetic herbicides, Rizvi et al. (1980) surveyed the effects of ethanolic extracts of leaves and seeds of over 50 species of weeds and crop plants against seed germination of the weed *Amaranthus spinosus*. Extracts of coffee (*Coffea arabica*) were most inhibitory, but extracts of seeds or leaves of twelve other species caused 30 to 50% inhibition, and eight other species caused 20 to 30% inhibition in the same concentrations. Caffeine was identified by Rizvi et al. (1981) as the most powerful phytotoxin in the coffee plant. A 1200-ppm aqueous solution completely inhibited seed germination of *A. spinosus*, but it had no effect on seed germination or subsequent growth of black gram, a common crop plant in India.

**Table 2** Early summer weed production as influenced by residues of fall- or spring-killed cover crops

| Cover crop | Time of kill | Kill method | Weed dry weight* ($g/m^2$) |
|---|---|---|---|
| Sorghum × Sudangrass hybrid | Fall | Frost | 31.2 bc |
| Sudangrass | Fall | Frost | 78.0 a |
| Sorghum | Fall | Frost | 54.1 b |
| Oats | Fall | Frost | 47.8 b |
| Rye | Spring | Desiccant | 28.6 cd |
| Wheat | Spring | Desiccant | 13.2 d |
| Barley | Spring | Desiccant | 28.3 cd |
| None | Spring | Desiccant | 81.5 a |

* Means with different letters are significantly different at $P = 0.05$ by Duncan's multiple-range test.

*Source:* From Putnam et al. (1983), with permission of Plenum Pub. Corp.

## II. WEED CONTROL BY PLANTS

### *Cultural Practices and General Weed Control*

As I pointed out earlier in this chapter, Alan Putnam and several of his colleagues were among the earliest scientists to experiment with the possibility of using allelopathy in weed control. They continued this research until Dr. Putnam's retirement from Michigan State University. Putnam et al. (1983) experimented further with the use of allelopathic mulches, cover crops, conventional and no-till farming, and nonallelopathic mulches. They demonstrated again that cover crops that were allowed to freeze or were killed by desiccants were very effective in helping control the population of weeds (Table 2). They also determined the effects of residues of several cover crops on various garden and field crops, since this information is critical in choosing the cover crops to be used with different crop plants (Table 3). Even a quick perusal of Table 3 makes it clear that most of the allelopathic cover crops cannot be used with cabbage, lettuce, and tomato. In fact, none of those tested could be used with lettuce and only one could be used with cabbage. The researchers found that when tillage is eliminated, there is a shift of weed species from primarily dicotyledons to monocotyledons, particularly to grasses such as species of *Panicum, Echinochloa,* and *Setaria.* Among weeds that decreased greatly in density were *Amaranthus, Ambrosia, Chenopodium,* and *Portulaca.*

**Table 3** General response of several annual crops to residues of spring–seeded cover crops in no–tillage systems

| | Indicator crop response* | | | | | | | |
|---|---|---|---|---|---|---|---|---|
| Cover crop | Cabbage | Carrot | Corn | Cucumber | Lettuce | Pea | Snap bean | Tomato |
| Barley (spring) | – | + | + | + | – | + | + | – |
| Barley (winter) | – | + | + | + | – | + | + | – |
| Corn | – | + | + | + | – | + | + | + |
| Oats | – | – | + | + | – | + | + | + |
| Rye | – | + | + | + | – | + | + | + |
| Sorghum | – | – | – | + | – | + | + | – |
| Sorghum × Sudangrass | + | – | – | + | – | + | + | – |
| Wheat | – | + | + | + | – | + | + | + |

*Plus indicates germination and growth equal or superior to that in no–tillage controls without residues; minus indicates significant reduction in germination, growth, or yield.

*Source:* From Putnam et al. (1983), with permission of Plenum Pub. Corp.

Putnam et al. (1983) reported also that companion planting of rye or wheat in the fall in orchards in Michigan did not interfere with tree growth. They desiccated the wheat or rye in May before competition began and found that the residues gave effective weed control for up to 60 days. After two seasons, growth of cherry trees at Traverse City, Michigan, was slightly increased by the presence of rye or sorghum residues, and all management systems with cover crops provided growth equal to or better than that in cultivated or repeatedly sprayed controls.

Leather (1983) found that germination of wild mustard (*Brassica kaber* var. *pinnatifida*) seeds was inhibited by undiluted aqueous leaf extracts of two varieties of cultivated sunflower (*Helianthus annuus*) and the native annual sunflower (*H. annuus*), but was stimulated by more dilute extracts from the same sources. He found also that seedling growth of velvetleaf, jimsonweed, morning glory, and mustard was inhibited by leachates of Hybrid 201 sunflower leaf and stem tissue. Root exudates inhibited velvetleaf and jimsonweed growth but stimulated growth of morning glory. Rape (*Brassica napus*) tissue leachates inhibited growth of velvetleaf and morning glory but did not affect wild mustard. Leather combined his greenhouse and laboratory experiments with a five-year field study, with sunflower and oat grown in rotation. Weed density increased in all plots but the amount of increase was significantly less in plots with sunflower than in control plots (Fig. 3). There was no significant difference between weed densities in the plots with the three hybrid sunflowers. His results definitely indicated that interference by certain crops in rotation limited weed growth, and the combined results of allelopathic studies with field studies strongly suggested that at least part of the effects were due to allelopathy.

Leather (1987) continued his earlier experiments by combining the use of the herbicide S-ethyl dipropyl carbamothioate (EPTC) with growth of sunflower hybrid 8941 in field tests. He found that weed biomass was reduced equally in plots planted with sunflower, whether or not the herbicide was applied in each of four years. This was somewhat of a surprise result to Dr. Leather, because his goal was to determine if weed control by sunflower could be extended throughout more of the year by combining a herbicide treatment with sunflower planting.

Panasiuk et al. (1986) investigated the allelopathic interaction between sorghum and 10 species of weeds: *Lepidium sativum, Echi-*

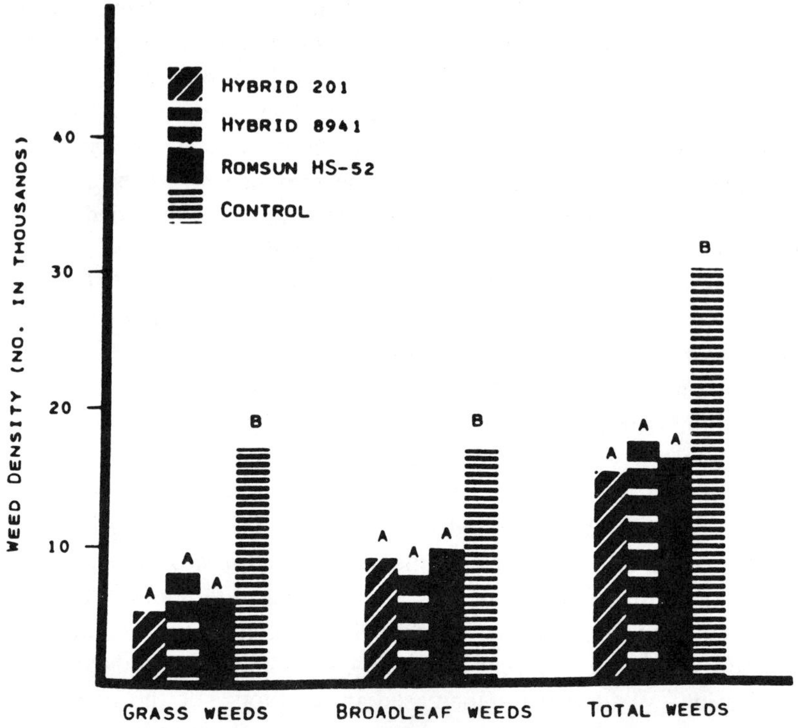


FIGURE 3 Average weed density over five growing seasons in field plots of three hybrids of sunflower in rotation with oat. Vertical bars within each group having the same letter are not significantly different at the 5% level by the Duncan's multiple-range test. *Adapted from Leather (1983), with permission of Plenum Pub. Corp.*

*nochloa crusgalli, Rumex crispus, Setaria viridis, Sorghum halepense, Ipomoea pandurata, Amaranthus retroflexus, Rumex acetosella, Abutilon theophrasti,* and *Sinapis arvensis. L. sativum* was treated as a weed, although it is grown as a crop plant in some areas; the sorghum cultivar used was Bird-a-Boo. Germination of weed seeds in petri dishes with germinating sorghum seed resulted in a range of effects—from slight inhibition to slight stimulation of weed seeds, depending on the species. The subsequent growth of the radicle and hypocotyl or the coleoptile of the weeds was significantly inhibited by the sorghum. When weeds were interplanted with sorghum and grown under greenhouse conditions, the inhibitory effects on some weed species were still obvious after two months of growth. Aqueous leachates from pots planted with sorghum alone or from a system in which sorghum roots protruded into water had strong allelopathic activity. It was obvious, therefore, that allelopathic effects of sorghum started when the seeds first germinated and continued during growth of the seedlings. It was also clear that the causative allelochemicals were very water soluble. In addition to the effects on growth of the weed seedlings, weeds germinated in the presence of sorghum generally were less healthy in appearance, with higher incidences of necrosis and chlorosis.

Shilling et al. (1985) conducted a very thorough series of experiments to determine the effects of tillage versus no-till, rye and wheat mulches, and chemicals in the mulches on weed control under laboratory and field conditions. Rye mulch (aboveground residue) caused a significant increase in the percent control of grass species even when the soil was tilled. Rye root residue also caused a significant improvement in grass control, although not as great as that provided by the addition of mulch. All three broad-leaf weeds, *Amaranthus retroflexus, Chenopodium album,* and *Ambrosia artemisiifolia,* responded similarly to tillage, showing an increase in density and biomass. Addition of rye mulch, however, reduced biomass of *C. album, A. retroflexus,* and *Ambrosia* by 96, 78, and 39%, respectively. The growth of these weeds was also reduced by the rye root residues.

Shilling et al. (1985) found that wheat mulch reduced growth of *Ipomoea lacunosa* and *I. purpurea* greatly in corn fields, but there was no effect of wheat mulch on *Ipomoea* spp. in soybean fields. They stated that the age of the wheat mulch at the time of treatment may have caused this difference, because the wheat straw was more mature and no longer green when it was used in the soybean plots. Later tests demonstrated that the level of phytotoxicity of wheat straw was much higher when the straw was cut while still green. To help explain the suppressive effects of the rye and wheat mulches on appropriate weeds, the scientists extracted and identified several phytotoxic allelochemicals and tested them against weed species involved in field experiments. They isolated and identified ß-phenyllactic acid (ßPLA) and ß-hydroxybutyric acid (ßHBA) from field-grown rye and found that ßPLA and ßHBA at a concentration of 8 m*M* inhibited *C. album* hypocotyl growth by 68 and 30%, respectively. Both compounds inhibited *C. album* root growth by 20% at 2 m*M*. *A. retroflexus* hypocotyl growth was inhibited 17% by ßPLA at 0.8 m*M* and 100% at 8 m*M*, and its growth was inhibited 59 and 39% at 2 m*M* by ßPLA and ßHBA, respectively.

Ferulic acid was the most phytotoxic compound identified from wheat mulch. At 5 m*M*, germination and root length of *I. lacunosa* were inhibited 23 and 82%, respectively. *Sida spinosa* germination and root growth were inhibited 85 and 82%, respectively, with carpels present on the seed. Shilling et al. discovered that ferulic acid was decarboxylated to a more toxic styrene derivative, 4-hydroxy-3-methoxystyrene, by a bacterium living on the carpels of *S. spinosa*. It was concluded that at least part of the suppressive effects of rye and wheat mulch on weeds in various field crops was probably due to allelopathy. Moreover, the results indicated that it may be possible to supplement, if not reduce, the herbicides used in no-till cropping systems with the proper choice and management of various cover crops and plant residues.

Forney et al. (1985) carried out a very thorough study at Blacksburg, Virginia, of weed suppression in no-till alfalfa by prior cropping of a sorghum-sudangrass hybrid (SSH) (*Sorghum bicolor* × *Sorghum sudanense*) or foxtail millet (FM) on land that had one of three types of land use history. The three previous uses were tall fescue (*Festuca arundinacea*) sod; old, weedy red clover; and no-till corn with a winter cover crop of rye. The annual forage grasses were planted (no-till) separately into each of the plots with the different histories; alfalfa was planted (no-till) into all the plots in late summer or early fall of the same year. Prior to planting, each plot received surface applications of the fertilizers deemed desirable in the area. Both grasses (SSH and FM) established well in no-till conditions—except for FM in the case where excessive mulch at planting time impeded seed placement. Desiccation of the mulches a few days prior to planting time eased this problem. Both grasses suppressed weeds and enhanced subsequent alfalfa establishment. Their data indicating successful weed control in the spring following fall planting of alfalfa were striking (Table 4). It was also obvious that plots in which tall fescue or rye was previously grown had enhanced weed control; both plants have strong allelopathic potential. These results were even more amazing when it is considered that 96 weed species were found at the study sites during the investigation. It was concluded that prior cropping of SSH or FM is a valuable option for growers wishing to establish alfalfa without tillage in the region of the study, especially when summer forage is useful. It is very evident from the species involved and the types of experiments performed that at least part of the weed control was due to allelopathy, with competition also having a strong effect.

Smeda and Putnam (1988) investigated the effect of fall-planted

**Table 4** Effects of prior cropping of summer annual grasses on dry weight of weeds harvested in early spring from fall-planted alfalfa plots at sites differing in land use history at Blacksburg, Virgina, in 1982 and 1983*

| | Harvest, April 19, 1982 Land use history | | | Harvest, April 26, 1983 Land use history | | |
|---|---|---|---|---|---|---|
| | Tall fescue sod | Weedy clover | Rye cover | Tall fescue sod | Weedy clover | Rye cover |
| Prior crop | Weed dry weight ($g/m^2$)† | | | Weed dry weight ($g/m^2$)† | | |
| None | 94 a | 246 a | 653 a | 177 a | 354 a | 645 a |
| Sorghum-Sudangrass hybrid, regrowth suppressed by paraquat | 91 a | 145 b | 31 b | 235 a | 359 a | 161 b |
| Sorghum-Sudangrass hybrid, no paraquat | 127 a | 160 b | 44 b | 186 a | 327 a | 166 b |
| Foxtail millet | 110 a | 140 b | 70 b | 142 a | 344 a | 88 b |

* Weeds in three 225-$cm^2$ areas were harvested from each replication.

† Means within a column followed by the same letter are not significantly different at the 5% level according to Duncan's multiple-range test.

*Source:* From Forney et al. (1985), with permission of Weed Sci. Soc. of America.

Wheeler rye, Yorkstar wheat, and Barstoy barley on weed suppression and yield of Midway and Guardian strawberries (*Fragaria* × *ananassa*) at two areas in Michigan. Part of each cover crop, along with the unseeded control plots, was treated with fluazifop-butyl in late May. Weed biomass was assessed for winter annual weeds on June 10, and for other weeds in early August. Spring and midsummer weed biomass was significantly reduced, compared with unseeded controls, by all cover crop treatments. Winter barley provided early spring weed suppression, but by midsummer, weed growth was similar to the unweeded control. Weed suppression increased with cover crop residue and was consistently high with increased cover crop seeding rates. Rye and wheat gave similar amounts of weed suppression throughout the growing season, with weed growth being reduced by 90 to 95% in early spring, decreasing to 55 to 85% control by midsummer. Strawberry yields were not significantly affected by the cover crops, although they appeared somewhat reduced. The conclusion was that winter annual and early summer annual weed suppression by rye and wheat cover crops beyond a critical weed-free period could delay or reduce the need for chemical weed control until postharvest renovation. Here again, the weed suppression probably resulted in part from allelopathy and in part from competition on the part of the cover crops.

Einhellig and Rasmussen (1989) investigated the effects of cropping with corn, sorghum, and soybeans on the following year's weed crop on the Gary Young farm in northeastern Nebraska. There was no history of herbicide or commercial fertilizer application in the study area, and during the study years, crops were planted using a no-till approach; no herbicide, other pesticide, or fertilizer was applied. Data from the first year were collected from a field with Nora silty-clay loam; the other two years' data were collected from two fields with Moody silty-clay loam soil. The first field had 29% clay and a pH of 7.1, whereas the field used in the last two years had 27% clay and a pH of 6.9. Weed reduction in areas previously cropped in grain sorghum was obvious with visual inspection. Sampling data showed very striking differences in weed abundance due to the prior years' cropping (Fig. 4). Percentage weed cover was significantly lower early in the year after sorghum, and midsummer weed biomass was well below that following corn and soybeans. In fact, weed biomass in June and July following corn was two to four times that in grain sorghum plots of the previous year. The inhibitory effects of grain sorghum were

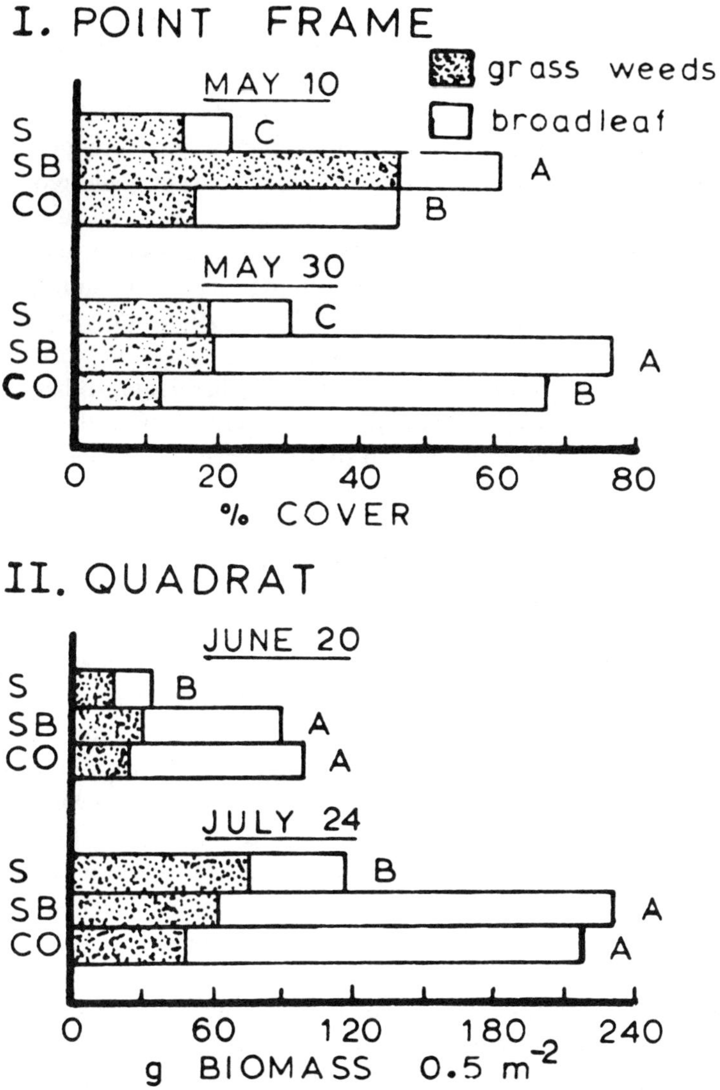


FIGURE 4 Effects of the prior year (1985) crop on weed abundance in 1986. S = grain sorghum; SB = soybeans; CO = corn. Bars within each sampling date having different letters are significantly different, $P < 0.05$, ANOVA, with Duncan's multiple-range test. *Adapted from Einhellig and Rasmussen (1989), with permission of Plenum Pub. Corp.*

primarily on broad-leaf weeds. No obvious differences were noted in the weed species present after the three crops. According to the authors, the known allelopathic potential of grain sorghum provided a logical explanation for the sorghum-mediated weed inhibition, Einhellig and Rasmussen concluded that "the planned use of a grain sorghum crop as a provision for weed management could reduce reliance on herbicides."

Worsham and Blum (1992) predicted that growers in the United States will increasingly use no-till crop production to attain the conservation requirements of the 1985 and 1990 Food Security Acts. They stated further that they have successfully grown crops in North Carolina in killed heavy mulches of rye without herbicides, other than a nonselective one to kill the rye. To test this possibility further, another experiment was started in the fall of 1989. Rye, subterranean clover, crimson clover, and hairy vetch were planted on a clay-loam soil, using both no-cover and conventional tillage treatments. Corn, soybeans, tobacco, and cotton were planted into the cover crops about two weeks after desiccating them with paraquat. After planting the summer crops, preemergent herbicides were used on one-half of each plot. Weed control ratings and soil samples were taken about four weeks after planting and soybean grain yields were determined as appropriate. Another, similar test was started in the fall of 1990, and corn or cotton was planted about two weeks after killing the cover crops in the early summer of 1991. *Amaranthus retroflexus, A. spinosus, A. hybridus,* and *Chenopodium album* were controlled 80 to 100% up to four weeks after planting into killed cover crops of rye and subterranean clover, without use of postemergent herbicides. Crimson clover and hairy vetch covers gave less control. The overall sequence for the control of the three pigweed (*Amaranthus*) species was as follows: rye, subterranean clover > crimson clover > hairy vetch, no cover > conventional till—all without preemergent herbicide. Control was 100% in all plots treated with preemergent herbicide. Regressions of environmental data against weed control suggested that pigweed control increased with increasing phenolic acid concentration and pH and responded in a convex manner to water and mulch. Worsham and Blum suggested that the effect of pH was likely related to the higher solubility of phenolic acids at pH 6 than pH 5. They concluded that, even though research indicated it is possible to substitute cover crop for preemergent herbicides in many cases in no-till cropping systems, much research needs to be done on factors responsible for the weed suppression and on improving the weed control effectiveness of allelopathic crop plants.

According to Obiefuna (1989), one of the major constraints in plantation agriculture in the tropical rain forest is rapid and adverse weed growth. Over the years manual weeding has remained the only effective means of weed control in plantains. Obiefuna decided to investigate the effect of intercropping musa AAB plantains with 2500, 5000, and 10,000 melons (*Colocynthis citrullus*) per hectare on delayed weeding for three, five, and seven months, respectively, after planting. The melons were planted on the same day as the plantains. Leaf production was relatively uniform in all intercropped plantains and controls until the third month after planting, at which time marked variations developed. Planting melons between the plantains significantly reduced weed growth so that weeding could be delayed up to seven months with the highest melon density. Melons planted at 5000/ha were equally effective at weed suppression and even accelerated plantain growth. At the low melon density, surface cover was poor about 40 days after planting, resulting in the development of areas of weed growth from the time of planting. The yield of melons significantly decreased with increasing melon population. The plantains established best when intercropped with 5000 melons/ha, but most plantains were suppressed beyond this level. The concentration of 5000 melons/ha also permitted delay in weeding for seven months, thus making that the best combination. No conclusions were drawn concerning the causes of the weed control, but it was implied that competition was responsible. No experiments were conducted, however, concerning this point.

Niggli et al. (1989) investigated the effects of several types of mulches on weed control, fruit yield, and fruit quality of Boskoop, Golden Delicious, and Gloster apples during an eight-year period in Germany. They reported that rape straw, fresh bark of conifers, and fresh bark of oak all gave excellent control of annual weeds. On the other hand, materials that are often rich in nutrient elements, such as bark compost, litter compost, and apple peelings, did not control weeds sufficiently. The researchers found that rape straw and oak bark reduced large fluctuations in annual yields of Boskoop apples, and all bark mulches improved fruit quality.

The final paper I am going to discuss here is not exactly appropriate to the subject at hand, but the results were so striking and so applicable to aquatic weed management that I feel more scientists should be aware of it. Elakovich (1989) determined the allelopathic effects of 16 species of aquatic plants native to the southeastern United States against lettuce seedlings and the aquatic plant *Lemna minor.* The plants used in the experimentation were *Brasenia schreberi, Cabomba carolina, Ceratophyllum demersum, Eleocharis acicularis, E. obtusa, Hydrilla verticillata, Juncus repens, Limnobium spongia, Myriophyllum aquaticum, M. spicatum, Najas guadalupensis, Nymphaea odorata, Nymphoides cordata, Potamogeton foliosus, Sparganium americanum,* and *Vallisneria americana.* Aqueous extracts which were made by blending 200 g of drip-dry plant material in 200 mL of deionized water and sterilized by filtering through 0.45-<μ Millipore filters, were tested against growth of lettuce seedlings and against growth of *L. minor.* The entire plant was used in each case in making the extract, except for *N. odorata,* which was divided into two portions: leaves in one and rhizomes and roots in the other. Six species inhibited lettuce seedling growth by more than 77%. The six species, in order of decreasing inhibition were *N. odorata* rhizomes and roots, *J. repens, V. americana, B. schreberi, C. demersum, E. acicularis,* and *N. odorata* leaves. In the *L. minor* assay, extracts of four species inhibited frond production by 68% or more. In order of decreasing inhibition, the species were *N. odorata* leaves; *M. aquaticum, N. odorata* rhizomes and roots; *C. carolina* and *B. schreberi* whole plants. The combined results showed that *N. odorata* leaves inhibited radicle growth of lettuce seedlings by 78% and inhibited frond production of *L. minor* by 98%. Moreover, *B. schreberi* inhibited lettuce seedling radicle growth by 82% and *L. minor* by 68%. Elakovich concluded that these two species are excellent candidates for use in aquatic weed management.

#### *Biological Control of Specific Weeds*

Anaya et al. (1990) pointed out that *Ipomoea tricolor* is commonly grown as a cover crop in the sugarcane fields of the state of Morelos, Mexico. The peasants use this practice before sowing and during the fallow period because they know this plant will eliminate all other weeds in two or three months. Anaya and her colleagues decided to determine how effective this plant is in weed control in scientific tests. Aqueous extracts of *I. tricolor* inhibited radicle growth of *Amaranthus leucocarpus* by 66% and radicle growth of *Echinochloa crusgalli* by 55%. Ethyl acetate and chloroform extracts were relatively active also. Isolation and identification of the active allelochemicals demonstrated that they were a mixture of glycosides, with jalapinolic acid as the aglycone glycosidically linked in the 11 position to an oligosaccharide composed of glucose, rhamnose, and fucose, which also combines with the carboxyl group of the aglycone to form a macrocyclic ester. These scientists concluded that "The practical use of *Ipomoea tricolor* as a weed controller in traditional agroecosystems, as well as the allelopathic activity of resin glycosides, offer promise as natural herbicides."

Saxena (1992) stressed the fact that the coupling of terrestrial and freshwater ecosystems is mediated through the process of litter fall. She emphasized further that the organic input of terrestrial origin from slopes of hills and nearby plains comes into water bodies through litter blow and litter flow, in addition to throughfall and seepage from the soil. These concepts led her to a biological control method for water hyacinth (*Eichhornia crassipes*). She decided to determine the effects of each of three known allelopathic plants—*Phragmites karka* (emergent) and *Jatropa curcas* and *Lantana camara* (terrestrial)—on the growth of water hyacinth. Plastic pots were divided by wire netting and cut pieces of one of the plants were allowed to decompose in one compartment, while water hyacinth was grown in the second compartment. Water hyacinth was completely killed by the decomposing residue of *L. camara* after nine days, and growth of the newly formed leaf of hyacinth was decreased by both *P. karka* and *J. curcas.* Because *L. camara* killed the water hyacinth, further tests were run with that plant only. Tests of various parts of *L. camara* showed that the flower extract was most inhibitory to hyacinth. These tests also demonstrated that the toxic allelochemicals of *Lantana* affect the roots of hyacinth first, causing them to shrink and then blacken within two to three days. Leaves start inrolling, blackening, and dying after four to five days. The phytotoxic agent(s) was still active after four weeks of decomposition. It certainly appears possible, therefore, that *L. camara* could be used in some fashion in the biological control of the pernicious weed water hyacinth, and perhaps other undesirable aquatic weeds.

*Asphodelus tenuifolius* is a common weed in wheat fields in Gujarat, India, according to Rao and Pandya (1992). These scientists decided to determine if wheat has a negative impact on *A. tenuifolius* and found highly significant inhibitory effects by wheat up to 60 days in age. They used IR, MS, and NMR to identify the chief inhibitory compound from wheat as untriacontance. This is a hydrocarbon, aliphatic in nature, with the following structural formula:

$$\left.\begin{array}{l} H_3C\text{-}CH_2\text{-}(CH_2)_{13} \\ \\ H_3C\text{-}CH_2\text{-}(CH_2)_{13} \end{array}\right> CH_2$$

This compound strongly inhibited seed germination and root and shoot elongation of *A. tenuifolius* at 240 μg/mL. Young wheat plants (20 days old) were much more inhibitory to growth of the weed than were older plants. Rao and Pandya stressed the fact that the effectivenes of untriacontance in controlling *A. tenuifolius* under field conditions needed to be tested, even though the compound was very active under laboratory conditions.

Johnston (1989) reported that spiny burrgrass (*Cenchrus longispinus* and *C. incertus*) is a serious summer-growing weed in light sandy soils in New South Wales, Victoria, and South Australia. There is a severe risk of soil erosion using herbicides and cultivation, so Johnston decided to determine if certain other plants would survive in areas covered with spiny burrgrass and help eliminate those species. Two areas were chosen in New South Wales, and plots were sown in 1980 with alfalfa (lucerne, CUF101); *Eragrostis curvula* (Consol, 4660, or 809); or buffel grass (*Chloris ciliaris,* American or Biloela). Control plots were also a part of the experiment. Some plots were fertilized and some were not. In June 1984, Consol and 4660 lovegrass still persisted from the 1980 planting; it was noteworthy that fertilizer N and P depressed lovegrass plant density. Lucerne, buffel grass, and 809 lovegrass did not do well, even with extra plantings. In March 1985, there were few spiny burrgrass plants in the plots first planted in Consol and 4660 lovegrass. Spiny burrgrass plants that did survive lacked the vigor, mass, and seed production of plants in control plots. By February 1987, Consol and 4660 lovegrass were spreading, especially in unfertilized plots. Thus, despite three extremely dry summers and uncontrolled grazing, Consol and 4660 lovegrass virtually controlled spiny burrgrass in four years. Johnston concluded that Consol and 4660 lovegrasses can be recommended for controlling spiny burrgrass on infertile, sandy-textured soils in low-rainfall environments in New South Wales and adjacent areas in Australia.

Harrison and Peterson (1986) reported that yellow nutsedge (*Cyperus esculentus*) grown in soil from sweet potato field plots accumulated less dry matter than did plants grown in soil from adjacent plots. Growth of yellow nutsedge was less when grown in a potting medium containing decomposing Regal or SC 1149–19 sweet potato plants, in comparison with plants grown in potting medium alone. When a sweet potato potting medium was incubated at 25°C and tested weekly with an alfalfa growth bioassay, inhibition was found to be high initially but decreased with time and was no longer present after 12 weeks. In another test, a methanolic extract of the periderms of sweet potato storage roots was applied to vermiculite, after which the vermiculite was dried and placed in test tubes. A yellow nutsedge tuber in an early stage of germination was placed in each test tube, and the tubes were placed in the light (111 $\mu Em^{-2}\ s^{-1}$). After one week, root and shoot dry weights and lengths were determined. Total root length, number of roots longer than 2 cm, and root dry weight were reduced by an extract concentration equivalent to 2.5 mg dry weight of periderm extracted per milliliter. Root growth inhibition was accompanied by root swelling, absence of secondary roots, and browning of root tips. At 10 mg/mL, root growth was almost completely inhibited. Extracts of SC 1149–19 were much less inhibitory than were extracts of Regal periderms, with significant root inhibitions of yellow nutsedge occurring only at 20 mg/mL. The ethanol extract did not affect nutsedge growth at the same concentrations as did the methanol extracts, even though it contained the same phenolic compounds. The concentrations of the phenolics were much lower, however, in the ethanolic extract. It is notable that at least 13 phytotoxins were present in the periderm extract, but they were not identified. The fact that field soil in which sweet potatoes previously grew was inhibitory to the growth of yellow nutsedge definitely indicates the allelopathic potential of sweet potatoes and suggests they could be included in weed control systems for selected weeds.

Peters and Mohammed Zam (1981) observed that some tall fescue (*Festuca arundinacea*) genotypes growing in a breeding nursery had crabgrass (*Digitaria sanguinalis*) infestations, whereas other genotypes had none. They collected the top 2.5 cm of surface soil in which the various tall fescue genotypes were growing and planted crabgrass seeds in those soils. The crabgrass germination in the greenhouse varied with genotypes in a fashion similar to that observed in the field. Next, plants from six of the parental genotypes of Missouri-96 tall fescue were used to make aqueous shoot leachates (10 g/100 mL distilled water). All the leachates were tested against seed germination and seedling growth of Dawn birdsfoot trefoil and Kenland red clover. All fescue genotype leachates significantly inhibited birdsfoot trefoil seed germination at 10 days; after 14 days, three genotypes caused significantly less germination than did the control. Leachates of all genotypes significantly reduced germination of red clover at both 10 and 14 days. Shoot and root growth of red clover was significantly inhibited by all genotype leachates of tall fescue. Root growth of birdsfoot trefoil was significantly inhibited by 10 g/100 mL leachates of two of the genotypes, as was shoot growth. Overall results indicated that tall fescue genotypes can be selected for control of large crabgrass in the fescue stands. Some of the genotypes could probably be used as cover crops, mulches, or in crop rotation systems to aid in weed control. Obviously, genotypes that did not inhibit the crops involved would have to be used. Tall fescue also appears to be an excellent candidate for genetic studies on allelopathy.

Leafy spurge (*Euphorbia esula*), which is rapidly invading rangelands in the upper Great Plains of the United States, is toxic to rangeland cattle and phytotoxic to palatable forage species and to crop plants (Hogan and Manners, 1990). Selleck (1972) observed that *Antennaria microphylla,* a small, low-growing perennial was able to invade dense stands of leafy spurge. Manners and Galitz (1986) identified three biologically active phenolic compounds in *A. microphylla*—hydroquinone, arbutin, and caffeic acid—and demonstrated that these compounds inhibited leafy spurge seed germination and radicle elongation. Hydroquinone was considered to be the most important allelochemical of the allelopathic interaction in the field. Hogan and Manners (1990) subsequently decided to study the chemical interactions of leafy spurge and *A. microphylla* in tissue culture. They found that media and media extracts of callus cultures of *A. microphylla* inhibited leafy spurge callus and suspension culture by 50 and 70%, respectively, compared with the control growth. Moreover, the media and extracts inhibited lettuce and leafy spurge root elongation by 64 and 77%, respectively. Hydroquinone was biosynthesized by callus and suspension cultures of *A. microphylla*. Exogenously added hydroquinone in a concentration of 0.5 m*M* was found to be inhibitory to leafy spurge suspension cell cultures and was only partially chemically transformed to the nontoxic, water-soluble monoglucoside, arbutin. Hogan and Manners concluded that these tests confirmed the chronic involvement of hydroquinone in the allelopathic interaction between *A. microphylla* and leafy spurge.

Daroesman (1981) stated that alang-alang (*Imperata cylindrica*) ranks high as one of the 10 worst weeds of the world; it now covers between 16 and 30 million hectares in Indonesia, mostly in Sumatra and Kalimantan. The Indonesian government has sponsored numerous transmigration projects to move families from the overpopulated urban areas to unoccupied areas in other parts of the country. Much of the land that has been made available for transmigration is alang-alang land. Each family is allotted 5 ha with a small area (0.5 to 1 ha) cleared for farming. Clearing with large tractors and plows is far too expensive, so much of the clearing has been done by overplanting with a fast-growing legume such as *Stylosanthes guyanensis.* This adds nitrogen through biological nitrogen fixation in addition to eliminating the alang-alang. It was assumed for a long time that, because of its rapid growth after the tops of the alang-alang were initially burned, the legume eliminated the grass by competition. Daroesman pointed out, however, that it is not known whether the legume eliminates the *Imperata* by smothering, by nutrient or water competition, or by allelopathic effects.

Dr. Roger Montgomery of the Hunting Technical Services in Jakarta, Indonesia, informed me in August 1989 that three other leguminous species were then being recommended as cover crops to be sown immediately after the burning of the alang-alang. The three species were *Centrosema pubescens, Calopogonium muconoides,* and *Pueraria phaseoloides.* Dr. Montgomery stated that he did not know why *Stylosanthes guyanensis* was discontinued. It is noteworthy that *S. guyanensis* (Sajise and Lales, 1975) and *Centrosema pubescens* (Bowen, 1961) have been shown to be allelopathic. In fact, *S. guyanensis* was reported to inhibit alang-alang for up to eight months in root-divider experiments. I have seen no information on possible allelopathic effects of *P. phaseoloides* or *C. muconoides,* and they need to be investigated. It is significant that Abdul-Rahman and Habib (1989) reported that *Imperata cylindrica* is a serious weed in Iraq as well and that decomposed alfalfa roots and their associated soil produced a 51 to 56% reduction in alang-alang seed germination. The alfalfa also reduced root and shoot growth of alang-alang by an average of 88%. Decayed and undecayed mixtures of alfalfa roots and soil (0.015:1 w/w) inhibited the production of new plants from alang-alang rhizomes by 30 and 42%, respectively. Moreover, root exudates of alfalfa significantly reduced root and shoot dry weights of alang-alang seedlings when alfalfa and alang-alang were grown together in nutrient culture.

Caffeic, chlorogenic, isochlorogenic, *p*-coumaric, *p*-OH-benzoic, and ferulic acids were identified in root exudates and residues of alfalfa; 126 μg of phenolics per gram of soil were found in alfalfa root residues after six months of decomposition in soil. It seems highly desirable that the three legumes that have been shown to have allelopathic effects on alang-alang should be investigated for control of that weed after burning in the transmigration areas of Indonesia. It appears likely that any control demonstrated by these species occurs because of a combination of allelopathic and competitive effects.

Liebl and Worsham (1983) decided to determine if the phytotoxic effects of plant residues on crop growth could also be responsible for the observed weed reduction in no-till cropping systems. They found that an aqueous extract of field-grown wheat reduced germination and root length of pitted morning glory (*Ipomoea lacunosa*) and common ragweed (*Ambrosia artemisiifolia*). Moreover, the inhibitory effect was increased by about 70% when the bioassays were conducted in the light. The hydrolyzed extract of wheat straw was fractionated with TLC; a compound that was isolated and found to have the greatest bioactivity on morning glory seed germination was ferulic acid. Ferulic acid ($5 \times 10^{-3}$ *M*) inhibited germination and root length of morning glory by 23 and 82%, respectively, and *Sida spinosa* with carpels by 85 and 82%, respectively. Crabgrass germination was inhibited by 100%. Root and shoot biomass of morning glory was reduced 52 and 26%, respectively, when it was grown in sand and watered with a $5 \times 10^{-3}$ *M* solution of ferulic acid. Ferulic acid was found to undergo decarboxylation in the presence of prickly sida fruit carpels, forming a styrene derivative, 2-methoxy-4-ethenylphenol. The researchers found that the decarboxylation of the ferulic acid was caused by a bacterium isolated from the carpels of *S. spinosa*.

Enache and Ilnicki (1990) pointed out that subterranean clover (*Trifolium subterraneum*) appeared to have potential as a successful living mulch because it is a winter annual legume with prostrate, nonrooting stems. The seeds germinate in late summer or early fall, and the plant grows vegetatively until becoming dormant in the winter. It resumes growth in early spring and flowers in late spring, and the seeds mature in the fruit at or below the soil surface. This causes the plant to reseed and eliminates the need to reestablish it. The clover dies soon after the seeds are produced, leaving a dense, dead mulch for the next two months, which does not compete with the main crop. Enache and Ilnicki decided, therefore, to test *T. subterraneum* for weed control as a living mulch, along with dead rye mulch, no mulch, and three tillage practices—conventional, minimum, and no-till. The crop planted in all cases was corn, and all living mulch plots had good ivyleaf morning glory (*Ipomoea hederacea*) control. The researchers stated that the subterranean-clover living-mulch control was a direct effect of the mulch and/or an allelopathic effect. Little control of fall panicum (*Panicum dichotomiflorum*) was obtained in the first year, but the living-mulch combinations effectively controlled fall panicum in 1987 and 1988, without any need for reseeding the clover. Weed biomass was reduced significantly by all living-mulch combinations, but all other combinations had higher weed biomass than did the living mulch. Corn silage and grain yields from the no-tillage plus living-mulch treatment were similar to or higher than those obtained with the conventional tillage plus no-mulch treatment.

As was pointed out in Chap. 2, *Parthenium hysterophorus* is a pernicious weed in fields and waste areas in many parts of India. Joshi and Mahadevappa (1986) noted that *Cassia uniflora* (= *C. sericea*) had gradually and effectively invaded many areas where parthenium had once vigorously grown. Subsequently, Joshi (1990, 1991a,b,c) and Ramachandra and Monteiro (1990) reported on several factors concerning the usefulness of *C. uniflora* in the biological control of parthenium. Joshi (1990) calculated the costs of mechanical, chemical, and biological control of parthenium and found that it was far cheaper to use biological control. There were numerous additional monetary returns from use in grazing, N fixation, haymaking, and human nutrition. Ramachandra and Monteiro (1990) reported that the seeds of *C. uniflora* contain 13.8% fiber, 19.5% crude protein, and 50.2% carbohydrate. Moreover, the protein digestibility was high (85.3%). Joshi (1991a) found that aqueous leachates of various parts of *C. uniflora* were very inhibitory to seed germination of parthenium, with seed leachates being most inhibitory (14%). In those cases where the seeds germinated, the growth of the seedlings was markedly retarded. The field plots in which *C. uniflora* was allowed to complete its life cycle had extremely small populations of the winter-generation parthenium plants compared with the parthenium plots. Joshi concluded that *C. uniflora* replaces parthenium through interference chiefly at three levels. It initiates replacement by inhibiting seed germination and seedling growth allelopathically. It outcompetes the summer generation of parthenium and causes a reduction in seed output. And it prevents the establishment of the winter generation of parthenium in its area of growth.

Aqueous leachates (1 g fresh wt/mL) of *Eucalyptus* sp. leaves collected from the botanical gardens at Chandigarh, India, were found to be very allelopathic to two-month-old seedlings of *Parthenium hysterophorus* when sprayed on the plants during three consecutive evenings (Kohli et al., 1988). Total chlorophyll was considerably reduced and the percent cell survival was markedly reduced. The RNA and water-soluble carbohydrate concentrations were also markedly reduced by the aqueous leachates. It was concluded that *Eucalyptus* sp. leaves contain allelochemics that affect growth of parthenium seedlings by inhibiting synthesis of RNA and carbohydrates.

The major plant parasite of tobacco in Andhra Pradesh, India, is *Orobanche cernua* var. *desertorum,* according to Krishnamurthy and Chandwani (1975). Tobacco is attacked by *Orbanche* before the tobacco has emerged from the ground, with resultant loss of yield and quality. The scientists investigated the ability of 17 crop plants to serve as trap crops (i.e., to stimulate seed germination of *Orboanche* without being suitable hosts themselves for the maturation of the parasite). They inoculated 500-mL aliquots of heavy black soil with 5000 seeds of *Orobanche* and placed each of the inoculated aliquots in a pot, along with 12 seeds of one of the crop species. After three weeks, the extent of infection of the roots, plus the percent germination of the *Orobanche* seeds, was determined. The crops tested were *Amaranthus gangeticus,* peanut, pigeon pea, chili, cotton, *Hibiscus sabdariffa,* tobacco cultivar Delcrest, M.T.U.- 17, rice, green gram (*Phaseolus aureus*), black gram (*P. mungo*), moth bean (*P. aconitifolius*), castor bean, til, Italian millet, sorghum, cowpea, and corn. The percentage germination of the *Orobanche* seed varied with the crop, with the maximum (69%) being obtained with chili and the minimum (1%) with *A. ganeticus.* Five crops—chili, sorghum, cowpea, moth bean, and *H. sabdariffa*—stimulated germination between 50 and 70%. Five crops stimulated 25 to 49% germination, and the rest fell between 1 and 24%. The control had only 0.88% germination. *A. gangeticus* and rice did not become infected at all with *Orobanche,* but infection was observed in all other cases. Sesamum was the only crop, however, in which pinhead-sized infections were observed in addition to the filamentous-stage infections. It was also observed that the filamentous infections in all other crops except tobacco started shrivelling and dying. The plants that stimulated germination of *Orobanche* between 25 and 70% would appear to be excellent possibilities for use in trap cropping to help control *Orobanche.*

Witchweed (*Striga asiatica*) is an important root parasite of agronomically important grasses and legumes. The seeds of this weed will germinate only under natural conditions, in the presence of chemical stimulants exuded from the roots of host plants. Strigol is one germination stimulant of witchweed seeds that was isolated and identified from root exudates of cotton. According to Hsiao et al. (1981), dl-strigol was first synthesized in 1974 and witchweed seeds are stimulated to germinate between 85 and 100% by concentrations of $10^{-14}$ to $10^{-6}$ *M* strigol. A $10^{-16}$ = *M* solution stimulates germination to 50%. Hsiao et al. (1981, 1983) decided to examine the conditioning and germination processes caused by dl-strigol (ST) and natural stimulants present in corn root exudates (STM). The overall goal was to control witchweed by stimulating its seed germination in the soil at times when suitable plants were not present to serve as hosts. The researchers found that witchweed seeds conditioned in water for 7 to 14 days required $10^{-10}$ and $10^{-12}$ *M* ST, respectively, to induce maximum germination (ca. 80%). It was noteworthy that conditioning seeds in ST or STM had adverse effects on the response of the seeds to the same compounds after conditioning. Exposure of witchweed to ST or STM before the appropriate pregermination threshold is reached apparently inhibits conditioning.

Hsiao et al. (1983) added a known amount of dl-strigol to the surface of a sandy-loam soil and leached the soil daily with 1.27 cm of simulated rainfall. After 21 days, 88% of the applied strigol remained in the top 2.5 cm of soil, about 6% in the 2.5 to 7.5-cm zone, and less than 1% at depths between 7.5 and 30 cm. Even with the lowest rate of application, sufficient strigol was leached to the zone between 22.5 and 30 cm to cause most of the witchweed seeds to germinate. Moreover, no significant degradation of the strigol occurred in moist soil in 21 days. The researchers concluded that dl-strigol has a potential to be an effective tool for a witchweed control or eradication program.

Netzly et al. (1988) isolated and identified four p-benzoquinones in the hydrophobic root exudate of IS8768 grain sorghum. The p-benzoquinones in the dihydroquinone form were found to be germination stimulants of witchweed. The major p-benzoquinone was sorgoleone; the others were structurally similar to sorgoleone but had different molecular weights. It is significant that sorgoleone is herbicidal to numerous weeds, and witchweed has apparently adapted dihydroquinones of this compound and close relatives as host-specific germination stimulants. The authors suggested that the dihydroquinone forms of the sorgoleones provide a new molecular model, simple in structure, which may be adapted to form more stable and still-active compounds to be used in the eventual eradication of witchweed.

Carson (1989) carried out two intercropping trials in 1985 and 1986 in The Gambia to compare the effects of interrow intercropping, intrarow intercropping, and sole cropping of grain sorghum and peanuts (groundnuts) on emergence of *Striga hermonthica* shoots on sorghum. He found that intrarow intercropping reduced the density of witchweed significantly when compared with sole cropping of sorghum. He pointed out that the result was apparently due to the lower soil temperature in the intracropping and/or to the peanuts having a trap-cropping effect. In 1985 and 1986 the grain yields of sorghum and the pod yields of peanuts were reduced somewhat below the yields obtained where sorghum and peanuts were the sole crop plants.

Four sesquiterpene lactones that share structural features of the lactone rings of strigol were found to have germination-stimulating activity against witchweed seed (Fischer et al., 1989a,b). Stimulation of germination by dihydroparthenolide reached a level of approximately 70% between $10^{-7}$ and $10^{-9}$ *M*; this is comparable to activities of strigol and some of its analogues. Confertiflorin and parthenin significantly increased witchweed germination at $10^{-4}$ *M*, and parthenin and desacetylconfertiflorin increased witchweed germination at $10^{-5}$ *M*. The structure of the lactone ring of dihydroparthenolide is similar to the butenolide ring of strigol, except, according to these scientists, it lacks the double bond. Moreover, most strigol analogues that have shown activity share the butenolide ring. Fischer et al. concluded that dihydroparthenolide could be used in witchweed control to stimulate germination in the absence of a host, as has been suggested for other stimulants. Fischer et al. (1990) subsequently tested 24 natural and synthetic sesquilactones against witchweed seed germination. All of the 10 germacranolides tested induced 40 to 65% germination across a range of $10^{-5}$ to $10^{-9}$ *M*. The highest activity occurred with a mixture of two eudesmanolides—santamarin and reynosin. The researchers pointed out that it is the spatial arrangement of the germacrane-eudesmane skeleton, in conjunction with the five-membered lactone ring, that is important in stimulation of witchweed seed germination.

Vail et al. (1990) tested 15 synthetic terpenoids, which were similar in structure to one of the four rings of the strigol molecule, against germination of broomrape (*Orobanche ramosa*). Nine were found to be active, and five of them contained ester groups. Several of the compounds were later tested against witchweed seeds, and the results were almost qualitatively identical. The researchers pointed out that monocyclic compounds with chemical structures similar to two of the rings of strigol have now been shown to have significant activity as germination stimulants of broomrape and witchweed.

## III. WEED CONTROL BY MICROORGANISMS

Boyette and Walker (1985) reported that sprays containing conidia of *Fusarium lateritium* and fungus-infested sodium alginate–kaolin clay granules controlled *Abutilon theophrasti* and *Sida spinosa* in corn, cotton, and soybeans in greenhouse studies. Moreover, there was no reduction in yields of the three crops. Postemergence foliar applications of the fungus controlled *A. theophrasti* by 40% and *S. spinosa* by 80% in the field in one test year, and in the second test year, the applications controlled the former by 27% and the latter by 34%. Preemergence application of the fungus in the granule form gave slightly better control of the same weeds in both years, except for a slightly lower control percentage of *Sida* in the first year. *F. lateritium* was the first pathogen to be evaluated as a biological control agent for more than one weed species, and it obviously proved successful. Because most mycoherbicides have a high degree of specificity, these findings could be significant from a commercial standpoint.

Many annual weeds persist in fields because of the ability of the seed to remain dormant for rather long periods of time (Kremer and Schulte, 1989). Chemical stimulation of imbibition and germination of dormant seed in the soil should help eliminate this possibility. Kremer and Schulte tested several chemicals, with and without the pathogen *Fusarium oxysporum,* for germination stimulation of *Abutilon theophrasti* seeds. The chemicals tested were $KNO_3$, $NaN_3$, gibberellic acid ($GA_3$), ethephon, and AC94377 (all germination stimulants); butylate, EPTC (S-ethyl-dipropyl-carbamothioate), and trifluralin (all herbicides); and the insecticides, carbofuran and cloethocarb. In the laboratory studies, *Abutilon* germination was increased, and hard seed numbers were decreased by all test chemicals except $KNO_3$, butylate, and trifluralin. Nonviable seeds were increased significantly only with trifluralin, and the proportion of hard seed was decreased only when *Fusarium* was combined with butylate. Overall growth of *Fusarium* and coincident infection of seedlings was enhanced with all chemicals except $NaN_3$ and trifluralin. The increases in nonviable seed and inhibition of radicle growth observed when *Fusarium* was added to certain treatments may have resulted from chemically induced susceptibility of the *Abutilon* seed and seedling tissues.

In greenhouse studies, Kremer and Schulte found that no treatment effectively reduced *Abutilon* emergence in the absence of *F. oxysporum.* Ethephon and carbofuran significantly reduced emergence, however, when combined with *Fusarium.* All chemicals except $NaN_3$ increased the number of nonviable seed when combined with *F. oxysporum.* Nonviable seed recovered from soil treated with ethephon, AC94377, butylate, and carbofuran plus *Fusarium* were heavily infected with mycelia. The researchers concluded that an integrated approach—combining low rates of soil-applied chemicals and selected microorganisms—might be useful in rapidly depleting the weed seed bank and decreasing seedling emergence and vigor.

*Colletotrichum* spp. cause diseases of many plants; *C. gloeosporoides* f. sp. *aeschynomene* is already sold as a mycoherbicide under the brand name of Collego. Silman et al. (1991) pointed out that *C. truncatum* is a pathogen of *Sesbania exaltata,* which is a pernicious weed for cotton and rice crops. A serious problem in using this microorganism for weed control is that it sporulates very poorly in liquid culture, and it is therefore very difficult to obtain sufficient conidia to inoculate *S. exaltata* in large fields. Silman et al. compared four culture systems, plus a great many media factors, against spore production. The four systems were dialysis membranes, liquid shake flasks, solid particles with humidity control, and solid particles without humidity control. The researchers' conclusion was that all systems gave spore counts of the same order of magnitude, and thus no choice of culture methods could be made from their tests. It was concluded that much more work was necessary.

*Ageratina adenophora* (crofton weed, Asteraceae) is rapidly invading grasslands and forested areas in the Natal mistbelt and Transvaal, as well as along stream banks in southwestern Cape Province,

South Africa. Morris (1989a) investigated the possibility of using the leaf-spot fungus *Phaeoramularia* sp. for the biological control of crofton weed in South Africa. He first inoculated *A. adenophora* and several other closely related species of *Ageratina* with the pathogen and found that *A. adenophora* was the only host in which symptoms developed. External hyphal growth and stomatal penetration took place on leaves of all species, but hyphae ceased growth in the sub-stomatal chambers of all species other than *A. adenophora.* Morris stated that this fungus has contributed to the partial control of crofton weed in Australia and appears to have promise in South Africa. He has received permission for field trials in South Africa, and some trial releases have been made. Kluge (1991) also stated that this fungal pathogen was released in 1987 and 1988 in some areas of South Africa and caused some defoliation of crofton weed. He pointed out, however, that by 1991 the pathogen had had no obvious impact on the populations of the weed.

The genus *Amaranthus* contains many species of weeds, some of which are considered among the worst in the world (Mintz et al., 1992). A fungus, *Aposphaeria amaranthi* was isolated from an unidentified species of *Amaranthus* in Arkansas. Preliminary studies demonstrated that it was pathogenic to several species of *Amaranthus.* Plants of *Amaranthus albus,* up to the 12-leaf stage, were sprayed with suspensions of *A. amaranthi* from $10^4$ to $10^7$ conidia/mL. When the spraying was done after a 24-hour dew period, 100% mortality of the *A. albus* seedlings occurred. The range of dew temperatures conducive to 100% mortality was 20 to 28°C when the *Amaranthus* seedlings had four leaves. Over 70% mortality was achieved with plants having eight leaves, but disease severity declined after plants began developing axillary buds. The highest level of control under field conditions was 99%, and this occurred when seedlings were sprayed to runoff with concentrations of $6 \times 10^6$ conidia/mL. Mintz et al. stated that more extensive host range tests and screening of additional isolates would be beneficial in determining the potential of this pathogen as a broad-spectrum mycoherbicide for several different species of *Amaranthus.*

Jones et al. (1988) inoculated nutrient-amended peat with a conidial suspension of *Gliocladium virens*; after five days the substrate colonized by the fungal hyphae was air-dried and ground to a powder. This powder was added to a mixture of 50% peat and 50% coarse sand, and the final mixture placed in individual 200-mL pots. The fungus-peat mixture was added to 16, 8.7, 4.5, or 2.3% of the total volume. An equivalent amount of peat moss alone was added to the controls. Fifteen crop plants and 17 weeds were tested for seedling emergence (percent of control) and dry weight. The weeds tested were *Avena fatua, Solanum ptycanthum, Convolvulus arvensis, Poa annua, Setaria glauca, Phalaris minor, Rumex crispus, Plantago lanceolata, Hypochoeris radicata, Amaranthus retroflexus, Portulaca oleracea, Amsinckia intermedia, Chenopodium album, Taraxacum officinale, Picris echioides, Senecio vulgaris,* and *Sonchus oleraceus.* Herbicidal activity was correlated with fungal production of viridiol, which had a broad spectrum of activity. Viridol was particularly effective against broad-leaf species and was less effective in monocot control. Emergence of most weeds was reduced by more than 90% of the control at application rates of 8.7% or less. The dry weights of the seedlings were drastically reduced in most cases. Applications of 4.5% reduced root and shoot weights of *A. retroflexus* by 93 and 98%, respectively. Crop plants were affected at higher treatment levels, but the toxicity was easily avoided by applying the fungus out of the root zone of the crop. Viridiol was detected three days after incorporation of the fungus and reached its peak concentration on days 5 and 6 at approximately 25 μg viridiol/100 mL of soil, based on an application rate of 11%. It declined to a nondetectable concentration by the end of two weeks. Jones et al. pointed out that this mycoherbicide might be particularly useful in soils where high levels of organic matter tie up synthetic chemical herbicides.

Hasan (1991) reported that *Asphodelus fistulosus* (onion weed) is widespread in South Australia where it invades pastures, making them unsuitable for grazing. He found a rust fungus, *Puccinia barbeyi,* which severely infects onion weed. The rust is monoecious, microcyclic, and multiplies primarily by aecial and telial stages. Several members of the Liliaceae were tested and found to be apparently resistant to the rust. It was concluded, therefore, that the rust is probably specific to *Asphodelus* spp. and has good potential for the biological control of onion weed in Australia.

Walker and Connick (1984) grew individual cultures of the fungi *Alternaria macrospora, A. cassiae, Fusarium lateritium, Colletotrichum malvarum,* and *Phyllosticta* sp. The mycelia were harvested, homogenized for 30 seconds, diluted to a 1:3 (v/v) with aqueous 1.33% Na-alginate, and pelleted in the presence of 0.25 *M* $CaCl_2$ at 40°C to give herbicidal products with extended shelf-life and sustained-release properties. The scientists tested granular formations of *A. cassiae* containing mycelium, Na-alginate, and kaolin and found that it controlled sicklepod by about 50% in soybeans at 500 kg/ha.

Daigle and Cotty (1991) reported that a formulation with a pH of 6.5 containing 0.1 to 1% Tween 80, 0.02-*M* potassium phosphate buffer, and 1% dehydrated potato dextrose broth best promoted germination of spores of *Alternaria cassiae. Cassia obtusifolia* (sicklepod) plants were sprayed with the above-described test solution at the two to three true-leaf stage, incubated in the dark at 100% relative humidity and 28°C for six hours, and placed in a growth chamber at 30°C. After two days, the researchers found that the formulation induced germination of *A. cassiae* quite well and improved infection of the sicklepod seedlings.

Spotted knapweed (*Centaurea maculosa*) is the chief weed in western Montana. Stierle et al. (1989) found that *Alternaria alternata* is a host-selective pathogen of that weed. Additionally, they found that the extracts of cultures of that fungus contained three classes of phytotoxins: diketopiperazines, tetramic acids, and perlenequinones. The three most active compounds identified were maculosin, tenuazonic acid, and alterlosin II, all of which were phytotoxic to knapweed. Maculosin was specific to spotted knapweed, but the other two were phytotoxic to several species tested. At a concentration of $10^{-3}$ *M*, tenuazonic acid was phytotoxic to 18 of the 19 test plants.

The search to find effective biological agents to control *Centaurea maculosa* and *Cirsium arvense* in Montana indicated that the pathogen with the most potential (*Sclerotinia sclerotiorum*) was one of the most nonspecific plant pathogens identified (Miller et al., 1989). Instead of seeking further for a more specific pathogen, Miller et al. chose to use a genetic approach to make *S. sclerotiorum* environmentally safe. Ascospores of the wild type of this pathogen were irradiated with UV light from a GE G8T5 (15-watt) source for 80 seconds, which permitted about 2% ascospore survival. Ascospores requiring yeast extract for growth were isolated, and protoplasts of the wild type and the mutant type were produced. Protoplast regeneration of the wild type and of the mutant type was determined; the mutant type was found to require either yeast extract or cytosine to develop. Several potential host plants were inoculated with either the wild type or the mutant type isolates. Four of the seven were not infected by the mutant type unless an exogenous cytosine source was also applied. On the other hand, all potential hosts were infected by the wild type without an addendum. The mutant fungus would be a good candidate for biological weed control, since infection would be limited to the area of cytosine application, thus reducing the threat to beneficial plants.

Sands et al. (1990) continued the type of research done by Miller et al. (1989), seeking to genetically limit the virulence or survival of a deadly pathogen instead of searching for other lethal pathogens with a narrow host range. In their 1990 investigation, Sands et al. developed three classes of mutants from *Sclerotinia sclerotiorum:* auxotrophic mutants that attacked hosts only when applied along with an exogenous source of the required nutrient; mutants unable to form sclerotia which just limits long-term survival; and mutants with reduced virulence and/or host ranges. Of 3000 ascospores treated with a mutagenic agent and screened for auxotrophy, only one showed dependence for pyrimidines. As was discussed above, this mutant fails to grow on minimal media and is avirulent unless an exogenous source of pyrimidines (i.e., cytosine) is present. After two years of in vitro culture, that mutant had not reverted to an autotrophic type. One nonsclerotial isolate was found to be virulent on all susceptible hosts tested, but it had slightly reduced virulence and killed fewer host plants. This mutant failed to overwinter or to produce ascocarps in field trials. A second nonsclerotial mutant was avirulent on all eight susceptible hosts tested. The researchers suggested that this mutant may be useful in research with protoplast fusion or genetic engineering techniques to obtain virulent, nonsclerotial mutants. The third class of mutants had reduced host ranges, apparently because of reductions in virulence. These mutants may be useful as biocontrol agents and in determining the causes of virulence and host-range specificity.

*Centaurea solstitialis* (yellow starthistle) is a very common and aggressive weed in the rangelands and pasturelands of the western United States. Natural pathogens are being investigated for possible use in biological control of that pest. Bennett et al. (1991) found one good prospect, a rust, *Puccinia jaceae,* which was collected on *C. solstitialis* in Turkey. The scientists investigated the effects of dew, plant age, and leaf position on susceptibility of the weed to the rust. The largest number of pustules developed on the starthistle with 12 or more hours of dew at 20°C; the most pustules developed on plants four to six weeks after planting, and susceptibility of older plants decreased with age. Inoculation with *P. jaceae* caused dry weights of rosettes and roots to be significantly less than those of controls. Root biomass of plants inoculated at four or six weeks after planting was between 30 and 53% less than root weights of controls. The researchers concluded that *P. jaceae* has the potential to become established in and damage yellow starthistle in North America.

According to Mortensen (1986), the goal in an effective biological control system is not to eliminate the weed but to keep the control agent and the host plant in an equilibrium with the population of the host plant at a low level. Some allowance must be made for minor attacks at times on nontarget organisms. Mortensen gave the example of the possible use of *Puccinia jaceae* in the biological control of *Centaurea diffusa* (diffuse knapweed) in British Columbia, Canada. Seedlings of safflower (*Carthamus tinctorius*) were attacked by the rust, but adult plants were resistant. Mortensen pointed out that it is unlikely that the rust developing on diffuse knapweed in the spring would be able to produce sufficient numbers of urediospores to attack safflower while it is still susceptible. *P. jaceae* has little chance of surviving on safflower and therefore appears to have good potential for biological control of diffuse knapweed. Generally, pathogens must be applied annually for successful biological control, so mass production of pathogen propagules and special storage and application techniques are required. Mortensen concluded that, because of the many special requirements, involvement of private industry is essential for commercialization of microbial herbicides.

*Datura stramonium* (jimsonweed) is a serious weed in many parts of the world, and its seeds are toxic to humans and livestock. Boyette et al. (1991) isolated *Alternaria crassa* from leaves of an infected jimsonweed collected near Southaven, Mississippi. The micro-organism was subsequently shown to be restricted in host range to jimsonweed and two tomato varieties; controlled tests indicated that it can infect and kill jimsonweed seedlings over a wide range of environmental conditions. Boyette et al. therefore decided to evaluate conidial and mycelial formulations of *A. crassa* for jimsonweed control under field conditions in parts of Arkansas and Mississippi. They found that this weed was successfully controlled by *A. crassa* from 1985 to 1987 (the duration of the tests). Conidial applications of *A. crassa* provided an average control of 96% in Mississippi and 87% in Arkansas. Mycelial formulations were less effective overall than were conidial formulations, but during periods of favorable environmental conditions, weed control by the mycelial formulations was similar to that provided by conidial preparations. Apparently, *A. crassa* holds real promise as a mycoherbicide for controlling jimsonweed.

Boyette and other colleagues (Abbas et al., 1991) isolated *Fusarium moniliforme* from greenhouse-grown jimsonweed plants in Mississippi in 1990. They applied aqueous suspensions of spores and mycelia ($>1 \times 10^7$ propagules/$mL^{-1}$) to potted soil in which two- to four-week-old jimsonweed plants were growing. Local lesions or mosaiclike patterns appeared on the leaves and growth was inhibited. When fungal suspensions were applied in higher dosages by soil drench or subirrigation, similar symptoms appeared. Moreover, when the fungus was grown on an autoclaved rice medium and the ground mixture was applied to jimsonweeds, the plants wilted in 24 hours and died after 48 hours. Cell-free filtrates of the fungal-rice medium also killed one to 2- and 3- to 4-week-old plants after 24 to 72 hours. A major toxin isolated and identified was fumonisin $B_1$, which was present in fermented rice at a concentration of 400 μg/g. Jimsonweeds treated with crude or cell-free extracts, or with purified aqueous fumonisin $B_1$ solutions, developed a soft rot that diffused along leaf veins. These scientists concluded that *F. moniliforme* may be superior to *Alternaria crassa* as a biological herbicide against jimsonweed because it does not require a dew period as does *A. crassa* and it grows and sporulates more rapidly.

If the chemical pesticides are not compatible with the fungal agent spores, it is counterproductive to apply chemical herbicides, fungicides, and/or adjuvants in fields where fungal biological control agents are used. Grant et al. (1990) found, for example, that commercial formulations of all grass weed chemical-control agents tested completely inhibited spore germination of *Colletotrichum gloeosporioides* f. sp. *malvae,* which is used in the control of *Malva pusilla* (round-leaved mallow). At recommended rates, however, all herbicides used for broad-leaf weed control—including 2,4-D ester, 2,4-D amine, benazolin, bentazon, clopyralid, cyanazine, cyanizine plus MCPA, dicamba, dicamba plus MCPA, dicamba plus mecoprop and MCPA, dicamba plus 2,4-D plus mecoprop, MCPA amine, MCPA-K, MCPA-Na, and metribuzin—caused no more than a 20% reduction in germination of the same kinds of spores. The fungicide triadimefon had no effect on *C. gloeosporioides malvae* spore germination at the recommended rate of application. Dicloran reduced germination by more than 50% and growth was distorted. Spore germination of the pathogenic fungus was inhibited by more than 90% by benomyl, carbathiin, chlorothalonil, iprodione, mancozeb, and thiophanate-methyl, although germination increased as the applied concentration of the fungicides was decreased below the recommended rate. Spore germination was totally inhibited by all concentrations of ferbam, thiram, and captan. Several adjuvants inhibited spore germination of the fungal control agent; but ammonium sulfate, Assist, Bio-Veg, CD-407, Tween 20, starch, sucrose, and water stimulated spore germination. Such tests should definitely be run for all chemicals planned for use in systems where any biological control agents are to be used.

Figliola et al. (1988) isolated two leaf-spotting fungal pathogens, *Bipolaris setariae* and *Piricularia grisea,* from badly infected *Eleusine indica* (goosegrass). Subsequent tests demonstrated that both pathogens were 100% effective in infecting goosegrass at 28°C, given a 72-hour dew period. The disease index increased as the dew duration increased at all temperatures tested. The optimum temperature for *B. setariae* was 24 to 28°C, and it was 28°C for *P. grisea.* Representative species of the Fabaceae, Malvaceae, Poaceae, and Solanaceae were inoculated with either *B. setariae* or *P. grisea* and placed in a dew growth chamber at 28°C. The researchers found that infection was limited to the Poaceae, and all the inoculated goosegrass plants were dead 20 days after inoculation with either fungal pathogen (Table 5). The limited host range tests indicated that neither fungus poses a threat to dicotyledonous crop plants, and thus both have real promise in the biological control of goosegrass.

*Euphorbia esula* (leafy spurge) is a perennial weed of the grasslands of North America and many croplands (Yang et al., 1990). Chemical control is both costly and difficult, so biological control methods are being sought. *Alternaria angustiovoidea* is known to be a pathogen of some leafy spurge plants and has been considered as a biological control agent of that weed. It is noteworthy, however, that 10 of 14 collections of leafy spurge showed resistance to that pathogen; it was not known whether the plants that were not infected in early tests might become infected in a longer dew period. Nor was it known whether an allelochemical is produced by the fungus and/or whether that substance might be allelopathic to leafy spurge. These questions were investigated by Yang et al., who found that a single dew period of 48 hours was required to obtain severe infection and kill the leafy

**Table 5** Efficacy of *Bipolaris setariae* and *Piricularia grisea* on 4-week-old goosegrass plants

| Population treatment* | % plants with symptoms | Average disease index† | % plants dead at day 20 |
|---|---|---|---|
| Inoculated with *B. setariae* | 100 | 94 | 100 |
| Anderson: Inoculated | | | |
| Control | 0 | 0 | 0 |
| Marlboro: Inoculated | 100 | 91 | 100 |
| Control | 0 | 0 | 0 |
| Inoculated with *P. grisea* | 100 | 100 | 100 |
| Anderson: Inoculated | | | |
| Control | 0 | 0 | 0 |
| Marlboro: Inoculated | 100 | 100 | 100 |
| Control | 0 | 0 | 0 |

* Plants were inoculated to the point of runoff with a suspension containing 20 000 spores/ml and were given a 72-hours dew period.
† Disease index was evaluated five days after inoculation. See text for disease-index evaluation method.
*Source:* From Figliola et al. (1988), with permission of Weed Sci. Soc. of America.

spurge plants. Twenty-two of 25 collections of leafy spurge were found to be susceptible to the pathogen when inoculated weeds were kept in dew chambers at 20 to 25°C for 48 hours after inoculation. Moreover, minor infection occurred on globe artichokes, corn, cowpeas, safflower, and zinnias under the same conditions. The researchers discovered also that this pathogen produced phytotoxins that caused chlorosis and wilting of leaves on cuttings of leafy spurge placed in the culture filtrate. The phytotoxins are not host-specific because the partially purified toxin also caused chlorosis in *Lemna obscura*. The phytotoxins were not identified.

*Euphorbia geniculata*, a widely distributed weed in India, was first introduced from South America (Lakshmanan et al., 1990). When a leaf and stem blight occurred on that weed on the campus of Tamil Nader Agricultural University during the fall of 1987 and 1988, the causative fungus was identified as *Cochliobolus carbonum* (or *Bipolaris zeicola*, or *Helminthosporium carbonum*). None of 19 crop plant species tested was infected by this pathogen. Lakshmanan et al. pointed out that the highly virulent nature of this fungus on *E. geniculata* and the limited host range suggested that the pathogen has promise for use as a biocontrol agent for *E. geniculata*.

According to Morris (1989b), *Hakea sericea* is a major weed in Cape Province, South Africa, and a strain of *Colletotrichum gloeo-*

*sporioides* occurs in many areas infested by *H. sericea*. That pathogen causes gummosis and death of *Hakea,* so Morris invęstigated the possibility of using it in the biological control of the weed. He sprinkled a dried formulation of *C. gloeosporioides* on wheat bran onto plots of young *Hakea* seedlings in southwestern Cape Province. After rainy periods, the fungus sporulated profusely on the bran, and rain splash dispersed the conidia. The seedlings became infected and died back from the stem tips. Application of the bran inoculum during early winter when seedlings of *Hakea* were in the cotyledonary to 20-leaf stage caused death of 93 to 98% of the seedlings, compared with death of 30 to 36% of the seedlings when the bran inoculum was applied at later stages of seedling development.

*Heliotropium europaeum,* a summer annual of Mediterranean and Middle Eastern origin, has become naturalized in Australia and is a serious weed in pastures (Hasan and Aracil, 1991). Conventional methods of control are expensive and environmentally unsafe, and thus Hasan and Aracil and others have investigated the possibility of biological control. The rust fungus, *Uromyces heliotropii,* is a macrocyclic rust that attacks all aerial parts and growth stages of the weed and causes serious damage and, frequently, death. The pathogen is spread by urediniospores during the summer, and toward the end of the summer unicellular teliospores are produced. Some of the teliospores overwinter and produce basidiospores in the spring, which infect new seedlings of *Heliotropium.* The infected plants produce aeciospores, which infect common heliotrope, which produce urediniospores, which carry on the cycle again during the summer. Because infection of heliotrope is not common during some years in southern Europe, Hasan and Aracil investigated the longevity and germination of the teliospores that cause the overwintering of the pathogen. They found that teliospores of *U. heliotropii* remained viable when stored for one year at temperatures between –8 and 20°C and carried infection over from one growing season to another under a Mediterranean climate. They found also that the rust rapidly killed infected heliotrope plants and reduced or prevented seed production. They concluded that the pathogen is capable of growing and surviving in Mediterraneanlike climates in Europe and Australia and is sufficiently virulent to cause significant field infection in both regions.

*Hydrilla verticillata* (hydrilla) is one of the worst pests of waterways in tropical and subtropical parts of the world (Joye, 1990). The high cost and environmental dangers associated with the use of synthetic herbicides make it highly desirable to find a biological control agent for hydrilla. A pathogenic isolate of *Macrophomina phaseolina* was collected from hydrilla in Lake Houston in 1987, and Joye began an investigation of this pathogen as a possible biological control agent of hydrilla. Repeated greenhouse and field tests demonstrated that this fungus greatly reduced the biomass of hydrilla within three to four weeks of inoculation. In 1988 and 1989, enclosures were established in a dense hydrilla stand in the Shelton Reservoir near Houston, Texas. Hydrilla in the enclosures was inoculated in September 1988 and October 1989 with a suspension of *M. phaseolina* to produce a final concentration of approximately $1 \times 10^4$ cfu/mL. After four weeks, the remaining biomass was collected, dried at 100°C, and weighed. The reduction in biomass of hydrilla, compared with control enclosures, was 61.3% in 1988 and 58% in 1989. Untreated plants in the controls remained healthy and vigorous. Results through 1990 indicate that *M. phaseolina* is an excellent prospect for the biological control of *H. verticillata.*

*Prunus serotina* (black cherry) was introduced into the Netherlands in 1920 to improve the understory and litter in planted pine forests (de Jong et al., 1990). It has become a pest rather than an asset, and control or elimination of black cherry by chemical or mechanical means has not proved satisfactory. *Chondrostereum purpureum* is a common saprophyte in wood of deciduous trees and a pathogen of black cherry, cultivated plum and cherry, and various ornamentals of the genus *Prunus.* De Jong et al. used epidemiological and air-pollution theories to determine the risk of infection of nontarget species by the use of *C. purpureum* in the attempted biological control of *P. serotina.* The experimental site was a 45-year-old stand of *Larix leptolepis* 17 m tall in the municipality of Ede, heavily infested with black cherry. Saplings of *P. serotina* 2 to 3 m tall were cut off at a height of 0.25 m, and agar pieces with mycelium from various isolates of *C. purpureum* were applied to the fresh wounds. The incidence of stumps with basidiocarps, the number of basidiocarps per stump, and the mean size of the basidiocarps were assessed in the two falls following inoculation. Checks of nontarget trees were made at various places away from the inoculated area; it was found that 500 m distant from the test area the risk to nontarget trees was high but was negligible 5000 m downwind. In general, results of biocontrol tests were highly satisfactory. In an experiment near Heerlen in 1986, 61% of 321 inoculated trees died within two years, whereas 56% of glyphosate-treated trees and 1% of 142 control trees died within those two years. De Jong et al. (1991) continued their work on modeling the escape of basidiospores of *C. purpureum* from a larch forest in which this pathogen was being used in the biological control of *P. serotina*. The downwind escape percentage was found to vary from 19 to 45%; the upwind percentage varied from 7 to 23%. Number of spores released from basidiocarps and escape percentages were then used as inputs for a Gaussian plume model to calculate movement of spores into *Prunus* spp. orchards.

*Sorghum halepense* (Johnsongrass) is a pernicious weed in several crops in the United States and is considered one of the 10 worst weeds in the world. Selective postemergence herbicides are often satisfactory to control this weed in dicot crops, but control in grass crops such as corn and sorghum is still a problem (Chiang et al., 1989a). Diseased Johnsongrass leaves were collected during a two-year period in North Carolina in a search for leaf-infecting pathogens that might have potential for biological control of this weed. Seven pathogenic fungi were identified and were investigated further for virulence, dew requirements, and sporulation capacity. Of these seven, *Exserohilum turcicum, Colletotrichum graminicola,* and *Gloeocercospora sorghi* were selected for still more tests. Multiple isolates of each were compared and found to differ in virulence and host specificity. *G. sorghi* was found to be more host-specific than the other two fungi; it was compatible only with *Sorghum* spp., whereas *E. turcicum* and *C. graminicola* were compatible with *Sorghum* and corn. Dicotyledonous species were immune to all three fungi.

Chiang et al. (1989b) selected four virulent foliar fungi known to infect Johnsongrass to investigate further their potentials for biological control of that weed. The four pathogens selected were *Exserohilum turcicum, Colletotrichum graminicola, Gloeocercospora sorghi,* and *Bipolaris halepense.* The greatest leaf injury (greater than 90%) was induced by *E. turcicum* with $2 \times 10^5$ conidia/mL and a 24-hour dew period after inoculation. Sequential inoculation with this pathogen at 15 and 20 days after emergence resulted in injury to more leaves than a single inoculation at 15 days. The researchers found no synergistic or antagonistic effects in combinations of *E. turcicum* with *C. graminicola* or *G. sorghi.* It is noteworthy that development of new leaves from inoculated plants was reduced by 30% or less at 14 days after inoculation compared with uninoculated plants. No kill was observed in any fungal tests. The percentage leaf area injured was generally greater when weeds were inoculated seven days after emergence than when inoculated at an older stage. It was concluded that failure to kill Johnsongrass plants in these tests did not rule out the possibility that this weed can be weakened and made less competitive with mycoherbicides such as those tested. On the basis of the researchers' overall studies, it appears that one or more of the pathogens might be very promising for Johnsongrass control in dicotyledonous crops but may not be safe in corn and sorghum crops.

*Taraxacum officinale* (dandelion) is an herbaceous perennial weed that is distributed worldwide but is most prevalent in temperate and subtemperate regions (Riddle et al., 1991). Chemical control with chlorophenoxy herbicides has been the accepted method of dandelion control, but the discovery of dioxin in certain formulations of 2,4-D and results of some studies that suggest that 2,4-D is a carcinogen have caused considerable concern about the continued use of 2,4-D and related synthetic herbicides for dandelion control. Riddle et al. investigated 60 isolates of *Sclerotinia sclerotiorum* and six isolates of *S. minor* for their potentials as biological control pathogens for dandelions. They found significant negative correlations between the relative virulence of the isolates and the dry weights of inoculated dandelion plants. Moreover, they found positive correlations between the relative virulence of isolates and reduction in numbers of dandelion plants in turf grass swards infested with inocula of the isolates. Twenty-three isolates of *S. sclerotiorum* were ranked as highly virulent against dandelion, twenty-nine isolates of *S. sclerotiorum* and five of *S. minor* were moderately virulent, and eight isolates of *S. sclerotiorum* and one of *S. minor* were weakly virulent. In a turf grass sward treated in 1987 with four applications of heat-killed seed of perennial ryegrass (100 $g/m^2$ $application^{-1}$) infested with isolate R30 of *S. sclerotiorum* followed by six applications at the same rate in 1988, there was an 80.7% reduction in the number of dandelion plants in the grass in August 1988. Populations of dandelions in untreated swards increased by 22.2% in the same time period. Application of isolate R30 along with 25 $g/m^{-2}$ of dandelion seed to a Kentucky bluegrass sward reduced establishment of dandelion seedlings by 85.5%. Necrosis or discoloration did not develop on Kentucky bluegrass, creeping bentgrass, annual bluegrass, or quackgrass when inoculated with isolate R30.

*Xanthium spinosum* (spiny cockleburr) is a widespread annual weed throughout much of the world, especially in temperate regions (Auld et al., 1988). It is particularly important in Australia and is susceptible to 2,4-D. However, that herbicide cannot be used close to broad-leaved crop plants that are also very susceptible to it. Therefore Auld et al. investigated the possibility of using biological control against *X. spinosum*. The fungus *Colletotrichum orbiculare* was found as the causative agent of anthracnose on *X. spinosum* in several places in southeastern Australia. It usually caused only leaf and stem lesions, but a few plants were apparently killed by the pathogen. The researchers therefore decided to determine the fungus' potential as a biological control agent for spiny cockleburr. The pathogen was grown on irradiated carnation agar until it sporulated, at which time spores were collected and suspended in sterile water. Concentrations of spores were determined using a haemocytometer. Ten 6-week-old cockleburr plants, each in a separate pot, were sprayed to complete wetness with different concentrations of spores of the pathogen. Ten plants were sprayed with distilled water as controls, and all treatments and controls were compared statistically when 50% of the plants in the most effective treatment were dead. The threshold for disease production was between $10^3$ and $10^4$ spores/$mL^{-1}$. All plants inoculated with more than $10^6$ spores/$mL^{-1}$ died within 21 days. Plants were killed after dew periods of 4 to 48 hours following inoculation, but 70% of the treated plants eventually died even without dew. Spore production for use in inoculation was no problem, because sporulation in submerged culture occurred rapidly, with a plateau of maximum production beginning at 16 days after inoculation. *C. orbiculare* is known as a pathogen of Cucurbitaceae, but the cucurbits growing in most cockleburr-infested areas in southeastern Australia are also weeds and are not likely to be an impediment to the widespread use of *C. orbiculare* as a biological control agent of spiny cockleburr in Australia.

## IV. POTENTIAL OR PRESENT HERBICIDES FROM PLANTS AND MICROORGANISMS

*Drechslera avenae* is a fungal pathogen of cultivated oats as well as wild oats (*Avena fatua*), *Arrhenatherum* spp. (oatgrass), and other weedy graminae. Sugawara and Strobel (1986) isolated a phytotoxin from the culture medium of this pathogen, which produces sunken lesions on cultivated and wild oats. They identified the compound as (-)-dihydropyrenophorin and found that it was active against the oat species and a variety of other plants at $3.2 \times 10^{-4}$ *M*. They were surprised to find that the phytotoxin was active against Johnsongrass at a concentration of $6.5 \times 10^{-8}$ *M*. It appears that the phytotoxin has excellent potential as a herbicide against several plants and particularly against Johnsongrass.

Putnam (1988) reviewed our general knowledge of allelochemicals from plants as possible herbicides and concluded that "[I]t seems doubtful that higher plant products will be active enough to be packaged and sold as herbicides unless some unique compounds are discovered that are more active than those now known." On the other hand, certain allelochemicals from higher plants may be chemically modified to become excellent herbicides. Putnam pointed out that commercial cinmethylin is adapted from cineole, a terpene contained in relatively high concentrations in numerous plants. Moreover, higher plants are being used widely in weed control, or at least in partial weed control, in many different ways, as discussed in Sec. II of this chapter.

*Streptomyces* strains isolated from soil produce many phytotoxins with strong biological activity—anisomycin, bialophos, cycloheximide, glufosinate, herbicidins A and B, herbimycins A and B, and toyocamycin among others (Heisey and Putnam, 1986). In their screening for the production of bioactive chemicals by microorganisms, Heisey and Putnam isolated and identified two very active herbicides, geldanamycin and nigericin, which were synthesized by *S. hygroscopicus.* Both compounds reduced garden cress radicle elongation by 50% at concentrations of 1 to 2 ppm and caused nearly complete inhibition at 3 to 4 ppm. Geldanamycin caused radicles to turn brown and disintegrate, but nigericin did not cause visible necrosis. Geldanamycin is an ansamycin antibiotic that is structurally identical to herbimycin B except for a methoxyl group on carbon 17. Geldanamycin is moderately active against fungi and bacteria and is a potent inhibitor of DNA synthesis. Nigericin is a polyether antibiotic that influences the transport of alkali cations across several types of membranes and impedes photophosphorylation. It inhibits gram-positive bacteria, mycobacteria, and certain pathogenic fungi.

Putnam and his colleagues continued their search for microbial isolates that could produce worthwhile herbicides (Mishra et al., 1988). They isolated and evaluated metabolites from 906 microbial isolates, which included 266 isolates from *Streptomyces* and 502 isolates from non-*Streptomyces* actinomycetes representing 18 genera, 28 unidentified aerobic actinomycetes, 70 fungi, and 40 isolates of eubacteria. All the isolates came from 18 soil samples from cultivated fields, indoor potted plants, and parks or lawns, with and without "fairy rings." Overall, metabolites from 72 isolates significantly inhibited germination of cress seeds. About 18% of all *Streptomyces* and *Nocardiopsis* isolates and 13% of *Actinoplanes* isolates were toxic to cress seeds. Other inhibitors included three isolates of *Actinomadura*; one isolate each from *Micromonospora, Micropolyspora, Streptosporangium, Streptoverticillium,* and *Bacillus,* and two isolates of unidentified actinomycetes. The herbicidal fungi included two isolates of *Penicillium* and one isolate each of *Aspergillus, Scopulariopsis,* and *Paecilomyces* species. About half the isolates inhibitory to cress were also inhibitory to *Echinochloa crusgalli.*

Rizvi et al. (1988) tested three terpenes—citral, citronellol, and geraniol—against seed germination and seedling growth of *Amaranthus spinosus* and tomatoes. All were inhibitory to seed germination and seedling growth of *A. spinosus.* Geraniol was most inhibitory, with 3-m*M* concentrations completely preventing germination and plumule growth and a 3-m*M* concentration completely inhibiting radicle growth. Neither compound inhibited seed germination, radicle growth, plumule growth, or biomass production of tomatoes. These results indicated that geraniol has some promise as a herbicide.

Rhizobitoxine is an allelochemical produced by certain strains of *Bradyrhizobium japonicum* in soybean nodules (Owens et al., 1972) and by *Pseudomonas andropogonis* (Mitchell, 1989). Anonymous (1969) pointed out that rhizobitoxine is an effective herbicide in amounts as low as 3 oz/acre. Owens (1973) reported that rhizobitoxine and the commercial herbicide, amitrole, were approximately equal in phytotoxicity to various weed and crop species. Crabgrass was moderately sensitive to rhizobitoxine, but Kentucky bluegrass was very tolerant of it. Amitrole had virtually the same toxicity to both of these grasses. The narrow difference in sensitivity between crabgrass and Kentucky bluegrass to commercial herbicides has limited their use on bluegrass lawns. Thus rhizobitoxine appears to have some promise in crabgrass control in such lawns.

*Drechslera indica* is a fungal pathogen of *Portulaca oleracea* (purslane), spiny amaranth, and rape (Kenfield et al., 1989). Two active compounds were isolated and identified from organic extracts of liquid cultures of *D. indica*. They were curvulin and O-methylcurvulinic acid, both polyketides. These compounds were tested against 18 plant species; curvulin, applied at 16 nmol/leaf, affected only purslane and spiny amaranth, whereas O-methylcurvulinic acid, applied at that rate, affected those weeds plus cucumber, soybean, and *Poa annua*. It was concluded that the necrogenic activity of these compounds warrants further research concerning their potentials as herbicides.

I pointed out in Sec. III of this chapter that pelleted mycelia of *Alternaria cassiae* gave relatively good control of sicklepod (Walker and Connick, 1984). Hradil et al. (1989) isolated and identified four phytotoxins in the ethyl acetate extract of the culture filtrate of *A. cassiae*. The identified phytotoxins were stemphyperylenol, stemphyltoxin II, alterperylenol, and altertoxin I. All were tested for the formation of necrotic areas on five weed and crop species 72 hours after application of 1.0 μg of a compound to a seedling. Stemphyperylenol was very active against crabgrass and slightly active against sicklepod, corn (B73), and timothy. Stemphyltoxin II was slightly active against sicklepod, crabgrass, timothy, and soybean and a bit more active to corn. Alterperylenol was somewhat active against crabgrass and timothy and moderately active against corn and soybean. Altertoxin I was very slightly active against sicklepod and soybean, but strongly active against corn. It is noteworthy that all the identified phytotoxins had no, or very slight, herbicidal activity against any of the test plants.

Three herbicidal compounds—3-hydroxybenzyl alcohol, 2-methylhydroquinone, and (+)-epiepoformin—were isolated and identified from the culture medium of the fungus *Scopulariopis brumptii* (Huang et al., 1989). The fungus was previously isolated from the rhizosphere of a potted asparagus plant. When the identified compounds were sprayed on several weeds and soybean, at 4.48 kg/ha, all showed some herbicidal action, with epiepoformin being most active. Epiepoformin completely killed *Amaranthus retroflexus* and gave 88% control of *Sinapis alba;* 3-hydroxybenzyl alcohol gave 78% control of *A. retroflexus* and slight activity against two other weeds; and 2-methylhydroquinone had slight activity against four weeds and soybeans. The crude extract of the culture medium gave complete control of *A. retroflexus* and *Sida spinosa* and slight to moderate activity against five other species. Potential control of this fungus as a biocontrol agent is still being studied.

*Avena fatua* (wild oats) is one of the most troublesome annual grasses in wheat and other cereals in many parts of the world; infestation by that weed causes extensive crop losses (Pérez, 1990). 2,4-Dihydroxy-7 methoxy-1,4-benzoxazin-3-one (DIMBOA) is the main hydroxamic acid of wheat, and that allelochemical and its decomposition product, 6-methoxy-benzoxazolin-2-one (MBOA), inhibited root growth of wild oats by 50% at concentrations of 0.7 and 0.5 m*M*, respectively. On the other hand, MBOA stimulated root growth of cultivated oats (*Avena sativa*) at 1.5-m*M* concentration or lower. Pulse experiments with DIMBOA indicated that it decomposed to MBOA in *A. fatua* seeds within 48 hours. Pérez suggested that further work on the release, dispersal, and uptake of hydroxamic acid should be undertaken before definite proposals are made concerning the use of hydroxamic acids in weed control.

Alpha-terthienyl (α-T) is produced in several species of the Asteraceae, and it is a potent phototoxin. Lambert et al. (1991) investigated its use as a herbicide in pot and field experiments. They found that it acted as a contact herbicide on corn and broad-leaf weeds and produced necrotic lesions under solar or solar-simulating illumination. It was considerably less active as a herbicide than some commercial synthetic ones, such as atrazine. For example, α-T caused only a 50% growth inhibition of *Amaranthus retroflexus* and *Chenopodium album* at 15 kg/ha.

Bialaphos is a product of *Streptomyces viridochromogenes* that is easily metabolized and highly biodegradable (Jobidon, 1991). It has been reported to have a half-life of 20 to 30 days in soils. Jobidon therefore decided to investigate the possibility of using this natural herbicide in the control of red raspberry (*Rubus idaeus*) in forest plantations of *Picea mariana* (black spruce) in Quebec. Approximately three weeks after bialophos was applied in July, August, or September, virtually all raspberry plants were killed. The current-year foliage of black spruce was affected more than was older foliage, and applications of bialaphos in June and July did more harm to spruce foliage than did the August application. Bialaphos treatments in August with rates ranging from 1.0 to 2.0 kg/ha of the active ingredient had only slight or no adverse effects on spruce survival, foliar injury, and growth increment. Raspberry density and height were both negatively correlated with photosynthetically active radiation reaching the black spruce seedlings.

Vulgamycin is an antibiotic that was isolated from *Streptomyces hygroscopicus* and is identical to enterocin isolated from cultures of *S. candidus* var. *enterostaticus.* Babczinski et al. (1991) tested vulgamycin against three crop plants—cotton, barley, and corn—and against eight weed species—*Cassia tora, Amaranthus retroflexus, Chenopodium album, Galinsoga parviflora, Helianthus annuus, Ipomoea* sp., *Sinapis alba,* and *Setaria viridis.* The crop plants were not significantly affected up to 500 g/ha, but all the weeds were strongly affected by 250 and 500 g/ha. In fact, all but two species were completely killed by 250 g/ha.

Scacchi et al. (1992) isolated a strain of *Streptomyces* sp. SD-702 from soil, the spent broth of which was very active against dicotyledonous plants. They subsequently found that the active principles in the broth were blasticidin S (B1-S) and 5-hydroxymethyl-blasticidin S (H-M-Bl-S). Barley, rice, and garden cress were all markedly inhibited by both compounds, with garden cress being the most sensitive. The compounds were tested against three monocot weeds—*Echinochloa crusgalli, Bromus inermis,* and *Lolium multiflorum*—and against five dicot weeds—*Vigna sinensis, Stellaria media, Ipomoea purpurea, Convolvulus arvensis,* and *Veronica persica*—all at 1 kg/ha. The monocots were affected only slightly, but all the dicots were markedly inhibited by Bl-S, and all except *Stellaria media* were inhibited by 70 to 100% by H-M-Bl-S. These compounds appear to have considerable promise in weed control but need more research to determine which weeds can be controlled consistently and in which crops they are safe.

**Conclusions** The information presented in this chapter makes it very clear that the proper use of allelopathic plants and microorganisms, or of natural products produced by plants and microorganisms, could decrease the use of many expensive and often environmentally unsound synthetic herbicides. This could be accomplished fairly soon and with very little additional research, particularly with an integrated use of allelopathy and synthetic herbicides. It has been demonstrated many times by numerous investigators that preventing or slowing of weed growth for a few weeks after crop emergence is sufficient to prevent significant loss of yield in many crops. It is unwise in many cases to keep fields weed-free until harvest. The U.S.D.A. should make the implementation of this recommendation one of its high-priority goals.

CHAPTER 5

# ALLELOPATHY IN BACTERIAL AND FUNGAL DISEASES OF PLANTS

## I. ALLELOPATHY IN THE DEVELOPMENT OF PLANT PATHOGENS

SPORES OF MOST pathogenic fungi are produced in or on infected host tissue, yet these spores usually do not germinate while they remain at the site of production. One cause of this phenomenon is production by the spores of fungistatic agents that are excreted into the water around the spores (Bell, 1977). These agents usually assure dispersal of viable ungerminated spores, and some spores produce germination stimulators that counteract the germination inhibitors after the spores are disseminated.

Survival of most fungal pathogens depends on the production of resting propagules, since they generally spend relatively long periods of time away from living host plants. According to Bell (1977), there is considerable evidence that allelochemicals may stimulate the production of sclerotia and chlamydospores by the fungal mycelium.

It was pointed out by French and Gallimore (1971) that French and his colleagues had previously reported that over 30 compounds (which included long-chain fatty aldehydes, alcohols, ketones, many noncyclic and cyclic terpenes, saturated and unsaturated hydrocarbons, and isoprene) were stimulatory to germination of uredospores of *Puccinia graminis* var. *tritici.* French and Gallimore (1971) tested 43 additional compounds, of which 31 stimulated germination of uredospores of *P. graminis tritici* and 12 did not. The volatile stimulatory compounds tested included sulfur derivatives, alcohols, ketones, nitrile, hydrocarbons, aldehydes, and amines. The nonvolatile stimulatory compounds included two amides, two sulfonates, and one alcohol; twelve compounds were inactive. The lowest stimulatory concentration of the volatile liquid compounds was 0.0001 μL/2 mL water; the lowest stimulatory concentration of the nonvolatile solid compounds was 0.0031 mg/2 mL water. French and Gallimore concluded that the unique physical and chemical properties of some of the stimulators may enable them to be used in controlling spore germination and infection of the host under various environmental conditions.

French and Gallimore (1972) subsequently investigated the possibility of stimulating *P. graminis tritici* uredospore germination while still in the pustules, using *n*-nonanal, 1-nonanol, octanol, nonylamine, or 1-nonene. They first tested each of the compounds in various concentrations against several pustules of Race 56 uredospores in Conway cells for an overnight period. Ten parts per million of nonanol caused massive germination and growth of germ tubes, with the mass of germ tubes appearing like a bit of cotton. The tests were then repeated in dew chambers with intact wheat plants. A small amount of the test compound (0.1 to 0.2 mL) was placed on a dry filter-paper disc supported on a glass cylinder in the middle of the chamber. After an overnight exposure to 0.1 mL of nonanal, virtually all pustules on treated plants were cottony white from the massive germination of spores in the pustules. All of the test compounds listed at the beginning of this paragraph actively induced spore germination in the pustules. After an overnight exposure, virtually no uredospores were visible or transferable to agar from germinated pustules because of the cottony mass of germ tubes. The collapsing germ tubes may provide entrance for microbial invasion, which could impede the growth and development of uredospores inside the pustules, in addition to providing a physical barrier to spore dispersal. Such an induction of germination in the pustules may offer a new method of controlling the spread of rust in the field, if the germination stimulator could be economically delivered and then retained at the site long enough to be effective.

The compound *n*-nonanal was identified in uredospore distillates of *Puccinia graminis, P. coronata, P. sorghi,* and *P. recondita,* and *P. graminis* was chemically induced to germinate by that aldehyde (French et al., 1975). French et al. investigated the possible stimulation of uredospore germination of *P. coronata, P. sorghi,* and *P. recondita* by n-nonanal and other related compounds. *P coronata avenae* was found to induce crown rust of oats, *P. recondita* caused leaf rust of wheat, and *P. sorghi,* induced common corn rust. All the compounds tested stimulated uredospore germination of all three species (Table 6). Stimulation of *P. sorghi* uredospores was less than that for the other two species, and spore concentrations had to be higher to elicit significant stimulation. In most tests, nonanol and octanol were slightly more effective and active than the other test compounds. 1-Nonanol stimulated uredospore germination of all four test species over a temperature range of 6 to 30°C. Uredospores of *P. coronata, P. recondita,* and *P. sorghi* germinated well in the pustules when exposed to the appropriate concentration of the appropriate stimulators. No uredospores germinated in the control pustules. The demonstrated extension of chemical stimulation of germination to other species of pathogens suggests that a broad spectrum of activity for these stimulators exists in the physiology of rust-spore germination.

Overwintering spores of fungi, such as teliospores of rusts and oospores of some other fungi, usually germinate slowly, and few compounds are known to speed germination (French, 1990). French found that onion and garlic volatiles slightly stimulated germination of teliospores of *Puccinia punctiformis.* He subsequently tested various allyl and isothionate compounds known to occur in *Allium* spp., plus some related compounds, against teliospores of *P. punctiformis,* the causal pathogen of Canada thistle rust. The teliospores produce basidiospores that infect Canada thistle and cause a serious disease of that pernicious weed. For that reason, it is important to study the possibility of using that pathogen as a biocontrol agent of Canada thistle. Low germination rates of the teliospores may be a limiting factor in using that pathogen, unless some way can be found to increase the rate. French tested various concentrations of linear, saturated alkyl isocyanates and found that only the nonyl, decyl, and dodecyl isothiocyanates significantly stimulated germination of the teliospores. No activity was found in the $C_4$, $C_5$, $C_6$, and $C_7$ derivatives for up to 21 days. At 250 μL/L, octadecyl-isothiocyanate stimulated germination by 12%. Nonyl-isothiocyanate stimulated germination by 15% at 10 μL/L at 28 days, but no activity was observed above 25 μL/L. Control germination was 0%. Dodecyl-isothiocyanate increased germination to 80% at 250 μL/L at 28 days. Mixtures of $C_9$-, $C_{10}$-, and $C_{12}$- isothiocyanates in a 1:1:1 ratio were less effective than any of the compounds alone.

The $C_{9,}$ $C_{10,}$ and $C_{12}$ isothiocyanates are the first known com-

**Table 6** Effect of nonanal, nonanol, octanol, and methylheptenone on germination of uredospores of *P. coronata*, *P. recondita*, and *P. sorghi*

| Concentration (ppm) | % germination ± standard error | | | | | | | | | | | |
|---|---|---|---|---|---|---|---|---|---|---|---|---|
| | *P. coronata* | | | | *P. recondita* | | | | *P. sorghi* | | | |
| | Nonanal | Nonanol | Octanol | Methyl-heptenone | Nonanal | Nonanol | Octanol | Methyl-heptenone | Nonanal | Nonanol | Octanol | Methyl-heptenone |
| 1000 | 0 ± 0 | 0 ± 0 | 0 ± 0 | 88 ± 1.5 | 0 ± 0 | 0 ± 0 | 0 ± 0 | 4 ± 1.8 | | | 0 ± 0 | |
| 500 | 0 ± 0 | 0 ± 0 | 0 ± 0 | 94 ± 1.8 | 0 ± 0 | 1 ± 1.7 | 2 ± 1.5 | 85 ± 2.0 | | | .2 ± 0.2 | |
| 250 | 0 ± 0 | 1 ± 1.0 | 72 ± 5.6 | 95 ± 1.6 | 12 ± 2.5 | 1 ± 1.0 | 38 ± 1.6 | 97 ± 1.0 | 62 ± 2.9 | 1 ± 0.8 | 50 ± 3.9 | 47 ± 2.0 |
| 100 | 90 ± 3.5 | 70 ± 2.4 | 87 ± 1.7 | 94 ± 1.5 | 89 ± 2.2 | 84 ± 1.0 | 82 ± 4.3 | 95 ± 0.6 | 58 ± 2.4 | 41 ± 5,0 | 62 ± 4.7 | 48 ± 6.0 |
| 50 | 96 ± 0.9 | 91 ± 0.8 | 88 ± 1.6 | 71 ± 5.5 | 86 ± 1.0 | 92 ± 1.8 | 82 ± 1.3 | 96 ± 1.2 | 69 ± 4.4 | 51 ± 1.1 | 61 ± 2.0 | 53 ± 6.8 |
| 25 | 98 ± 0.4 | 83 ± 1.9 | 87 ± 2.1 | 67 ± 6.0 | 82 ± 3.3 | 94 ± 1.6 | 83 ± 3.0 | 96 ± 0.5 | 63 ± 2.0 | 45 ± 2.9 | 63 ± 7.3 | 60 ± 8.5 |
| 5 | 89 ± 3.4 | 91 ± 0.6 | 87 ± 6.0 | 9 ± 3.8 | 17 ± 4.5 | 86 ± 5.7 | 79 ± 1.7 | 51 ± 13.1 | 46 ± 7.7 | 50 ± 3.8 | 65 ± 1.6 | 32 ± 4.7 |
| 0.5 | 6 ± 0.5 | 45 ± 6.1 | 39 ± 6.1 | 2 ± 1.9 | 1 ± 0.4 | 36 ± 9.4 | 14 ± 1.5 | 12 ± 2.6 | 24 ± 2.5 | 35 ± 8.5 | 59 ± 2.8 | 27 ± 9.3 |
| 0.05 | 4 ± 2.1 | 1 ± 0.4 | 22 ± 10.1 | 1 ± 0.5 | 6 ± 1.9 | 9 ± 1.3 | 4 ± 2.0 | 31 ± 8.3 | 17 ± 3.8 | 32 ± 4.5 | 37 ± 6.5 | 23 ± 7.7 |
| 0 | 1 ± 0.6 | 3 ± 0.9 | 10 ± 2.8 | 1 ± 0.3 | 4 ± 1.2 | 1 ± 0.3 | 4 ± 1.0 | 21 ± 4.8 | 20 ± 3.0 | 19 ± 2.8 | 43 ± 2.5 | 23 ± 2.7 |

*Source:* From French et al. (1975), with permission of Amer. Chem. Soc.

pounds found to stimulate germination of teliospores of *P. punctiformes.* They are of known structure and are commercially available. It is therefore possible that they may be useful in increasing teliospore germination of Canada thistle rust in biocontrol procedures.

It is evident from the information presented above that volatile compounds play important roles in the morphogenesis of fungal pathogens of plants. Hamilton-Kemp and Andersen (1986) decided to determine what volatile compounds are produced by Arthur 71 wheat at different stages of plant development. They harvested wheat from the field at different stages and isolated volatiles by steam distillation-extraction in a water-recycling still operated at reduced pressure. The volatiles were separated further by column chromatography, and the fractions were identified by MS. The researchers identified 35 volatile compounds in Arthur 71 wheat. Prominent groups of compounds identified were the unsaturated $C_9$ aldehydes and alcohols and the $C_6$ aldehydes and alcohols. The same compounds were found in fresh and frozen wheat, but the $C_6$ compounds were isolated in amounts up to several thousandfold greater from fresh than from frozen plants of the same age. On the other hand, freezing caused an increase of two- to threefold of most of the $C_9$ volatiles, except for the 2-enols which increased more than tenfold.

A comparison of the volatile compounds in culms and leaves indicated that the $C_9$ compounds, nonanal and nonanol, were present in greater amounts in the leaves by two- to fivefold. On the other hand, there was a marked increase of 10- to 20-fold in the 3-enols in the culm. On the whole, most of the volatile compounds identified from wheat were the same compounds or were related to those found by French and his colleagues to stimulate spore germination of various fungal pathogens.

Goel and Jhooty (1987) found that teliospores of *Urocystis agropyri,* which causes flag smut of wheat, remain ungerminated in the soil until the wheat is sown in the next crop season. They decided to determine the nature of the stimulus that induces germination of those spores. They grew host plants (wheat) and nonhost plants—corn, rice, sorghum, cotton, mungbean, and peanuts—in a growth chamber. When the plants reached the selected stage, they were harvested and used to determine their effects on teliospore germination of *U. agropyri.* One test involved placing some of the plant material of a given plant at the bottom of a Von Tiegham cell; presoaked teliospores were then put in a drop of water hanging from the lid at the top. The other method involved placing the possible plant stimulus at the bottom of a desiccator of 2-L capacity, and the presoaked teliospores were mounted in a drop of water on a microscope slide placed on a perforated screen. The lids were sealed using a lanolin paste and each apparatus was incubated at 18 to 20°C for 18 hours.

Leaves, stems, roots, and seeds of wheat stimulated germination of the teliospores of *U. agropyri.* Different cultivars of wheat had similar effects, including resistant and susceptible cultivars. The nonhost plants also stimulated germination, indicating that the stimulant was nonspecific. Ethyl acetate, ethyl ether, ethanol, chloroform, acetone, carbon dioxide, and an aqueous solution of 2-chloroethyl phosphoric acid (ethephon, a source of ethylene) were tested against the teliospores; only ethephon stimulated teliospore germination. Ripening banana fruit and lemon fruit are also sources of ethylene, and they also stimulated teliospore germination of *U. agropyri.* Goel and Jhooty concluded that ethylene may play an important role in triggering the germination of teliospores of this pathogen and may thus regulate a host-parasite relationship by direct effect on the pathogen.

Two diseases of sweet potato are caused by *Sclerotium rolfsii*—sclerotial blight, which develops on sprouts and roots in plant production beds, and circular spot, which develops on storage roots just before harvest (Clark, 1989). Sweet potatoes are propagated by storing roots from the previous crop for several months and then bedding the whole roots in field beds. These roots are covered with soil, and transplants are harvested from the beds by cutting above the soil surface; the beds are maintained to allow regrowth of sprouts and later harvest of transplants. Mycelia of *S. rolfsii* usually grow profusely over the surface of the sweet potatoes in the beds and invade the plants most often near the point of emergence of the shoots. *S. rolfsii* has been shown to react dramatically to the presence of certain volatile chemicals. In the presence of the stimulatory volatiles, groups of hyphae erupt through the rind of the sclerotia; they do not require an outside food source to infect certain hosts. Clark decided to study the effects of host volatiles on the growth of *S. rolfsii* and the development of sclerotial blight. In preliminary experiments, all isolates of *S. rolfsii* responded similarly to sweet potato root tissue, so only two isolates were selected for the project, WJM-8 and 84-75.

Clark reported that *Fusarium solani*—infected storage roots of sweet potato increased eruptive germination of sclerotia of *S. rolfsii* to a greater extent than did butanol, which is a known stimulant of eruptive germination. On the other hand, healthy storage root tissue of sweet potato did not stimulate eruptive germination. It is notable that storage root tissue infected with *Erwinia chrysanthemi, Macrophomina phaseolina, Diplodia gossypina, Ceratocystis fimbriata, Plenodomus destruens,* and *Trichoderma* spp. also stimulated greater eruptive germination of sclerotia than did butanol or healthy tissue. On the other hand, hyphal growth of *S. rolfsii* was greater following germination in the presence of healthy storage roots of sweet potato and was chiefly directed toward the healthy tissue. Moreover, secondary sclerotia were initiated more frequently in control dishes and in dishes with healthy root tissue. Sclerotial germination on natural soil was increased by the presence of both healthy and either *F. solani*–infected or *Erwinia chrysanthemi*–infected storage root tissue. The frequency of infection, however, of sweet potato stem segments was increased only by the infected tissue. Knowledge of the sources and roles of the volatiles in development of sclerotial blight may be used to improve disease control. Careful selection of mother roots free of diseases such as fusarium root rot will reduce these diseases and may also reduce subsequent sclerotial blight.

Two species of cotton, *Gossypium barbadense* and *G. hirsutum,* attracted *Pythium dissotocum* zoospores to the root caps when seedling roots were placed in suspensions of the zoospores. Moreover, the roots were infected after the attraction (Goldberg et al., 1989). Roots of neither cotton species attracted *P. catenulatum* zoospores, and no roots became infected. Isolated root cap cells had exactly the same effects as did intact roots in the cotton experiments. Zoospores of *Pythium dissotocum* and *P. catenulatum* were attracted to all or some part of the seedling roots of the following species: cucumber, *Agrostis palustris,* lettuce, tomato, corn, spinach, and *Salicornia* sp. All roots were infected. No attempts were made to identify the specific attractant(s) in any case.

I previously pointed out in this chapter that probably most fungal propagules are released from the host in a dormant condition and have to be exposed to a stimulant before germination occurs. It appears now that stimulants most commonly released into the soil environment that alleviate fungistasis are components of seed and root exudates released during seed germination and root development (Nelson, 1990). The specific soluble molecules involved in stimulating propagule germination of soil-borne fungal pathogens are largely unknown but are thought by many researchers to involve primarily sugars and amino acids. Some resting spores respond only to $C_{16}$-$C_{18}$ fatty acids, and many respond chiefly to volatile stimulants, as we have already discussed. Ethanol is the chief volatile stimulant evolved from germinating cotton seeds and, according to Nelson, is active at nanomolar to subnanomolar concentrations.

Sporangia of *Pythium ultimum* produced on common synthetic media behave very differently in their response to sugars and amino acids than do those produced on diseased tissue (Nelson, 1990). By leaching soil columns with water or less polar solvents such as methanol, ethanol, or acetone, soluble exudate stimulants can be extracted from nonsterilized soil in which cotton seeds were germinated. It seems that aqueous-soil leachates yield the highest level of stimulatory activity when seed exudates are extracted 48 hours after planting. Interference with the production and activity of exudate stimulants is a promising approach to inhibiting pathogen activity and achieving biological control.

Plant-microbe interactions must ultimately reach a stage in which the microbe detects a susceptible host. Chemical signals are extremely important in a great many allelopathic actions and reactions, but unfortunately most researchers in allelopathy have completely overlooked these signals. *Agrobacterium tumefaciens* causes crown gall disease in a wide variety of dicotyledonous plants, yet the importance of chemical signals in the development of this disease has been overlooked until relatively recently. The bacterium causes tumors on its host by passing transfer-DNA into the host plant. The t-DNA includes genes that encode enzymes of auxin and cytokinin biosynthesis, and these growth regulators transform the host plant cell (Spencer and Towers, 1988). Detection of susceptible host cells and early stages of tumor formation are chiefly controlled by a set of virulence (*vir*) genes. Stachel et al. (1985) identified two active signal compounds, acetosyringone and hydroxyacetosyringone, from transformed tobacco root cultures. They suggested that these signals enable *A. tumefaciens* to recognize susceptible tissue and to initiate transformation of the plant tissue. According to Stachel et al., these two signal compounds had not been previously identified as natural components of plant cells, indicating that they are not widespread. The researchers' overall conclusion was that acetosyringone and hydroxyacetosyringone are specifically synthesized and exuded by metabolically active, wounded cells.

Spencer and Towers (1988) questioned the conclusions of Stachel et al. concerning the widespread role of acetosyringone and hydroxyacetosyringone in the attraction of *A. tumefaciens* and the induction of its virulence genes. Spencer and Towers tested the effects of some cinnamic acid derivatives, chalcones, and lignin precursors (sinapyl alcohol and coniferyl alcohol) as inducers of the virulence genes in *A. tumefaciens.* All had at least slight inducing effects and most had considerable effects. Sinapyl alcohol, coniferyl alcohol, syringaldehyde, methylsyringate, methylferulate, and 5-OH-methylferulate had almost as much activity as did acetosyringone. It was concluded that two basic structural features are generally required to confer activity: guaiacyl or syringyl substitution on a benzene ring and, with the exception of the monolignols, a carbonyl group on a substituent para to the hydroxy substituent on the ring. In light of the many dicotyledonous plants that serve as hosts of the crown gall pathogens, these conclusions seem more logical than those of Stachel et al.

Until 1990, acetosyringone and hydroxyacetosyringone remained the only *A. tumefaciens* virulence-gene-inducing compounds identified from host plant tissues (Spencer et al., 1990). In that year, Spencer et al. reported the isolation of a *vir*-inducing phenolic compound from several *Vitis* cultivars. This compound was identified as a syringic acid, methyl ester.

Anthracnose caused by *Colletotrichum gloeosporioides* is the most widespread and serious disease of *Stylosanthes guyanensis* (or *S. guianensis*). Isolates of *C. gloeosporioides,* obtained from the same anthracnose lesions on leaves of *S. guyanensis,* showed considerable variation in pathogenicity, and conidia of weakly pathogenic isolates germinated more rapidly and to a higher percentage on detached leaves of *Stylosanthes* (CIAT 136) than did pathogenic isolates (Lenné and Brown, 1991). These scientists also found that conidia of pathogenic isolates germinated poorly on the relatively resistant CIAT 136 *Stylosanthes* in comparison with germination on the more susceptible CIAT 1283. Moreover, leachates from leaves of CIAT 136 suppressed germination of all four test isolates of *C. gloeosporioides.* In subsequent work, leaf leachates from CIAT 136 suppressed germination of all four isolates of *C. gloeosporioides.* Leachates from leaves of CIAT 1283 significantly enhanced germination of conidia of the pathogenic isolates on leaves of both accessions. On the other hand, germination of conidia of the weakly pathogenic isolates in leaf leachate from CIAT 1283 was similar to germination in water.

Lenné and Brown isolated seven species of bacteria from leaves of *S. guyanensis* and several species each of *Bacillus* and *Pseudomonas* that were not identified. Of these, two isolates of *Bacillus subtilis,* two nonfluorescent *Pseudomonas* spp., *Pseudomonas stutzeri,* and *Alcaligenes faecalis* inhibited radial growth of colonies of two pathogenic isolates of *C. gloeosporioides* to a significantly greater extent than they did the growth of two weakly pathogenic isolates. According to Lenné and Brown, the antagonistic activity of these bacteria appeared to be due to antibiotic production.

Much of the literature on volatile stimulants of spore germination for 43 species of fungi was reviewed by French (1992). He discussed some of the physical and chemical properties of the volatiles, sources of information, germination in pustules, natural occurrence, mode of action, and potential usefulness.

*Pythium ultimum* is a pathogenic fungus that causes preemergence and postemergence damping-off and root rot of numerous crop plants. Several crops have been protected against *Pythium* damping-off by seed treatment with fluorescent pseudomonads, so Paulitz (1991) decided to investigate the mechanism by which the protection occurred. A suspension of *Pseudomonas putida* (NIR) was adjusted to approximately $10^9$ cells per milliliter in sterile distilled water and mixed with a solution of pelgel as a sticker. Laxton Progress peas and Maple Arrow soybeans were inoculated with the *P. putida* suspension before being placed in moist, sterile sand. Some of the seeds of each were placed in separate petri dishes. A filter paper was moistened with sterile distilled water and placed in each petri dish lid, after which a soil-sporangia inoculum of *Pythium ultimum* was added to the dishes. Three treatments were used: no seeds in the petri dish, seeds treated with pelgel only, and seeds treated with *P. putida* plus pelgel. After 24 hours at 26°C, the soil inoculum was removed from the lid of each petri dish and the total hyphal length was determined. Tests were also run for the production of volatile antifungal compounds by *P. putida* and for effects of ethanol and acetaldehyde on the stimulation of growth of *Pythium ultimum.*

When an inoculum of *P. ultimum* was exposed to volatiles from germinating seeds of peas or soybeans for 24 hours, an extensive network of hyphae grew out of the soil. The presence of *Pseudomonas putida* significantly reduced the hyphal growth of *P. ultimum* exposed to volatiles from the germinating seeds. Hyphal growth was also stimulated by 7.6 μL of ethanol or by 9.5 μL of acetaldehyde. Emergence of soybeans treated with *P. putida* and planted in infested soil was similar to that of seeds planted in noninfested soil. When pea seeds were treated with *P. putida,* significantly less ethanol was detected at 18 hours compared with the nontreated soil. Acetaldehyde was significantly lower in *P. putida*–treated soil at all sampling times and was below the detection limit in many samples. *P. putida,* in a culture with ethanol as the sole carbon source, completely metabolized 15 to 500 μL of ethanol in 48 hours. Paulitz concluded that *Pseudomonas putida* (NIR) may reduce damping-off by *Pythium ultimum* by using volatile exudates from germinating pea and soybean seed. He concluded also that the volatile seed exudates may serve both inductive and nutritional functions in the pathogenesis of damping-off caused by *P. ultimum.*

Marine phycomycetes are common and widespread pathogens of microscopic and macroscopic algae that periodically cause significant damage to crops such as *Porphyra* (Kerwin et al., 1992). Species of *Pythium* have been most often reported as infecting the red algae, but this may be due to the economic importance of these algae. Kerwin et al. isolated 30 strains of *Pythium marinum* from infected blades of *Porphyra nereocystis* collected in Friday Harbor, Washington. All strains of *P. marinum* isolated from *P. nereocystis* readily entered sexual and asexual reproductive cycles. The C-8 strain was chosen for further investigations because it had a slightly greater tendency to enter the sexual cycle than did other strains. Vegetative growth was found to be significantly greater at 12, 20, and 25°C than at 4°C, with maximum mass occurring at two to three days in shake culture, when inoculated with 1 part mature culture to 100 parts culture medium. Zoospores were readily induced in *Pythium* following dilution of the culture with seawater. When *Porphyra* discs were exposed to *Pythium* zoospores, encystment on *Porphyra perforata* was greater than on *P. nereocystis.* For a given zoospore density, encystment increased progressively from 4 to 20°C and from 5- to 60-minute exposure. Zoospores encysted on thalli of red algae other than *Porphyra,* but no encystment occurred on brown or green algae.

The initial evidence supported the existence of chemical signals that induced zoospore encystment on red algae cuticles. Heat and base hydrolysis stability and acid lability of the encystment signal implicated algal cuticle carbohydrates. Many carbohydrates were tested, but only agars that were a complex mixture of methylated and/or sulfated galactans derived from several red algae consistently induced encystment or attachment of *Pythium marinum* zoospores. Sequential extraction of carbohydrates from *Porphyra perforata* supported the existence of separate signals for zoospore encystment and appressorium induction. Amphotericin B induced 100% encystment in less than four minutes when added to zoospore suspensions at 3 μg/mL. Because amphotericin B binds to sterols, this test suggested the existence of sterols at or near the surfaces of zoospore membranes. At 20°C oospores with well-defined cell walls appeared three days following exposure to zoospores, whereas a comparable stage of development in oospores did not occur for six weeks in *Porphyra* discs held at 4°C.

Seven volatile organic compounds known to be produced by baker's yeast (*Saccharomyces cerevisiae*)—acetaldehyde, ethylacetate, ethanol, n-propanol, isobutanol, 2-methylbutanol, and 3-methylbutanol—were tested singly and in various combinations against spore germination of five test pathogens: *Helminthosporium oryzae, Cercospora personata, Cunninghamella blakesleeana, Colletotrichum capsici,* and *Alternaria solani* (Saksena and Tripathi, 1987). When they were tested singly, all the compounds inhibited spore germination of all species except *C. blakesleeana.* Only ethanol and ethylacetate inhibited spore germination of that species. All combinations tested also inhibited spore germination of all species except *C. blakesleeana.* In that case, only four combinations inhibited spore germination. Saksena and Tripathi suggested the possibility of controlling the prevalence of the tested pathogens in the soil by yeast. It seems rather strange that these compounds were inhibitory to spore germination of all the tested pathogens when, in many other tests, some of them have been shown to be stimulatory to spore germination of so many other fungal pathogens. I suspect the quantity (2 mL) of each pure compound used in the tests by Saksena and Tripathi may have been so high that it was in the inhibitory range, that is, above the stimulatory concentration.

The most important residue-associated pathogen of wheat in Kansas and other Great Plains states is *Pyrenophora tritici-repentis,* an ascomycete. Pfender (1988) noticed that *P. tritici-repentis* was sometimes found at much reduced levels in the presence of secondary saprophytes. *Limonomyces roseipellis,* a basidiomycete, was occasionally isolated from ascocarps of *P. tritici-repentis* in the field. Pfender felt that *L. roseipellis* might be a good candidate for biological control of the pathogen, so he tested the ability of that basidiomycete to suppress ascocarp formation of *P. tritici-repentis* in the laboratory by using nonsterile straw precolonized during normal growth of the pathogen. Some pieces of wheat (TAM 105) infected with *Pyrenophora* were inoculated with *L. roseipellis* and some were not. Two environmental conditions were used for incubation, and the number of ascocarps produced in the presence of *L. roseipellis* in all cases was less than in the respective control. Overall, *L. roseipellis* reduced total spore production more than ascocarp numbers. It is notable that these results occurred even though *P. tritici-repentis* was firmly established on the wheat straw at the time of inoculation with the basidiomycete.

*Phymatotrichum omnivorum* is the causal pathogen of phymatotrichum root rot, which infects over 2000 dicotyledonous species (Zuberer et al., 1988). This fungus occurs commonly in alkaline soils, where it can persist for many years at a depth generally of 30 to 45 cm. Zuberer et al. produced sclerotia in the laboratory and buried relatively large numbers of them in three Houston black-clay soils at a 45-cm depth. One area was prairie, one a continuous cotton field, and one was in a rotation of cotton, corn, and sorghum. Some of the sclerotia were dug up monthly over a 12-month period and analyzed for bacteria and actinomycetes. Numbers associated with sclerotia ranged from $10^6$ to $10^9$ cells per gram of sclerotia plus adherent soil, which was two to three orders of magnitude greater than numbers from soil at the same depth but free of sclerotia. Up to 26% of the isolates from the sclerotia inhibited growth of *P. omnivorum.* No marked seasonal variation in microbial numbers was noted during the investigation. It is noteworthy that no appreciable differences in numbers of positive isolates occurred in the various sites. The overall results suggested that the sclerotial associates of *P. omnivorum* may help limit the occurrence or spread of that fungus.

I previously discussed numerous research projects in which volatile compounds were demonstrated to stimulate the germination of spores of certain pathogenic fungi. In higher concentrations, several of the volatiles were found to be inhibitory to germination of the same spores. Moreover, I have previously pointed out that spores of many fungal pathogens often fail to germinate when they are in contact with their parent cultures. McKee and Robinson (1988) maintained that, to the time of their publication, sporostasis was linked with the production of volatile metabolites only by indirect evidence. Even though their statement could be strongly contested, they carried out a careful project, which did provide more direct evidence to link sporostasis in *Geotrichum candidum* with the self-production of volatile signals. They found that when agar discs containing arthrospores of *G. candidum* were placed in petri dishes containing uninoculated agar, there was excellent arthrospore germination, whereas similar discs with arthrospores on them showed no germination when they were placed in dishes with living mycelia of *G. candidum.*

*Pyronema domesticum* does not produce apothecia nor grow well in the presence of other fungi, and this may be the reason it often forms apothecia at the site of a recent fire or on sterilized media. Because of its rather unusual responses to various environmental stimuli, Moore-Landecker (1988) decided to determine what effects volatiles emanating from various microorganisms and substrates would elicit. Moore-Landecker and Shropshire (1984) found that volatiles produced by that pathogen prevent apothecial formation by it, but adsorption by charcoal eliminates that inhibition. Volatiles from *Chaetomium globosum, Trichoderma,* and *Trichophaea abundans* had little or no effect on apothecial numbers or development. Numbers of apothecia were strongly reduced by volatiles from *Cyathus stercoreus* and *Poronia aedipus.* Volatiles from *Dipodascopsis uninucleatus* depressed development of ascogenous hyphae. Volatile compounds from *Rhizopus nigricans, Schizophyllum commune,* and *Sordaria fimicola* also prevented apothecial formation in *P. domesticum.*

The four bacteria tested—*Bacillus subtilis, Micrococcus luteus, Proteus vulgaris,* and *Streptomyces griseus*—all interfered with the development of the ascogenous system in *P. domesticum.* Volatiles from germinating pea, bean, and corn seed also prevent production of apothecia by *P. domesticum.* Apothecia do not form on *P. domesticum* when exposed to soil or wood containing their normal microbial flora. These findings make it very clear why *P. domesticum* invades primarily after burns and other phenomena that cause an area to become relatively bare.

*Talaromyces flavus* is a potential biocontrol agent of *Verticillium dahliae* and *Sclerotinia* spp. It is known to produce an antibioticlike compound that kills microsclerotia of *V. dahliae* (Kim et al., 1988). These facts caused Kim and his colleagues to determine what the antibioticlike compound was and how it brought about the death of the microsclerotia. They were able to isolate the active factor by HPLC and identified it as glucose oxidase, which acted in conjunction with glucose to produce hydrogen peroxide ($H_2O_2$), which inhibited germination of the microsclerotia of *V. dahliae* in a concentration of approximately 12 μg/mL.

## II. ALLELOPATHY AND THE PROMOTION OF INFECTIONS

As early as 1948, Cochrane postulated that some root rots are induced by toxic action of plant residues. Patrick et al. (1963) observed in field studies in the Salinas Valley in California that sunken or discolored lesions were often present on roots of lettuce and spinach plants when their roots grew in contact with or close to fragments of plant residues. No known primary pathogen was consistently obtained when isolations were made from the lesions; usually, only common soil saprophytes were found. The researchers therefore suggested that toxins produced in the decomposing residues conditioned the roots to invasion by various low-grade pathogens.

Patrick and Koch (1963) found that the extent and severity of black root rot of tobacco caused by *Thielaviopsis basicola* were much greater when tobacco roots were exposed to allelopathic extracts of decomposing plant residues before inoculation. Additionally, those scientists found that, after treating the roots with the allelopathic extracts, the pathogen was equally damaging to susceptible and resistant varieties of tobacco.

Toussoun and Patrick (1963) reported that the occurrence of root rot of bean caused by *Fusarium solani* f. *phaseoli* was increased greatly if the bean roots came in contact with the allelopathic extracts of decomposing plant residues prior to being inoculated with the pathogen. Thus, for several years, many plant pathologists have believed that allelochemicals play causal roles in at least some plant diseases.

Kochhar et al. (1980) demonstrated that fescue (*Festuca arundinacea*) leaf leachate inhibited growth of white clover (*Trifolium repens*) seedlings. Subsequently, Kochhar et al. (1982) experimented with effects of multiple factors on growth of white clover seedlings. They found that inoculation of clover with *Rhizoctonia solani* reduced seedling growth, and clover treated with *R.solani* and ozone ($O_3$) had a greater reduction in biomass when stressed by leachate from fescue leaves. This demonstrated again the effect of allelochemicals in the promotion of infections.

Both allelopathy and infections of asparagus by one or more pathogens (*Fusarium oxysporum* f. *asparagi* and *F. moniliforme*) have been implicated in the asparagus decline and replant problem. Peirce and Colby (1987) reported that asparagus root leachate depressed asparagus seedling emergence in a sterile vermiculite medium. In a medium inoculated with *Fusarium oxysporum* f. *asparagi,* the effect was magnified. When results were compared with a sucrose solution with the same concentration of soluble solids as present in the asparagus root leachate, it appeared that the reduced emergence of asparagus seedlings may have been due in part to enhancement of *F. oxysporum* growth by the energy source in the medium. However, when the seeds were germinated in the asparagus root leachate until radicle emergence and then rinsed in water and transferred to *F. oxysporum*—inoculated medium, emergence was still reduced in comparison with the controls. Thus depression of emergence was apparently related both to an allelochemical predisposing young radicles and/or hypocotyls to increased *F. oxysporum* infection and to stimulation of *F. oxysporum* in the rhizosphere by the soluble-solids content of the root exudate. These experiments indicated again that allelopathy can promote infection of seedlings.

## III. CROP ROTATION AND PLANT DISEASES

I previously discussed the importance of the rotation of crops in preventing autoallelopathic effects and also in preventing allelopathic effects of certain crops on other crops (Chap. 1, Sec. II). Many farmers have known for a great many years that crop rotation can help control certain plant diseases. Unfortunately, there have been many pressures on U.S. farmers in the past 40 years or so to plant the same crop year after year. Additionally, there has been very little sponsored research done concerning the importance of rotation in preventing plant diseases and other agricultural problems. In many ways, the U.S.D.A. has failed to be the innovator in agricultural research, supposedly one of its important missions.

The chinampa agricultural system, developed in the Valley of

Mexico by the Aztec Indians, has been maintained for about 2000 years (Lumsden et al., 1987). In that system farmers employ methods that balance productivity with sustained long-term production. The chinampas were constructed in swamps and shallow lakes by heaping soil into ridges, thus leaving canals that drained the fields and also provided water for crops in drier times. The fields are maintained by incorporating large amounts of plant residues, animal manures, and sediment from the canals into the soil. The crops appeared to be relatively free of soil-borne plant diseases compared with crops in nearby agricultural areas employing more "modern" techniques, including the use of herbicides and fungicides. Large numbers of crop plants are intermingled each year in the chinampa and are varied from year to year.

Lumsden et al. (1987) decided to investigate comparable soils cultivated under the two systems, to determine the physical and biological properties of those soils in relation to damping-off diseases caused by *Pythium* spp. The incidence of these diseases caused by natural populations of *Pythium spp.* was found to be much lower in the chinampa than in soils under the modern system of cultivation (Table 7). Moreover, the populations of *Pythium aphanidermatum* and *Rhizoctonia solani* were significantly lower in the chinampa than in the Chapingo system, a modern system of cultivation. Disease caused by an introduced inoculum of *P. aphanidermatum* was suppressed in the chinampa soils. Moreover, the biological factors associated with disease suppression in the traditional system were found to be an increased soil dehydrogenase activity, a reduced rate of germination of *P. aphanidermatum* oospores, and higher populations of *Pseudomonas* spp. and saprophytic *Fusarium* spp. Physicochemical factors generally associated with suppression of damping-off included lower pH and higher concentrations of Ca, K, Mg, and organic matter.

The Chinese have become more interested in recent years in allelopathy and crop rotation (Li, 1988). They are now attempting to grow more cotton and are having a high incidence of fusarium wilt. Li decided to determine if soil from a peppermint (*Mentha haplocalyx* var. *piperascens*) field would suppress the wilt. He found that the peppermint soil decreased cotton *Fusarium* wilt, from 68 to 82% in the control to 50 to 65% in the test. When severely infested soils were treated with 0.5 to 1% peppermint shatter or with 0.025 to 0.05% peppermint oil, the incidence of the disease decreased by 44 to 74%.

**Table 7** Populations of *Pythium aphanidermatum, P. ultimum* and related species, and *Rhizoctonia solani* in modern Chapingo agricultural soil and traditional chinampa agricultural soil, and the incidence of damping-off of radish and cucumber seedlings

| Soil | *P. aphanidermatum** (propagules/$g^{-1}$)† | *P. ultimum* and *Pythium* spp.* (propagules/$g^{-1}$)† | R. solani *colonization**‡ % | *% damping-off** Radish (20°C) | Cucumber (30°C) |
|---|---|---|---|---|---|
| Chapingo | 8.1 a | 254 b | 6.2 a | 56.7 a | 33.3 a |
| Chinampa | 0.0 b | 3530 a | 5.3 a | 19.2 b | 14.2 b |

* Values in each column followed by the same letter are not significantly different at $P = 0.05$, according to Duncan's multiple-range test.
† Numbers estimated on five batches of soil with use of selective media for *P. aphanidermatum* and *P. ultimum*.
‡ Percentage colonization (average of three batches of soil) of beet seed added to soil, incubated three days, recovered and plated on water agar.

*Source:* From Lumsden et al. (1987), with kind permission from Pergamon Press Ltd.

Fumigation of *Fusarium* spores in a petri plate with a 2000 to 10,000-ppm solution of peppermint oil completely inhibited spore germination. Li concluded that peppermint-cotton rotation would probably control cotton fusarium wilt, or at least markedly decrease its incidence. Obviously, considerably more research needs to be done on this problem.

The basal stem rot disease of several vegetable crops caused by *Sclerotium rolfsii* alone virtually prevents vegetable production in the savanna zone of Nigeria (Wokocha, 1988). A suitable rotation of crops is known to diminish the populations of soil-borne plant pathogens to relatively harmless levels in some situations. Thus Wokocha investigated the effects of several rotation systems on the populations of viable sclerotia of *S. rolfsii* in Nigeria. After five years of each rotation, soil samples were collected from six field and counts of sclerotia were made. Two of the fields had significantly higher sclerotial counts than did the other four fields. One field with a high count had been monocropped with tomatoes for three years prior to sampling, whereas the other field with a high count, had tomatoes following peanuts in the two years before the count. Numbers of sclerotia were lowest in fields in which susceptible crops followed corn in the two years before sampling or in which corn or sorghum was the last crop prior to sampling. This experiment demonstrated that monocropping with tomatoes or planting tomatoes after peanuts may result in a buildup of the sclerotia of *S. rolfsii* in fields in the Nigerian savanna.

**Conclusion** It is obvious that the number of papers reviewed here concerning the effects of crop rotations on the incidence of diseases is rather small, but the positive results and the logic behind such experiments suggest that this is a very fertile field for further research.

CHAPTER 6

# ALLELOPATHY IN THE BIOLOGICAL CONTROL OF PLANT DISEASES: HOST PLANTS A–M

## I. CONTROL OF PLANT PATHOGENS BY PLANTS

WHILE SCREENING LEAVES of 150 plant species for fungitoxicity against *Drechslera oryzae,* which causes the leaf-spot disease of rice, Chaturvedi et al. (1987) found that *Adenocalymma allicea* (Bignoniaceae) had by far the greatest activity. The active fraction was an essential oil that was yellow in color, had a strong garliclike odor, and according to gas-liquid chromatography, was a mixture of 10 major and 35 minor components. The minimum inhibitory concentration of the oil was 500 ppm, a concentration that killed the pathogen in 30 minutes. The toxicity of the oil was thermostable up to 100°C, the maximum temperature tested, and was still active up to 21 days. Several tests of the oil were compared against some synthetic fungicides; the oil was found to be ten times more active than Blitox-50 and Karathane and four times more active than Dithane M-45, Ceresan, and Hinosan-50. The oil did not harm rice seedlings, even at 1000 ppm. The oil was also very fungitoxic to many fungal pathogens other than *Drechslera oryzae* (Table 8). In fact, it completely inhibited mycelial growth of all the tested fungi at 1000 ppm and mycelial growth in about half of them at 500 ppm. In preliminary field trials, the oil was found effective in the control of the leaf-spot disease of rice, thus suggesting its possible use as a fungicide under field conditions.

Because of the serious side effects of many synthetic chemicals and because of their costs, developing countries have become very

**Table 8** Fungitoxic spectrum of oil extracted from leaves of *Adenocalymma allicea*

| Test fungi | % inhibition of mycelial growth at | | |
|---|---|---|---|
| | 500 ppm | 700 ppm | 1000 ppm |
| *Alternaria raphani* Groves & Skolko | 100 | 100 | 100 |
| *A. solani* (Ellis & Mart) Jones & Grout | 100 | 100 | 100 |
| *Aspergillus flavipes* Thom. & Church | 45 | 65 | 100 |
| *A. flavus* Link. | 35 | 45 | 100 |
| *A. fumigatus* Fres. | 40 | 50 | 100 |
| *A. japonicus* Saito | 65 | 75 | 100 |
| *A. niger* van Tiegh. | 30 | 60 | 100 |
| *A. terreus* Thom. | 41 | 78 | 100 |
| *A. versicolor* (Vuill.) Tiraboschi | 46 | 81 | 100 |
| *Cephalosporium sacchari* Butler | 100 | 100 | 100 |
| *Cladosporium cladosporioides* (Fres.) de Vries | 100 | 100 | 100 |
| *Colletotrichum lindemuthianum* (Sacc. & Magn.) Bor. & Cav. | 100 | 100 | 100 |
| *C. tinosporae* Sydow | 100 | 100 | 100 |
| *Fusarium moniloforme* Scheldon | 100 | 100 | 100 |
| *F. solani* (Mart.) Sacc. | 100 | 100 | 100 |
| *Macrophomina phaseolina* (Tassi) Good. | 100 | 100 | 100 |
| *Paecilomyces varioti* Bainl. | 100 | 100 | 100 |
| *Penicillium decumbens* Thom. | 39 | 100 | 100 |
| *P. viridicatum* Westling | 40 | 100 | 100 |
| *Rhizoctonia solani* Kuhn | 100 | 100 | 100 |
| *Telaromyces* sp. Benjamin | 100 | 100 | 100 |

*Source:* From Chaturvedi et al. (1987), with permission of Butterworth Heinemann Ltd.

interested in finding natural fungicides to control plant pathogenic fungi. Many medicinal plants have been found to possess remarkable fungicidal and bactericidal properties, so Ram (1989) tested leaves, roots, seeds, or bark of 26 plant species against *Sclerotium rolfsii,* the causal agent of chickpea collar rot. Generally, he added 2 and 5% of the selected part or parts to 1 kg of unsterilized sandy loam soil from the Banares Hindu University farm in which *S. rolfsii* had been allowed to grow for six days under good conditions. In the tests of effectivity of the natural fungicides, 10 chickpea seeds were planted in each pot of soil containing a selected natural fungicide. Observations were recorded of the preemergence infection percentage at the end of 14 days and of the postemergence collar rot percentage after 30 and 45 days. Data were statistically analyzed and most treatments were found to be significant, at least at the 0.05 level. After considering the efficacy, availability, and cost of the different treatments, Ram listed the plants in the following order (most to least effective): *Vitex negundo, Cucuma amada, Hydnocarpus laurifolia, Hemidesmus indicus, Ferula assa-foetida, Saussurea lappa*, and *Cassia tora. Aegle marmelos* was found to be effective in controlling *Sclerotinia sclerotiorum*, which causes stem rot of chickpea.

There have been some reports in the literature stating that aqueous extracts of garlic (*Allium sativum*) bulbs and leaves are inhibitory to some plant pathogenic fungi. Singh et al. (1990) decided to test the compound ajoene from garlic bulbs to see what effects it would have on spore germination of some plant pathogenic fungi. Ajoene was dissolved in 87% glycerol (20 mg/mL) and diluted further with sterile distilled water to give 25, 50, 75, and 100 μg/mL. Each concentration of ajoene was tested against germination of 200 conidia of *Alternaria* sp., *A. solani, A. tenuissima, A. triticina, Curvularia* sp., *Colletotrichum* sp., *Fusarium lini, F. oxysporum, F. semitectum,* and *F. udum.* Appropriate controls without ajoene were also tested. Most of the species of *Fusarium* were completely inhibited by relatively low concentrations of ajoene, but *F. lini* was relatively resistant. *A. solani* was the most resistant of the species tested.

Stem and leaf blight of pigeon pea (*Cajanus cajan*) in India caused by *Phytophthora drechsleri* f. sp. *cajani* is difficult to control under field conditions because the pathogen is naturally resistant to metalaxl, the most common fungicide used in India to control diseases caused by *Phytophthora* (Singh et al., 1992). Singh et al. therefore tested ajoene to determine whether it would be effective. Ajoene inhibited mycelial growth of *P. drechsleri* f. sp. *cajani* at the higher concentrations tested (15 to 25 ppm). Considerably smaller concentrations (1 to 5 ppm) inhibited sporangial formation and sporangial germination. On the other hand, higher concentrations (15 to 20 ppm) of ajoene were required to inhibit zoospore germination and the number of germ-tube branches produced per zoospore. It appears therefore, that ajoene from garlic bulbs may be a successful chemical for the control of *Phytophthora* leaf blight of pigeon pea in India.

Wang et al. (1990) identified two antifungal compounds in a dichloromethane extract of the whole plant of *Artemisia borealis.* The compounds were the polyacetylenes, heptadeca-1,9 (Z), 16-trien-4, 6-diyn-3,8-diol, and heptadeca-1,9(Z), 16 trien-4,6-diyn-3-ol. Only 1.25 μg of each compound inhibited growth of the fungal pathogen *Cladosporium cucumerinum.*

Even though India is the largest producer of groundnuts (peanuts) in the world, the average yield is low because of diseases (Ghewande, 1989). Late leaf spot (*Phaeoisariopsis personata*) and rust (*Puccinia arachidis*) together cause 70% loss in pod yield. Ghewande decided to experiment with low-cost protection technology that would be within reach of small and marginal farmers. He carried out an experiment during two rainy growing seasons with Gang I Spanish peanuts, a variety highly susceptible to both late leaf spot and rust. Treatments used were 2% aqueous leaf extracts of neem (*Azadirachta indica*), mehandi (*Lawsonia inermis*), Mexican daisy (*Tridax procumbens*), and karanj (*Pongamia pinnata*); the chemicals used were carbendazim 0.05% + mancozeb 0.2% and NCP-75 0.3%. Each of these treatments was sprayed on separate plots three times at two week intervals. Each of the treatments controlled both the diseases and caused higher yields than given by the controls. Carbendazim plus mancozeb gave the maximum control of both diseases, but among plant species, neem and mehandi gave the best control and showed good prospects for use by the small farmers.

*Eucalyptus* sp. (blue gum) and neem leaves have both been reported to contain antimicrobial allelochemicals, so Singh and Dwivedi (1990) tested the volatile and nonvolatile fractions of the leaf distillates of the two species, plus the volatile and nonvolatile fractions of neem oil against germination of the sclerotia of *Sclerotium rolfsii*, a fungal pathogen that causes foot rot in barley. The volatile fractions of blue gum leaf distillate reduced sclerotia germination by 60%, and the nonvolatile fractions reduced it by 40%. On the other hand, the volatile fractions of neem leaves reduced sclerotia germination by 40%, and the nonvolatile fractions reduced it by 10%. Neem oil volatiles reduced sclerotia germination by 55%, the nonvolatiles by 60%. Viability of the sclerotia after treatment with the various distillates for five days was as follows: blue gum leaf distillate, 20%; neem leaf distillate, 55%, and neem oil, 8%. It appears that blue gum distillate and neem oil have very good prospects in the control of *Sclerotium rolfsii.*

The cucurbitacins are tetracyclic triterpenoids that are present in many members of the Cucurbitaceae, and cucurbitacins I and D have been shown relatively recently to inhibit induction of extracellular laccase formation by *Botrytis cinerea,* which causes a type of fruit rot of cucumber. Bar-Nun and Mayer (1990) hypothesized that laccase formation is part of the infective process by *Botrytis* and that prevention of laccase production might afford protection of the host against the invading fungus. They tested this hypothesis by applying extracts of *Ecballium elaterium* (squirting cucumber), a ready source of cucurbitacins, or authentic cucurbitacin I to cucumber fruit or cabbage leaves prior to, or during, inoculation with *Botrytis.* This treatment prevented infection of the tissue, the infecting fungus being restricted to the site of infection (Fig. 5). The results using the partially purified extract and the pure cucurbitacin were indistinguishable. Similar tests were conducted using the inner leaves of white cabbage, and again the results were similar. It was concluded from the results of all experiments that the ability of cucurbitacin I to inhibit induction of laccase by *Botrytis* was responsible for its protective effect.

In preliminary tests, Salama et al. (1988) determined the effects of water extracts of leaves of *Eucalyptus rostrata* on germination of sclerotia of *Sclerotium cepivorum,* which causes the white rot disease of onion in Egypt. They found that soaking 30-day-old sclerotia for 24 hours in even a 1:15 solution (w/w, leaf powder/distilled $H_2O$) completely prevented germination of the sclerotia. Addition of water extracts of *E. rostrata* leaves to dishes containing *S. cepivorum* slowed growth considerably, decreased the number of sclerotia formed, but did not markedly affect the sizes of the sclerotia. Addition of *E. rostrata* leaves to sandy-clay soil (1:2, w/w) previously inoculated with *S. cepivorum* and planted in onions almost completely prevented infection of the onion plants for 12 weeks.

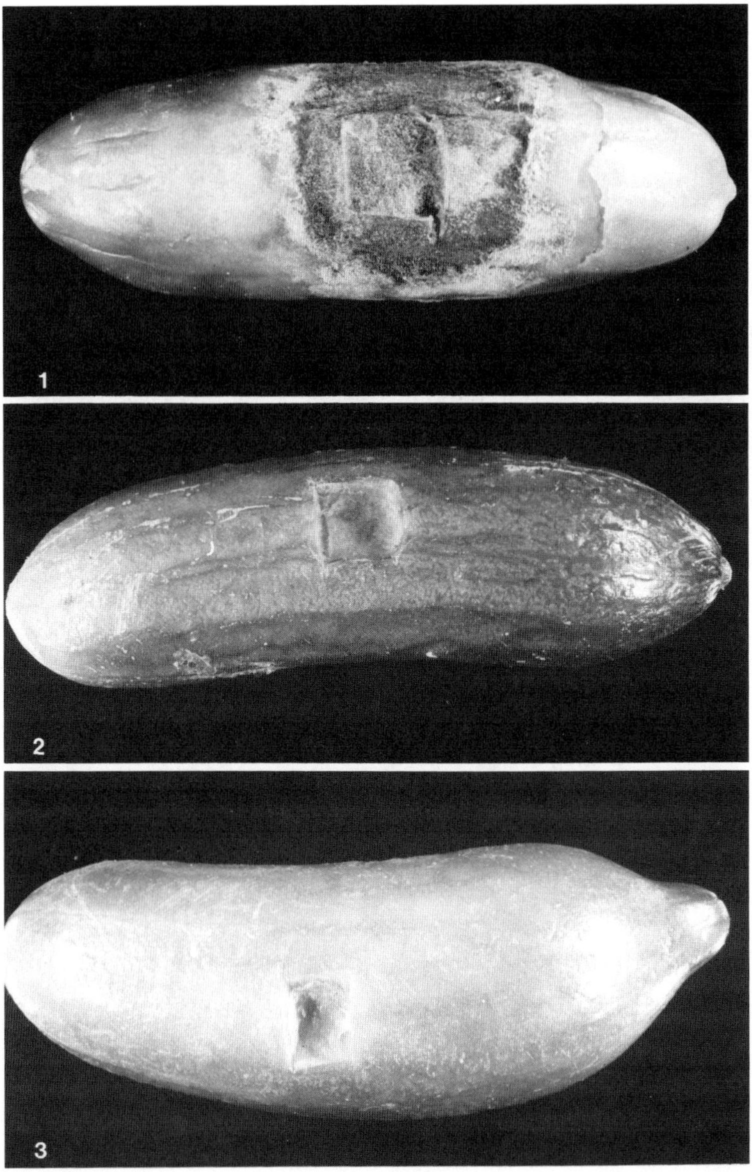


FIGURE 5 (1) Cucumber fruit infected with *Botrytis cinerea*. Photograph taken 120 hours after inoculation. Note spread of fungus. (2) Cucumber fruit pretreated with *Ecballium* extract prior to inoculation. Photograph taken 120 hours after inoculation. Note absence of fungal infection. (3) Cucumber fruit pretreated with cucurbitacin prior to inoculation. Photograph taken 120 hours after inoculation. *From Bar-Nun and Mayer (1990),* with kind permission from Pergamon Press Ltd.

Leaf juices of 18 plant species were screened against four pathogenic fungal species—*Pythium debaryanum, Fusarium oxysporum, Rhizoctonia solani,* and *Sclerotium rolfsii* (Kumar and Tripathi, 1991). The plants involved were four species of the Asteraceae, two of the Brassicaceae, one of the Myrtaceae, two of the Fabaceae, one of the Acanthaceae, one of the Annonaceae, two of the Verbenaceae, two of the Rutaceae, one of the Umbelliferae, one of the Papaveraceae, and one of the Papilionaceae. The juices were made by taking 20 g fresh weight of leaves, macerating the leaves in 20 mL of distilled water, and filtering out the pulp. The tests were made against the pathogens by adding the juice of a plant to a petri dish containing a sufficient amount of the agar medium to make a final dilution of 1:1, 1:3, and 1:5 of the juice with respect to the medium. The plates were then inoculated with the selected pathogen. Of the 18 plant species, only *Eupatorium cannabinum* leaves completely inhibited the mycelial growth of all four test pathogens at a minimum dilution of 1:1. The leaf juices of *Saraca indica* and *Adhatoda vasica* exhibited complete toxicity at all dilutions against *P. debaryanum* and *F. oxysporum.* The juices of *Adhatoda vasica* and *Crotolaria medicaginea* at 1:1 dilution were toxic to *R. solani* and *S. rolfsii,* respectively. The leaf juices of *Murraya koenigii, Aegle marmelos,* and *Crotolaria medicaginea* at all dilutions completely inhibited *P. debaryanum.* The leaf juice of *Solidago canadensis* at 1:3 dilution completely inhibited mycelial growth of *S. rolfsii,* and at 1:1 dilution was completely toxic to *P. debaryanum.* Damping-off of pea seedlings was successfully controlled by treating the host seed with leaf juice of *E. cannabinum,* which is a common herb in the temperate Himalayas and Khasi Hills. The plant is also cultivated in various gardens as an ornamental, and medicinally, it is used to cure jaundice and scurvy and to treat sores and ulcers. It was concluded that its use in controlling damping-off would be safe, convenient, and nonpollutive.

Asafoetida is a natural product obtained from the plant, *Ferula assa-foetida;* it is used in the Ayurveda and Unani systems of medicine. Chaurasia and Ram (1992) obtained the resin from a local market in Varanasi, India, and made a 10,000-ppm standard asafoetida solution, which they sterilized with a 0.22-μm filter and diluted to nine solutions between 50 and 7000 ppm in small flasks. Each flask was inoculated with either *Rhizoctonia solani* or *Sclerotinia sclerotiorum.* Growth was recorded after 7 and 14 days at 25°C. The weight of the mycelium in each flask was determined; after 14 days' growth, the number of sclerotia produced was also recorded. Even the lowest concentration (50 ppm) caused growth inhibition of each pathogen, but *S. sclerotiorum* was affected the most. *R. solani* was completely inhibited by 7000 ppm, and *S. sclerotiorum* was completely inhibited by 2000 ppm. No sclerotia were produced by either fungus in a concentration of 1000 ppm after 14 days. It appears that even 1000 ppm could be used to control these two pathogens; because neither produced sclerotia at that concentration, the survival of the pathogens in the soil would be reduced.

Based on previous results by Ram and his colleagues using asafoetida, Ram and Jalali (1992) investigated the fungicidal effect of that allelochemical on *Ascochyta rabiei,* the causal pathogen of chickpea blight. The spore germination of *A. rabiei* was completely inhibited at 1000 ppm of asafoetida. Mycelial growth was inhibited at 2500 ppm. Fungicidal efficacy was tested on 60-day-old chickpea plants by spraying them with 500 to 3000-ppm concentrations of asafoetida 24 hours before inoculation with *A. rabiei.* Data on infection and plant girdling were recorded 10 and 20 days after inoculation. The researchers found that spraying with 3000 ppm of asafoetida before inoculation reduced the incidence of the disease by 50%. Postinoculation spraying with asafoetida was not effective.

Essential oils were collected by hydrodistillation from 10 angiosperms—*Aegle marmelos, Alpinia galanga, Artemisia vulgaris, Callistemon lanceolatus, Curcuma longa, Eucalyptus citriodora, Foeniculum vulgare, Lippia alba, Rosa indica,* and *Salvia plebeia.* All these oils were tested in a concentration of 3000 mg/L against 15 species of storage fungi. The leaf oils of *F. vulgare* were completely toxic to *Aspergillus flavus.* The oils of *A. galanga, C. longa, E. citriodora, L. alga,* and *R. indica* were moderately toxic (above 50% but below 96%), whereas the oils of *A. marmelos, A. vulgaris, C. lanceolatus,* and *S. plebeia* had low toxicity (below 50%). Fifteen fungi were tested; the oil of *F. vulgare* inhibited the mycelial growth of six fungi completely at 1500 mg/L, eight fungi at 2000 mg/L, and eleven fungi at 2500 mg/L. Moreover, the oil was not toxic to seed germination and seedling growth of *Cyamopsis tetragonoloba* (guar). Because there is little chance it would have any adverse effect on humans (this plant is used as a vegetable and a spice), it appears that the essential oil of *F. vulgare* may be an ideal fungal toxicant for the protection of seed of guar against storage fungi.

A member of the tribe Inuleae (Asteraceae), *Helichrysum decumbens* grows in maritime areas of southeastern Spain (Tomás-Lorente et al., 1989). In a project to find new sources of fungicides, Tomás-Lorente and colleagues isolated and identified three antifungal phloroglucinols in chloroform extracts of *H. decumbens.* The phloroglucinols were externally deposited on the plant surfaces and were found to prevent growth of *Cladosporium herbarum* in a bioassay.

*Lonchocarpus castilloi* (Leguminosae) is a large tree that grows in the northern part of Chiapas, Mexico, the southern part of Tabasco, and the Yucatán Peninsula. Gómez-Garibay et al. (1990) isolated and identified five flavonoids from a petrol extract of the heartwood of that tree. The substances consisted of four methyl furan auranols and α-hydroxydihydrofuran chalcone, named castillenes A–E. All were yellow oils and solutions (0.3 mg/mL) were made in $Me_2CO$. In antifungal bioassays, 0.5 mL of a given test solution was poured into a petri dish, after which 6 mL of hot, sterilized, growth medium was added. Control plates were treated with solvent (0.5 mL) only. All plates were allowed to stand overnight in a sterile hood to evaporate the solvent. Next, the plates were inoculated with a 6 mm plug of *Lenzites trabea* mycelium and incubated at 25°C for seven days, after which the growth of the fungus in each plate was measured. The inhibition by each flavonoid was as follows: castillen A, 21.5%; castillen B, 10.1%; castillen C, 20.2%; castillen D, 33.9%; castillen E, 68.2%.

Volatiles from crushed Mountain Pride tomato leaves significantly inhibited growth of spores from two fungal pathogens, *Alternaria alternata* and *Botrytis cinerea,* and Hamilton-Kemp et al. (1992) investigated the compounds responsible for the inhibition. Two major groups of volatile compounds known to be produced by crushed tomato leaves are the $C_6$ and $C_9$ aldehydes and the terpene hydrocarbons. The $C_6$ aldehydes, E-2-hexanal and hexanal, and $C_9$ aldehydes, E-2-nonenal and nonanal, all inhibited hyphal growth of both fungal pathogens. However, the terpenes, 2-carene and limonene, had no effect on mycelial growth of the test fungi.

Methyl esters of three fatty acids were isolated and identified from methanol extracts of the tops of the grass *Micanthus sinensis* and were found to be inhibitory to the rice blast fungus, *Pyricularia oryzae* (Mori et al., 1987). The three fungal inhibitors were methyl palmitate, methyl linoleate, and methyl linolenate. Moreover, linoleic and linolenic acids have been found as antifungal factors against *P. oryzae* in rice plants treated with probenazole (a controlling agent against rice blast and bacterial leaf blight of rice). It was concluded from a test of 44 fatty acids that a free monocarboxylic fatty acid with a carbon length of 10 to 13 is the basic structure required to induce antifungal activity against *P. oryzae.*

Menetrez et al. (1990) isolated and identified the leaf surface compounds from *Nicotiana tabacum* and tested their influence on germination and germ-tube morphology of *Peronospora tabacina,* the tobacco blue mold pathogen, and on *Alternaria alternata,* the tobacco brown spot pathogen. Of the compounds identified, germination of sporangia of *P. tabacina* was significantly decreased by *cis*-abienol, but that same compound did not inhibit sporangial germination when combined with sucrose esters or hydrocarbons at a combined concentration of 10 $\mu g/cm^2$. Germination of sporangia was completely inhibited by α- and ß-duvatrienediols. None of the leaf surface chemicals affected germination of *A. alternata* conidia.

Later, cuticular components were extracted from *Nicotiana tabacum, N. glutinosa,* and 24 other species of *Nicotiana* growing in field plots at the Georgia Coastal Plain Experiment Station (Kennedy et al., 1992). The glucose and sucrose ester fractions were separated and tested against germination of the sporangia of *Peronospora tabacina* taken from *N. tabacum.* The tests were run on leaf discs cut from the blue-mold-susceptible *N. tabacum* cv. TI1406. Germination was inhibited by greater than 20% when the sporangia were exposed to sugar esters isolated from *N. acuminata, N. lenthamiana, N. attenuata, N. clevelandii,* and *N. miersii* and accessions 10 and 12 of *N. bigelovii.*

Other compounds that were separated from leaf washings of the 26 species of *Nicotiana* were the α- and ß-4,8,13,-duvatriene-1,3-diols; (13-E)-labda-13-ene-8 α, 15-diol; (12Z)-labda-12,14-diene-8-α-ol (*cis*-abienol); (13-R)-labda-8,14-diene-13-ol (manool); 2-hydroxymanool; a mixture of (13R)-labda-14-ene- 8 α, 13 diol (sclareol); and (13-S)-labda-14-ene- 8 α, 13 diol (episclareol). Estimated $IC_{50}$ values (50% inhibition values) against germination of *P. tabacina* sporangia were 3.0 $\mu g/cm^2$ for α-DVT-diol, 2.9 $\mu g/cm^2$ for ß-DVT-diol, 0.4 $\mu g/cm^2$ for labdenediol, and 4.7 $\mu g/cm^2$ for the sclareol mixture. Manool, 2-hydroxymanool, and *cis*-abienol at application rates up to 30 $\mu g/cm^2$ had little or no effect on sporangium germination. Plant breeders and geneticists are able now to manipulate the composition of trichome exudates, and this capability, together with the knowledge of the antimicrobial activity of the exudate compounds, could enable the development of genotypes with enhanced disease resistance. Tewari and Nayak (1991) tested ethanol extracts of fresh leaves of *Ocimum sanctum, Piper betle, Nyctanthes arbor-tristis,* and *Citrus limon* against *Cochliobolus miyabeanus, Pyricularia oryzae,* and *Rhizoctonia solani,* all of which cause diseases of rice. All four leaf extracts reduced the radial growth of the pathogens on petri plates, but *O. sanctum* and *P. betle* were most inhibitory.

Another experiment was carried out to test the efficacy of the leaf extracts against the effects of the three pathogens on living rice plants. HR 12 and Benibhog rice plants were inoculated with either *Pyricularia oryzae* or *Cochliobolus miyabeanus* 20 days after transplanting, and Karuna rice was inoculated with *Rhizoctonia solani* 20 days after transplanting. An aqueous suspension of each leaf extract (2.5 g/L) was sprayed to runoff, 24 hours after inoculation. A standard fungicide (Bavistin 1 g/L), instead of leaf extracts, was also sprayed on some inoculated plants for comparison. These tests were run on three occasions. All leaf extracts were somewhat effective against all pathogens, but *P. betle* and *O. sanctum* were most effective in controlling the diseases (Table 9). In fact, one or the other of these two was more effective in most tests than the commercial fungicide, Bavistin, and the spray cost was less than one-third that of Bavistin.

Damping-off is a serious problem in forest tree nurseries in the southeastern United States. The common fungi associated with that disease in slash pine (*Pinus elliotii*) are *Rhizoctonia solani,* binucleate *Rhizoctonia* spp., *Pythium aphanidermatum,* and *Fusarium* spp. Huang and Kuhlman (1991a) found that a growth medium of 100% milled pine bark (60% *Pinus taeda* and 40% *P. echinata*) completely controlled damping-off caused by *P. aphanidermatum* and reduced damping-off caused by *R. solani* by 50%. It did not protect, however, against the binucleate *Rhizoctonia.* The researchers subsequently formulated a soil amendment (SF-21) from 750 g of milled pine bark, 35 g of $(NH_4)_2SO_4$, 10 g of triple superphosphate, 30 g of $CaCl_2$, 25 g of KCl, 150 g of $Al_2(SO_4)_3$, and 750 mL of glycerine. When SF-21 was added at 1% (w/w) to soil, it controlled more than 50% of damping-off caused by *P. aphanidermatum, R. solani,* and binucleate *Rhizoctonia.*

When SF-21 was added to field soil at the rate of 2400 kg/ha before sowing slash pine seeds, it significantly reduced postemergence damping-off caused by *R. solani, P. aphanidermatum,* and *Fusarium*

**Table 9** Glasshouse tests of extracts against the three fungal diseases of rice*

| Spray | Occasion 1 | | | Occasion 2 | | | Occasion 3 | | |
|---|---|---|---|---|---|---|---|---|---|
| | *P. oryzae* | C. miyabeanus | R. solani | P. oryzae | C. miyabeanus | R. solani | P. oryzae | C. miyabeanus | R. solani |
| *P. betle* (2.5 g/L$^{-1}$) | 1.2 | 4.0 | 2.8 | 1.0 | 4.0 | 2.7 | 2.6 | 3.3 | 3.6 |
| *O. sactum* (2.5 g/L$^{-1}$) | 2.4 | 4.5 | 4.9 | 2.8 | 4.4 | 4.9 | 2.0 | 4.8 | 5.4 |
| *N. arbor-tristis* (2.5 g/L$^{-1}$) | 7.0 | 7.5 | 6.0 | 7.0 | 7.0 | 6.0 | 6.4 | 7.1 | 5.5 |
| *C. limon* (2.5 g/L$^{-1}$) | 7.0 | 7.0 | 7.8 | 6.8 | 7.0 | 7.3 | 7.2 | 8.3 | 7.9 |
| Bavistin 50 WP (1.0 g/L$^{-1}$) | 1.6 | 1.9 | 3.5 | 1.1 | 1.9 | 4.0 | 1.9 | 2.5 | 3.8 |
| Control | 8.2 | 8.0 | 8.1 | 8.5 | 8.2 | 8.0 | 8.7 | 8.5 | 8.0 |
| LSD ($P = 0.05$) | 0.8 | | | 2.3 | | | 0.8 | | |

* Figures represent scores for disease on 0–9 scale, as described in IRRI's SES (1980).

*Source:* From Tewari and Nayak (1991), with permission of Butterworth Heinemann Ltd.

*moniliforme* var. *subglutinans* by 36 to 38, 25 to 28, and 12 to 22%, respectively, in two tests. Amendment with SF-21 also increased the number of healthy seedlings produced per unit area and the height of the slash pine seedlings. It is noteworthy that populations of *R. solani* and *P. aphanidermatum* were reduced by 50 to 90% four weeks after treatment of the plots. Several antagonistic fungi were stimulated in soil treated with SF-21, and numbers of colony-forming units were increased greatly and remained high for over 50 days.

Huang and Kuhlman (1991b) subsequently demonstrated that the mechanism by which SF-21 inhibited damping-off pathogens of slash pine seedlings was related to the stimulation of the antagonistic fungi. They found that soil amended with 1% (w/w) SF-21 did not immediately inhibit hyphal growth and stem-segment colonization by *R. solani,* but seven days after amendment it significantly inhibited that fungus. Moreover, the population of the antagonistic fungi increased significantly four days or slightly more after application of the SF-21. The density of the antagonistic population was correlated with suppression of stem colonization by *R. solani* in the amended soil. Additionally, stem-segment colonization by *R. solani* increased 30 to 40% after amended soil was autoclaved, but the inhibitory effect returned seven days after infestation with *Trichoderma harzianum* or *Penicillium oxalicum* or both. Increasing the population of *T. harzianum* or *P. oxalicum* or both in amended soil at day 0 to about the same concentration as that in amended soil at day 7 immediately rendered the soil suppressive to stem-segment colonization by *R. solani.*

Soil amended with SF-21 and incubated for seven days produced inhibitory substances that diffused into a synthetic medium and inhibited hyphal growth of *R. solani* and *P. aphanidermatum.* Diffusates from nonamended soil did not have a suppressive effect. When the pH of amended soil was adjusted from 4.3 to 5.8, the suppressive effect on colonization by *R. solani* was completely nullified, but the soil remained partially inhibitory to germ-tube elongation by *P. aphanidermatum.* Huang and Kuhlman found, moreover, that the population density of *Trichoderma* spp. and *Penicillium* spp. in amended soil was negatively correlated with pH values from 4 to 6. They concluded that the effects of SF-21 on various pathogens differ. The effects on *R. solani* are indirect through proliferation of antagonistic species such as *Trichoderma* spp. and *Penicillium* spp. The effects on *P. aphanidermatum* are direct by the inorganic and organic components of SF-21 and indirect by reduced soil pH and stimulation of the antagonistic soil population.

*Salvia splendens* (red sage or scarlet salvia) is a member of the mint family and an ornamental plant. Many mints are considered medicinal plants. Quereshi et al. (1989) collected fully grown, flowering plants from the beds in which they were grown, then washed, dried, and cut the plants into three parts; aerial vegetative parts, flowers, and roots. Each part (500 g) was soaked with continuous agitation, in ethanol, acetone, or water (1.7 L each) for five days. Solvents were decanted, then concentrated under reduced pressure at room temperature to a semisolid mass. Each extract of each part was tested against 13 gram-positive bacteria and 14 gram-negative bacteria. Aqueous extracts had very little activity, whereas ethanolic and acetone extracts often had considerable activity. The root extracts had maximum activity in all solvents. It is obvious that *S. splendens,* like other members of the Labiatae, possesses considerable antibacterial activity. This is true of gram negative organisms as well as of gram-positive ones.

*Scutellaria violaceae* and *S. woronowii* were grown at the Kew Royal Botanic Gardens; clerodin and jodrellin A were isolated and identified from *S. violacea,* and jodrellin B was isolated and identified from *S. woronowii* (Cole et al., 1991). The three neo-clerodone diterpenoids were then tested against the plant pathogenic fungi *Fusarium oxysporum* f. sp. *lycopersici* and *Verticillium tricorpus.* Spore germination of *V. tricorpus* was delayed by clerodin and jodrellin B, and growth of *F. oxysporum* and *V. tricorpus* was inhibited in a dose-dependent manner by all three compounds. Clerodin completely inhibited spore germination of *F. oxysporum* for 18 hours at 100 ppm and spore germination of *V. tricorpus* at 25 to 100 ppm for 42 hours; jodrellin B inhibited spore germination of *V. tricorpus* for 66 hours at 100 ppm. Clerodin inhibited growth of both pathogens at 50 to 100 ppm for 66 hours, and jodrellin B inhibited growth of *F. oxysporum* for 66 hours at 50 to 100 ppm and growth of *V. tricorpus* at 25 ppm or above for 66 hours.

Essential oils were extracted from 12 plant species by hydrodistillation, and the oils were tested at a concentration of 5000 ppm against the mycelial growth of *Pythium aphanidermatum (*Kishore and Dwivedi, 1991). The plants involved were *Artemisia nelagrica, Chrysanthemum indicum, Citrus reticulata, Cirrus sinensis, Pinus* sp., *Ju-*

*niperus communis, Anethum graveolens, Foeniculum vulgare, Seseli indicum, Caesulia axillaris, Tagetes erecta,* and *Callistemon lanceolatus.* All oils were inhibitory to the growth of *P. aphanidermatum,* with the inhibitory range being from 41 to 100%. *T. erecta* was the only plant causing a 100% inhibition of fungal growth. The *Tagetes* oil was completely inhibitory to *P. aphanidermatum* at 2000 ppm; moreover, it was completely inhibitiory to 18 of 25 pathogenic fungi at 2000 ppm and to all 25 pathogenic fungi at 3000 ppm. Additionally, *Tagetes* oil controlled the damping-off of tomato seedlings by *Pythium* when the seeds were treated with 2000 ppm (32.6% control) or 3000 ppm (49.2% control). It was suggested that the control of tomato damping-off might be considerable at higher concentrations without being toxic to the tomato plants.

Thompson Seedless grape berries were sampled weekly from bloom through harvest and the exudates were obtained by sequential immersion in distilled water, ethanol, and diethyl ether (Padgett and Morrison, 1990). The water solutions of the exudates contained sugars, malic acid, K, and $N_2$; the ethanol extracts contained primarily phenolic compounds; the ether extracts contained primarily lipid and phenolic compounds. Because *Botrydis cineria* is the primary causal agent of bunch rot in grapes, the exudates were tested against growth of that pathogen. The water extracts promoted mycelial growth, and the ethanol and ether extracts collected in the first three weeks after bloom strongly inhibited mycelial growth of *B. cinerea.* The inhibitory effects of these fractions decreased later in the season. It has been known for some time that grape berries are relatively resistant to *Botrytis* infection until after they start to ripen. The data of Padgett and Morrison therefore suggest that the changes in composition of grape berry exudates with time may play a significant role in the early-season resistance and late-season susceptibility to *B. cineria* infection.

The soil-borne plant pathogen, *Sclerotium rolfsii* can survive for long periods in soil and is highly resistant to microbial attack; thus Hadar and Gorodecki (1991) decided to test germination of sclerotia of that pathogen in compost. The composted grape marc (CGM) used in their experiments consisted of grape skin, seeds, and stalks left over after wine processing. The materials were composted in windrows for six months, the windrows being turned every two weeks for three months and then allowed to cure for three months. Sclerotia of *S. rolfsii* were collected, dried in a dessiccator for 48 hours, and then placed on the surface of either peat moss or the grape compost in petri dishes. After four days, they were checked for germination; those that had not germinated were then tested to see if they were still viable. After 48 hours, 80 to 90% germination occurred in peat but no germination occurred in CGM. Sclerotial viability decreased from 100% to less than 10% within the first 40 hours in GCM, but remained close to 100% in peat.

Sclerotia of *S. rolfsii* that did not germinate in GCM were dipped in 1% sodium hypochlorite for 30 seconds, washed in sterile water, dried on tissue paper, and placed on various media. In this fashion, *Penicillium* spp. were isolated with high frequency from sclerotia incubated in GCM. Their antagonistic ability was examined by inoculating sclerotia with *Penicillium* sp. and placing the inoculated sclerotia on peat moss. After 48 hours, only 5 to 15% of the inoculated sclerotia had germinated. All other sclerotia that did not germinate were attacked by the antagonistic fungus, and sporulation of *Penicillium* sp. was apparent on the sclerotia after 48 hours. Inhibition of sclerotial germination and reduction of sclerotia viability by antagonists decrease both the inoculum potential of the pathogen and the disease severity.

Ginger is obtained from the rhizomes of *Zingiber officinale;* it is one of the most familiar spices and one of the most frequently prescribed Oriental drugs. Endo et al. (1990) isolated and identified four new curcumine derivatives from ethanol extracts of dried fresh ginger rhizomes: gingerenone A, gingerenone B, gingerenone C, and isogingerenone B. Gingerenone A caused complete inhibition of elongation of invading hyphae of *Pyricularia oryzae* at concentrations above 10 ppm and reduced appressorial formation by 50% at 10 ppm.

## II. CONTROL OF PLANT PATHOGENS BY MICROORGANISMS: HOST PLANTS A–M

Host plants are arranged alphabetically by scientific names in this section and pathogens are arranged alphabeticaly under host names, followed by an alphabetical arrangement of antagonists under each pathogen. In those cases where specific hosts were not discussed, I have arranged the pathogens and antagonists in a similar fashion under the category of "general hosts." When several pathogens and/or several antagonists are discussed in a given paper, I have arranged the pathogens and antagonists alphabetically in reviewing that paper.

Strains of *Bacillus subtilis* have been used to control several diseases associated with such vascular wilt pathogens as *Verticillium dahliae* (Hall and Davis, 1990). Such pathogens are confined to the vascular tissue and are unlikely to be affected directly by surface applications of a bacterial antagonist. Hall and Davis therefore used a rifampicin-resistant strain of *B. subtilis* to determine if it could move through the xylem and remain alive in *Acer saccharinum* (silver maple) and *Acer saccharum* (sugar maple). When plants were in full leaf, stems were fitted with rubber serum caps that held 5 mL of an aqueous cell suspension of *B. subtilis.* A small scalpel was used to make a wound 1 mm deep on each side of the stem to be inoculated; the serum caps were refilled twice daily over a five-day period. Distilled water was used for the controls. Near the end of the dormancy period following each growing season, samples of 20 trees of each species treated with *B. subtilis* and of 20 controls were collected and examined. Similar rifampicin-resistant *B. subtilis* were recovered from the xylem tissue after 12 and 24 months. After two years, these bacteria were recovered 30 to 72 cm and 5 to 54 cm, respectively, from the inoculated site in silver and sugar maple. Thus *B. subtilis,* a potential antagonist to *Verticillium dahliae* can survive and be transported into newly formed xylem tissue of both silver and sugar maple.

*Gaeumannomyces graminis* var. *avenae* (take-all) severely affected the golf greens sown with *Agrostis palustris* (creeping bent, cv. Pencross) but had a lesser effect on the fairways planted with *Agrostis palustris* and *Festuca rubra* subsp. *rubra* on a golf course at Benodet, France, in 1987 (Sarniguet and Lucas, 1992). Both species were severely damaged near the take-all patch margin but the central part was recolonized by susceptible *F. rubra* subsp. *rubra.* Counts of rhizosphere bacteria showed that the total population was nearly the same in all zones across the patches, but the ratio of fluorescent *Pseudomonas* spp. to total bacteria was 1:22, 1:15.4, 1:3.5, and 1:2.9, respectively, in the disease-free area, the front of the patch, the damaged part of the patch, and the recolonized central part. Moreover, in the last zone, 44 to 82% of the fluorescent *Pseudomonas* spp. were antagonistic in vitro to the take-all fungus, whereas only 12 to 34% from the disease-free area were antagonistic. Thus the development of take-all induced quantitative and qualitative changes in populations of fluorescent pseudomonads.

Top dressings made from cornmeal-sand mixtures infested with strains of *Enterobacter cloacae* were spread over *Agrostis palustris–Poa annua* greens at Ithaca, New York, that were infested with the dollar-spot pathogen, *Sclerotinia homoeocarpa* (Nelson and Craft, 1991). The researchers found that strains EcCT-501 and E1 of *E. cloacae* significantly reduced the dollar-spot disease when compared with untreated spots. Monthly applications of strain EcCT-501 resulted in 63% disease control; this strain was as effective as iprodione or propiconazole in reducing that disease. Disease suppression due to EcCT-501 and E1 was noticeable up to two months after application. Suppression of dollar spot was more effective when *E. cloacae* was applied as a preventive treatment rather than a curative treatment. Nevertheless, strain EcCT-501 significantly reduced dollar spot even when applied to severely infected turf. In two years of treatment, recoverable populations of *E. cloacae* immediately after treatment were approximately $10^7$ to $10^9$ cfu per gram dry weight of thatch each year. The numbers declined each year with time, but they remained greater than $10^4$ cells per gram for up to 13 weeks and were detectable in the spring of 1989 after application in the summer of 1988.

A similar experiment was conducted on *Agrostis palustris* greens infected with the dollar-spot disease at Saskatoon, Canada, but here the researchers used many different potential antagonists (Goodman and Burpee, 1991). Four of 27 isolates tested in the greenhouse were significantly antagonistic to the dollar-spot organism. Five organisms tested in the greenhouse were used in field tests, those selected ranging from very suppressive to dollar spot to nonsuppressive. In the field tests, maximum disease intensities (percentage of plot area blighted) were 5, 18, or 44% respectively, in plots treated with sand-cornmeal infested by *Fusarium heterosporum, Acremonium* sp., or an unidentified bacterium, compared with 84% in plots not top-dressed. In the second year of the tests, treatments at two-week intervals with sand-cornmeal infested with *F. heterosporum* at 400 $cm^3/m^2$ limited disease intensity of dollar spot to 3% compared with 18% in nontreated plots. Both greenhouse and field tests suggested that *F. heterosporum* (the isolate) produced substances toxic to *S. homoeocarpa.*

Typhula blight of turf grasses is a common disease caused by *Typhula ishikariensis* and *T. incarnata* in areas where snow cover exceeds 90 continuous days (Lawton and Burpee, 1990). Field studies were conducted from 1985 to 1988 at Guelph, Ontario, to determine the effects of *T. phacorrhiza* (T016) as an antagonist in the control of

Typhula blight of *Agrostis palustris.* Applications of 50, 100, 200, or 400 g/m$^2$ of colonized grain ($1.75 \times 10^3$, $3.5 \times 10^3$, $7.0 \times 10^3$, and $1.4 \times 10^4$ cfu, respectively) and 100, 200, and 400 g/m$^2$ of heat-killed colonized grain were made to the surface of 1 m$^2$ plots of creeping bentgrass on Dec. 4, 1985. Remaining plots were left untreated or received an application of PCNB17G (a fungicide) at 30 kg/ha. Snow covered the plots on December 4, 1985, and melted on March 25, 1986 (111 days of snow cover). There was a highly significant decline in disease intensity in 1986, 1987, and 1988 as a result of the applications of isolate T016 of *Typhula phacorrhiza* in the fall of 1985 and 1986.

As the concentration of isolate T016 increased, the time required for turf grass to recover to less than 3% necrotic tissue decreased significantly in 1986 and 1987. A significant increase in the number of sclerotia of *T. phacorrhiza* was detected in plots in the spring of 1986, 1987, and 1988 in response to increases in concentration of grain colonized by isolate T016 applied in the fall of 1985 and 1986. The increases in concentrations of grain colonized by isolate T016 in the fall of 1985 and 1986 caused a significant decrease in number of sclerotia of the pathogens, *T. i. ishikariensis* plus *T. incarnata* in plots in the spring of 1986, but not in 1988. Results of the experiments indicated that a rate of 200 g/m$^2$ of grain colonized by *T. phacorrhiza* ($7.0 \times 10^3$ cfu) is required to provide suppression of Typhula blight equivalent to suppression of that blight by the fungicide PCNB. Lawton and Burpee suggested, however, that incorporation of highly disease-suppressive isolates of *T. phacorrhiza* into alginate pellets may improve the efficacy of biological control.

*Sclerotinia sclerotiorum* causes yield losses in many economically important crops, including *Apium graveolens* (celery) and *Lactuca sativa* (lettuce). Tests were therefore conducted by Budge and Whipps (1991) to determine the effects of *Coniothyrium minitans, Trichoderma harzianum* (HH3), and *Trichoderma* sp. (B1) as antagonists against *S. sclerotiorum.* In control plots, over 80% of the celery and 90 and 60% of the lettuce in the first and second crops, respectively, were infected at harvest. Only the *C. minitans* treatment in the first lettuce crop decreased disease and increased marketable yield. Moreover, *C. minitans* reduced the number of sclerotia recovered at harvest in the celery and first lettuce crops and decreased sclerotial survival of the pathogen over the autumn fallow periods following the celery and second lettuce crop. *C. minitans* survived in soil for over one year and spread to infect sclerotia in virtually all other plots. *Trichoderma* spp. tested had no effect on disease and very little effect on survival of sclerotia of the pathogen, even though they could be recovered from the soil for the duration of the experiments.

*Cercospora arachidicola* is the causal agent of Cercospora leaf spot of *Arachis hypogaea* (peanut). Knudsen and Spurr (1987) investigated the survival and efficacy of several bacterial strains for control of *C. arachidicola* in the field over a 12-week period. The bacteria selected were *Bacillus thuringiensis* (HD-1), *B. thuringiensis* (HD-521), *B. cereus* var. *mycoides* (Ox-3), and *Pseudomonas cepacia* (Pc 742). They were applied—as aqueous suspensions of wettable powders or as dusts—individually to Florigiant peanuts at three locations in North Carolina and Virginia at biweekly intervals. Leaflets were sampled periodically to determine survival of applied bacteria and development of the Cercospora leaf spot. Numbers of *Bacillus* spp. declined between applications, 90 to 99% during most two-week intervals. Populations of *P. cepacia* (Pc 742, wettable powder or dust formulations) were more variable; reductions ranged from approximately 10% to over 99%. Analysis of variance with treatments blocked by location indicated that both disease severity and area under the disease progress curve differed significantly among treatments and among locations. Disease severity was significantly less in plots treated with fungicide or with three of the biocontrol agents (HD-1; Pc 742, wettable powder; and Pc 742-D, dust) than in unsprayed plots. The means for the area under the disease progress curves for all biocontrol agents and fungicide treatments were significantly lower than they were in unsprayed plots. Overall, the *P. cepacia* strain controlled the disease slightly more effectively.

Rust diseases are widely distributed among peanuts in India; in susceptible varieties they can cause up to 50% losses (Ghewande, 1990). Ghewande therefore decided to determine the effects of the fungi *Acremonium persicinum, Eudarluca caricis, Penicillium islandicum, Tuberculina costaricana,* and *Verticillium lecanii* on the peanut rust caused by *Puccinia arachidis.* All the antagonists and their culture filtrates inhibited urediniospore germination and reduced rust development by varying amounts. The maximum effects in laboratory tests and in the field were caused by *V. lecanii* and *P. islandicum.* The inoculation of antagonists and rust on the same day gave better control of rust than did inoculations on a different day. Sprays of culture

**Table 10** Effect of culture filtrates of mycoparasites on the development of rust under field condition

| Culture filtrates of | Mean pustules per leaf: Number of pustules | Mean pustules per leaf: Number of pustules ruptured | Mean pod yield (g/plant) |
|---|---|---|---|
| *Acremonium persicinum* (CMI) | 125 | 20 | 13.78 |
| *A. persicinum* (IARI) | 155 | 27 | 10.5 |
| *Eudarluca caricis* | 71 | 9 | 15.75 |
| *Penicillium islandicum* | 65 | 7 | 13.5 |
| *Tuberculina costaricana* | 105 | 20 | 16.17 |
| *Verticillium lecanii* | 56 | 9 | 17.5 |
| Control (water spray) | 157 | 28 | 9.75 |
| C.D. ($P$ = 0.05) | 22.37* | 8.35* | 4.03* |

* Significant at 1% level.

*Source:* From Ghewande (1990), with permission of Taylor and Francis Ltd.

filtrates of these antagonists significantly reduced rust disease under field conditions (Table 10). Thus they obviously produce metabolites that reduce the in vivo development of *P. arachidis.*

Other antagonists have also been tested against *Puccinia arachidis* on peanuts (Govindasamy and Balasubramanian, 1989). This rust was significantly reduced by *Trichoderma harzianum* on detached leaves of TMV.7 peanuts. Pretreatment with *T. harzianum* conidia caused a greater inhibition of germination and germ-tube growth of *P. arachidis* than did mixed/posttreatment. After *T. harzianum* application, many of the fully developed pustules harbored the antagonist. By contrast, unopened pustules yielded fewer colonies of *T. harrzianum* when plated on an appropriate medium. In a subsequent experiment, a phenollike, Folin-positive antifungal compound was isolated from an ether-chloroform extract of the growth medium of *T. harzianum.* This compound is apparently released by germinating conidia of this antagonist and probably explains why rust uredospores infected with *T. harzianum* do not germinate at all, even after repeated washings. Moreover, when infected uredospores were smeared on detached peanut leaves, they did not germinate and induce infection.

Another serious disease of peanut in India is root rot caused by *Rhizoctonia solani,* and Savithiry and Gnanamanickam (1987) tested several antagonistic strains of *Pseudomonas fluorescens* for biological control of that disease. One of the strains was originally isolated from citrus (pfc) and another from peanut (pfp). The *R. solani* used in the tests was isolated from peanut plants. In plate assays, mycelial growth of *R. solani* was strongly reduced by both strains of *P. fluorescens,* with inhibition zones of 2.5- to 4-cm diameters being observed. Siderophores of *P. fluorescens* prevented germination of sclerotia of *R. solani,* as did cell suspensions of *P. fluorescens.*

In greenhouse experiments, Jalgaon 24 and TMV2 peanut plants inoculated with *R. solani* wilted and died in 10 days, but plants bacterized with *P. fluorescens* (pfc) before they were inoculated with *R. solani* survived. The bacterized plants had restricted lesions induced by *R. solani* on lower hypocotyl regions, whereas the nonbacterized plants had longer and spreading lesions, which often girdled the hypocotyls of such plants. In field tests, TMV2 peanuts grown from seeds treated with a suspension ($10^8$ cfu/mL$^{-1}$) of *P. fluorescens* (pfc) grew taller and appeared greener than those plants that were nonbacterized. The number of pods and fresh weight of pods were 41.1 and 59.0% higher, respectively, than in control plants. Moreover, the number of branches of bacterized plants was 14.2% higher than for nonbacterized plants. It therefore appears that in India certain strains of *P. fluorescens* can help control root rot of peanuts caused by *R. solani;* they can also enhance growth and yields, even in plants free of disease.

Ganesan and Gnanamanickam (1987) found that *Pseudomonas fluorescens* can also help control root and stem rot of peanuts caused by *Sclerotium rolfsii.* This pathogen was isolated from infected peanut plants in two fields in Tamilnadu, India. Fluorescent pseudomonads were isolated from peanuts, rice, citrus, and cotton. These were screened further, and only biotype II strains of *P. fluorescens* were used in tests. In plate assays, mycelial growth of *S. rolfsii* (three strains) was strongly reduced by four strains of *P. fluorescens.* Clear inhibition zones ranging from 2.5- to 5.5-cm diameter were observed. The citrus strain of *P. fluorescens* caused the largest inhibitory zone, indicating it was the most effective antagonist. Mycelial growth inhibition was also observed when cell-free siderophore plugs of *P. fluorescens* were used. The siderophore plugs also prevented germination of the sclerotia of *S. rolfsii.* Cell suspensions of the citrus and peanut strains of *P. fluorescens* also reduced the survival of sclerotia of *S. rolfsii.* After two months, only 0 to 10% of the sclerotia suspended in bacteria germinated, compared with 70% germination of sclerotia suspended in sterile distilled water. In greenhouse experiments, all peanut plants inoculated with *S. rolfsii* wilted within seven days, but 99% of plants inoculated with *P. fluorescens* before they were inoculated with *S. rolfsii* survived. Thus it is evident that efficient strains of *P. fluorescens* can effectively control the root and stem rot of peanut plants caused by *S. rolfsii*.

*Asparagus officinalis* is subject to several *Fusarium* diseases, some of which can cause several serious diseases, including the root rot and stem wilt caused by *Fusarium oxysporum* f. sp. *asparagi*. Smith et al. (1990) pointed out that the actinomycete *Streptomyces griseus* var. *autotrophicus* (ATCC 53668) had been found to produce a strong antifungal antibiotic, faeriefungin. They decided therefore to test the organism and the antibiotic for control of the root rot and stem wilt disease of asparagus caused by *F. o. asparagi*. The antagonist completely inhibited the growth of *F. o. asparagi* on all media assayed but did not show any protection to plants on the basis of the variables measured. In fact, all of the *S. griseus* treatments reduced asparagus growth in the absence of the pathogen. The root-dip treatments challenged with the pathogen did not differ from the treatments with the pathogen alone. In the greenhouse experiments, inoculation with the pathogen reduced growth for all variables measured. *S. griseus* incorporation into sterile soil provided no reduction in disease, as evidenced by all variables measured.

In contrast, the antibiotic produced by *S. griseus* did not adversely affect asparagus seed germination or plant growth, but it did decrease the severity of disease caused by *F. o. asparagi* in sterile asparagus plants grown on water agar at 25 ppm; it also increased fibrous root weight of plants grown in the absence of the pathogen. The antibiotic, faeriefungin, had a minimum inhibitory concentration of 12.5 ppm against the pathogen. The reasons for the contrasts in results between the experiments carried out with the antibiotic and those with the organism (*S. griseus*) that produces it are not clear from the investigation.

Phytophthora root rot is very common in nurseries in Western Australia; therefore Hardy and Sivasithamparam (1991) tested a composted *Eucalyptus* bark (CEB) medium and a pine bark mix based on a nurseryman's mix (NM) to determine their relative effects in suppressing different *Phytophthora* species. The pine bark mix consisted of *Pinus radiata* bark, peat, sand, and *Eucalyptus marginata* sawdust. The composted *Eucalyptus* bark medium consisted of composted *E. calophylla* and *E. diversicolor* bark and a coarse river sand (3:1).

The pathogens tested were *Phytophthora drechsleri,* isolate GE5 from *Banksia grandis; P. citricola,* isolate GE18 from *Azalea indica; P. nicotianae* var. *nicotianae,* isolate GE16 from *Eutaxia obovata; P. cinnamomi,* GE25 from *Pimelea ferruginea;* and *P. cryptogea,* GE28 from *Grevillea drummondii.*

Banksia (*Banksia occidentalis*) was used as the indicator host plant for *Phytophthora cinnamomi,* and waratah (*Telopea speciosissima*) was used as the indicator host species for the other four *Phytophthora* species. The selected pathogens were introduced into the containers with the chosen indicator plants and media. Disease symptoms included blackening of the crown region and wilting of seedlings. Diseased plants were removed and appropriate tests were run to confirm that the symptoms and death were due to the introduced pathogens. Disease progress curves were plotted for each medium and *Phytophthora* sp. as mean percentage mortality. The CEB medium was suppressive in decreasing order to *Phytophthora cryptogea, P. nicotianae* var. *nicotianae, P. drechsleri,* and *P. cinnamomi* infections of waratah and *B. occidentalis.* The suppressiveness of the CEB medium appeared to be biological, since it became conducive to root rot after steam sterilization. The NM medium was conducive to all *Phytophthora* species tested. In the NM medium, *P. citricola* was the most pathogenic of the five *Phytophthora* species tested, but when the medium was steamed, the pathogenicity was virtually the same for all *Phytophthora* species. The suppression in the CEB medium was shown to be one that involves fungistatic effects. Thus it is likely that plants conducive to the disease will show signs of infection when planted at sites conducive to the disease.

Damping-off of *Beta vulgaris* (sugar beet) is commonly caused by *Aphanomyces cochlioides, Phoma betae, Pythium ultimum,* and *Rhizoctonia solani* in Switzerland (Walther and Gindrat, 1988). *Gossypium hirsutum* (cotton) is commonly infected with *Pythium ultimum* and *Rhizoctonia solani* also. Walther and Gindrat selected a strain of *Chaetomium globosum* (Cg-1) and a strain of *Pseudomonas* sp. (Ps-4) from 300 isolates of fungi, actinomycetes, and bacteria, tested under growth-chamber conditions for their ability to control damping-off of sugar beet caused by *P. ultimum.* The researchers tested the selected antagonist strains against damping-off of KWS Mono sugar beet plants and Acala cotton. Seed treatment with ascospores of *C. globosum* (Cg-1) reduced damping-off of sugar beets caused by seed-borne

*Phoma betae* and soil-borne *Pythium ultimum,* or *Rhizoctonia solani* in growth-chamber experiments. Seed treatment with *Pseudomonas* sp. (Ps-4) controlled *P. betae* and *P. ultimum,* but not *R. solani.* Coating cotton seeds with ascospores of *C. globosum* controlled damping-off of cotton caused by *P. ultimum* and *R. solani.* In some experiments, biological control was equally or more effective than seed treatment with captan.

Nondormant oospores of *P. ultimum,* pycnidiospores of *P. betae,* and ascospores of *C. globosum* germinated at rates of 80 to 92% on cellophane membranes covering sterilized soil. When soil was inoculated with *C. globosum* or *Pseudomonas* sp., germination of *P. ultimum* and *P. betae* was strongly inhibited, whereas ascospores of *C. globosus* germinated well.

Ascospores of *C. globosus* germinated well even after storage on seeds for 420 days at 20°C. The germination percentage after 2.5 years at 20°C was 93. Thus there is no problem with storing this antagonist for a prolonged period. *Pseudomonas* sp. (Ps-4) on filter discs revived after four months at 20 to 30°C, and may have survived longer. If root colonization occurs, long-term protection may be possible. *C. globosus* appears to be an especially promising antagonist for control of sugar beet and cotton damping-off.

Gordon-Lennox et al. (1987) did a very similar but less extensive investigation on selected antagonists against damping-off in sugar beets and cotton. They screened 224 microbial species for antagonistic action and selected *Chaetomium globosum, Pseudomonas* sp., and *Pythium oligandrum* for further studies. Their results were less extensive but relatively similar to those of Walther and Gindrat (1988).

A more detailed study was conducted later on effects of several strains of *Chaetomium globosum* (Cg-1, Cg-3, Cg-13, Cg-14, Cg-20, Cg-29, Cg-40, and Cg-43) on damping-off of sugar beets caused by *Pythium ultimum* (P-71) obtained from Ciba-Geigy AG in Switzerland (DiPietro et al., 1992). The *C. globosum* strains were obtained from Dr. Gindrat at Changins, Switzerland. A spontaneous variant of Cg-13 was isolated early in the investigation and was designated as ADP-13. In a study of the efficacy of *C. globosum* strains in reducing damping-off, only Cg-13 was effective. The inhibition of mycelial growth of *P. ultimum* by culture filtrates of *C. globosum* varied with the strain. Filtrates of strain Cg-13 from all growth temperatures were inhibitory to *P. ultimum.* With Cg-43, maximum activity was detected in filtrate at 30°C. Two metabolites with antifungal activity against *P. ultimum* were isolated from liquid cultures of *C. globosum*. These were 2-(buta-1, 3-dienyl)- 3-hydroxy- 4-(penta-1, 3-dienyl)-tetrahydrofuran (BHT) and chaetomin, an epidithiadiketopiperazine. BHT was produced by six of eight tested strains of *C. globosum*, and chaetomin was produced by five of nine tested strains. Additionally, chaetomin was extracted from pasteurized soil inoculated with strain Cg-13, an effective biocontrol strain, but not from pasteurized soil inoculated with ADP-13, the spontaneous mutant of Cg-13 that is unable to suppress *Pythium* damping-off (Fig. 6). The activity of chaetomin was comparable to the fungicide metalaxyl, a hundred times as inhibitory to mycelial growth of *P. ultimum* as BHT, and five to ten times higher than gliotoxin, a different epidithiadketopiperazine. The results indicated that chaetomin production in soil plays an important role in the antagonism of *C. globosum* against *P. ultimum*. Except for the mutant strain ADP-13, however, all the nonproducing strains still gave some degree of protection against *Pythium* damping-off. The authors suggested that this may have been due to the production of other antifungal compounds, such as an unidentified compound produced by Cg-43.

On the basis of previous research, which indicated that *Pythium oligandrum* might be a suitable antagonist against some fungal pathogens, Walther and Gindrat (1987) decided to test it against damping-off of sugar beets caused by *Phoma betae, Pythium ultimum,* and *Rhizoctonia solani*. Preliminary tests suggested that results were rather poor, apparently because of a constitutive dormancy of the oospores of *P. oligandrum*. The researchers selected four strains of *P. oligandrum* (Po-1, Po-4, Po-8, and Po-9) and found that soaking the oospores for several days in sterile distilled water (SDW) enhanced germination in three of the four isolates. Moreover, treatment of the water-soaked oospores with an equal volume of a 10% aqueous myo-inositol solution for five minutes also aided germination of the oospores. After the oospores were dried at room temperature, germination was tested on an agar medium at 15°C and 75 to 85% relative humidity in darkness. The myo-inositol treatment increased germination of the oospores during the first 20 days of storage at 15°C. In tests of biocontrol of damping-off of sugar beets, oospore suspensions of *P. oligandrum* were adjusted to $10^6$ oospores/mL. Two milliliters were used for dipping 120 sugar beet seeds, which were subsequently air-dried at room temperature and stored for 3 to 38 days before planting in containers

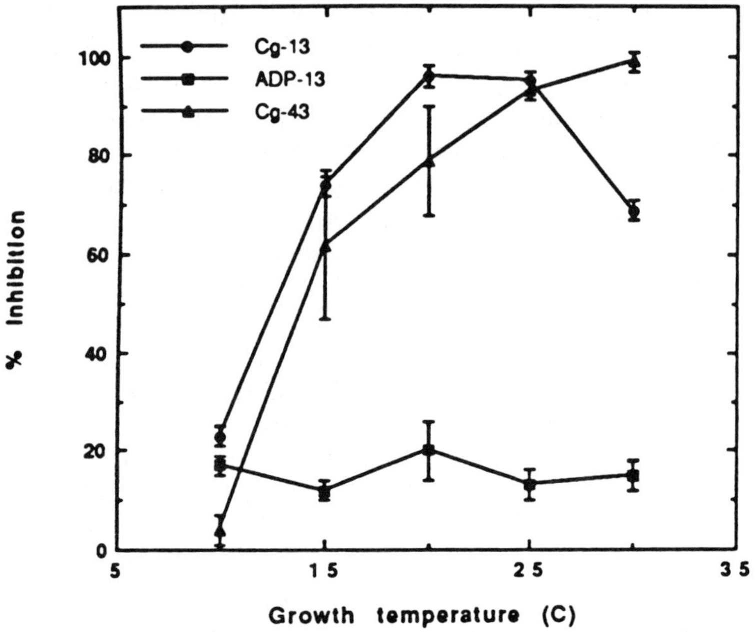


FIGURE 6 Inhibition of mycelial growth of *Pythium ultimum* by culture filtrates of *Chaetomium globosum* strains Cg-13, ADP-13, and Cg-43 grown in 1% malt extract broth at different temperatures. Culture filtrates were incorporated in malt extract agar. Each point represents the means of three replicate plates from each of three experiments. Error bars represent standard errors. *Adapted from DiPietro et al. (1992), with permission of Amer. Phytopathological Soc.*

of a horticultural mixture inoculated with *P. ultimum* or *R. solani.* Emergence and post-emergence damping-off were observed for 21 days after planting. Killed seedlings were discarded or occasionally plated on PDA to check the cause of damping-off.

Seed dressing with oospores of *P. oligandrum* controlled pre- and postemergence damping-off of sugar beets caused by *P. ultimum* and *P. betae.* No significant control of damping-off caused by *R. solani* was observed, however. Water flooding of oospores and treatment with myo-inositol before seed coating enhanced the effectiveness of *P. oligandrum* in controlling *P. ultimum* and *P. betae.* Walther and Gindrat concluded that efficacy of *P. oligandrum* appears to depend on the rate of germination of oospores at the time of sowing and that the efficacy of the activated oospores was comparable to that of captan, a commercial fungicide.

Martin and Hancock (1987) independently investigated the antagonistic role of *Pythium oligandrum* against *P. ultimum,* the causative agent of *Pythium* damping-off in sugar beets. They did this because of the involvement of *P. oligandrum* in soils naturally suppressive to *P. ultimum.* In seed treatments, sugar beet seeds (V137H8) were shaken with sufficient oospores of *P. oligandrum* to apply approximately 12,500 oospores per seed. These were dried and stored in plastic bags at 4°C until used. Concentrations of *P. oligandrum* in field and fumigated soil were adjusted by amendment with autoclaved soil that had been inoculated by adding culture-grown oospores of *P. oligandrum* and moistening the soil to field capacity. After air drying, the soil was ground to pass through a 0.5-mm-mesh sieve and stored at 4°C until used. Greenhouse trials were used to determine the effectiveness of *P. oligandrum* in controlling damping-off of sugar beets caused by *P. ultimum.* After planting the oospore-treated seeds, the researchers recorded emergence and percentage of surviving plants on alternate days for 18 days after planting.

Coating sugar beet seeds with oospores of *P. oligandrum* controlled preemergence damping-off caused by *P. ultimum* equivalent to treating the seeds with the fungicide fenaminosulf. This protection was evident within 24 hours of planting and coincided with seed colonization by *P. oligandrum.* The parts of the seed and seedlings colonized by *P. oligandrum* had a lower frequency of colonization by *P. ultimum* than did untreated seeds. Thus early colonization by *P. oligandrum* apparently protects against subsequent infection by *P. ultimum.* A similar colonization of crop debris by *P. oligandrum* in suppressive soils in California prevented subsequent colonization by *P. ultimum.* Pelleting of sugar beet seeds with oospores of *P. oligandrum* caused no necrosis of any part of the seedling. On the other hand, pelleting the seeds did cause a reduction in the root length of plants grown in fumigated soil, though it did not reduce shoot growth. Coating seeds of 12 crop plants from six families with oospores of *P. oligandrum* had no detrimental effects on seedlings planted in fumigated soil. Moreover, observation by microscope indicated that none of the plants showed signs of root necrosis or root invasion by *P. oligandrum.*

A large-scale production of oospores of *P. oligandrum* in liquid culture is now possible, and this allows controlled quantities of oospores to be applied to seeds rather easily by the use of film-coating techniques. McQuilken et al. (1990) pointed out that the successful use of such techniques would allow a rapid increase in the use of biological control procedures. These scientists experimented with commercially coated seeds of sugar beet (cv. Amethyst) and cress (*Lepidium sativum* cv. Curled). Coating was done by Germain's (UK) Ltd., Kings Lynn, Norfolk, England, employing two techniques—pelleting and film coating. Fungicides and *P. oligandrum* oospores were incorporated separately, either into seed pellets using the EB3 pellet process or coated onto seed surfaces using a film-coating finder system. Prior to coating, sugar beet seeds were steeped for 12 hours at 25°C, either in water or in 0.2% aqueous thiram solution. Thiram steeping controls seed-borne *Phoma betae,* and all sugar beets sown in Britain are now treated routinely with this fungicide. Thiram was also added as a dust to the EB3 pellet, and thiram-treated seeds were then treated with hymexazol or metalaxyl. Hymexazol controls both *Pythium* spp. and *Aphanomyces cochlioides,* whereas metalaxyl controls only *Pythium* spp. Cress seeds were not treated with fungicides.

Cress seeds and some of the water-steeped sugar beet seeds were treated with 4.0 or 4.5 g of oospore inoculum to achieve $35 \times 10^3$ or $50 \times 10^3$ oospores per seed, respectively. Other cress and sugar beet seeds were pelleted and film-coated without additives as controls. Pelleting and film-coating processes both produced similar numbers of oospores of *P. oligandrum* per seed, achieving 86 to 94% and 76 to 86%, respectively, of the target rate for cress and sugar beets. Oospores obtained from the treated seeds had a similar germination to that of those in the inoculum used to coat the seeds. In sand artificially infested with *P. ultimum, P. oligandrum* pellets gave increases in seedling stand and shoot dry weight equivalent to the fungicide drench. Film coating with *P. oligandrum* increased seedling stand and shoot dry weight, but was not as effective as the fungicide treatment. In potting compost artificially infested with *P. ultimum, P. oligandrum* seed coatings gave significant increases in seedling stand and shoot dry weight, but none of these was as effective as the fungicide drench. Seedling stand of the pelleted *P. oligandrum* treatments in potting compost was consistently less than that in sand, although there was no difference with the film-coating treatment. In sand artificially infested with *R. solani,* all *P. oligandrum* coatings gave significant increases in seedling stand and shoot dry weight, but none was as effective as the tolclofos-methyl drench treatment.

The use of *P. oligandrum* as a biological control agent of damping-off diseases has now reached a stage requiring field testing on numerous sites. This fungus has a worldwide distribution and can control most of the major damping-off diseases of sugar beets; it can be simply produced and coated on seed, and it is nonpathogenic to over 15 species of crop plants.

Lewis and Papavizas (1987) investigated the effect of age, concentration, and time of addition to the soil of germlings of an isolate of *Trichoderma hamatum* (TRI-4) on the survival and saprophytic growth of *Rhizoctonia solani* in soil; the influence of soil pH, moisture, and incubation temperature on the antagonist-pathogen interactions; and the efficacy of germlings in different soil types and against several isolates of *R. solani*. The isolates of *R. solani* investigated were R-23 from cotton; R-18, R-53, and R-103 from tomatoes; R-34 from snap beans; and R-108 from potatoes. Some isolates were obtained in Maryland, some in Mississippi, and some in South Carolina. Pathogen-infested beet seeds were used as the inoculum to simulate natural conditions. Preparations of *T. hamatum* (TRI-4) were made on wheat bran mixed with water and were used in most tests. A spore suspension was added to the bran mixture to provide $1 \times 10^5$ conidia/$g^{-1}$ of bran, and the cultures were incubated for three days at 25°C before being added to the soil. This preparation, designated as germling, contained young, actively growing hyphae with no evidence of conidia or chlamydospores. In one experiment, different periods of incubation were used, giving everything from conidia on the bran through the germling stage to the stage with large amounts of conidia and chlamydospores.

A loamy sand of pH 6.4 and 0.4% organic matter was used in most experiments. However, seven other soils with different characteristics were also used; these included a silty-clay loam of pH 6.3 with 1.4% O.M., a loam of pH 6.2 and 5.0% O.M., a sandy-clay loam of pH 6.3 and 1.0% O.M., a sandy-clay loam of pH 4.5 and 3.2% O.M., a sandy loam of pH 6.9 and 1.2% O.M., a sandy loam of pH 5.1 and 0.2% O.M., and a sandy loam of pH 6.7 and 0.3% O.M. The soils came from various parts of Maryland and New Jersey.

In most experiments, germlings of TRI-4 were added to soil one week after the infested beet seeds to provide 0.5% bran and $5 \times 10^3$ propagules/g of soil. During the first week of incubation there was only 10 to 15% survival of the pathogen in beet seeds in soil amended with the inoculum, whereas there was 95% survival of the pathogen in beet seeds from nonamended soil. Germling and eight-day-old preparations of *T. hamatum* were very effective in preventing growth of the pathogen from a food base into soil. Conidia with bran (0 days old) and mature, 40-day-old preparations did not significantly reduce pathogen survival or prevent growth of the pathogen in the soil.

Germlings of *T. hamatum* (TRI-4) reduced survival and saprophytic growth of five isolates of *R. solani* to less than 9% within one week. Survival and saprophytic growth of *R. solani* (R-23) in seven soils other than the loamy sand used in most experiments were also reduced by the germlings. Addition of germlings to soil reduced survival of the pathogen to less than 10% in one week. There were no significant differences in survival of the pathogen between the seven soils. Germlings of the antagonist reduced pathogen survival under all conditions of temperature, moisture, and pH tested, except under high moisture at 32°C. The greatest reduction of the pathogen (91 to 100%) was observed at all soil moisture levels at 15 and 23°C and at pH 5.0 and 6.4. The range of environmental conditions under which *T. hamatum* controlled the population of *R. solani* was truly impressive. This type of information concerning other potential antagonists and pathogens could certainly help simplify the development of practical and effective biological disease control methods.

*Sclerotium rolfsii* is the most destructive pathogen of sugar beets in India, causing 15 to 59% reduction in root yield in different cultivars (Upadhyay and Mukhopadhyay, 1986). It can be controlled by pentachloronitrobenzene (PCNB) at 15 to 20 kg/ha. This amount is excessively expensive and causes soil and ground water pollution. Upadhyay and Mukhopadhyay therefore investigated the possibility of using *Trichoderma harzianum* as a biological control agent for that disease pathogen. *T. harzianum* (IMI-238493) was isolated from sugar beet field soil and used in all tests. When *S. rolfsii* and *T. harzianum* were inoculated on the same petri plate, *T. harzianum* attacked and lysed the mycelium and sclerotia of *S. rolfsii.* The types of interactions between the organisms were hyphal coiling, entry through haustoria-like structures, and direct entry into the hyphae and sclerotia of *S. rolfsii.* When *T. harzianum* was applied to *S. rolfsii*–infested soil in the form of sorghum culture, it gave as high as 76 and 88% disease control of sugar beet seedlings (Table 11). The degree of control in-

**Table 11** Biocontrol of *S. rolfsii* (SR) by *T. harzianum* (TH) in sugar beet seedlings grown two successive times in the same soil

| | Growth cycle 1 | | Growth cycle 2 | |
|---|---|---|---|---|
| Inoculum ratio (SR:TH) | Seedling mortality (%) | Disease control (%) | Seedling mortality (%) | Disease control (%) |
| 1:1 | 58.78 (50.09)* | 19.12 | 63.33 (52.86) | 14.46 |
| 1:2 | 54.23 (47.42) | 25.17 | 42.77 (40.74) | 41.72 |
| 1:3 | 24.62 (30.97) | 61.05 | 26.31 (30.63) | 64.76 |
| 1:4 | 21.80 (27.59) | 68.46 | 19.90 (25.94) | 73.03 |
| 1:5 | 17.93 (24.52) | 76.03 | 7.71 (15.46) | 88.33 |
| SR 2 g/kg (Check) | 76.33 (65.26) | 0.00 | 77.26 (65.43) | 0.00 |
| LSD 0.05 | 13.9 | | 2.10 | |

* Arc sin — percentage transformation.

*Source:* From Upadhyay and Mukhopadhyay (1986), with permission of Taylor and Francis Ltd.

creased with an increase in the amount of *T. harzianum* applied. The IMI-238493 isolate of *S. rolfsii* was tolerant of the fungicide PCNB, and at low concentrations PCNB improved disease control when integrated with *T. harzianum* treatment. Integration of PCNB and *T. harzianum* under field conditions significantly reduced the incidence of *Sclerotium* root rot (76% disease control in sugar beets) and increased the root, green foliage, and sucrose yield.

Brown girdling root rot in *Brassica napus* (rape) caused by *Rhizoctonia solani* has been a serious problem in the Peace River region of Alberta, Canada, so Dahiya et al. (1988) investigated the possibility of using a biological control organism isolated from a healthy rape field at Agriculture Canada Research Station, Beaverlodge, Alberta. They challenged numerous isolates against *R. solani* and found the most active antagonist to be a fluorescent pseudomonad identified as *P. fluorescens.* Three antibiotics inhibitory to *R. solani* were isolated and identified as pyocyanin, pyrrolnitrin, and phenazine carboxamide. Pyocyanin also inhibited growth of other fungi associated with the rape seedling disease complex, *Fusarium roseum* and *Pythium ultimum.* Pyrrolnitrin and phenazine carboxamide inhibited *Gaeuman-*

*nomyces graminis* (take-all fungus of wheat), *Alternaria brassicae,* and *Botrytis cinerea,* whereas *F. roseum* was only partially inhibited and *Pythium ultimum* was not affected. Treating rape seed at planting time with *P. fluorescens* in *R. solani*–infested soil prevented root rot.

Elsherif and Grossmann (1991) investigated the effects of 141 isolates of fluorescent pseudomonads against *Plasmodiophora brassicae* on *Brassica oleracea* (Capitata group, Chinese cabbage), of 44 isolates against *Rhizoctonia solani* on beans, and of five isolates against *Tilletia caries* on wheat. The researchers found that 19 of the 141 isolates of *Pseudomonas fluorescens* and *P. putida* significantly reduced the percentage of Chinese cabbage plants damaged by *P. brassicae,* 1 of 44 isolates suppressed the damping-off of beans caused by *R. solani,* and 1 of 5 reduced the incidence of the smut caused by *T. caries* on wheat. The method of infesting the system with the antagonists determined somewhat their effects on the pathogens, particularly in the case of Chinese cabbage and beans.

*Cajanus cajan* (pigeon pea, or red gram) is a very important crop in tropical countries, and wilt caused by *Fusarium udum* seriously reduces its yield. Upadhyay and Rai (1987) therefore investigated the interactions between the rhizosphere microorganisms of pigeon pea with the goal of detecting potential antagonists against *F. udum. Aspergillus niger, A. terreus,* and *Penicillium citrinum* were the most antagonistic fungi tested against *F. udum.* Among the actinomycetes, *Streptomyces griseus* and *Micromonospora globosa* were most inhibitory against *F. udum. F. udum* was a strong antagonist against some fungi, but these effects were not of any consequence in controlling the pigeon pea disease caused by *F. udum.*

Volatile allelochemicals were produced by some of the potential antagonists, but they had much less activity against *F. udum* than did the nonvolatiles produced. The culture filtrates of *A. niger, P. citrinum,* and *Micromonospora globosa* were most effective in inhibiting growth of *F. udum* at a 20% concentration. A complete inhibition of the conidial germination of the test pathogen was caused by the culture filtrates of *A. niger, Myrothecium roridum,* and *P. citrinum.* The culture filtrates of *A. niger, A. luchuensis,* and *P. citrinum* completely inhibited chlamydospore germination of *F. udum,* followed by a highly significant inhibition by filtrates of *Aspergillus flavus, A. terreus, Alternaria alternata, Neocosmospora vasinfecta, Myrothecium roridum, Streptomyces griseus,* and *Micromonospora globosa.*

In the investigation by Upadhyay and Rai, species antagonistic under cultural conditions were also antagonistic to the test pathogen in soil. Species such as *A. niger, A. flavus, A. terreus,* and *Micromonospora globosa* highly suppressed conidial and/or chlamydospore germination of *F. udum* and also the growth of that pathogen when inoculated in soil. The researchers concluded therefore, that these antagonistic species may be used for the biological control of the wilt disease of pigeon pea.

*Castanea sativa* (chestnut) cultivation is becoming more popular in Italy and some other countries, and interest in controlling chestnut blight caused by *Cryphonectria parasitica* is increasing. No satisfactory control method has been developed to this time so attention is being directed more toward biological control. Turchetti and Maresi (1991) investigated the use of four hypovirulent (H) strains of *C. parasitica* as possible antagonists against five virulent (V) strains collected from four regions of Italy. To test for vegetative compatibility, V strains were transferred to fresh plates of an appropriate medium and paired with H testers in all possible combinations of H and V strains. H strains were paired with each other.

Inoculations were carried out on *Castanea* sprouts with a combination of 4 H strains, with all V-H strain combinations, and with each V strain. One year after inoculation, infected bark tissue was collected from each type of inoculation, and reisolations were made from the cankers to verify the morphological characters of the cultures.

Field inoculation trials were made on blight-free stump sprouts in a chestnut stand near Florence, Italy. The set of four hypovirulent strains was tested for its ability to limit V-strain growth. Two types of controls were set: wounds without any inoculum and wounds inoculated with a V strain and surrounded by bore holes filled with sterilized bentonite. These inoculations were examined after one year.

Overall, the four H strains showed vegetative compatibility, but the V strains were mutually incompatible with conversion between the H strains and all five V-compatibility groups. Cankers from V strains receiving no treatment grew until they girdled the sprouts and killed them. Uninfected sprouts inoculated with the four H strains survived. The V-infected sprouts with H strains applied around the infections had reduced canker size. The mycelia of the H strains growing in the bark anastomized, and callus formed around the edge of the inoculations. Artificial inoculations with four hypovirulent strains of *C. parasitica* in combination and with virulent infections treated with the same H combination both produced healing cankers. Treatments to cure existing infections are available and can start or increase the natural spread of hypovirulence in chestnut stands.

An important soil-borne disease of *Chrysanthemum* sp. in Taiwan is the basal stem rot caused by *Rhizoctonia solani* (Tschen, 1991). Tschen tested, by various methods, several potential antagonists isolated from Taiwanese soils against *R. solani*. In an early test, the basal portions of chrysanthemum stems were coated with culture filtrates of eight different antagonists—*Penicillium* sp. CF111, *Paecilomyces marquandii* CF302, *P. marquandii* CF407, *Gliocladium deliquescens* F-92, *Bacillus* sp. CB9, *Bacillus* sp. CB20, *Bacillus cereus* CB22, and *B. subtilis* F29-3. *P. marquandii* CF407 and *Gliocladium deliquescens* F-92 significantly reduced the stem rot disease caused by *R. solani* when the coated chrysanthemums were planted in sand inoculated with the pathogen.

In another experiment, 16 antagonists were grown in sawdust compost that was later used as antagonist inocula in sand beds containing *R. solani*. Chrysanthemums were subsequently planted in the beds, with beds free of *R. solani* serving as controls. All 16 of the antagonists significantly reduced the incidence of root rot caused by *R. solani* (Table 12). *Trichoderma harzianum* TVCN1 and TVCN2, *Paecilomyces marquandii* CF302, and *Penicillium* sp. CF111 gave the best control, however. In a test of three delivery systems for the antagonists, raw sawdust was the poorest system; sawdust compost and rice husks were roughly equivalent.

*Cicer arietinum* (chickpea), an important crop in many parts of the world, has become a promising alternative crop for the nonirrigated areas of the Palouse region of eastern Washington and northern Idaho (Kaiser et al., 1989). A very important soil-borne disease of this crop is seed rot and preemergence damping-off caused by *Pythium ultimum*. Greenhouse and field tests were carried out with chickpea accession PI48870 (USA) that is highly susceptible to pythium seed rot and preemergence damping-off. Seven strains of fluorescent pseudomonads (six strains of *P. fluorescens* and one of *P. putida*) and *Penicillium oxalicum* were tested as potential antagonists to *P. ultimum*.

Two *Pseudomonas fluorescens* strains (Q29z-80 and M8z-80) increased emergence and yields of chickpea in some field trials in eastern Washington over a three-year period. In some trials, these seed treatments were equivalent to those due to captan, metalaxyl, or *Pen-*

**Table 12** Control of *R. solani* in chrysanthemum by sawdust compost containing antagonists

| Antagonist | Disease incidence (%)* |
|---|---|
| *Penicillium* sp. CF111 | 9.2 efgh |
| *Paecilomyces marquandii* CF302 | 8.0 fgh |
| *Paecilomyces marquandii* CF407 | 15.6 cd |
| *Gliocladium deliquescens* F-92 | 14.1 def |
| *Penicillium vermiculatum* F-60 | 23.3 bc |
| *Trichoderma viride* TD | 23.3 bc |
| *Trichoderma koningii* T2 | 25.2 b |
| *Trichoderma viride* T20 | 13.8 def |
| *Trichoderma pseudokoningii* T33 | 25.4 b |
| *Trichoderma harzianum* TVCN1 | 6.8 h |
| *Trichoderma harzianum* TVCN2 | 7.9 gh |
| *Trichoderma koningii* T12 | 12.3 efg |
| *Bacillus* sp. CB9 | 22.4 bc |
| *Bacillus* sp. CB20 | 20.6 bcd |
| *Bacillus cereus* CB22 | 13.5 defg |
| *Bacillus subtilis* F29–3 | 9.7 efgh |
| Control | 42.3 a |

* Values within the column not followed by the same letter are significantly different at 5% level, according to Duncan's multiple-range test.

*Source:* From Tschen (1991), with permission of Plant Prot. Soc. of Republic of China.

*icillium oxalicum* treatments. The large variation in results of the treatments indicates that there is a strong environmental effect on the action of the antagonists. Protection with these biological agents needs to be improved so that acceptable levels of control are attained over a wider range of environmental conditions. Combining the fluorescent pseudomonads with *Penicillium oxalicum* might enhance their biological control.

Worldwide there are large decay losses of citrus because of *Penicillium digitatum* (green mold) and *P. italicum* (blue mold). Moreover, isolates resistant to fungicides used to control these fungi have been found in the major producing areas of the world (Wilson and Chalutz, 1989). This makes the need for biological control methods for these decay organisms even more urgent.

Potential antagonists were isolated from *Citrus limon* (lemon) harvested in unsprayed groves in Florida. One hundred and four isolates were obtained and tested against mold on lemon and on *C. paradisi* (grapefruit), using inocula of the potential antagonists with cell counts ranging from $10^7$ to $10^9$ cfu/mL. Grapefruits and lemons previously washed in a 2% hypochlorite solution and then dried were wounded six times with a dissecting needle. The wounds were filled with broth containing the test antagonist. After the wound site dried, each wound was challenged with 20 μL of an inoculum of either *Penicillium digitatum* or *P. italicum* at a concentration of $10^4$ spores/mL. The fruit was incubated for four days at room temperature in a plastic container, with some sterile water added to maintain a high humidity. The wound sites were examined for 14 days to determine the effects of the potential antagonists. After the initial screening, four isolates were selected for further testing on the basis of their antagonism against both *P. digitatum* and *P. italicum.* The four selected inhibited infection by either fungus by 50% or more 11 days after inoculation.

Two of the isolates selected were bacteria, *Pseudomonas cepacia* and *P. syringae,* and two were yeasts, *Debaryomyces hansenii* and *Aureobasidium pullulans.* Only one of the isolates, *P. cepacia,* produced antibiotic zones of inhibition against the two pathogens. Compared with the control, all four of the selected isolates demonstrated significant control of the two decay organisms. *P. cepacia* gave the best protection, however, against both species of *Penicillium* over a two-week period. Actually, all four selected isolates were equally antagonistic during the first week of lesion development, but not after about two weeks. It was concluded that all four isolates selected after initial tests have potential as biocontrol agents against *Penicillium* fruit rots of citrus and that their efficacy can probably be enhanced by changing the preparation and method of application of each.

*Corchorus capsularis* (jute) is an important crop in parts of India, Bangladesh, and a few other countries. At least some of its cultivars are susceptible to root rot and stem rot caused by *Macrophomina phaseolina.* Bhattacharyya et al. (1985) explored the possibility of using *Aspergillus versicolor* in the biological control of that pathogen. The *Aspergillus versicolor* used in their experiments was isolated in their laboratory; the jute used was cultivar JRO-632, which is known to be susceptible to *M. phaseolina.* A soil or soil-plus-compost medium was inoculated with the antagonist, and the medium so inoculated was incubated at 25 ± 2°C for 10 days. It was then dried in aluminum trays to a semisolid cake, which was then infested with *M. phaseolina* by spreading a suspension of mycelia and sclerotia on the soil surface. Two-day-old jute seedlings were planted in the described inoculated medium, some with the pathogen and some without (control). The seedlings were then allowed to grow in the greenhouse for 15 days, after which they were removed, the roots were washed free of soil, and the extent of infection by *M. phaseolina* was determined.

The best inhibition of *M. phaseolina* (50%) by the antagonist, *A. versicolor,* occurred in a soil medium in which the pH was set initially at 4.0. On the basis of published reports, the authors speculated that this may have been due to an increased production of the antibiotic mycoversilin at that pH. When a soil-compost medium was used in the experiment, a mixture of 10% compost and 90% soil gave the best biological control of *M. phaseolina* by the antagonist. Again the authors used published reports to speculate that the greater control by *A. versicolor* in a medium with added compost may have been due to a greater production of mycoversilin. Unfortunately, no tests were run for the presence or quantity of mycoversilin. The evidence obtained did indicate, however, that *A. versicolor* is a good biological control antagonist against *M. phaseolina,* the causative agent of root rot and stem rot of jute.

Fire blight on ornamentals caused by *Erwinia amylovora* is rather common in several parts of the world, and it is a rather serious problem on *Cotoneaster bullatus* (cotoneaster) in Germany. Isenbeck and Schulz (1986) previously had success in the biological control of the fire blight organism, and they decided to investigate the activity of several bacteria isolated from blighted ornamentals. They obtained several yellow-pigmented, rod-shaped bacterial strains with antagonistic activity against *E. amylovora.* They selected four isolates for additional studies: active strains C/1 and C/81 from blighted shoots of cotoneaster; E.H. 112y from an apple shoot; and an inactive strain, C/5, from cotoneaster. They found that a strong pH decrease occurs in the media of antagonistic isolates, whereas the pH increases in the medium of the inactive strain. No inhibitor formation could be measured in buffered media; it was found that it is not only the production but also the inhibitor action that is optimal under acid conditions. When three-day-old culture filtrates were titrated to the initial value of 6.6 and tested in a plate assay, hardly any inhibitory activity occurred. The inhibitors from the isolates were not destroyed by autoclaving for 15 minutes. In each case the inhibitor was soluble in water and methanol; slightly soluble in ethanol; and not soluble in chloroform, acetone, hexene, or acetic acid ethyl ester.

Fire blight could be somewhat reduced in *Cotoneaster bullatus* by shoot-tip inoculation using culture filtrates of the antagonistic bacteria. Use of living antagonists in the shoot-tip inoculation, however, gave considerably better fire blight control. It therefore appears that the permanent *de novo* synthesis of the antibiotic under natural conditions is necessary for effective fire blight control.

Considerable evidence exists in the literature that many nonpathogenic isolates of *Fusarium oxysporum* exert considerable antagonistic effects against pathogenic forms of this species. Paulitz et al. (1987) decided to determine if some of the nonpathogenic isolates of *F. oxysporum* could protect *Cucumis sativus* (cucumber) against *Fusarium* wilt in a conducive soil. They obtained nonpathogenic isolates of *F. oxysporum* from surface-disinfested, symptomless cucumber roots grown in two raw (nonautoclaved) soils. The isolates were screened for pathogenicity and antagonistic activity against *Fusarium* wilt of cucumber in raw soil infested with *F. oxysporum* f. sp. *cucumerinum.*

Twenty-five nonpathogenic isolates of *F. oxysporum* from each soil were chosen for their ability to reduce the incidence of *Fusarium* wilt of cucumbers caused by *F. oxysporum cucumerinum.* Three isolates (C1, C5, C10) significantly reduced the rate of increase in disease incidence. In fact, these isolates delayed the development of 50% disease incidence by 3 to 6 days in the first trial and 6 to 13 days in the second trial. Three isolates did not affect the infection time. C5 was added to raw soil infested with various inoculum densities of *F.o.c.* and reduced the infection rate at all inoculum densities of the pathogen. Paulitz et al. pointed out that some biological antagonists work well in steamed or sterilized soil but are not successful in field trials because they are not adapted to the environment in which the pathogen lives. They stated that use of the antagonist that occupies the same niche as the pathogen could overcome that obstacle.

In a subsequent paper (Park et al., 1988), the same scientists investigated the interactions between *Pseudomonas putida* and nonpathogenic isolates of *Fusarium oxysporum* in the biological control of *Fusarium* wilt of cucumber. They reported that the combination of the two types of antagonists was effective in reducing that disease of cucumber when added to the soil together at pH 6.7, but the antagonists were not effective when added separately. The scientists found that the combination was most effective in nearly neutral to alkaline soils (pH 8.1). The population of fluorescent pseudomonads increased significantly in the cucumber rhizosphere in the presence of a nonpathogenic isolate of *F. oxysporum* in an alkaline soil (pH 8.1). The researchers speculated that the addition of nonpathogenic strains of *F. oxysporum* increased root exudation of the cucumber, thereby enhancing the activity of added pseudomonads in the rhizosphere through their production of siderophores, which increases competition for the iron necessary for effective germination of spores of the pathogen and penetration of the host.

Two new strains (T-68 and Gh-2) of *Trichoderma* spp. were tested against *Fusarium* wilt of *Cucumis melo* (muskmelon) caused by *Fusarium oxysporum* f. sp. *melonis* and of *Gossypium barbadense* (Pima cotton, S5) caused by *F. oxysporum* f. sp. *vasinfectum* (Ordentlich et al., 1991). Their efficacies were also compared with the antagonistic activity of *Trichoderma harzianum* (T-35). The muskmelon pathogen was isolated from diseased muskmelon and the cotton pathogen from diseased cotton in commercial fields in central Israel. In initial tests, a disc containing a *Trichoderma* strain was placed on a petri plate inoculated 48 hours previously with one of the pathogens. Four or five days later, the plates were observed for overgrowth or other interactions. Both strains T-68 and Gh-2 of *Trichoderma* spp. completely overgrew each of the pathogens, effectively preventing their growth. On the other hand, *T. harzianum* (T-35) did not overgrow the pathogens, but it did prevent them from colonizing the plates.

In tests concerning the biological control of *Fusarium* wilt by the *Trichoderma* isolates, one of these was added to soil infested with either the muskmelon or cotton pathogen, and the appropriate plants were placed in the soil. After 14 days, the disease incidence was observed in the greenhouse. Disease symptoms on cotton plants were reduced from 75% wilted seedlings in the inoculated control to 5, 7, and 7% by T-35, T-68, and Gh-2, respectively. The percentage of wilted melon seedlings was not affected by T-68 or Gh-2, whereas T-35 reduced the disease incidence from 87 to 17%. The scientists found no correlation between enzymatic activity (chitinase and glucanase) and the biocontrol capability against *Fusarium* wilt on muskmelon and cotton.

*Cucumis sativus* (cucumber) damping-off is a serious disease caused by *Pythium ultimum*. Paulitz and Baker (1987) studied its possible biological control by *Pythium nunn*, which is nonpathogenic to cucumber. Isolates of *P. nunn* (N3) and *P. ultimum* (N1) used in previous studies were selected for these experiments. Addition of *P. nunn* to aerated, steamed soil infested with *P. ultimum* resulted in disease and pathogen reduction. The population densities of *P. ultimum* and disease incidence decreased as the initial inoculum density of *P. nunn* increased. Moreover, disease reduction decreased as the initial inoculum density of *P. ultimum* increased. Suppression lasted longer when the *P. nunn* inoculum was produced on 1% rolled oats than when produced on 0.3% ground bean leaves. Cucumbers grown in steamed soil infested with *P. ultimum* and *P. nunn* had greater root dry weight than did plants grown in soil infested with *P. ultimum* alone. The overall evidence indicated that *P. nunn* requires organic substrates instead of living fungal hosts for disease suppression to occur.

Paulitz et al. (1990) continued the investigation of Paulitz and Baker (1987) on *Pythium* damping-off of cucumber, but used an additional antagonist, *Trichoderma harzianum* (T-95), along with *Pythium nunn*. The isolates of *P. nunn* and *P. ultimum* were the same as those used in the previous investigation. In the steamed soil without a bean-leaf amendment, and infested with *P. ultimum,* disease in the control exceeded 80% in each cucumber (Marketer Long) planting. *P. nunn* applied to the soil did not reduce damping-off in the first or second planting, but was effective in the third. In raw soil, however, without bean-leaf amendment, *P. nunn* significantly reduced the disease incidence in the first and third plantings. Seed treatment with *T. harzianum* (T-95) reduced disease to less than 10% in each planting. No significant differences occurred in disease incidence between the treatment with T-95 alone or in combination with *P. nunn.* In Colorado soil treatments amended with bean leaves, *P. nunn* was as effective as the seed treatment with T-95 or the combination of the two. The biocontrol agents reduced disease to less than 10% by the third planting.

In the experiment performed in the Oregon soil mix, seed treatment with T-95 was not as effective in reducing disease as it was in the Colorado soil mix. In the soil infested with *P. ultimum,* the initial damping-off was more severe than in the Colorado mix, up to 100% at the first planting. The only biological control treatment that significantly reduced disease in steamed soil infested with *P. ultimum* at the first planting was the *P. nunn* plus T-95 treatment. In the absence of the bean-leaf amendment, *P. nunn* alone did not reduce damping-off in subsequent plantings. When bean leaves were added to the raw soil mix, damping-off was reduced by 27 and 57%, respectively, by the *P. nunn* and *P. nunn* plus T-95 treatments when compared with the control. In the Colorado soil without leaf amendment, *P. nunn* applied alone resulted in an increase in shoot dry weight of cucumbers only in the steamed soil infested with *P. ultimum.* In soil amended with bean leaves, *P. nunn* enhanced shoot biomass in both steamed-infested and raw-infested soil. Seed treatment with T-95 stimulated shoot dry weight over the control in all Colorado soil treatments, except for raw soil amended with bean leaves. The combination of *P. nunn* and T-95 resulted in greater shoot dry weight than either applied alone in raw-infested soil without bean leaf and in steamed-infested soil amended with bean leaves. Again in the Oregon raw soil, the combination of both biocontrol agents stimulated cucumber growth more than did either alone. This is another of the numerous experiments that demonstrate that two or more compatible biological control agents can be combined to give additional control of a pathogen.

We have seen that fluorescent pseudomonads suppress several diseases caused by soil-borne plant pathogens. One of the mechanisms involved in disease suppression is the production of antifungal metabolites (antibiotics) (Maurhofer et al., 1992). Maurhofer et al. were aware that *Pseudomonas fluorescens* strain CHAO suppresses several plant diseases and produces several antibiotics, two of which are pyoluteorin (Plt) and 2,4-diacetyl-phloroglucinol (Phl). They decided therefore to develop a strain of P.f. CHAO that produces larger amounts of these two antibiotics and compare the two strains for suppression of *Pythium ultimum* on *Cucumis sativus* (Chinesische Schlange), cress (*Lepidium sativum* Gartenkresse einfach), and corn (*Zea mays* var. *saccharata*). They developed a new strain of CHAO by transferring recombinant cosmids from CHAO using *Escherichia coli* and triparental mating using the helper plasmid pME497. The transconjugants were screened for enhanced inhibition of *P. ultimum.* The scientists selected CHAO/pME3090 as a transconjugant that produced significantly more Plt and Phl on malt agar and significantly more Plt on King's B agar. In the absence of the pathogen, the two strains had no effect on the prospective host plants. In a gnotobiotic system the recombinant strain protected cucumber plants against disease caused by *P. ultimum* more than did the wild type CHAO. Both strains protected cress and sweet corn from *P. ultimum,* but in the presence of the pathogen, fresh weights of cress and corn protected by strain CHAO/pME3090 were lower compared with those of plants protected by strain CHAO (Table 13). The recombinant strain (but not strain

CHAO) also reduced the growth of cress and sweet corn in the absence of the pathogen. Cucumber, cress, and corn were all inhibited in growth by the antibiotics Plt and Phl, but cucumber was much less sensitive to Phl than were the other plants. It was inferred, therefore, that the enhanced pyoluteorin production might be responsible for the increased ability of CHAO/pME3090 to suppress damping-off of cucumber and that that strain's detrimental effect on cress and sweet corn might be due to the phytotoxic effects of both antibiotics.

Filippi et al. (1987) previously found that *Bacillus subtilis* M51 protected *Dianthus caryphyllus* (carnation) against fusarium wilt caused by *F. oxysporum* f. sp. *dianthi.* They therefore decided to investigate the relation between protection against the fusarium wilt of carnation and colonization of the bacterial antagonists on the roots of that plant. They first determined the antibiotic resistance of several isolates of M51 and selected a strain (designated MZ51) resistant to selenomycin (25 μg/mL). Roots of 2000 carnation cuttings were inoculated with MZ51 cells on Waksman agar supplemented with 25 μg/mL of selenomycin. One thousand of these cuttings were planted in soil naturally infested with *F. o. dianthi,* and the other 1000 were planted in soil free of the pathogen. Field trials were conducted similarly at two field stations in Italy.

The presence of the MZ51 strain on the roots was determined at five day intervals, from planting to the ninetieth day. Up to the thirty-fifth day, MZ51 colonies were present on all carnation root fragments examined; after that, the cuttings showed a rapid decrease in the percentage of roots with bacteria present. They reached very low values after 90 days, and differences were seen from one part of a root to another. Root apices remained inoculated somewhat longer than other areas, except for the area next to the stem, where the antagonists remained on all cuttings. It was inferred that the antagonistic effect of MZ51 was a temporary one, with all cuttings being protected for 60 days. After that period, there was no difference from the controls.

Van Peer et al. (1990) reported that fusarium wilt of carnation was reduced some by the iron source and also by the antagonist *Pseudomonas* sp. strain WCS417r. For example, fusarium wilt caused by *F.o.d.* was reduced when Fe-EDDHA was used instead of Fe-DTPA, and the reduction was increased by addition of the antagonist WCS417r in Pallas carnation, which is moderately resistant to *F.o.d.,* but not in

**Table 13** Influence of increased pyoluteorin (Plt) and 2,4-diacetylphloroglucinol (Phl) production in *Pseudomonas fluorescens* strain CHAO on the suppression of damping-off of cucumber, cress, and sweet corn caused by *Pythium ultimum*

| Plant tested | Microorganisms added | | Root fresh weight[‡§] (mg) | Plant fresh weight[‡§] (mg) | Fluorescent pseudomonads[§] ($10^8$ cfu/g root) |
|---|---|---|---|---|---|
| | Bacteria* | *P. ultimum*[†] | | | |
| Cucumber | None | – | 547 a | 1,147 ab | 0.0 a |
| | CHAO | – | 544 a | 1,175 a | 1.0 b |
| | CHAO/pME3090 | – | 498 a | 1,138 ab | 0.4 b |
| | None | + | 71 d | 91 d | 0.0 a |
| | CHAO | + | 158 c | 471 c | 18.0 c |
| | CHAO/pME3090 | + | 394 b | 948 b | 1.8 b |
| Cress | None | – | 1,032 a | 2,696 a | 0.0 a |
| | CHAO | – | 1,045 a | 2,627 a | 14.0 b |
| | CHAO/pME3090 | – | 462 b | 1,317 b | 30.0 b |
| | None | + | 47 d | 176 d | 0.0 a |
| | CHAO | + | 618 b | 1,578 b | 23.0 b |
| | CHAO/pME3090 | + | 247 c | 751 c | 49.0 b |
| Sweet corn | None | – | 1,489 a | 2,111 a | 0.0 a |
| | CHAO | – | 1,317 a | 1,917 a | 1.1 b |
| | CHAO/pME3090 | – | 982 b | 1,430 b | 1.2 b |
| | None | + | 576 d | 929 c | 0.0 a |
| | CHAO | + | 952 b | 1,471 b | 1.2 b |
| | CHAO/pME3090 | + | 665 c | 1,010 c | 1.7 b |

* CHAO is the wild-type strain of *P. fluorescens,* and CHAO/pME3090 is the Plt- and Phl-overproducing transconjugant of strain CHAO. Bacteria were added as described in "Materials and Methods."

† *P. ultimum* was added as described in "Materials and Methods."

‡ Fresh weight per plant for cucumber and sweet corn; fresh weight per flask for cress.

§ Means for the same host plant within the same column followed by the same letter are not significantly different at $P = 0.05$ according to the Student's *t*-test considering one independent experiment as a repetition. Cucumber: each value is the mean of three experiments, with three replicates per experiment and one flask with three plates per replicate. Cress: each value is the mean of six experiments, with five replicates per experiment and one flask with plants grown from 0.2 g of seeds per replicate. Sweet corn: each value is the mean of six experiments, with three replicates per experiment and one flask with five plants per replicate.

*Source:* From Maurhofer et al. (1992), with permission of Amer. Phytopathological Soc.

the susceptible cultivar Lena. Higher concentrations of the siderophore fusarine produced by *F. o. diantha* WCS816 were found when Fe-EDDHA was used as the iron source instead of Fe-DTPA. Germ-tube length of *F. O. diantha* was less with Fe-EDDHA than with Fe-DTPA. It seemed that EDDHA had a toxic effect on the conidia of *F. o. diantha* strain WCS816, in addition to limiting the Fe supply. The scientists concluded that the reduced fusarium wilt due to Fe-EDDHA was caused by a limited iron supply and that the fluorescent pseudomonad reduced the infection even more because of its strong competition for iron.

Soils suppressive to fusarium wilts have been recognized in the United States and France for several years, and it is known that two different microbial populations, nonpathogenic *Fusarium* spp. and fluorescent *Pseudomonas* spp., are involved in the suppressiveness (Lemanceau et al., 1992). Lemanceau et al. decided to investigate the effect of pseudobactin 358 (an antibiotic) produced by *Pseudomonas putida* WCS358 on the combined action of *P. putida* WCS358 and the nonpathogenic *F. oxysporum* Fo47 on fusarium wilt of carnations. These combined antagonists efficiently suppressed fusarium wilt of that plant; the reduction attained was significantly greater than that obtained by each antagonist alone. *P. putida* WCS358 had no effect on disease severity when inoculated on its own, but it significantly increased the control achieved with the nonpathogenic *F. oxysporum* Fo47b10. On the other hand, a siderophore-negative mutant of WCS358 had no effect on disease severity even when inoculated with Fo47b10 (Fig. 7). The researchers inferred that because the cfu densities of the two antagonists were the same at the root level, the difference between the two *P. putida* strains in reducing disease must have been due to the production of pseudobactin 358 by the wild-type WCS358 strain. Moreover, the production of the antibiotic by WCS358 must have been responsible for the increased disease reduction by Fo47b10 combined with WCS358 compared with that by Fo47b10 alone.

*Phytophthora nicotianae* var. *parasitica* is the chief fungus responsible for foot rot of citrus in Spain; therefore Tuset et al. (1990) decided to investigate the possible biological control of foot rot on *Citrus sinensis* (sweet orange) and *C. deliciosa* (willowleaf mandarin) with *Myrothecium roridum*. Initial tests against *P. n. parasitica, P. syringae,* and *P. capsici* demonstrated that *M. roridum* produced potent toxins in vitro to those three species of *Phytophthora*. In vivo tests with

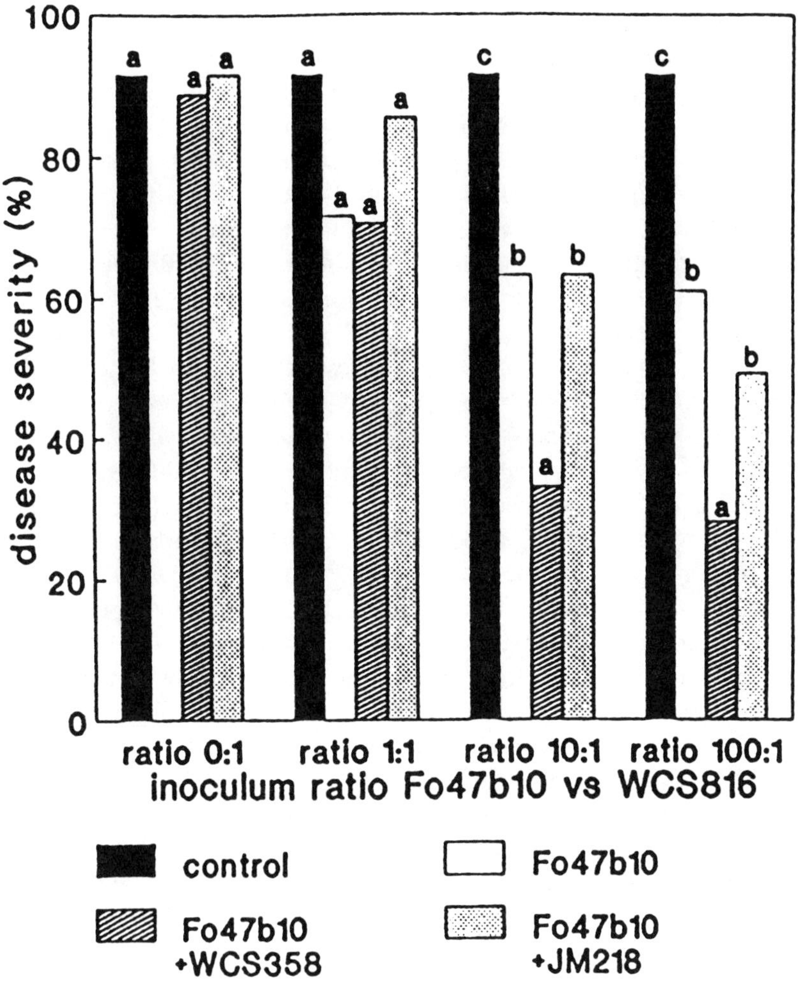


FIGURE 7 Comparison of the effects of *Pseudomonas putida* WCS358 and the sid-negative mutant JM218 on fusarium-wilt severity for different inoculum ratios of nonpathogenic *Fusarium oxysporum* Fo47b10 to pathogenic *F. oxysporum* WCS816. For the same Fo47b10/WCS816 inoculum ratio, means designated with the same letter are not significantly different ($P = 0.05$). *Adapted from Lẹmanceau et al. (1992), with permission of Amer. Soc. for Microbiology.*

*Citrus sinensis* or *C. deliciosa* (1) inoculated with *M. roridum* alone, (2) first inoculated with *M. roridum* and subsequently with *P. n. parasitica,* and (3) inoculated with *P. n. parasitica* alone demonstrated that *M. roridum* was an excellent antagonist against *P. n. parasitica* in both species of *Citrus.* Only 32.6% of the sweet orange plants were infected when *M. roridum* was inoculated into the bark before the pathogen, and 26.2% of the willowleaf mandarin plants were infected when similarly inoculated (Table 14). On the other hand, 100% of the controls inoculated only with *Phytophthora n. parasitica* were infected in both citrus species.

Dandurand and Menge (1992) investigated the influence of *Fusarium solani* on citrus root rot caused by *Phytophthora parasitica* (= *P. nicotianae* var. *parasitica*) and *P. citrophthora. Citrus paradisi* (grapefruit), *Citrus sinensis* (sweet orange), and *Poncirus trifoliata* × *C. sinensis* (Troyer citrange) were the plants used in the tests. Inoculation of citrus with either *F. solani* and *P. parasitica* or *P. citrophthora* increased root rot compared with inoculation with *P. parasitica* or *P. citrophthora* alone, when plants were inoculated with *Phytophthora* by dipping their roots in zoospore suspensions and subsequently transplanted into soil infested with *F. solani.* Root rot was not increased by simultaneous coinoculation of *P. parasitica* and *F. solani,* however, or when plants were inoculated first with *F. solani.* Nor was root rot increased when citrus plants were inoculated with *P. parasitica* 30 days after transplanting into soil infested with *F. solani.* Root dipping into zoospore suspensions of *P. parasitica* and transplanting into soil infested with *F. solani* reduced feeder root length by 62% and root weight by 61% when compared with inoculation by *P. parasitica* alone. When citrus roots were inoculated with *P. citrophthora* and transplanted into soil infested with *F. solani,* feeder root length was reduced by 68%, but feeder root weight was not significantly reduced when compared with that of plants inoculated with only *P. citrophthora.* It was concluded that the role of *F. solani* in the development of root rot of citrus is secondary to that of *P. parasitica* or *P. citrophthora,* but that there is sufficient evidence to suggest that *F. solani* can contribute to citrus root rot.

*Penicillium digitatum* is a major postharvest pathogen of citrus. An effective biocontrol organism would be of great value. Huang et al. (1992) tested the effects of the potential antagonist, *Bacillus pumilus* (B-PRCA-1), on the control of *P. digitatum* in vitro and on Valencia

**Table 14** Control by *Myrothecium roridum* of foot rot due to *Phytophthora nicotianae* var. *parasitica* in citrus seedlings under glasshouse conditions. Data recorded 21 days after inoculation with *P. n. parasitica* and 30 days after inoculation with *M. roridum*

| | Pathogen (0.4-cm agar discs inoculated into bark of basal part of stem) | | | | | |
|---|---|---|---|---|---|---|
| | *M. roridum* alone | | *M. roridum* then *P. n. parasitica* | | *P.n. parasitica* alone | |
| Citrus species | % affected | Mean lesion area ($cm^2$) | % affected | Mean lesion area ($cm^2$) | % affected | Mean lesion area ($cm^2$) |
| Sweet orange (*C. sinensis*) | 0 | – | 32.6 | 2.2 | 100 | 3.6 |
| Willowleaf mandarin (*C. deliciosa*) | 0 | – | 26.2 | 1.7 | 100 | 2.3 |

*Source:* From Tuset et al. (1990), with permission of Blackwell Scientific Pub. Ltd.

oranges (*Citrus sinensis*), Washington navel oranges, and Lisbon lemons (*C. limon*). The scientists also compared the effects of *B. pumilus* with the fungicides benomyl and imazalil.

*B. pumilus* had a strong effect in vitro, producing large inhibition zones in the *P. digitatum* growth on petri plates, significantly larger zones than did benomyl and imazalil. In in vivo tests, inoculated sites on Valencia oranges were infected by *P. digitatum;* in the water control, benomyl gave poor control, whereas *B. pumilus* and imazalil completely inhibited *P. digitatum* decay. *B. pumilus* was quite inhibitory to *P. digitatum* on Lisbon lemons at all concentrations ($1.9 \times 10^7$ to $10^{10}$ cfu/mL). Again on Washington navel oranges, all antagonist concentrations significantly controlled *P. digitatum* infections. Thus *B. pumilus* proved to be an excellent biocontrol agent for *P. digitatum* on citrus.

The large financial losses due to the citrus fruit rot caused by *Penicillium digitatum* are very obvious; witness the many research projects related to that disease organism. Another of these projects concerned the possible biocontrol of *P. digitatum* by *Trichoderma viride.* These two organisms were collected from citrus, the first from decaying and the second from healthy fruit. Navelina oranges were used as the source fruits for the tests.

Navelina oranges protected with aqueous suspensions of *T. viride* spores ($2.5 \times 10^6$ to $2.5 \times 10^9$/mL) had an increase in resistance toward *P. digitatum.* The fruits inoculated with *P. digitatum* did not produce lesions in five days when *T. viride* was applied 48 or 72 hours before inoculation with *P. digitatum.* Moreover, fruits inoculated with *T. viride* alone did not exhibit any disease symptoms. Thus, in this experiment, *T. viride* proved to be an effective antagonist to *P. digitatum* on citrus fruits.

Fire blight of *Crataegus monogyna* (hawthorn) and other rosaceous plants in Britain is caused by *Erwinia amylovora,* for which there is no suitable chemical control agent allowable for use (Wilson et al., 1990). Wilson et al. collected 136 isolates of presumed *Erwinia herbicola* from flowers and leaves of hawthorn. Many suppressed symptoms of fire blight, and 34 isolates considered more suppressive than *E. herbicola* isolate Eh252 (a known suppressive isolate) were selected for further tests.

The 13 most suppressive isolates of *E. herbicola* gave significant control of blossom blight of crataegus six days after inoculation in the first trial. In a second test, all isolates except two gave significant control of the disease. No isolate gave better control than Eh252 in either test, however. Two isolates of *E. herbicola* were compared with Agrimycin 17, Copac E, and the experimental bactericides S-0208 and JF4387. All gave significant control of blossom blight at five days after inoculation. The amount of control given by one of the isolates of *E. herbicola* at $10^8$ cells/mL did not differ significantly from that provided by Agrimycin 17, Copac E, or JF4387; even at $10^7$ cells/mL, there was no significant difference from the protection provided by JF4387 or Copac E. No isolate gave the protection afforded, however, by the experimental bactericide S-0208.

Six isolates were tested for control of shoot blight of hawthorn; all significantly reduced the disease index in two tests. The isolates that were most effective in the control of blossom and shoot blight of hawthorn were identified in the AP120E Profile Index as belonging to the *E. herbicola–Enterobacter agglomerans* complex.

*Balansia cyperi*–infected leaves of *Cyperus rotundus* did not harbor *Rhizoctonia solani* but uninfected leaves did, and the leaf tissue harboring *R. solani* showed no symptoms of disease (Stovall and Clay, 1991). Extracts of *B. cyperi* mycelia showed no inhibitory effects on growth of *F. oxysporum, R. oryzae, R. solani, T. harzianum,* or *Alternaria alternata.* Moderate to strong inhibition of *Cladosporium cladosporiodes* was caused by four different solvent extracts (hexane, chloroform, ethyl acetate, and ethyl ether), whereas *Penicillium glabrum* was inhibited slightly by ethyl acetate and ethyl ether extracts only. Solvent extracts of *Balansia cyperi*–infected leaves inhibited growth of most test fungi, including *F. oxysporum, R. oryzae,* and *R. solani.* Extracts of uninfected purple nutsedge leaves inhibited only the last two species, and to a lesser extent. It was concluded that infection of *C. rotundus* by *B. cyperi* may deter infection of its host by other, more pathogenic fungi.

*Armillaria luteobubalina* is the most pathogenic and widespread of five *Armillaria* spp. described from Australia (Pearce, 1990). Studies outside of Australia have shown that inoculation of tree stumps with saprotrophic fungi can significantly reduce inoculum levels of root-rot fungi. Pearce carried out a very comprehensive laboratory experiment to screen wood-decay fungi for their ability to replace or restrict growth of *A. luteobubalina* on *Eucalyptus diversicolor* (karri). Four isolates of *A. luteobubalina* were selected for the tests, and 128 isolates of wood-decay fungi were tested for antagonism against *A. luteobubalina.* About eight major kinds of interactions occurred between the fungal isolates and the *A. luteobubalina* strains in petridish tests. Some of the test isolates rapidly overgrew the *A. luteobubalina,* severely limiting its growth. Twenty-one possible antagonists were selected for further tests on karri-stem sections. Pearce found a close correlation between the relative antagonistic effect on agar and on the karri blocks. For example, *Coriolus versicolor* consistently replaced *A. luteobubalina* on karri wood, just as it did on agar plates. It appears, therefore, that *C. versicolor* would be a good prospect for biocontrol of *A. luteobubalina* under field conditions.

There are many research projects and publications that concern the biological control of widely occurring pathogens in which specific hosts are not selected for specialized study. I have arranged these papers alphabetically according to pathogen, and I discuss them below under the heading "general hosts."

Crown gall caused by *Agrobacterium tumefaciens* (or *Agrobacterium radiobacter* var. *tumefaciens*) occurs widely on many hosts in many countries. Rysheuvels et al. (1984) isolated 139 *Agrobacterium* strains from galls and soil in the fruit tree area around Gorsen, Belgium. Thirty isolates were identified as *A. tumefaciens:* eleven belonged to biotype 1, nine to biotype 2, four to biotype 3, and six to an undetermined biotype. When strain K84 of *A. radiobacter* (or *A. radiobacter* var. *radiobacter*) were tested on artificial culture against the isolates of *A. tumefaciens,* isolates from biotypes 1 and 2 were sensitive to K84, but their sensitivity was variable. Biotype 3 isolates were generally resistant to K84. When K84 was tested for antagonistic action against datura plants inoculated with one of three biotype 1 pathogens, or one of three biotype 2 pathogens, or one biotype 3 pathogen, K84 gave good control of biotype 1 or biotype 2 isolates but not of the biotype 3 isolate when inoculated with a concentration of $10^{10}$ K84 bacteria/mL and $10^{10}$ pathogens/mL. When $10^9$ isolates/mL were used in the inoculation of the pathogen, the $10^9$ concentration of K84 bacteria/mL gave good control of all biotype 1 and 2 isolates, but not of the biotype 3 isolate. It appears therefore that K84 is a good antagonist for the biocontrol of biotype 1 and 2 strains of *A. tumefaciens* if the concentrations of K84 are sufficiently high.

According to Pesenti-Barili et al. (1991) the use of strain K84 in the control of crown gall "is considered an outstanding example of effective microbial biological control." A better carrier for the storage and dispersal of K84 is desirable, however, and Pesenti-Barili et al. screened nine materials to select those with the best qualities. They found it was possible to preserve *A. radiobacter* cells on dry solid supports for a long time, provided the storage temperature is 4°C and the inoculum volume for $4 \times 10^9$ cfu/g is not less than 0.15 mL/g of carrier. However, a substantial carrier water content was necessary for room-temperature storage. At 21°C, the differences between most carriers were small. Porosil MP, expanded clay, and kaolin showed the best results, with storage times greater than three months. Vermiculite provided the longest storage time (563 days) at 4°C, and Micro-cel provided the shortest (99 days). Good carriers at 4°C were often poor carriers at 21°C and vice versa. In addition to its long storage time, vermiculite also assured full and immediate biological activity in the prevention of crown gall. It was concluded that vermiculite is suitable for a new formulation of *A. radiobacter* K84 cells for biological control of crown gall.

Many soils used for the cultivation of fruit trees and ornamental shrubs in Poland are contaminated with *Agrobacterium tumefaciens.* According to regulations there, nursery material with galls on the collar or main roots cannot be sold, and it is estimated that about a half million nursery trees and roses are destroyed annually because of crown gall. Sobiczewski et al. (1991) therefore investigated the efficacy of *A. radiobacter* K84 to control crown gall caused by a highly pathogenic strain of *A. tumefaciens* on rootstocks of fruit trees and roses. Hard-pruned roots of one-year-old generative rootstocks of fruit trees were treated by immersion for five minutes in a water suspension of K84 bacteria and left to dry for 10 to 15 minutes. The roots were then immersed for five minutes in a water suspension of a highly pathogenic strain of *A. tumefaciens* AT-4, biotype 2, at a concentration of about $10^8$ to $10^9$ cells/mL. Treatment with a water suspension of Miedzian 50 (50% copper as copper oxychloride) was used for comparison with the K84 treatment. Treated rootstocks were planted immediately in the field, and galls on collars and roots were evaluated after six months. There were 96 rootstocks of each species per treatment (8 replicates of 12 plants). In a separate trial, rootstocks of fruit trees and roses were treated similarly, but instead of being inoculated with the pathogen, they were planted in the field where a high intensity of crown gall had been observed every year. The treatments with K84 were highly effective in all cases. Treatment with copper oxychloride was usually less effective against crown gall and was highly phytotoxic in some species. The results were so striking that the pharmaceutical factory at Pabianice cooperated in the formulation of the biopreparation of K84 named Polagrocyna, which has now been introduced into commercial use.

Strains of *Agrobacterium* are classified into three biotypes on the basis of carbon source utilization and other biochemical tests (Ryder and Jones, 1991). The biotypes correspond to different chromosomal forms of the bacterium. The significance of the biotypes as causal agents of crown gall varies with location and plant host. In Spain, biotype 2 strains usually affect stone fruits, whereas biotype 1 strains are important pathogens on roses. In Australia the main pathogenic forms on almond and stone fruit are nopaline-agrocinopine A strains of biotype 2. Biotypes 3 strains are the most important pathogens on grapes.

As previously discussed, *A. radiobacter* strain K84 has been successfully used for the biocontrol of numerous pathogenic strains of *A. tumefaciens,* but not all. Production of the antibiotic agrocin 84 is the major reason for the success of strain K84, but the continued success of this strain has been jeopardized by the possibility of transfer of the agrocin plasmid pAgK84 to pathogenic agrobacteria, thus making them resistant to control. Through genetic engineering, Ryder and Jones (1991) constructed a transfer-deficient deletion mutant of pAgK84, which they designated strain K1026. This achievement is significant because the agrocin plasmid (pAgK1026) can no longer be transferred to other agrobacteria. The scientists compared the action of K1026 and K84 in controlling crown gall on almond planted on soil infested with biotype 2 *A. tumefaciens.* They found that K1026 was as effective as K84 in controlling crown gall on the almond plants seven months after the start of the test. K1026 was registered as a pesticide in Australia and has been available there in every state since 1988.

I mentioned previously that *Armillaria* spp. are very important pathogens of woody plants in forests. Several attempts have been made to use fungi in the biocontrol of *Armillaria* but the potential of bacteria as biocontrol agents of forest root pathogens is relatively unknown (Dumas, 1992). Dumas collected soil samples from 24 areas in the boreal mixed forest of North America that had low, moderate, and high concentrations of *Armillaria* infections. Dumas isolated 2462 bacteria

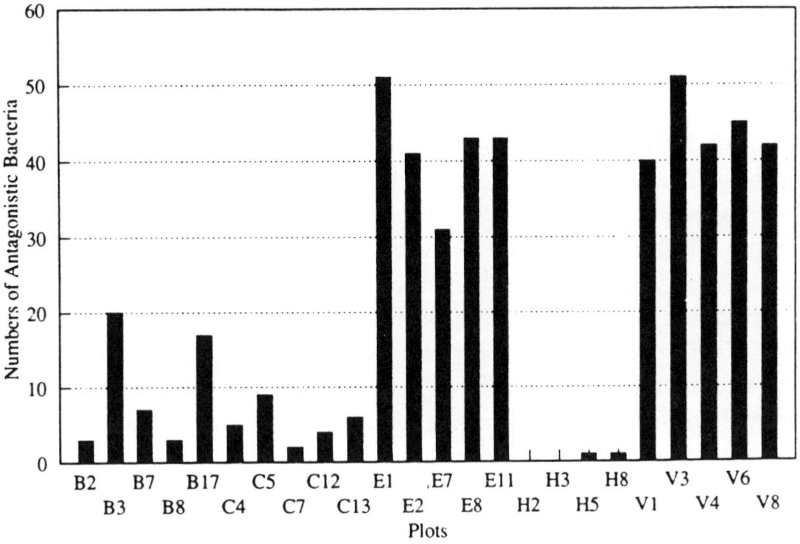


FIGURE 8 Number of inhibitory bacterial strains per plot isolated from the soils of the Boreal Mixedwood Forest. Letter/number combinations refer to specific plots. *Adapted from Dumas (1992), with permission of Paul Parey GmbH & Co. KG, Hamburg, Germany.*

from the soil samples, of which 507 inhibited the linear growth of *A. ostoyae*. Bacteria were isolated from all depths of the soil columns. No correlation was observed between the numbers of antagonists and soil depth, and there were large variations in numbers of antagonists isolated from various plots (Fig. 8).

The most frequent inhibitory bacteria isolated were the pseudomonads, of which *Pseudomonas fluorescens* was the most common. Other bacteria that inhibited the linear growth of *A. ostoyae* were species of *Bacillus, Enterobacter,* and *Serratia* and *Agrobacterium radiobacter. Serratia* spp. and *A. radiobacter* were the least frequently isolated bacterial inhibitors of *Armillaria*. Isolates of *P. fluorescens, Pseudomonas* spp., and *A. radiobacter* were the most variable in their inhibition, whereas *Bacillus* spp. were the least variable. *Enterobacter* spp. and *Serratia* spp. were only slightly inhibitory to *A. ostoyae.*

Because of the inconsistent ability of *A. ostoyae* to produce rhizomorphs in culture, *Armillaria gallica* was used as the test organism to determine the abilities of the bacteria to prevent rhizomorph formation. The abilities of the bacteria to inhibit rhizomorph formation was strain-dependent within a bacterial group. *P. fluorescens* and *Bacillus* spp. had the greatest influence on rhizomorph production. This is an important finding because the rhizomorphs are the infectious structures of *Armillaria* spp. It is now anticipated that the more active bacteria can be reintroduced in high concentrations in the boreal mixedwood forest of Ontario, Canada, to protect seedlings from infection by *Armillaria*. Dumas suggested that the variations in antagonistic activity between strains of the same species of bacteria may result from variations in production of antibiotics.

*Epicoccum purpurascens* is a widespread, red-pigmented fungus that survives saprophytically on senescent plant remains. It produces several antibiotics (Brown et al., 1987). *E. purpurascens* was found in initial tests on agar to inhibit growth of *Botrytis allii, Cochliobolus sativus, Rhizoctonia solani, Fusarium oxysporum* f. sp. *lini, Phytophthora* spp., and *Pythium* spp. Four isolates of *E. purpurascens* were tested; all of them inhibited growth of the pathogenic fungi listed above. One of them (I3) consistently inhibited opposing fungi more than the others did and was selected for further study. Six antifungal compounds were isolated from cultures of *E. purpurascens* (I3). Four of these compounds (epicorazines A and B and two unidentified compounds, X and Y) were produced at the same time at an early growth stage in a sucrose-plus-casamino-acid medium but were not detected in a glucose-plus-$(NH_4)_2HPO_4$ medium. Flavipin was identified from both media, and a third antifungal compound was produced at a later stage of growth in both media.

Mycelial growth of *Phytophthora* spp. was generally more sensitive to flavipin, whereas growth of *Pythium* spp. was more sensitive to the epicorazines A and B and to unknowns X and Y. When hyphae of *E. pupurascens* and *Phytophthora* spp. or *Pythium* spp. hyphae approached each other on plates or gel slides, the hyphae of the pathogen became swollen and coiled before contact, probably a result of antibiosis. *E. purpurascens* produced extracellular ß, 1–3 glucanase, and also cellulase, which allowed the hyphae to penetrate the mycelium of *Phytophthora* spp. or *Pythium* spp. after contact. It appears therefore that *Epicoccum purpurascens* (I3) would be a good antagonist against several species of *Phytophthora* or *Pythium.*

Attafuah and Bradbury (1989) isolated a bacterium from the mealybug *Planococcoides njalensis* and identified it as a species of

*Pseudomonas.* It did not fit any known species of that genus so the researchers named it *P. antimicrobica.* The bacterium was tested for antimicrobial activity on agar plates and was found to be inhibitory to growth of all of the 33 fungi and 6 of the 8 bacteria against which it was tested. The action varied from fungistasis and bacteriostasis to complete lysis of fungal mycelia. The results demonstrated that *P. antimicrobica* produced an exogenous, water-soluble, thermolabile, inhibitory compound—or possibly more than one compound. This bacterium definitely has potential as a biocontrol organism against several pathogenic fungi and a few pathogenic bacteria.

*Pseudomonas syringae* occurs on many plants; some are pathogens, whereas others are weak pathogens or saprophytes (Harrison et al., 1991). A transposon-generated mutant of the wild-type strain of *P. syringae* 174(MSU16H) has been isolated and found to produce larger inhibitory zones on culture media than does the wild type. Harrison et al. isolated and identified a novel family of peptides from the medium of *P. syringae* and named the peptides pseudomycins. The pseudomycins are different from previously described antibiotics from *P. syringae,* including syringomycin, syringotoxin, and syringostatins. Pseudomycin A is the predominant peptide in a family of four. Pseudomycins A through C contain hydroxyaspartic acid, aspartic acid, serine, arginine, lysine, and diaminobutyric acid. The molecular masses of pseudomycins A through C are 1224, 1208, and 1252 Da, respectively. Pseudomycin D has a molecular mass of 2401 Da and is a more complex compound.

The pseudomycins were tested for antimycotic activity against the fungal pathogens listed in Table 15. A semipurified combination of the pseudomycins was used in the tests; nine of the fourteen plant pathogenic fungi were inhibited in growth, which certainly indicates a broad range of activity against pathogens. The dominant pseudomycin A was tested for activity against four monocots and four dicots, all common crop plants; except timothy, all the plants were sensitive to the antimycotic. Tomato and sickle pod (*Cassia tora*) were sensitive to application of 5 or 10 μg to a leaf puncture, but not to 1 μg.

Liao (1989) reported that 27 of 963 strains of epiphytic bacteria from fruits of bell pepper and tomato were antagonistic to *Erwinia carotovora* f. sp. *carotovora,* a soft-rot bacterium on several plants. Eight of the antagonistic organisms were identified as *Pseudomonas fluorescens,* sixteen as *P. putida,* and one as *Flavobacterium* sp.; two

**Table 15** Activity of pseudomycins (semipurified, Amberlite step) against several plant-pathogenic fungi. Pseudomycins (10 μl) were applied to PDA plates at 10 $\mu g/\mu L^{-1}$ and dried; the plate was then oversprayed with spores and mycelial fragments of the fungus being tested. Readings were made after five days. The experiment was replicated four times, with identical results each time.

| Fungus | Activity* |
|---|---|
| *Rynchosporium secalis* | + |
| *Ceratocystis ulmi* | + |
| *Cephalosporium gramineum* | – |
| *Pyrenophora teres* | – |
| *Pyrenophora graminea* | – |
| *Rhizoctonia solani* | + |
| *Botrytis allii* | – |
| *Sclerotinia sclerotiorum* | + |
| *Verticillium albo-atrum* | + |
| *Verticillium dahliae* | + |
| *Thielaviopis basicola* | + |
| *Fusarium oxysporum* | + |
| *Fusarium graminearum* | – |
| *Fusarium culmorum* | + |

* + = A zone of inhibition.
– = No zone of inhibition.

*Source:* From Harrison et al. (1991), with permission of Soc. of General Microbiology.

were not identified. One strain of *P. putida* (PP22) inhibited 24 strains of soft-rot bacteria in species of the genera *Erwinia, Pseudomonas, Xanthomonas,* and *Cytophaga,* one strain of *Pseudomonas solanacearum,* four pathogenic forms of *Xanthomonas campestris,* and five pathogenic forms of *Pseudomonas syringae. P. putida* PP22 produced iron-chelating siderophores and an antibacterial compound that was heat- and trypsin-resistant. It also suppressed the growth of *Erwinia carotovora* f. sp. *carotovora* on potato slices and survived on the slices and on roots of potato plants for several weeks. Strain PP22 reduced the severity of bacterial soft rot on potato tubers caused by *E. c. carotovora* and *Xanthomonas campestris* by 21 and 44%, respectively. Strain PP22 was able to produce an unidentified antibiotic as indicated above, but a Tu5-generated mutant of PP22 was unable to do so and was unable to act as an antagonist against *E. c. carotovora.*

*Trichoderma koningii* was isolated from a soil suppressive to the saprophytic growth of take-all of cereals caused by *Gaeumannormyces graminis* var. *tritici* and was grown in PDB (Simon et al., 1988). The activity of the metabolites in the medium after filter sterilization was determined against *G. graminis tritici,* and the results were striking. The remainder of the broth in which *T. koningii* (strain 7a, IMI No. 308475) grew was extracted with ethyl acetate and a very active compound was isolated by solvent partitioning and TLC, after which it was identified by GC-MS and NMR. The compound was determined to be 6-n-pentyl-2H-pyran-2-one. A very high yield (145 mg/L) was obtained from the *T. koningii* culture. Tests demonstrated that this antifungal compound inhibited growth of *Bipolaris sorokiniana, Fusarium oxysporum, G. graminis tritici, Phytophthora cinnamomi, Pythium middletonii,* and *Rhizoctonia solani.* It was concluded that the effect of *T. koningii* and/or its metabolite merits further study. It appears to have excellent promise against several soil-borne plant pathogens.

In a subsequent project (Dunlop et al., 1989), *Trichoderma koningii* (a strain different from 7a) produced a new antibiotic in a pure culture. This was 4,8-dihydroxy-2-(1-hydroxypeptyl)-3,4,5,6,7,8-hexahydro-2H-1-benzopyran-5-one. The compound and the broth containing the compound inhibited growth of take-all of cereals and of several other fungal plant pathogens.

Numerous fluorescent pseudomonads from soil have been shown to produce antibiotics that are recognized as the main reason for their antagonistic action against several plant pathogens (Shanahan et al., 1992). Shanahan and his colleagues isolated a fluorescent *Pseudomonas* strain (F113) from the rhizosphere of sugar beets and demonstrated that it inhibits several plant fungal pathogens. They isolated and identified an antibiotic, 2,4-diacetylphloroglucinol (DAPG), and developed a high-pressure liquid chromatographic assay to detect DAPG quantitatively in culture media and soil.

Strain F113 of *Pseudomonas* sp. inhibited the growth of *Escherichia coli* and the plant pathogenic fungi, *Pythium ultimum, Phoma beta, Fusarium oxysporum,* and *Rhizopus stolonifer.* An antibiotic-negative strain of F113 was constructed by mutating K113 with Tn5-lac, and Kmr colonies were spot-inoculated onto LB agar medium for inhibition assays, with *E. coli* as the test strain. One colony was obtained that did not inhibit *E. coli* and was designated strain F113G22. This strain was found to have lost its ability to inhibit *Pythium ultimum;* it also was unable to produce DAPG. It is now recognized that antibiotic production is an important characteristic of plant-disease suppression by many strains of *Pseudomonas* spp.

According to Ordentlich et al. (1992), Ordentlich and Chet (unpublished) isolated a strain of *Trichoderma harzianum* that secreted substances into the growth medium that inhibited growth of microorganisms and decreased germination of conidia and chlamydospores of *Fusarium oxysporum* f. sp. *melonis* and *F. o. vasinfectum.* The substances also reduced the occurrence of diseases caused by those *Fusarium* pathogens in melon and cotton plants. Three inhibitory compounds were isolated from the growth medium of their isolate of *T. harzianum.* The scientists identified the antibiotic produced in the largest amount as 3-(2-hydroxypropyl)-4-(2-hexadienyl)-2(5H)-furanone. This compound appears to have good possibilities for use in the biocontrol of several fungal pathogens of various hosts.

*Trichoderma* spp. have been shown (see above) to produce both volatile and nonvolatile compounds that inhibit the growth of pathogenic fungi. *T. reesei* (strain P-12, a mutant of *T. reesei* QM9414) was reported by Watts et al. (1988) to be active against several plant pathogenic fungi. They extracted the mycelium with 70% acetonitrile and separated out two compounds that had antifungal activity. One of these was identified as trichodermin, a previously identified antibiotic produced by *Trichoderma* spp. The other was not identified and was designated as $T_x$. Trichodermin was quite inhibitory to the growth of nine test pathogenic fungi, and $T_x$ was inhibitory to all except three of them. It was less inhibitory than trichodermin in all cases except against *Rhizopus stolonifer,* where it was more inhibitory than trichodermin. *T. reesei* P-12 thus appears to have excellent biocontrol potential against several fungal pathogens.

*Nigrospora oryzae* is a known producer of antibiotics, so Szewczuk et al. (1991) isolated a strain of *N. oryzae* from rye and cultured it for three weeks in a nutrient medium containing 3% malt extract, with maltose as the main component. The mycelium was then extracted with acetone, and the filtrate was extracted with chloroform. Both extracts contained the same crude product, but more was obtained from the mycelium. Three major fractions were obtained from the extract by TLC and preparative chromatography. All three were tested separately against spore germination of *Fusarium culmorum,* and all were found to be inhibitory. However, fraction III was most inhibitory. When the organic extract of the mycelium of *N. oryzae* was tested against six species of *Fusarium* and *Botrytis cinerea,* both spore germination and mycelial growth were inhibited in all species. A concentration of the extract as small as 0.1 $\mu g/cm^3$ was effective against

*F. culmorum, F. gramminearum, F. oxysporum,* and *F. equiseti.* It appears therefore that the strain of *N. oryzae* used in these tests may find use in the biocontrol of some fungal pathogens.

Brown and Hamilton (1992) found that *Zygorrhynchus moelleri* produces indole-3-ethanol (IEt) as a major metabolite in cultures containing casein hydrolysate as the nitrogen source. They reported that IEt inhibited germination of zoospores of *Phytophthora cinnamomi,* oogonia of two *Pythium* species, and conidia of *Fusarium oxysporum* f. sp. *lini,* with $ED_{50}$ values between 1.8 and 12.7 μg/mL. They also found that mycelial growth of the same fungi and of *Rhizoctonia solani* and *Sclerotinia sclerotiorum* was suppressed by that compound. Thus *Z. moelleri* may have potential as a biocontrol agent.

We have pointed out several times that seed rot and damping-off due to *Pythium* spp. are the cause of considerable economic loss. Thompson and Burns (1989) investigated several microorganisms as potential biocontrol agents of these diseases. On the basis of preliminary screening tests, *Penicillium claviforme* (IMI44744) was selected as the most promising *Pythium ultimum* antagonist. Samples of *P. claviforme* culture filtrate were filter-sterilized and incorporated into molten cornmeal agar in petri dishes. After solidifying, the agar was inoculated with *P. ultimum;* subsequent measurements of *P. ultimum* colonies demonstrated that the filtrate of *Penicillium claviforme* was very inhibitory to growth of *P. ultimum.*

Sugar beet seed pellets with filtrate of *Penicillium claviforme* incorporated germinated well, and seedling emergence and establishment in compost infested with *P. ultimum* were significantly improved. The filtrate of *P. claviforme* was somewhat less effective in the above tests than the synthetic fungicide hymexazol, but the efficacy of the two in suppressing a milder infection was the same.

Melouk and Akem (1987) investigated effects of the filtrate of *Penicillium citrinum* on the mycelial growth of several plant pathogens. They obtained the *P. citrinum* from the sclerotia of *Sclerotinia minor* in a field planted to peanuts and allowed it to grow for six to eight weeks in Czapek-Dox broth. The mycelium was then filtered out and the filtrate was tested against *Sclerotinia minor, S. major, Sclerotium rolfsii,* and *Rhizoctonia solani.* The filtrate inhibited the mycelial growth of all test fungi. Inhibitors in the filtrate were extracted with ethyl acetate, and the active component was tentatively identified as citrinin. This antibiotic was found to be the active component in the filtrate against the mycelial growth of *Sclerotinia minor, S. major,* and *Sclerotium rolfsii.*

Ordentlich et al. (1987) isolated 203 bacteria from roots of bean, peanut, and chickpea plants grown in soil infested with *Sclerotium rolfsii.* All were tested against disease incidence in beans caused by *S. rolfsii.* One isolate giving the best control in early tests was selected for further tests employing *Rhizoctonia solani* on beans, *Pythium aphanidermatum* on cucumbers, and *S. rolfsii* on beans. The isolate, identified as *Serratia marcescens,* was found to reduce disease incidence in beans caused by *S. rolfsii* by up to 75% and damping-off of beans caused by *R. solani* by 50%. It was not effective against *P. aphanidermatum* in tomatoes. A rifampicin-resistant natural mutant of *S. marcescens,* which was as effective as the wild type, was used to study the colonization of roots of Brittle Wax beans. When the bacteria were mixed to all profiles of the soil before planting beans, the biocontrol populations on the root system were highest at the base and the tip of the root. When the mutant of *S. marcescens* was applied to the upper part of the soil, however, the bacteria became established near the root base, declining toward the root tip, where they increased again.

Application of *S. marcescens* as a drench or in a drip system was most effective in reducing disease incidence (75%). Spraying, mixing with soil, or seed coating reduced incidence of disease caused by *S. rolfsii* by 55, 44, and 32%, respectively. Germination of sclerotia dipped in an *S. marcescens* suspension and placed on soil, and of nontreated sclerotia placed on soil infested with *S. marcescens,* was decreased by 36 and 20%, respectively, after five days of incubation. It appears, therefore, that *S. marcescens* is promising as an antagonist against diseases of certain crops caused by *S. rolfsii* and *R. solani.*

Ten isolates of *Chaetomium* spp. were obtained from sclerotia of *Sclerotinia sclerotiorum* buried in soil in the Okinawa prefecture in Japan (Nakashima et al., 1991). Eight of the ten isolates inhibited growth of the mycelium of *S. sclerotiorum.* Seven of these isolates were identified as members of *Chaetomium trilaterale* var. *diporum* and the other three as members of *C. trilaterale* var. *cupreum.* The isolate of *C. t.* var. *diporum* designated as RC-5 was used for the production of an antifungal substance. This substance, found to have the molecular formula of $C_{31}H_{35}O_8N$, was very inhibitory (0.5 ppm) to mycelial growth of *S. sclerotiorum* (90% by hour 156). On the other hand, a 10-ppm solution of the antibiotic did not affect ascospore germination of *S. sclerotiorum. This antibiotic action suggests that C. t. diporum* has good possibilities for antagonistic action against *S. sclerotiorum.*

Zhou and Reeleder (1990) isolated a strain of *Epicoccum purpurascens* from a lettuce leaf and exposed it to shortwave UV light. A resultant isolate (M-20-A) was grown on media containing the fungicides iprodine or vinclozolin, and fungicide-tolerant strains were obtained. Sporulation was improved, when compared with the wild type, in some of the strains obtained. The researchers found that strains differed in their tolerance to iprodine and vinclozolin but were not tolerant to the fungicide benomyl. In in vitro tests, the fungicide-tolerant strains R4000, 16B, and 7A inhibited *Sclerotinia sclerotiorum* more than the wild type or M-20-A did. Moreover, the fungicide resistant strains exhibited greater control of white mold of beans caused by *S. sclerotiorum* in the greenhouse than did *E. purpurascens* M-20-A. It is very important to consider other factors—such as plant colonization ability—when selecting strains for use in the field. Nevertheless, selecting antagonistic strains that are tolerant to some of the more commonly used fungicides is an excellent goal.

Roberts and Lumsden (1990) investigated the mechanism of action of *Gliocladium virens* as a biocontrol agent of *Pythium ultimum.* Most of their research was done with *G. virens* (G20) and *P. ultimum* (PuZs1 and PuM×L1) from the U.S.D.A.'s Biocontrol of Plant Diseases Laboratory culture collection. Culture supernatants from *G. virens* grown in 5% bran extract contained gliotoxin, laminarinase, amylase, carboxymethylcellulase, chitinase, and protease. The researchers reported that the supernatant inhibited sporangial germination and mycelial growth of *P. ultimum* before and after removal of the enzyme activity by heating or by size fractionation. Moreover, supernatants from *G. virens* grown in a medium in which *G. virens* did not produce the enzymes listed above, still strongly inhibited sporangial germination and mycelial growth of *P. ultimum.* Through several tests, the scientists determined that gliotoxin was the only compound present in the size-fractionated medium that inhibited sporangial germination and mycelial growth of *P. ultimum.* It appears from these data that gliotoxin and perhaps other antibiotics produced in different media are responsible for the antagonistic activity of *G. virens* against *P. ultimum.*

Bin et al. (1991) pointed out that biocontrol efficiency is likely to increase with increasing use of biological control organisms. They suggested therefore that studies of factors affecting growth and proliferation of biocontrol agents are not only desirable, but necessary. They decided to determine some of the effects of the antagonist *Pseudomonas fluorescens* (2-79$RN_{10}$) on another biological control antagonist, *Trichoderma harzianum* (ThzIDI), under different conditions. *P. fluorescens* (2-79$RN_{10}$) inhibited growth of ThzIDI on two types of agar plates; 2-79$RN_{10}$ populations remained about the same when inoculated into steamed soil along with ThzIDI, and populations of ThzIDI increased over the same period. Treatment with 2-79$RN_{10}$ resulted, however, in significantly lower numbers of ThzIDI compared with controls. Neither the effect of matric potential nor the interaction between 2-79$RN_{10}$ and matric potential was significant for ThzIDI numbers, colony radius, or hyphal density. Time × treatment interactions were not significant either. Addition of 2-79$RN_{10}$ at either $3 \times 10^4$ or $3 \times 10^7$ cfu/g significantly reduced radial growth of hyphae and hyphal density. In two years of field experiments using raw or steamed soil, populations of 2-79$RN_{10}$ decreased gradually after one to two weeks and did not reduce the ability of *Trichoderma* spp. to colonize sclerotia of *S. sclerotium.* Colonization of sclerotia by *Trichoderma* spp. after nine weeks was significantly higher in steamed soil when ThzIDI was added, which suggested a possible inhibition of ThzIDI by indigenous soil microorganisms or utilization by ThzIDI of nutrients released by steaming the soil. The overall results caused Bin et al. to infer that under some conditions, high populations of antagonistic bacteria in soil suppressed a fungal biocontrol organism, but that the suppressive effect was reduced or eliminated when a high bacterial population was not present.

Mycorrhizal fungi have been reported since 1964 to have antagonistic activity against certain pathogens (Rice, 1984). Related research has continued in recent years in greater depth, particularly in relation to *Pisolithus tinctorius* (Kope and Fortin, 1989, 1990; Kope et al., 1991; Tsantrizos et al., 1991; Suh et al., 1991). Kope and Fortin (1989) tested filtrates of 16 ectomycorrhizal fungi against mycelial growth of 23 plant pathogens and 1 ectomycorrhizal fungus, *Sphaerosporella brunnea.* Filtrates of seven of the ectomycorrhizal fungi inhibited at least one or more test fungi, with *Pisolithus tinctorius* (71%) and *Tricholoma pessundatum* (46%) having the most widespread activity. The *P. tinctorius* filtrate caused hyphal lysis of *Truncatella hartigii, Rhizoctonia praticola,* and the ectomycorrhizal fungus *S. brunnea.*

Kope and Fortin (1990) reported that a metabolite secreted by *Pisolithus tinctorius* lysed conidia of several phytopathogenic fungi, lysed the hyphae, and inhibited germination of the conidia. Charcoal adsorbed the inhibitor and lysing metabolite. Later, two antifungal compounds were isolated from the culture medium of *P. tinctorius* (or *P. orhizus*) and were identified as *p*-hydroxypbenzoylformic acid and (R)-(-)-*p*-hydroxymandelic acid, pisolithin A and pisolithin B, respectively (Kope et al., 1991). The concentration necessary for 50% germination inhibition ($GI_{50}$) of conidia of *Truncatella hartigii* was 67 μg/mL for pisolithin A and 71 μg/mL for pisolithin B. Both compounds were more active than polyoxin B in inhibiting the mycelial growth of *T. hartigii,* a phytopathogen. Tsantrizos et al. (1991) substantiated the work of Kope et al. (1991) by isolating the same antibiotics from *P. pisolithus* and testing the activity of the compounds against several phytopathogenic fungi.

Deb (1990) isolated *Aspergillus candidus, A. flavus, A. niger, A. versicolor, Fusarium oxysporum, F. solani, Penicillium oxalicum, P. rubrum, Trichoderma harzianum, T. koningii,* and *T. viride* from the rhizosphere of *Glycine max* (soybean) and determined their activity in vitro against *Sclerotium rolfsii.* Most of the listed fungi inhibited growth of *S. rolfsii. Trichoderma viride* and *T. koningii* had volatile activity, whereas all three *Trichoderma* species and *A. flavus* showed nonvolatile activity against *S. rolfsii.* The culture filtrates also had inhibitory activity against growth of *S. rolfsii,* indicating probable antibiotic activity. Nonsterile culture filtrates had more activity on growth of *S. rolfsii* than did sterile ones, but they did not affect soybean seed germination.

Deb continued the work described above with Dutta (Deb and Dutta, 1991), and they reported that all the fungi tested decreased infection of soybeans by *S. rolfsii* in infested soil when each of the antagonists was added individually to the soil. *Trichoderma harzianum* and *T. koningii* gave better control of the foot rot disease of soybeans caused by *S. rolfsii* than did any of the other nine potential antagonistic fungi. *T. harzianum* or *T. koningii* reduced the population of *S. rolfsii* by 50% when added to the *S. rolfsii*–infested soil, and each increased soybean seed yield by several grams per plant compared with the control.

Howell (1991) studied the biological control by *Gliocladium virens* of damping-off of *Gossypium hirsutum* (cotton) caused by *Py-*

*thium ultimum.* He obtained 20 strains of *G. virens* (Gl-G20) and cultured all of them on substrates of millet, wheat, sorghum, soybean, cotton, wheat bran, rice, rice hulls, or oats. Most of the *G. virens* strains grown on millet were ineffective as biocontrol agents of cotton damping-off caused by *P. ultimum* (P-1); however, strains G-4, G-9, and G-12 were very effective when grown on millet. Their efficacy varied, however, with the growth substrate. Strains grown on millet, rice hulls, or wheat bran gave good disease suppression. Untreated cotton seed or seed coated with millet preparations of *G. virens* and planted in uninfested soil emerged and produced healthy radicles without lesions.

Analyses of the strain-substrate extracts indicated that G-1, G-3, G-4, G-8, G-9, and G-12 produced viridin, viridiol, and gliovirin; G-2, G-5, G-6, G-10, G-11, G-13, G-15 through G-18, and G-20 produced viridin, viridiol, and gliotoxin, but not gliovirin. All strain-substrate combinations with efficacy as biocontrol agents of pythium damping-off were able to produce the antibiotic gliovirin. This antibiotic was very active against *P. ultimum,* giving a 12-mm clear zone at a concentration of 0.5 μg/mL. Combination treatments with reduced concentrations of the fungicide metalaxyl and the biocontrol preparation of *G. virens* resulted in a synergistic effect that gave disease suppression equal to that of full-strength fungicide treatment. Pythium damping-off of cotton seedlings can be controlled by seed-coat preparations of *G. virens* in combination with the fungicide, and the amount of fungicide can thus be reduced greatly. This form of biological control could no doubt be used in many more cases with a minimum of additional research.

Ninety-six bacterial strains obtained from rhizosphere soils of cotton were tested for their effects on colonization of cotton seed by *Pythium ultimum* (Loper, 1988). Only nine of the strains protected cotton seed from colonization by *P. ultimum* at a level statistically similar to that of metalaxyl seed treatment. All nine strains were fluorescent pseudomonads, belonging to either *Pseudomonas fluorescens* or *P. putida. P. fluorescens* 3551 was selected for further study because it could be manipulated genetically; it protected cotton seed from colonization by *P. ultimum* in three different soil types with indigenous or introduced populations of *P. ultimum.* Strain 3551 also consistently increased cotton seedling emergence.

Strain 3793, a spontaneous, rifampicin-resistant derivative of strain 3551, was selected for further experimentation because it was fluorescent, it had the same growth rate as the parental strain, and its effect in seed treatment of cotton was statistically similar to that of strain 3551. Tn5 mutants of strain 3793 were obtained by matings with *Escherichia coli* strain SM10 (pSUP1011). Twenty-two of 8,017 matings were nonfluorescent; of the twenty-two, fourteen had single bands with homology to the λ:Tn5 probe following Southern analysis. The 14 strains with single bands were unable to grow on an iron-deficient medium, but their growth was restored when FeCl was incorporated in the medium. The $Flu^-$ mutants were deficient in siderophore production, as indicated by their iron-limited growth and lack of iron-regulated antibiosis against *E. coli* AN194. Significant differences were observed between six representative $Flu^-$ mutants with respect to cotton seedling emergence and that resulting from treatment with the parental strains 3551 or 3793. No consistent differences were observed between spermosphere populations of *P. fluorescens* 3793 and the $Flu^-$ mutants. It was concluded that the fluorescent siderophore production by *P. fluorescens* 3551 contributes to, but does not account for all, its antagonistic activity against *Pythium ultimum.*

Howie and Suslow (1991) also investigated the role of antibiosis in the antagonism of the fluorescent pseudomonad strain Hv37aR2 against *Pythium ultimum* in cotton. They found that the suppression of disease development by that strain is dominated by its biosynthesis of an antibiotic. Strain Hv37aR2 caused an average reduction of 70% in root infection and an average of 50% increase in emergence of cotton seedlings infested with *P. ultimum.* On the other hand, Afu isogenic mutants of Hv37aR2, deficient in the biosynthesis of an antifungal compound, colonized the spermosphere and rhizosphere of cotton as well as the parent strain but were not effective in preventing damping-off of cotton caused by *P. ultimum.* Howie and Suslow concluded that an essential component in the suppression of *P. ultimum* by strain Hv37aR2 was the biosynthesis of an antibiotic that corresponds to oomycin A.

Hagedorn et al. (1989) reported on a survey made during the 1983 field season at Starkville, Mississippi, of the rhizobacteria of cotton plants inventoried at different plant growth stages. Isolates (1000) were identified to genus level and characterized for repression of *Pythium ultimum* and *Rhizoctonia solani* in vitro. Fluorescent pseudomonads were the most numerous gram-negative rhizobacterial isolates, and they provided the largest number of isolates with fungal-repressive activity. Several other gram-negative bacteria were isolated, but in lower numbers. Cotton seedlings were initially colonized by many different bacterial genera. Populations quickly reached $10^8$ cfu/g of root tissue, but declined as the season progressed, as the root mass increased, and as the roots became woodier. There was no correlation between the proportion of rhizobacterial isolates that possessed fungal-repressive activity and the plant growth stage from which the isolates were obtained. There were considerably more isolates that repressed activity of *P. ultimum* than of *R. solani.*

Two isolates of *Trichoderma viride* (T-1-R9, TS-1-R3), three of *T. harzianum* (WT-6–24, Th-23-R9, Th-58), three of *T. hamatum* (TRI-4, Tm-23, 31–3), and three of *Gliocladium virens* (Gl-3, Gl-17, Gl-21) were used by Lewis and Papavizas (1987) in biocontrol surveys against *Rhizoctonia solani* (R-23) isolated from a diseased cotton plant and occasionally against strains R-18, R-53, and R-103 from tomatoes, R-34 from snap beans, and R-108 from potatoes (*Solanum tuberosum*).

A loamy sand (pH 6.4, 0.4% O.M.) was used in most experiments, but other soils were used at times, including a silty-clay loam (pH 6.3, 1.4% O.M.), a loam (pH 6.2, 5.0% O.M.), a sandy-clay loam (pH 6.3, 1.0% O.M.), a sandy clay loam (pH 4.5, 3.2% O.M.), and a sandy loam (pH 6.7, 0.3% O.M.).

Alginate pellets were made using conidial suspensions of the biocontrol fungi, sodium alginate, distilled water, wheat bran (ground), and a $CaCl_2$ gellant. The spherical, beadlike pellets were air-dried and stored at 5 and 25°C and were generally used within two days of preparation. In most experiments, alginate bran pellets containing fungal biomass of biocontrol isolates were added to soil with beet seed infested with *R. solani* at a rate of 1.0 g/100 g soil (dry weight equivalent). After one and three weeks, beet seeds were retrieved from an aliquot of soil and tested for survival of *R. solani.*

Pellets with eight (of eleven) of the biocontrol isolates reduced survival of *R. solani* on beet seeds by 34 to 78%. Pellets containing *T. harzianum* (Th-58) and *T. hamatum* (TRI-4) were the most effective. Populations of isolates proliferated in soil to $10^6$ to $10^{11}$ cfu/g from propagules within the pellets. Pellets with TRI-4 reduced pathogen survival and growth by more than 70% in six different soils and were effective against six *R. solani* isolates in a natural loamy sand. Small amounts of fungal biomass in pellets were as effective as large amounts in repressing the pathogen. Pellets prepared with four and

**Table 16** Effect of pellet formulations* of isolates of *Trichoderma* spp. and *Gliocladium virens* on cotton and sugar beet stands in soil infested with *Rhizoctonia solani*

| | Plant stand (%) at 3 weeks | |
|---|---|---|
| Isolate | Cotton† | Sugar beet† |
| Control (uninfested)* | 84 a | 56 a |
| Control (*R. solani*) | 0 c | 0 d |
| Control (*R. solani* and pellets) | 0 c | 0 d |
| *T. viride* | | |
| T-1-R9 | 0 c | 0 d |
| TS-1-R3 | 38 b | 25 c |
| *T. harzianum* | | |
| WT-6-24 | 0 c | 0 d |
| Th-58 | 0 c | 0 d |
| Th-23-R9 | 0 c | 0 d |
| *T. hamatum* | | |
| TRI-4 | 93 a | 44 ab |
| Tm-23 | 0 c | 6 d |
| 31-3 | 0 c | 4 d |
| *G. virens* | | |
| Gl-3 | 20 bc | 1 d |
| Gl-17 | 38 b | 8 d |
| Gl-21 | 80 a | 33 bc |

* Alginate pellets were added to soil at a rate of 1.0 g/100 g soil (dry weight equivalent). Pellets in control soil contained no fermenter biomass.

† Values in each column followed by the same letter are not significantly different according to Duncan's multiple-range test ($P = 0.05$).

*Sources:* From Lewis and Papavizas (1987), with permission of Blackwell Scientific Pub. Ltd.

three, respectively, of the eleven isolates prevented damping-off of cotton and sugar beets (Table 16). It was concluded that even though we possess effective biocontrol fungi, appropriate formulations of the microorganisms are critical for the implementation of biological control.

Lewis and Papavizas (1991) followed up their previous work on biocontrol of damping-off of cotton caused by *Rhizoctonia solani.* Again, they used *Trichoderma* spp. and *Gliocladium virens* preparations as biocontrol agents—this time in field tests over four growing seasons. The preparations tested were bran/germling, a powder (Pyrax/biomass), and alginate pellets containing milled biomass of the biocontrol fungi. The bran/germling preparation consistently prevented disease, reduced saprophytic activity of the pathogen, and caused proliferation of populations of the biocontrol fungi. In all four years, bran/germling preparations of *T. hamatum* TRI-4 and *G. virens*

Gl-21 significantly prevented disease, and in three of the years gave a plant stand comparable to that of the *R. solani*–noninfested plots. Pyrax/biomass preparations of Gl-21 prevented damping-off in two of the four years, but stands similar to the noninfested plots were never achieved. Pyrax/biomass preparations of most other isolates used were not effective in preventing the damping-off. Alginate pellets did not prevent disease in the two years they were applied to the soil.

The economically important diseases of *Helianthus annuus* (sunflower) in India are leaf spots caused by *Alternaria helianthi,* wilts caused by *Sclerotium rolfsii,* and occasionally root rots caused by *Rhizoctonia solani* and *Macrophomina phaseolina* (Hebbar et al., 1991). Hebbar et al. investigated the potential antagonistic effects of many bacteria isolated from diseased and healthy roots and tops of sunflower plants. The pathogens used in tests were *R. solani* and *M. phaseolina* from the Indian Institute of Horticultural Research, Hessarghatta; *A. helianthi* from the Agricultural University, Hebbal; and *S. rolfsii* from the Biological Control Centre, Bangalore.

Antagonistic bacteria from leaves were mainly actinomycetes and pigmented gram-positive bacteria, while those from roots and crowns were *Pseudomonas fluorescens-putida, P. maltophilia, P. cepacia, Flavobacterium odoratum,* and *Bacillus* sp. In soil bioassays *P. cepacia* strain N24 significantly increased seedling emergence when used as a seed inoculum in the presence of *S. rolfsii.* Those bacterial strains that had activity against a broad range of pathogens were found to colonize *H. annuus* roots well and to reduce disease and fungal wilt. On an average, only 30% of seedlings were diseased when treated with the antagonistic strains in the presence of the pathogen, whereas 60% were diseased in the presence of the pathogen alone. In field plots treated with *P. cepacia* N24, only 1 to 3% of the seedlings were wilted, whereas 14% were wilted in the presence of the pathogen alone.

*Fusarium culmorum* and *Bipolaris sorokiniana* are common pathogens on *Hordeum vulgare* (barley) and *Triticum aestivum* (wheat) under Finnish conditions (Tahvonen and Avikainen, 1990). *Streptomyces* spp. are common microorganisms in soil and are well known for their ability to produce antibiotics. Tahvonen and Avikainen decided, therefore, to determine the suitability of a commercial preparation of *S. griseovirides* (Mycostop) for the biological control of *F. culmorum* and *B. sorokiniana* on barley and wheat.

Seeds were dusted with 3 to 15 g of the Mycostop, which contained $10^8$ to $10^9$ cfu/g. The controls and comparison consisted of nondusted seed and chemical dusting with alkoxyalkylmercury (Ceresan) at a dosage of 2 g/kg of seed. Dusting of artificially infected seed with Mycostop reduced greatly the damage caused by *F. culmorum* and *B. sorokiniana*. Both chemical and biological control increased the dry weight of the shoots of both barley and wheat. *Streptomyces* dusting of uninfected barley and wheat seed, which naturally contained varying amounts of *Fusarium* spp. and *B. sorokiniana*, reduced foot damage in the sprouting experiments, as compared with undusted seed. The results were not as good, however, as those obtained with mercury dusting. Mycostop dusting reduced germination of barley seed but had no effect on wheat seed germination. There were no significant differences of the efficacy of control given by differing dusting doses (5 g/kg of seed vs. 10 to 15 g/kg). The effect of seed treatment remained stable for two to three weeks when the treated seeds were stored under dry conditions. According to the pot experiments described here, *Streptomyces griseovirides* (Mycostop) has potential use as a biological control agent against cereal pathogens.

*Erysiphe graminis* f. sp. *hordei* is a rather common fungal pathogen on leaves of barley and is the causative agent of powdery mildew of barley (Klecan et al., 1990). Yeastlike *Tilletiopsis* spp. are common epiphytes on leaves infected with powdery mildew, rust or smut. During the maintenance of stock cultures of *E. g. hordei*, an isolate of *Tilletiopsis* was found that seemed to interfere with the growth of that pathogen. The isolate was identified by Klecan et al. as *T. pallescens*. Three other species of *Tilletiopsis* were obtained for comparison in various experiments. An antagonistic relationship between *E. g. hordei* and *T. pallescens* was demonstrated on barley leaf segments. On a gross level, *T. pallescens* caused a pronounced reduction of mycelial expansion and spore production by *E. g. hordei; T. minor* was antagonistic to a lesser extent. *T. fulvescens* and *T. washingtonensis* had no effect on *E. g. hordei*. Low-temperature scanning and conventional transmission electron microscopy showed that hyphae of *E. g. hordei* were collapsed and degenerated by *T. pallescens*. A note was added in proof that a Dr. Skow had reported in 1986 that *T. albescens* was an effective biocontrol agent for powdery mildew of barley (Nordisk Jordbrugforskning 68: 331–332) and that a Dr. Hijwegen had demonstrated in 1986 that *T. minor* is an effective biocontrol agent for powdery mildew of cucumber (Neth. J. Plant Pathol. 92: 93–95).

Lynch et al. (1991) compared the efficacy of *Trichoderma harzianum* IMI 275950 (isolated from wheat straw), *T. viride* IMI 298375, *T. harzianum* WT, *Gliocladium virens* G20, and *Enterobacter cloacae* ATCC 29987 against damping-off of *Lactuca sativa* (lettuce cv. Ravel) caused by *Pythium ultimum* strain PuMXL (isolated from a lettuce plant with root rot). The tests were run in the greenhouse in pots containing a nonsterile compost potting mix.

No clear dose-response relationship was found in the induction of damping-off by *Pythium ultimum;* 50 sporangia/g of compost were sufficient to induce a substantial level of disease. There was no statistically significant difference in plant yield when 50 or 100 sporangia were applied per gram of potting mix. At all levels of pathogen inoculum, it was necessary to add the *T. harzianum* IMI 298375 preparation at 0.5% w/w compost to obtain good disease control. It was not possible, however, to overcome the lowering of shoot weight induced by the *P. ultimum. T. harzianum* WT, *G. virens* G20, and *E. cloacae* ATCC 29987 were also effective against damping-off of lettuce caused by *P. ultimum,* but *T. viride* IMI 298375 was not effective. There was no adverse effect on the biocontrol action of mixing the fungal strains with the bacterium (*E. cloacae*). *T. harzianum* IMI 275950 and *T. viride* IMI 298375 had an adverse effect on lettuce shoot growth, but the other test antagonists did not. There was no significant added benefit from mixing the antagonists over that achieved by adding them individually. The coexistence of the bacterial and fungal antagonists was revealed on the root surface and inner surface of the testa by scanning electron microscopy. This demonstrated the compatibility of the antagonists.

Lettuce is a major greenhouse crop in Finland during the winter months, and *Botrytis cinerea* is the most serious disease organism encountered, although bottom rot of lettuce caused by *Rhizoctonia solani* is becoming common in some areas (Tahvonen and Lahdenperä, 1988). Regulations in Finland prohibit the use of chemical control methods for plant diseases; thus biocontrol procedures are very important. Lettuce plants (var. Ostinata) were grown in new, fertilized *Sphagnum fuscum* peat, and biocontrol of *B. cinerea* and *R. solani* by *Streptomyces griseoviridis* was investigated. The standard preparation of *S. griseoviridis* contained 1 g of a conidial and mycelial mass/100 mL $H_2O$. In the experiments, seedlings were sprayed with dilutions made of that preparation. Either the seedlings or substrate was infested with either *B. cinerea* or *R. solani* before or after applying the spore suspension of the antagonist. Treatment of the lettuce seedlings with either of several *S. griseoviridis* strains significantly reduced yield losses caused by *B. cinerea* but had no effect on losses caused by *R solani.* It appears therefore that *Streptomyces* is a promising antagonist for damping-off of lettuce caused by *B. cinerea,* but the method used in treatment requires further development. I should add here that *B. cinerea* has been shown to be very sensitive to the antibiotic(s) excreted by several *Streptomyces* isolates.

We previously pointed out that some strains of *Trichoderma harzianum* were shown to produce several antibiotics, as was *T. koningii.* Claydon et al. (1987) found that two strains of *T. harzianum,* IMI 275950 and IMI 284726, each produced two volatile antibiotics: 6-*n*-pentyl-2H-pyran-2-one and 6-*n*-pentenyl-2H-pyran-2-one. The pentyl analogue was the major product and was very inhibitory to the mycelial growth of all 15 fungi used in tests, most of which were plant pathogens. Moreover, it was very inhibitory to all strains of *R. solani* used in tests. The antibiotic also markedly suppressed the rate of damping-off of lettuce seedlings by *R. solani.* It was concluded that the capacity of some *Trichoderma* species to produce antifungal alkyl pyrones may therefore be a useful property to exploit in the choice and formulation of these organisms as biological control agents.

Maplestone et al. (1991) tested *Trichoderma harzianum* isolates Thl, T35, T12, T95, WT, WT6; *T. viride* isolate IMI 298375; and *Gliocladium virens* G20 against *Rhizoctonia solani* anastomosis group 2, type 1 (RS2). The antagonist inoculum was a peat-bran system. Preliminary tests demonstrated that incorporation of autoclaved peat-bran inoculum of *Trichoderma* gave equal or better disease control than that of nonautoclaved inoculum. In subsequent tests, the *T. harzianum* inoculum was autoclaved and either used directly or extracted with ethyl acetate before incorporation in the plant-growth medium.

Six of the eight antagonists decreased damping-off and three of these increased yield of lettuce used in the tests, as compared with the *R. solani* treatment alone. Both autoclaved and nonautoclaved inoculum of *T. harzianum* Thl decreased disease and increased yield. Moreover, the ethyl acetate–extracted inoculum of Thl resulted in similar levels of biocontrol and improved plant growth, as did the autoclaved medium that was not extracted with ethyl acetate. The fact that the antagonistic action of the autoclaved *T. harzianum* in preventing the damping-off of lettuce was the same as that of the non-autoclaved inoculum suggests strongly that the antagonistic action was due to antibiotics produced by *T. harzianum.*

Adams and Fravel (1990) calculated that effective disease control at acceptable rates of addition is essential to the practical use of bio-control in field systems. *Sporidesmium sclerotivorum* was selected as a model system for the control of *Sclerotinia minor,* which causes leaf drop of lettuce. *S. sclerotivorum* is an obligate mycoparasite of the sclerotia of *Sclerotinia* spp., is a persistent biocontrol agent in soil, and reproduces on host sclerotia.

Field plots were established in an area at Beltsville, Maryland, that had been in turf grass for at least 10 years. The grass was removed and the plots were rototilled. Parris Island Cos lettuce seedlings were transplanted into the plots in early September, and about two months later each plant was inoculated with oat grains infested with *S. minor.* By the following March, the inoculum density of *S. minor* ranged from 54 to 122 sclerotia/100 g of soil. *S. sclerotivorum* was grown on vermiculite saturated with SM-4 medium. After sporulation, the entire substrate with spores and mycelium was mixed in a blender to pass through a 0.6-mm screen. The inoculum was added to water and sprayed on mature, diseased lettuce plants in May at rates of 0.2, 2, and 20 kg/ha. Immediately after application of the antagonist, the plots were rototilled to mix the diseased lettuce debris throughout the soil, after which the plots were sprinkler-irrigated.

Five consecutive crops of lettuce were grown in the plots from the fall of 1987 through the fall of 1989. Soil samples were collected about every two weeks and assayed for the population densities of sclerotia, including percentages of conidia infected by the antagonist. The samples were also assayed for the populations of macroconidia of the pathogen.

The population density of the sclerotia of the pathogen, *S. minor,* declined more rapidly in the plots treated with *S. sclerotivorum* than in the nontreated plots. During the winter of 1988–1989, the population density of the sclerotia of *S. minor* was below 30/100 g of soil in all treatments, and there were no significant differences in population densities of *S. minor* in the various treatments from that time until the end of the experiment. During the fall of 1987, only *S. sclerotivorum* at 20 kg/ha reduced the incidence of lettuce drop compared with the untreated control. In the spring of 1988, both the 20 and 2 kg/ha treatments had significantly lower disease incidence than did the untreated controls. In the fall 1988 crop, the incidence of lettuce drop was significantly less in all treatments of *S. sclerotivorum* than in untreated controls. Because of an increase in indigenous populations of *S. sclerotivorum* in the nontreated plots, there were no differences in disease incidence among treatments in the last two crops. By 1989, populations of *S. minor* were low in all plots because of the activity of the biocontrol agent. The estimated cost of the three rates of application was \$2, \$20, and \$200/ha, respectively. In New Jersey, the average annual loss due to lettuce drop is 10%, or \$1089/ha. If even 50% disease control would save the grower almost \$550/ha, it becomes obvious that application of *S. sclerotivorum* is economically beneficial.

Wilt of bottlegourd (*Lagenaria siceraria*) caused by *Fusarium oxysporum* has destroyed some 90% of that crop throughout the world since the beginning of the twentieth century (Gaikwad et al., 1987). These scientists decided to test two strains of *Bacillus subtilis* (Bs-13 and Bs-14) as biocontrol agents against *F. oxysporum*. Seeds of bottlegourd were soaked in a suspension of $8 \times 10^7$ cells/mL of Bs-13 or Bs-14, or Bs-13 and Bs-14, for 24 hours at 27°C, after which they were planted in sterilized sand in trays. *F. oxysporum* was grown for seven days on PDA and then flooded with sterile distilled water to make a spore suspension of $4 \times 10^6$ spores/mL. Twenty milliliters of this suspension was put into a vial, along with a seven-day-old gourd seedling. Observations were recorded five and seven days after the start of the test. Bottlegourd seedlings coated with antagonists did not show any plant reaction to *F. oxysporum,* whereas in the control, the gourd plants showed moderate to serious reactions.

Other coated bottlegourd seedlings were planted in soil infested with *F. oxysporum,* and the seedlings were checked every 7 days up to 35 days from planting. It was obvious that coating the seeds with Bs-13 markedly decreased the numbers of pathogen propagules per gram of soil (Fig. 9). Coating with Bs-14 had almost as marked an effect.

Tomato (*Lycopersicon esculentum*) plants commonly suffer from two fusarium diseases: the plant wilt caused by *Fusarium oxysporum* f. sp. *lycopersici* and crown and root rot caused by *F. oxysporum* f. sp. *radicis lycopersici.* The latter is now the most serious disease in the soilless culture of tomato widely practiced in western Europe (Lemanceau and Alabouvette, 1991). Both *Pseudomonas* spp. and *Fusarium* spp. have been widely tested as antagonistic microorganisms against

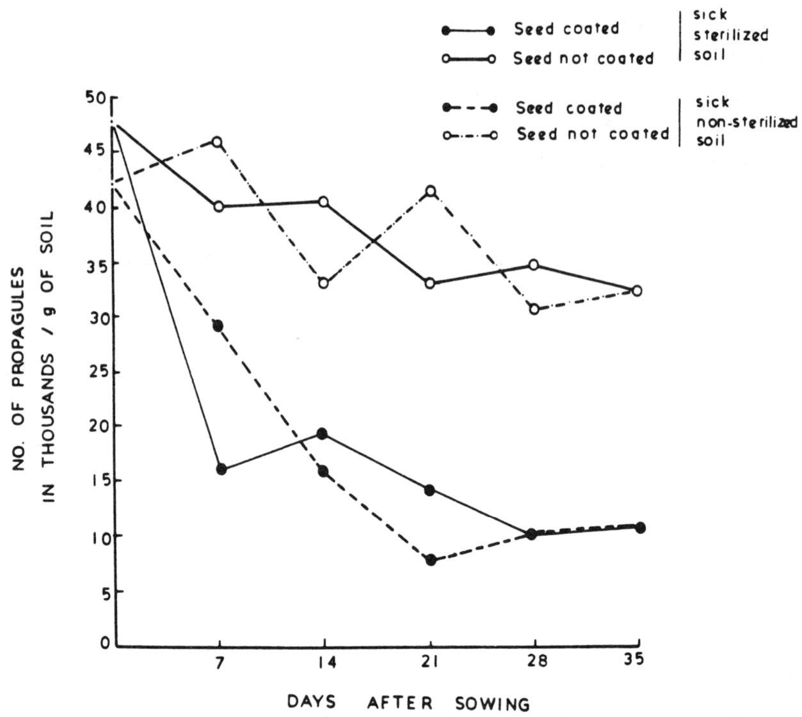


FIGURE 9 Effect of seed coating with antagonist-13 on soil population of *Fusarium oxysporum* in bottlegourd cropping. *Adapted from Gaikwad et al. (1987), with permission of Kluwer Academic Pub.*

certain plant diseases, so Lemanceau and Alabouvette decided to test the effects of the two together in the biocontrol of fusarium diseases of tomato. They had a large collection of bacterial strains to test, which made it impossible to use tomato in the original screening tests. *Linum usitatissimum* (flax) reacts similarly, so Regina flax was used in initial tests.

Two strains of pathogenic *Fusarium* were used: (1) a strain of *F. oxysporum* f. sp. *lini* (Foln 3) that induces wilt of flax was used in screening strains of *Pseudomonas* for ability to reduce fusarium wilt incidence and (2) a strain of *F. oxysporum* f. sp. *radicis lycopersici* (Forl 19) that induces crown and root rot of tomato was used in other experiments. The strain of nonpathogenic *F. oxysporum* (Fo47) used was isolated from the suppressive soil at Châteaurenard, France. In the screening experiment, both the pathogenic and nonpathogenic *Fu-*

*sarium* were introduced into the rockwool as a talc inoculum. Rockwool was infested by addition of a given volume of suspension to give $2 \times 10^7$ cfu/mL of substrate for the nonpathogenic *Fusarium* and $5 \times 10^2$ cfu/mL of substrate for the pathogenic strain. In the biological control tests conducted under commercial-type conditions, both types of *Fusarium* were introduced into the substrate as a suspension of microconidia from a five-day-old shake culture in malt extract (10 g/L). To remove nutrients before adding to the substrate, the conidia were washed three times in sterile, demineralized water.

Seventy-four strains of *Pseudomonas* were used in the initial tests, 71 from France, 2 from America, and 1 from Swiss soil. The bacteria were washed out of the growing medium and suspended in the nutrient solution used to water the growing substrate. The inoculum density was adjusted to $1 \times 10^8$ cfu/mL of substrate. All 74 strains of *Pseudomonas* were tested for their ability to reduce the incidence of fusarium wilt of flax when applied either alone or in association with one preselected nonpathogenic strain of *F. oxysporum* (Fo47). Most of the strains did not modify the percentage of wilted plants; however, 10.8% of them significantly improved the control due to Fo47. One of these effective bacterial strains (C7) was used for further tests, and two trials under commercial conditions indicated that C7 was effective in association with the nonpathogenic *Fusarium* Fo47 in controlling fusarium crown and root rot of tomato, even when each antagonistic organism was not effective alone.

The populations of *Trichoderma* spp. on roots and in the soil are often important in regulating the occurrence of several kinds of plant diseases, as we have pointed out numerous times. Therefore Dewan and Sivasithamparam (1988b) studied the frequency and identity of three species of *Trichoderma* on *Lolium rigidum* (ryegrass) and *Triticum aestivum* (wheat) roots in a field in Western Australia. Generally, *Trichoderma* spp. were isolated in greater frequency from wheat roots than from ryegrass. *T. hamatum* occurred in higher frequency, however, in ryegrass roots than in wheat roots, whereas *T. harzianum* occurred with greater frequency in roots of wheat. *T. koningii* was recovered with a higher frequency from roots of ryegrass at the seedling stage than from wheat, but the reverse was true at the tillering stage.

The take-all fungus (*Gaeumannomyces graminis* var. *tritici*) was present in both wheat and ryegrass roots and was severely pathogenic to both. *T. hamatum* and *T. koningii* reduced mortality of both wheat and ryegrass inoculated with the take-all fungus, whereas *T. harzianum* protected neither wheat nor ryegrass from that pathogen.

I mentioned previously that fusarium crown rot of tomato caused by *Fusarium oxysporum* f. sp. *radicis lycopersici* is a very serious disease worldwide and has been difficult to control. Sivan et al. (1987) investigated the potential of *Trichoderma harzianum* (T-35) obtained from a fusarium-wilted cotton plant against *F. o. radicis lycopersici* isolated from a diseased tomato plant. Biological control tests were conducted in artificially infested sandy-loam soil (pH 7.4, 0.3% O.M.) consisting of 82.3% sand, 2.3% silt, and 15% clay. Soil infestation with the pathogen was carried out using a microconidial suspension of *F. o. radicis lycopersici*. *T. harzianum* was applied as a conidial coating on seeds of 1684-Naama tomato. In field tests, *T. harzianum* was applied in the nursery as a wheat bran–peat preparation to the peat and vermiculite propagative mixture of tomatoes. After the end of the rooting period (30 days from planting), the tomatoes were transplanted to the field.

*Trichoderma* (T-35)-treated tomato transplants were better protected against fusarium crown rot than were untreated controls when planted in methyl-bromide fumigated or nonfumigated soil. The total yield of tomato plants in the *T. harzianum* plots was increased as much as 26.2% over the controls (Fig. 10). *T. harzianum* proliferated nicely in the rhizosphere. When tomato seeds were treated with conidia of *T. harzianum,* the antagonist was detected on root segments from plants sampled 20 weeks after planting.

Mónaco et al. (1991) examined the antagonistic potential of one isolate each of *Trichoderma aureoviride, T. koningii,* and *T. harzianum* against the pathogens *Sclerotium rolfsii, Fusarium solani, F. equiseti,* and *F. oxysporum*. All these fungi were isolated from soil of the La Plata area of Argentina.

Tomato seeds of La Plata line 7 were disinfected in sodium hypochlorite (5%) for 15 to 20 minutes and washed with tap water for 10 minutes. They were then divided into four groups; each of three groups was treated with one isolate of *Trichoderma* and the fourth group was left untreated. In an initial experiment, the antagonists were tested against seed germination on agar plates. *T. harzianum* stimulated seed germination during early stages, whereas, compared with the controls, *T. aureoviride* and *T. koningii* retarded germination. The difference in

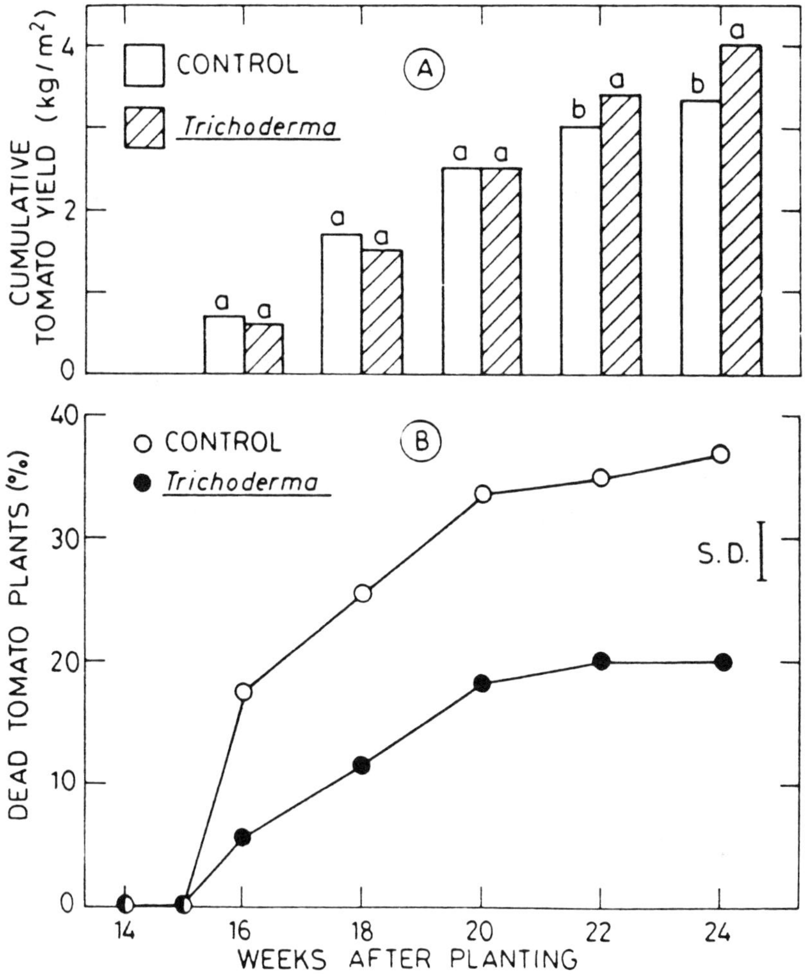


FIGURE 10 Biological control of Fusarium crown rot of tomatoes under field conditions (field experiment 1). *Trichoderma harzianum* was applied as a wheat bran–peat preparation introduced to the rooting mixture of tomato transplants (10%, v/v). (A) Cumulative yield of tomatoes; bar values on each harvest date marked with the same letter are not significantly different ($P = 0.05$). (B) Disease reduction determined 16, 18, 20, and 24 weeks after planting was significant ($P = 0.05$). Statistical comparisons were done by *t*-test. *Adapted from Sivan et al. (1987), with permission of Amer. Phytopathological Soc.*

percentage germination decreased some by the tenth day but was never completely overcome.

When *Trichoderma* spp. were applied to the seeds sown in soils infested with *Fusarium* spp. or *S. rolfsii,* they increased seedling emergence significantly in most treatments, as compared with the controls without antagonists. The antagonists were not equally effective against all pathogens, however. Unlike the laboratory tests, *T. aureoviride* was ineffective against *F. solani,* but all *Trichoderma* species were effective against *F. oxysporum. T. koningii* protected tomato seedlings against *S. rolfsii* significantly better than the protection given by *T. harzianum* and *T. aureoviride. T. harzianum* and *T. koningii* were significantly more effective against *F. solani* than was *T. aureoviride,* but all three antagonists were equally effective against *F. oxysporum.*

Bacterial wilt of tomato caused by *Pseudomonas solanacearum* is widespread in the world, particularly in tropical areas (Misaghi et al., 1992). Initial observations showed that some possible bacterial antagonists recovered from tomato roots could move along developing roots in the absence of percolating water. Misaghi et al. investigated the possibility that mobility of certain potential antagonists could enhance their biocontrol activity. From 65 root bacterial isolates of *Bacillus* sp. *P. fluorescens* and *P. putida* collected from the rhizosphere of field-grown tomato plants, Misaghi et al. selected 10 isolates that exhibited mobility (group 1) and 10 that did not (group 2). Four replicate tests were run to see if any changed categories (motile to nonmotile or vice versa). None changed. Group 1 isolates significantly reduced percent mortality caused by inoculation with *Pseudomonas solanacearum,* whereas group 2 isolates did not. All isolates from group 1 and 50% of the isolates from group 2 grew in the rhizosphere of tomato plants. Mobility was negatively correlated with percent plant mortality and positively correlated with the growth in the rhizosphere. Unfortunately, the classifications of the group 1 isolates and group 2 isolates were not given. The difference in antagonism may have been due to other characteristics of the type of bacteria in groups 1 and 2 rather than to mobility.

Fruit rot of tomato caused by *Rhizoctonia solani* is very serious in many areas of the world. Lewis et al. (1990) investigated the efficacy of *Trichoderma hamatum* (TRI-4, Tm-23) and *Gliocladium virens* (Gl-21) in preventing fruit rot caused by *R. solani* (isolates R-23, R-18, R-103, RMP). Field soil was incubated on a greenhouse bench for one week before addition of a mixed-millet inoculum consisting of the four *R. solani* isolates (100 g/20 kg soil, dry weight equivalent). This soil was then thoroughly mixed and distributed in planter flats. After one week, conidia, germlings, or pellets of the isolates were incorporated into the soil. Conidia were added in water to provide $10^5$ conidia/g of soil; germlings and pellets were added at a rate of 0.5% (w/w). Green, mature tomato fruits were placed on their sides in flats. Soil and fruits were kept damp by frequent light irrigation, and after ten days each fruit was examined for symptoms of fruit rot. In the greenhouse, germlings and alginate pellets of the individual antagonistic fungi significantly prevented fruit rot (> 80%), reduced pathogen inoculum (> 75%), and resulted in increased populations of the antagonists. In field tests the same types of preparations of the antagonists did not significantly prevent fruit rot, even though saprophytic activity of the pathogen isolates was reduced by the germlings in some tests. In greenhouse and field tests, conidia of the antagonists were generally ineffective in preventing disease, in reducing pathogen inoculum, and in stimulating antagonistic fungi to proliferate rapidly.

*Sclerotium rolfsii* is one of the most destructive soil-borne fungal pathogens of cultivated plants grown in the southeastern United States (Ristaino et al., 1991). Moreover, effective, economical disease control methods are still lacking for many hosts of that pathogen. Soil solarization and biocontrol have been studied some for control of the pest, but most studies of solarization have been made in arid climates. Ristaino et al. decided to test solarization and biocontrol with *Gliocladium virens,* alone and together, for control of *S. rolfsii* on Chico III tomato. Sclerotia of *S. rolfsii* (isolate Sr-DD-10) were obtained from Franklin County, North Carolina, and were cultured in autoclaved oats.

Field plot experiments were conducted in three successive years on Orangeburg sandy-loam soil (77% sand, 17% silt, 6% clay) in the coastal plain region of North Carolina. The inoculum of *S. rolfsii* was applied to main plots, solarization to subplots, and *Gliocladium virens* (Gl-21) pellets (W. R. Grace, Columbia, Maryland) at the same depth as the *S. rolfsii* to half of the subplots, whereas the remaining plots were not amended with the antagonist. Tomatoes were transplanted (six weeks after planting) into the field plots. Disease incidence was monitored occasionally for several weeks after planting each year. *G. virens* significantly reduced numbers of sclerotia of *S. rolfsii* by 100, 96, and 56% to depths of 30 cm in 1988, 1989, and 1990, respectively, whereas solarization alone reduced numbers of sclerotia by 62% in 1990. When disease pressure was high in 1988, disease incidence in plots planted immediately after treatment was reduced by 49% in solarized soils amended with *G. virens* before solarization, as compared with untreated controls. When disease pressure was low in 1990, disease incidence was reduced by either 77% in solarized soils or 53% in soils amended with *G. virens* alone, but the combined treatment did not significantly reduce disease the season after treatment. At shallow depth, introduced *G. virens* and indigenous fungi (*Trichoderma* spp.) did not survive solarization, but recovery of *Trichoderma* spp. increased when alginated pellets containing *G. virens* were added at depths of 20 and 30 cm in the soil. Plant dry weights were doubled, as compared with untreated controls, four weeks after solarization, but fruit yield of tomato planted the season after solarization was not affected by either solarization or the presence of *G. virens.* It was concluded that solarization for a six-week period during the warmest months of the summer between a spring- and fall-grown crop and treatment of solarized soil with *G. virens* could provide an additional management tool for southern blight in the coastal plain of North Carolina.

Some of the most important postharvest diseases of *Malus* × *domestica* (apples) are caused by *Penicillium expansum, Botrytis cinerea, Mucor* spp., *Gleosporium perennans,* and *Phialophora malorum* (Janisiewicz, 1988). When one considers that most fruits have a number of important diseases, controlling only one may favor another. Janisiewicz therefore decided to investigate the biocontrol of *B. cinerea* with an antagonist isolated from apple leaves and use it in combination with an antagonist against *P. expansum* to control both diseases simultaneously. In vivo screening of apples was done by treating wounded apples with an aqueous suspension of a potential antagonist and an aqueous suspension of pathogen spores ($1 \times 10^4$ spores/mL). Lesion diameters were measured after seven days.

The selection of antagonists was based on their good performance in screening on Golden Delicious apples. Combinations of two antagonists, each known to inhibit the development of *B. cinerea* or *P. expansum* on fruit, were applied in mixtures as a water suspension to wounds on Golden Delicious apples. This was followed by a challenge with an aqueous mixture of *B. cinerea* and *P. expansum* spores. Antagonist mixtures consisted of either two bacteria, two yeasts, or organisms from these two groups; 18 combinations were tested. The concentration of bacteria in the mixtures was approximately $3 \times 10^8$ cfu/mL; that of the yeast was $3 \times 10^6$ cells/mL. For each combination, 20 μL of the mixture of the suspension of *B. cinerea* and *P. expansum* ($1 \times 10^2$ spores of each per milliliter) was added. Lesions were measured after seven days at 24°C. From the four treatments showing good inhibition of both pathogens, one combination of isolates was selected for further study.

The selected combination of *Pseudomonas* sp. and *Acremonium breve* was tested against spore mixtures of *B. cinerea* and *P. expansum* on the apples. Tests on Golden Delicious apples just after harvest showed that total protection from *B. cinerea* ($1 \times 10^4$ spores/mL) was obtained with $1.5 \times 10^4$ cfu/mL, the lowest concentration of *A. breve* applied. The protection lasted for ten weeks at 24°C, the duration of the experiment. When *Pseudomonas* sp. (L-22-64) from an earlier study and *A. breve* were combined, they gave excellent protection against both *B. cinerea* and *P. expansum.* The concentration of the antagonists needed for control was low enough to be considered for commercial use.

In an experiment related to that described in the previous paragraphs, Janisiewicz and Roitman (1988) isolated numerous potential antagonists from apple leaves and fruits and found several to be antagonistic to *B. cinerea* or *P. expansum,* or both. The organism that produced the largest zone of inhibition on agar plates was identified as *Pseudomonas cepacia.* It gave good control of gray mold on both Golden Delicious apples and Bosc pears (*Pyrus*) and reduction in blue mold, caused by *P. expansum,* on both.

An effective antifungal compound that was isolated from *Pseudomonas cepacia,* identified as a pyrrolnitrin, inhibited growth of both gray and blue mold on agar at a concentration of 1 mg/L. Complete control of gray mold was obtained on apples and pears with a pyrrolnitrin concentration of 10 mg/L, at a pathogen concentration of $10^3$-$10^5$ conidia/mL. Blue mold was controlled at the same concentration of pyrrolnitrin at inoculum concentrations of $10^3$ conidia/mL on pears and $10^3$ and $10^4$ conidia/mL for apples. At 50 mg/L or higher, complete control of both diseases was obtained on both fruits at all tested inoculum concentrations.

Pusey (1989) isolated numerous microorganisms from stone fruits and tested them against *Monilinia fructicola,* the pathogen that causes brown rot of stone fruits. This was done by spraying solid media with bacterial isolates. Numerous bacteria were very inhibitory to *M. fructicola* in culture but had little or no effect against the pathogen when tested on fruit. This was not true of one bacterium, designated as B-3 and later identified as *Bacillus subtilis.* Even the cell-free filtrates from cultures of B-3 protected fruit from the pathogen, and the filtrates remained active even after autoclaving.

The biologically active substance was isolated from the filtrate and identified by mass spectroscopy, nuclear magnetic resonance spectrometry, and IR spectrometry. It was concluded that the major antifungal peptides produced by B-3 were the same as the iturin antibiotics previously reported. The iturins are cyclic peptides made up of seven alpha-amino acids and one beta-amino acid. The iturin antibiotics are active against few bacteria, but have a wide antifungal action. The antibiotic extract from B-3 inhibited growth of 15 of 19 fungal pathogens.

Suspensions of B-3 cells controlled brown rot on *Prunus persica* (peaches), nectarines, apricots, plums, and cherries. Brown rot on peach was retarded by B-3 cell suspensions of $10^6$ and $10^7$ cfu/mL. At $10^8$ cfu/mL, peach fruits never became infected by *M. fructicola* but eventually became infected by other fungi such as *Penicillium expansum.* Although infections developed more slowly on nonwounded fruit, results with B-3 were generally even more dramatic on nonwounded fruits. In a simulated commercial fruit-packing test, B-3 ($2 \times 7^8$ to $7 \times 10^8$ cfu/kg fruit) was equal to benomyl (1–2 mg/kg fruit) in controlling brown rot on peach. B-3 was also tested on several types of fruits inoculated with various postharvest fungi. In vivo activity was shown for apple rots caused by *M. fructicola, Botrytis cinerea,* and *Glomerella cingulata* and for gray mold of grapes caused by *B. cinerea.*

*Phytophthora cactorum* is the major cause of phytophthora crown and root rot of *Malus* × *domestica* (apple) trees in the Kootenay Valley of British Columbia, Canada (Utkhede and Smith, 1991). Twenty-one strains of *Bacillus subtilis* previously shown to produce diffusible antibiotics and one strain of *Enterobacter aerogenes* (B8) were tested, along with metalaxyl and monoammonium phosphate, in an eight-year field test in the Kootenay Valley for control of phytophthora crown and root rot of *Malus* × *domestica* (MacSpur budded on MM.106 rootstock).

The first application of chemical and biological materials was made at planting in the spring of 1983 and a second that fall. Treatments were then applied semiannually for three more years in early May and late September. In the years following, the treatments were applied only once a year, in early May. Soil around each tree was infested annually in mid-June with *Phytophthora cactorum* at the rate of about 2300 sporangia per tree. The soil was removed from the crown region to a depth of 4 cm and replaced immediately after application of the fungal pathogen or bacterial antagonist.

The fungicide metalaxyl prevented phytophthora crown and root rot of apple trees for the full eight years. Compared with the control, strain B8 of *E. aerogenes* also significantly reduced the number of trees infected with *P. cactorum.* Again compared with the control, Strains AB8, BACTX, NZB1, AB7, AB8, and BACT2 of *B. subtilis* significantly reduced the number of trees infected by *P. cactorum* and also increased the trunk diameter. It was concluded that the level of protection provided by these six *B. subtilis* strains under field conditions suggests that they could provide biological control of the crown and root disease of apple trees in the Kootenay Valley.

We have noted above that phytophthora crown and root rot of apple trees can be satisfactorily controlled by several strains of *Bacillus subtilis.* Roiger and Jeffers (1991) decided to see if *Trichoderma* spp. could control this disease on apple. Using two delivery methods for the antagonists, they tested 223 isolates of five species of *Trichoderma.* A viscous suspension of conidia in an aqueous gel was applied to roots, or a colonized mixture of peat and wheat bran was added to the soil. Of all the isolates that were tested twice, six delivered in peat-bran and five delivered as conidial suspensions increased survival time of seedlings compared with controls without *Trichoderma.* The 11 isolates were compared in a separate experiment, and isolate TW.055 of *T. virens* (or *Gliocladium virens*) in peat-bran was consistently superior to all other isolates. Seedlings treated with *T. virens* survived an average of 30 (of 44) days, whereas those in the control survived only 19 days. *T. koningii* isolates TW.105 and TW.189 in peat-bran caused apple seedlings to survive an average of 24 days, and isolate TW.138 of *T. harzianum* caused seedlings to survive 22 days. The isolates of *Trichoderma* that were successful in protecting apple seedlings against phytophthora crown and root rot in these tests should be investigated further as potential biocontrol agents in the protection of apple seedlings against such diseases.

Apple scab caused by *Venturia inaequalis* is a relatively serious problem in the apple-growing area around Lake Constance in Germany. Miedkte and Kennel (1990) investigated the possible biological control of the disease by a strain of *Athelia bombacina* and two strains of *Chaetomium globosum* (F6 and NRRL6296). At the beginning of leaf fall in October, green leaves naturally infected with *V. inaequalis* were collected from 10-year-old Golden Delicious apple trees. The infected leaves were sprayed on both surfaces with either *A. bombacina* or a strain of *C. globosum* until uniformly wet. The treated leaves were allowed to overwinter in special boxes on the orchard floor, where they were exposed to natural weather conditions.

In the *Athelia*-bran treatment, *A. bombacina* completely colonized the leaves and suppressed the pseudothecial formation and ascospore production of *V. inaequalis* by 100%. When added without nutrient, *A. bombacina* colonized leaves by 60 to 80% and reduced ascospore production of *V. inaequalis* by 60 to 70%. Both strains of *C. globosum* colonized the leaves completely, but strain F6 reduced ascospore production by only 30 to 40%, whereas NRRL6296 reduced ascospore production by about 65%. It was suggested that the greater efficacy of *C. globosum* NRRL6296 over F6 might have been due to a greater production of antibiotics, as reported by others. Of the three antagonists tested, however, *A. bombacina* was the best in preventing ascospore formation by *V. inaequalis.*

*Phytophthora megasperma* f. sp. *medicaginis* causes damping-off of *Medicago sativa* (alfalfa) seedlings in poorly drained soil in many parts of the world (Handelsman et al., 1990). Even resistant seedlings may be susceptible to this disease in early stages of development; thus effective biocontrol could be valuable. Handelsman et al. isolated 700 bacterial strains from the roots of field-grown alfalfa and screened them against Iroquois alfalfa seedlings inoculated with *P. megasperma* f. sp. *medicaginis.* Sufficient *P. m. m.* zoospores were added to culture tubes containing alfalfa seedlings to cause 100% mortality of control seedlings. Only one of the 700 isolates reduced seedling mortality to 0%; it was identified as *Bacillus cereus* and was designated as UW85. Subsequent tests demonstrated that both fully sporulated cultures of UW85 and sterile filtrates were effective in protecting seedlings from damping-off. In tests in a field infested with *P. m. m.,* coating seeds with UW85 significantly increased emergence of alfalfa. It appears therefore that UW85 has good potential as a biocontrol organism against damping-off of alfalfa caused by *P. m. m.*

*Cercospora moricola* causes a severe leaf-spot disease of *Morus indica* (mulberry) leaves, the chief source of food for silkworms. Using fungicides to control the pathogen often causes extreme residual toxicity to the silkworms. Sukumar and Ramalingam (1986) isolated 14 types of microorganisms, including both bacteria and fungi, from mulberry leaves and screened them for antagonism against *C. moricola.* All inhibited the pathogen at least slightly, but *Nigrospora sphaerica* and *Trichoderma viride* were considered moderate inhibitors, while *Cladosporium cladosporioides, Staphylococcus* sp., *Corynebacterium* sp., *Curvularia lunata, Pseudomonas maltophila* and *M. leucotrichum* were strong inhibitors. Even though *M. leucotrichum* completely inhibited growth of *C. moricola,* it behaved as a pathogen when inoculated in sufficient concentration and probably should not be used as a biocontrol agent. In vivo tests indicated, however, that *P. maltophila, C. cladosporioides,* and *C. lunata* significantly reduced the number of leaf-spots on mulberry leaves and showed promise as biocontrol agents against the leaf-spot disease.

Panama wilt of *Musa balbisiana* (banana) caused by *Fusarium oxysporum* f. sp. *cubense* has been a major cause of loss to growers. There are no effective chemical control methods for this disease, so Sivamani and Gnanamanickam (1988) investigated the possibility of using some native strains of *Pseudomonas fluorescens* as biocontrol agents against it. The wilt pathogen *F. oxysporum* f. sp. *cubense* was isolated from wilted banana trees in two states of southern India, and a race 1 and race 4 isolates were used. The *P. fluorescens* strains were from the local collection isolated from roots of rice (strain Pfr 13), peanut (Pfgn), and banana (Pfb) and from leaves of blackgram (Pfbg), citrus (Pfcp), and cotton (Pfco). Plate assays were used for initial screening of the possible antagonists against *F. o. c.* In in vitro tests, all six *P. fluorescens* strains reduced growth of the pathogen (TV15, race 1).

In vivo tests were conducted by dipping roots of *Musa balbisiana* seedlings in suspensions of *P. fluorescens* (Pfcp), the most active strain in in vitro tests, and then dipping in a conidial suspension of the pathogen *F. oxysporum cubense* (TV15, race 1). Some seedlings were dipped in just the spore suspension; control seedlings were dipped for one hour in sterile water. The banana seedlings treated with the antagonist (Pfcp) showed less wilting and internal discoloration due to the pathogen than did inoculated plants not treated with Pfcp. The antagonist also resulted in better root growth and plant height. It appears therefore that use of an appropriate strain of *P. fluorescens* and proper treating conditions could give good biological control of the Panama wilt of banana.

Another significant wilt pathogen of banana, eggplant, potato, and tomato in Asia and the South Pacific region is *Pseudomonas solanacearum.* Anuratha and Gnanamanickam (1990) screened 125 strains of fluorescent and 52 strains of nonfluorescent bacteria against *P. solanacearum* and found that *P. fluorescens* strain Pfcp and strains B33 and B36 of *Bacillus* spp. were most inhibitory. In subsequent tests in the greenhouse and field, they found that strain-Pfcp-treated banana, eggplant, and tomato plants were protected from wilt caused by *P. solanacearum* by 50, 61, and 95%, respectively, in the greenhouse and by 50, 49, and 36%, respectively, in the field. *Bacillus* strains B33 and B36 gave considerably less protection against the wilt. Plants treated with the antagonists after inoculation with the pathogen had increased height and biomass over the plants treated only with the pathogen and had close to the values for untreated control plants.

CHAPTER 7

# ALLELOPATHY IN THE BIOLOGICAL CONTROL OF PLANT DISEASES: HOST PLANTS N–Z

SINCE MANY EXPERIMENTAL METHODS were discussed in considerable detail in Chap. 6, fewer details will be given concerning many of the investigations covered in this chapter, where I conclude my treatment of biological control of plant pathogens by microorganisms.

Basal rot of *Narcissus pseudo-narcissus* is responsible for major losses in the bulb industry in Britain and elsewhere (Beale and Pitt, 1990). This disease is caused by *Fusarium oxysporum* f. sp. *narcissi;* no economic and effective control methods have been found for this pathogen. Beale and Pitt isolated many bacteria, actinomycetes, and fungi from stored *Narcissus* bulbs and soil at 10 field sites and at a long-established habitat of indigenous *Narcissus.* These were screened against *F. oxysporum narcissi* on agar plates. Twenty-nine fungi and one actinomycete were found to have considerable inhibitory activity against the pathogen.

The 30 promising antagonists were screened against thiabendazole, the most commonly used fungicide in *Narcissus* production; about two-thirds of them had some level of tolerance against the fungicide. Ten of the antagonists grew in the presence of 1, 5, or 10 μg/ml of the fungicide, some tolerated up to 100 μg/ml.

Seventeen antagonists to *F. oxysporum* f. sp. *narcissi* were evaluated for biocontrol activity in pots with Golden Harvest *Narcissus,* and seven that were thiabendazole-tolerant were evaluated in combination with the fungicide. Six antagonists gave disease levels significantly lower than infected controls. Two of these, a *Streptomyces* sp. (131)

and *Minimedusa polyspora* (128L), gave disease levels 56% lower than those for the infected control. The other effective antagonists—*Trichoderma hamatum* (094), *Penicillium rubrum* (082), *T. viride* (089L), and *T. harzianum* (092)—all reduced disease levels by over 30% of that recorded in the infected control. Dual treatments of fungal antagonists and thiabendazole, although resulting in lower levels of disease, did not significantly reduce levels over that of thiabendazole or the antagonist alone.

In a one-year field trial, bulb treatment with *Minimedusa polyspora,* a *Streptomyces* sp., and an isolate of *Trichoderma viride* significantly suppressed reductions in flower numbers, reductions in bulb yields, and percentage infection. The combination of thiabendazole with *T. viride* gave bulb yields significantly greater than either *T. viride* or thiabendazole alone, and integration of *M. polyspora* and thiabendazole gave significantly greater flower numbers and significantly fewer infections than either treatment alone. Again I emphasize that, with the development of appropriate methods, integration of a chemical treatment with an antagonist could probably give excellent control, with a considerably lower use of a fungicide or bacteriocide in many instances.

Bacterial wilt of *Nicotiana tabacum* (tobacco) and other plants in the Solanaceae is caused by *Pseudomonas solanacearum.* This pathogen is a real nuisance because it can survive for a long time in soil, even in the absence of the host plant (Aoki et al., 1991). It is well known that this disease is less prevalent when tobacco is grown in rice paddies; therefore Aoki et al. isolated several effective antagonists of the pathogen from paddy soil. Two antagonists were used, *Pseudomonas cepacia* (B5) and a mutant strain of B5 (B5-R) that is resistant to nalidixic acid, and one strain of the pathogen was selected, *P. solanacearum* (PS27), which came from the Leaf Tobacco Research Laboratory in Tochigi, Japan.

An antibiotic produced by *P. cepacia* (B5) was isolated and identified by IR, NMR, and MS as 2-keto-D-gluconic acid (2KGA). Two milligrams per 10 mL of a PS27 suspension culture completely eliminated all viable cells after two days. When B5 cells were added to a soil suspension of PS27 cells along with glucose or wheat bran, 2KGA was synthesized and equaled 10 mg/10 mL after two days with added glucose and 3 mg/10 mL after two days with added wheat bran. Moreover, no viable cells of PS27 remained after two days when glucose or wheat bran was added to the suspension along with the B5 antagonist. The production of 2KGA and the destruction of the pathogen in the soil suspension to which B5 and wheat bran or glucose were applied support the inference that 2KGA was involved in disease suppression by *P. cepacia* (B5). This is another of the many investigations indicating that suspected antagonists often reduce disease only if they produce antibiotics; at least they are more effective if they produce antibiotics.

Another disease of tobacco, black root rot, is caused by the fungal pathogen *Thielaviopsis basicola,* which can also survive in soil in the absence of host plants (Reddy and Patrick, 1992). This pathogen also attacks a wide variety of crops, and Reddy and Patrick had previously isolated fluorescent pseudomonads from soil that were antagonistic to *T. basicola.* They decided therefore to test the ability of one strain of *Pseudomonas* (RD:1) to suppress black root rot of tobacco under greenhouse conditions.

Tobacco plants with roots bacterized with *Pseudomonas* RD:1 were planted in nonsterilized fallow soil and in soil containing decomposing rye residues artificially infested with chlamydospores of *T. basicola.* Compared with nonbacterized controls, bacterization with RD:1 significantly increased shoot and root dry weights of tobacco seedlings in both fallow soil and in soil containing decomposing rye residues. RD:1 also became well established in the rhizosphere of tobacco seedlings and significantly reduced the severity of black root rot disease in both soils. Strain RD:1 also inhibited the mycelial growth of *T. basicola* on agar plates. It was concluded that RD:1 is potentially promising for the biocontrol of *T. basicola.*

About the same time Reddy and Patrick were investigating the possible role of *Pseudomonas* RD:1 on black root rot of tobacco caused by *T. basicola,* Keel et al. (1992) were investigating the efficacy and mechanism of action of *Pseudomonas fluorescens* (strain CHAO) as antagonist to the same disease and also to take-all of wheat caused by *Gaeumannomyces graminis* var. *tritici.* Strain CHAO was found to produce two antibiotics, 2,4-diacetylphloroglucinol (Phl) and 2,5-dichloro-2-(2',6'-dihydroxybenzoyl)-pyrole (pyoluteorin). No further work was done with the latter, but Phl was very inhibitory to growth of *T. basicola, G. graminis tritici* and *Fusarium oxysporum* f. sp. *lycopersici, F. o.* f. sp. *lini, Pythium debaryanum, P. ultimum,* and *Rhizoctonia solani.* It also was very inhibitory to several species of bacteria, especially *Bacillus subtilis, B. thuringiensis, Micrococcus luteus, Pseudomonas syringae* pv. *phaseolicola, P. s.* pv. *tabaci,* and *Staphylococcus aureus.* In contrast, strains of the *P. fluorescens-putida* group, including strain CHAO and its Phl$^-$ mutant CHA625, and *P. aeruginosa* were very insensitive to the antibiotic. Phl was found to be inhibitory to the growth of corn, cress, cucumber, flax, sweet corn, tobacco, tomato, and wheat.

Compared with strain CHAO, a *P. fluorescens* mutant of CHAO obtained by Tn5 insertion did not produce Phl, showed lowered inhibition of *T. basicola* and *G. g.* var. *tritici,* and had a reduced suppressive effect on tobacco black root rot and on take-all of wheat. When the mutant strain was complemented with an 11-kbDNA fragment of wild-type CHAO, it was again able to produce Phl, inhibit the fungal pathogens, and suppress the diseases. Phl was shown to be produced in the rhizosphere of wheat by strain CHAO and the complemented mutant but not by the Phl-deficient mutant (Table 17). These experiments support the importance of 2,4-diacetylphloroglucinol production by the *P. fluorescens* strain CHAO in the suppression of soil-borne plant pathogens in the rhizosphere.

Sheath blight of *Oryza sativa* (rice) is caused by *Rhizoctonia solani* and occurs throughout the world (Mew and Rosales, 1986). Inoculum of the pathogen is carried over from one crop to the next by sclerotia, infected weeds, straw, and stubble, and this disease has not been effectively controlled by host resistance. Mew and Rosales isolated bacteria from more than 300 randomly selected healthy and sheath-blight-infected rice plants and others from paddy water, sclerotia, and soil. All isolated bacteria were tested for inhibitory activity against strain LR-1 of *R. solani.* Inhibitory strains of bacteria were lyophilized for storage. The LR-1 strain of *R. solani* came from cv. IR36 of rice.

Ninety-one percent of the fluorescent bacterial isolates inhibited mycelial growth of *R. solani* (LR-1), whereas only 33% of the nonfluorescent isolates did so. The isolate In-b-150 slightly stimulated germination of seeds of rice cv. IR-36. When sclerotia of *R. solani* were soaked for different periods (10 minutes; 12, 24, and 48 hours; and 1 week) in an effective bacterial suspension, then allowed to infect rice plants grown to maximum tillering, disease incidence was reduced, compared with plants infected with sclerotia soaked in sterile distilled water. Isolate In-b-17, a nonfluorescent bacterium, seemed to have a

**Table 17** Production of 2,4-diacetylphloroglucinol (Phl) by *Pseudomonas fluorescens* strain CHAO and its derivatives in the rhizosphere of wheat grown under gnotobiotic conditions and the relationship between antibiotic production and suppression of *Gaeumannomyces graminis* var. *tritici*–induced take-all by the bacteria*

| Microorganisms added | | μg Phl†/g | | Plant fresh weight§ (mg) | Root fresh weight§ (mg) | Disease ratings§¶ | Fluorescent pseudomondads§ ($10^8$ cfu/g of root) |
|---|---|---|---|---|---|---|---|
| *P. fluorescens* | *G. graminis* | Rhizosphere‡ | Root | | | | |
| None | – | < 0.001 | < 0.01 | 598 ab | 320 a | 0 e | 0 b |
| | + | < 0.001 | < 0.01 | 318 d | 156 c | 3.1 a | 0 b |
| CHAO (Phl†) | – | 0.04 ± 0.020 | 0.94 ± 0.48 | 638 a | 332 a | 0 e | 1.1 a |
| | + | 0.10 ± 0.050 | 1.36 ± 0.16 | 606 ab | 323 a | 0.7 d | 1.2 a |
| CHA625 (Phl-) | – | < 0.001 | < 0.01 | 609 ab | 320 a | 0 e | 0.7 a |
| | + | < 0.001 | < 0.01 | 496 c | 249 b | 1.9 b | 1.1 a |
| CHA625/pME3128 (Phl†) | – | 0.01 ± 0.008 | 0.26 ± 0.14 | 631 a | 335 a | 0 e | 1.3 a |
| | + | 0.01 ± 0.003 | 0.19 ± 0.05 | 540 ab | 294 a | 1.3 c | 1.4 a |

* CHAO (a wild-type strain of *P. fluorescens*) and its derivatives, CHA625 (a Tn5-insertion Phl-negative mutant) and CHA625/pME3128 (a transconjugant of CHA625, restored in Phl production), and *G. graminis* var. *tritici* were added at $10^7$ cfu and as 1.25 mg, respectively, of colonized millet seed per cubic centimeter of artificial soil, 14 and 7 days, respectively, before planting.

† Phl was extracted from the roots and the adhering artificial soil of 125 plants per treatment; each value is the mean of three experiments.

‡ Rhizosphere weight includes fresh weight of the roots with the adhering rhizosphere soil.

§ Means within columns followed by the same letter are not significantly different at $P = 0.05$ according to Student's *t*-test. Each value is the mean of 6 independent experiments with 3 or 25 replicates (for Phl extraction from rhizosphere) per experiment and 1 flask (5 plants) per replicate. (See text.)

¶ Disease severity was rated on a 0–4 scale (0 = no disease, 4 = plants dead).

*Source:* From Keel et al. (1992), with permission of Amer. Phytopathological Soc.

greater effect on disease resistance than did fluorescent isolates In-b-24 and In-b-150. When bacterial suspensions (about $10^6$ cfu/mL) were sprayed on rice plants infected at the same time with sclerotia inserted in the tillers, the difference in disease incidence and lesion size was significant. Sheath-blight incidence was 100, 30, and 75%, respectively, on plants exposed to *R. solani* and to *R. solani* in combination with either isolate In-b-17 or In-b-24. Again, the nonfluorescent isolate In-b-17 appeared to protect the plants better than did the isolate In-b-24.

In a second planting in flats used to grow rice plants from seeds treated with bacterial antagonists in the first planting, plants grown from untreated seeds had reduced disease incidence. The bacterial antagonists appeared to have established themselves in the soil. In fact, the total bacteria counts from soil samples at 35 days after planting were considerably higher in soil from seed boxes where seeds had been treated with the antagonistic bacteria.

Strains of fluorescent and nonfluorescent bacteria isolated from rice rhizospheres in southern India inhibited mycelial growth of *R. solani,* affected sclerotial viability, and protected IR20 and TKM9 rice seedlings from infection by *R. solani* in greenhouse trials (Devi et al., 1989). In field plots, TKM and IR-50 rice plants raised from bacterized seeds had 65 to 72% less sheath blight than did plants from untreated seeds.

One of the most devastating diseases of rice is bacterial blight caused by *Xanthomonas oryzae* pv. *oryzae;* it occurs as a vascular wilt in the early stages of growth and as a leaf blight in later stages (Sakthivel and Mew, 1991). Some strains of *X. oryzae* pv. *oryzae* were previously reported to produce bacteriocins, proteinaceous antibacterial substances. Sakthivel and Mew isolated 144 strains of the pathogen from small lesions of bacterial-blight-infected leaves collected from rice fields of the IRRI at Los Baños, the Philippines. The isolates were tested for pathogenicity against IR24 rice, known to be susceptible to all *X. oryzae* pv. *oryzae* isolates known. All the isolates were next tested for bacteriocin production, after which selected strains were screened to identify the natural avirulent isolates. Next, selected isolates that showed broad-spectrum bacteriocin production and growth inhibition of the pathogens were exposed to N-methyl-N-nitro-N-nitrosoguanidine (NTG) mutagenesis. After treatment, isolates that produced no symptoms on susceptible rice strains were tested again for bacteriocin production. Antibiotic markers were then developed against streptomycin and rifampicin. The efficacy of the nonpathogenic-bacteriocin-producing bacteria against the bacterial blight disease was then tested under greenhouse and field conditions (screenhouse). The selected potential antagonists reduced bacterial blight incidence up to 31 to 99% in greenhouse tests and 11 to 73% in screenhouse tests. Bacterial leaf streak severity was reduced 4 to 20% in greenhouse tests; disease incidence was reduced in the screenhouse by 20 to 39%. The nonpathogenic-bacteriocin-producing mutants of *X. oryzae* pv. *oryzae* survived well epiphytically and maintained population levels up to $10^6$ to $10^7$ cfu/g of leaves of 34-day-old plants. These nonpathogenic bacteriogenic bacteria are well adapted for survival in the same ecological niches in which the pathogen survives because both belong to the same taxonomic group. They may therefore be more effective than other biocontrol organisms against rice bacterial diseases. It should be emphasized that this is another of the many examples where the success of an antagonistic organism is related to its ability to produce one or more antibiotics.

A rust of *Pelargonium* x *hortorum* (garden geranium) is caused by *Puccinia pelargonii-zonalis* and was first discovered in South Africa. Rytter et al. (1989) selected *Bacillus* spp. as a group for potential biological control of geranium rust. They isolated *Bacillus* spp. from geranium cultivars, including Salmon Queen, Schone von Grenchen, Pink Camellia, Springtime Irene, and Snowmass, whose reactions to the rust ranged from immune to highly susceptible. Additional isolations were made from a rust-infected plant of the cv. Snowmass. Preliminary tests confirmed that 12 isolates were *Bacillus* spp. A control *B. subtilis,* strain 13, obtained from Pennsylvania State University, was also used in all tests.

All strains were tested against urediniospore germination of *Puccinia pelargonii-zonalis.* Strain 3 of *B. subtilis,* isolated from a rust-infected geranium leaf, inhibited spore germination and reduced the incidence of rust pustules on inoculated leaves in the greenhouse. The inhibitory substance(s) was present in the culture filtrate of strain 3 and was retained in or on washed bacterial cells. The filtrate was most inhibitory in decreasing the number of pustules per unit of leaf area, followed by the washed bacterial cell treatment. Cells cultured and applied to geranium leaves in nutrient broth were more effective in reducing rust than was a culture filtrate, indicating that nutrients affected the pathogen-antagonist interaction. When *B. subtilis* (strain 3) was applied one to four days before inoculation with the pathogen, it was significantly more effective in protecting against the disease than when it was added on the same day as the pathogen. It was concluded that strain 3 of *B. subtilis* probably could not be used as the only control for geranium rust, because there is a zero tolerance for geranium rust in commercial production. It could, however, possibly be used alternately with fungicide applications, thus saving fungicide costs.

Root rot of *Persea americana* (avocado) is the major disease of that plant in California; it is caused by *Phytophthora cinnamomi* (Gees and Coffey, 1989). Gees and Coffey isolated bacteria and fungi from avocado feeder roots growing in a soil suppressive to *P. cinnamomi* and added them to various soil mixtures and natural soils infested with the pathogen to test the biocontrol potential of *Persea indica,* a relative of avocado. Other tests were conducted in five artificial soil mixes.

Thirty-six fungal isolates and 110 bacterial isolates were tested against infection of hosts by *Phytophthora cinnamomi.* The test hosts for potential bacterial antagonists were blue lupine (*Lupinus angustifolius*) and *Persea indica;* the test host for fungal antagonists was *P. indica.* The duration of tests was four weeks with blue lupine and four to six weeks with *P. indica.* No bacterial strain was found to be a good potential antagonist for root rot of avocado. Strain TW of *Myrothecium roridum* was the best and most consistent antagonist in controlling *P. cinnamomi* in several greenhouse tests with highly susceptible seedlings of *P. indica. M. roridum* was grown on a wheat-bran ture at 2.5% (w/v) two weeks before inoculation with zoospores of *P. cinnamomi.* In a UC-mixture with *P. indica* inoculated with zoospores of *P. cinnamomi, M. roridum* suppressed root infection by 50 to 94% compared with uninoculated controls. There was no significant difference in the disease reduction achieved by either *M. roridum* or the fungicide potassium phosphonate (2.5 mg/pot). Root infection ranged from 12 to 54% in three naturally infested field soils, compared with 58 to 93% for controls over the same four-week period. On a medium containing carbendazim, a fungicide-resistant mutant of strain TW, TWm14, was consistently isolated from the root tips of *P. indica* growing in infested soil four weeks after transfer. This indicated the rhizosphere competence of this strain. Current use of strain TW at a rate of 2.5% (w/v) bran inoculum will probably require application of 6.5 to

26.0 kg/ha, based on the recommended avocado planting density of 260 trees/ha.

I have pointed out frequently that *Sclerotium rolfsii* is a very destructive pathogen on many plants and has been the subject of many biocontrol experiments. Papavizas and Lewis (1989) screened 285 wild-type strains and mutants of *Gliocladium virens, Trichoderma hamatum, T. harzianum,* and *T. viride* against *S. rolfsii* in the greenhouse. Ten strains of *G. virens* and four strains of *T. harzianum* suppressed damping-off of *Phaseolus vulgaris* (snap beans) by 30 to 50% and blight (both diseases due to *S. rolfsii*) by 36 to 74%. All strains of *T hamatum* and *T. viride* tested as conidia were ineffective. In disease suppression, several strains of *G. virens* and *T. harzianum* used alone were equal to or more effective than double and triple mixtures of such strains. Of four formulations of *G. virens* tested, germlings, alginate-bran-fermenter biomass pellets, and Pyrax-fermenter biomass mixtures reduced disease considerably, and all three formulations were more effective than conidia in aqueous suspension. When strain Gl-3 of *G. virens* was added to soil as Pyrax-fermenter, biomass mixtures in amounts to provide $1.5 \times 10^3$ to $1.2 \times 10^4$ cfu/g of soil provided statistically significant protection of the host at all concentrations. The strains of the pathogen used also affected the antagonistic action. Strains Gl-3 and Gl-21 of *G. virens* suppressed disease caused by strain Sr-1 (small sclerotia) of *S. rolfsii,* gave partially effective control of Sr-116 (medium sclerotia), and were not effective against Sr-3 (large sclerotia), even though they colonized sclerotia of all three pathogen strains. It was concluded that deep-tank fermentation in inexpensive media for mass production of biomass of *Gliocladium* and *Trichoderma* is the technological approach most likely to be used in the United States and in other industrialized countries.

Root rot of *Phaseolus angularis* (adzuki bean, or red bean) caused by *Rhizoctonia solani* is a very destructive disease in Taiwan (Liu, 1991). Liu reported that *Trichoderma harzianum* (strain TC11) and *T. koningii* (strain T12) were strongly antagonistic to *R. solani* in screening tests. In order to improve biocontrol efficacy and survival in the field where a number of fungicides are sprayed, he induced TC11 and TC12 to become benomyl-resistant by use of UV light and chemical mutagens. Of many new strains produced, Liu selected one (T12-R33-D25) that was resistant to 1000 μg/mL of benomyl and still retained its lytic or antibiotic activity against *R. solani.* A rhizosphere-competent phenomenon and growth-promotion effect were observed in seed coating of adzuki bean or soil amendment of T12-R33-D25. In field trials at Pingtung and Kaohsiung, Taiwan, from 1986 to 1990, seed treatment or soil treatment prevented root rot of adzuki beans and significantly increased yields. Even though both delivery systems gave equally good results, seed coating took the least labor. This biocontrol method with strain T12-R33-D25 is now available through extension services to the bean producer in Taiwan.

In a search of a successful biocontrol of the pathogenic fungus *Fomes annosus,* Donnelly and Sheridan (1986) cultured several fungi, one at a time, on a plate with *F. annosus.* One fungus, *Trichoderma polysporum* isolated from the roots of *Picea sitchensis* (Sitka spruce), displayed an antagonism in the form of a demarcation line that was associated with increased production of aerial hyphae and deposition of crystalline material by both fungi at the line of mycelial contact. The crystals produced by *F. annosus* in the presence of *T. polysporum* were identified as 3-hydroxy-7,11,11-trimethyl-cyclopenta-(g)-benzopyran-1-one (formajorin D), a sesquiterpene isocoumarin previously isolated from cultures and fructifications of *F. annosus.*

The pigment produced by *T. polysporum* in paired cultures with *Fomes annosus* was separated into its components by column chromatography. The components were 1-hydroxyl-3-methyl-anthraquinone (pachybasin); 1-8-dihydroxy-3-methyl-anthraquinone (chrysophanol); and 1,6,8-trihydroxy-3-methyl-anthraquinone (emodin). These anthraquinones had been isolated previously from *T. viride* and *Phomea foveata.* The original pigment, the isolated anthraquinones, and their O-acetyl and O-methyl derivatives were tested against growth of two strains of *F. annosus.* The O-acetyl derivatives of the anthraquinones markedly inhibited the linear growth rate of both strains of *F. annosus.* It appears therefore that production of these compounds by *T. polysporum* when it comes in contact with *F. annosus* makes it a likely candidate for the biocontrol of *F. annosus.*

Duchesne et al. (1989a) carried out a time course of disease suppression by the ectomycorrhizal fungus *Paxillus involutus* against *Fusarium oxysporum* f. sp. *pini* on *Pinus resinosa. P. resinosa* seedlings were inoculated with the ectomycorrhizal fungus *P. involutus* and one day later with a spore suspension of the root pathogen *F. oxysporum pini.* Controls consisted of pine seedlings inoculated with plugs of sterile medium and *F. oxysporum pini.* Seedling survival was counted, sporulation of *F. o. pini* was measured, and ethanol extractions were made every day through day 8 and then on day 10 and day 14 after planting. The ethanol extracts were assayed for fungitoxic activity by measuring the germination of microconidia of *F. o. pini*.. Dead seedlings were first observed three days after planting in pathogen-inoculated controls, and sporulation of *F. o. pini* was reduced significantly three days after seedling inoculation with *P. involutus* when compared with controls lacking *P. involutus.* The fungitoxic activity of the rhizosphere of the seedlings inoculated with *P. involutus* was greater than that of the control rhizosphere and increased markedly two to four days after inoculation with the pathogen. It was inferred therefore that suppression of *Fusarium* root rot by *P. involutus* may be the result of antibiosis by this ectomycorrhizal fungus, since the time course of events definitely indicated it.

Duchesne et al. (1989b) followed up the previous study by identifying one of the fungitoxic or fungistatic compounds in the ethanol extracts of the rhizosphere of pine seedlings inoculated with *Paxillus involutus* as oxalic acid. Simultaneous inoculation of pine seedlings with oxalic acid and a spore suspension of *F. o. pini* protected the seedlings against *Fusarium* root rot and decreased sporulation of *F. o. pini* in the rhizosphere when compared with controls lacking oxalic acid. Moreover, there was a fivefold increase in oxalic acid production by *P. involutus* in tubes containing pine seedlings, compared with production in tubes lacking pine seedlings.

In still another follow-up investigation of the antagonistic action of the ectomycorrhizal fungus *Paxillus involutus,* Farquhar and Peterson (1990) followed the pine root infection by *F. o. pini* microscopically. They found that the presence of the pathogen within root tissues was significantly reduced within the first six days of infection after prior inoculation with *P. involutus.* Hyphae and conidia of the pathogen were swollen and filled with strange material when *P. involutus* was also present in the rhizosphere of the *Pinus resinosa* seedlings. Suppressed, ungerminated conidia of the pathogen had thick cell walls and large cytoplasmic inclusions. Thus different criteria also indicated the antagonistic nature of *P. involutus* against *F. o. pini* in pine seedlings.

In another investigation of the antagonistic role of *Paxillus involutus* against *Fusarium* root rot of *Pinus resinosa,* Farquhar and Peterson (1991) reported that colonization of primary roots of red pine seedlings by two isolates of *F. o. pini* in a sterile, soil-free system and a nonsterile rooting medium was suppressed for two months if the roots were exposed to *P. involutus* for one week before being inoculated with the pathogen. In contrast, roots not exposed to *P. involutus* were extensively colonized by *F. o. pini,* and host tissue vacuolation and disorganization occurred.

*Fusarium solani* f. sp. *pisi, Phoma medicaginis* var. *pinodella,* and *Mycosphaerella pinodes* are the fungal pathogens that generally cause foot rot of *Pisum sativum* (peas) in the United Kingdom (Bradshaw-Smith et al., 1991). Because foot rot is usually caused by a complex of fungi, any potential antagonist has to have a broad spectrum of inhibitory activity. On the basis of other reports, *Pythium oligandrum* (IMI 133857) seemed a likely choice to test. *F. solani* f. sp. *pisi* strain B2, *P. medicaginis* var. *pinodella* strain P2, and *M. pinodes* strain M13, known to be highly pathogenic to *Pisum sativum,* were isolated from infected pea plants.

*Pythium oligandrum* (IMI 133857) antagonized each of the three pathogens (one at a time) under several conditions. Direct observation on dual plates indicated that necrotophic parasitism occurred against each of the pathogens. The initial event in every interaction was contact. Following contact, the pathogen hyphae typically lysed within three minutes. The final stage of the interaction was growth of *P. oligandrum* through the parasitized host hyphae. The antagonist produced a volatile antibiotic that reduced the growth rate of *P. medicaginis* var. *pinodella* and *M. pinodes,* but not of *F. solani* f. sp. *pisi.* No nonvolatile antibiotics were found. *P. oligandrum* was thus shown to be an effective antagonist of the three major foot rot pathogens of peas, apparently through at least two main mechanisms of action.

Another serious disease of peas is pythium damping-off caused primarily by *Pythium ultimum* and *P. sylvaticum* (Parke, 1990). Parke isolated *Pseudomonas cepacia* strain AMMD (ATCC52796) from the rhizosphere of peas grown in the *Aphanomyces* root-rot nursery at Arlington (Wisc.) Agriculture Experiment Station. A rifampicin-resistant strain (at 100 μg/mL), which was a spontaneous mutant designated AMMDRI, was used in some studies concerning the antagonistic action of *P. cepacia* against *Pythium ultimum* and *P. sylvaticum.* Soil used in all experiments was Plano silt loam (pH 6.4), which is naturally infested with *Pythium* (429 propagules/g).

Nontreated seeds of Perfection 8221 peas, and seeds treated with *P. cepacia* (AMMD or AMMDRI) or metalaxyl fungicide, were compared for their effects on pea seedling emergence in a growth-chamber experiment at 16, 20, 24, and 28°C. Strain AMMD reduced pre-emergence damping-off caused by *Pythium ultimum* and *P. sylvaticum* by 47% in growth-chamber experiments. The antagonist was as effective as metalaxyl at all temperatures. Under controlled soil matrix potential, 100% of the seeds were infected 30 hours after planting in the control. Seed infection during the first 48 hours was reduced 44 to 60% by seed treatment with *P. cepacia* strain AMMDRI (resistant to rifampicin) or with metalaxyl. When strain AMMDRI was applied at log 4.9 cfu per seed, the doubling time of this strain during the first 24 hours after planting was 3.1 hours. This strain represented an increasing proportion of the bacterial population associated with the seed (Fig. 11); thus it was very successful in colonizing the young pea seedling. As I previously indicated, other strains of *P. cepacia* have been shown to produce antibiotics, so it is possible that they may have been responsible, at least in part, for the strong antagonistic action against *Pythium* in this case.

Parke et al. (1991) followed up the project just described with an investigation concerning the antagonistic effects of *Pseudomonas cepacia* (strain AMMD), *P. fluorescens* (PRA25), and *Corynebacterium* sp. (5A) on *Pythium* spp, the cause of pythium damping-off of peas, and on *Aphanomyces euteiches* f. sp. *pisi,* the cause of *Aphanomyces* root rot of peas. Perfection 8221 *P. sativum* seeds were treated with a suspension of each of the 85 potential antagonistic bacterial strains isolated from healthy-looking pea plants grown in field soils throughout Wisconsin. One strain was used for each test. Six days after planting, the seedlings were inoculated with *A. e. pisi* (isolate P4A-5 mL of $2 \times 10^4$ zoospores/mL/cone). Plants were harvested two weeks later and rated for disease severity.

Twenty-two of the 85 strains tested for control of aphanomyces root rot were selected for tests against both *Pythium* spp. and *A. e. pisi.* Nontreated seeds served as controls. Twelve of these 22 strains were selected for field trials in 1986; they were identified as species of *Pseudomonas, Bacillus, Corynebacterium,* and *Flavobacterium.* Seven of these were selected for further tests in 1987 at Arlington and Hancock, Wisconsin.

Three species—*Pseudomonas cepacia* (AMMD), *P. fluorescens* (PRA25), and *Corynebacterium* sp. (5A)—consistently increased emergence and yield in 1986 and 1987 field tests. These three were tested for

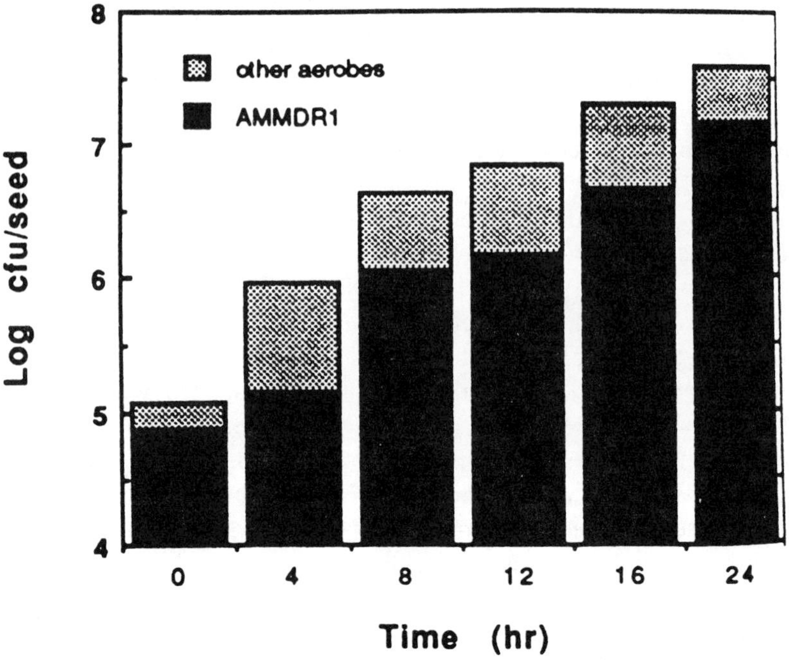


FIGURE 11 Population density of *Pseudomonas cepacia* strain AMMDR1 and other aerobic bacteria recovered from pea seeds during the first 24 hours after planting in naturally infested soil. Seeds were treated with *P. cepacia* strain AMMDR1 at an inoculum density of log 4.88 cfu/seed. *Adapted from Parke (1990), with permission of Amer. Phytopathological Soc.*

efficacy alone and in combination with captan during 1988 and 1989. In 1989, when *Aphanomyces* root rot was moderate to severe, seed treatment with the bacteria resulted in significant increases in emergence and yield at all field sites and in reduced severity at some sites, regardless of captan treatment. *P. cepacia* was the most effective antagonist, increasing emergence by 40% and yield by 48% when compared with captan alone. *P. fluorescens* without captan was also very effective. *Corynebacterium* sp. without captan increased emergence by 23% and yield by 12%, as compared with captan alone. It is evident that each of these antagonists is effective in the biocontrol of both *Pythium* damping-off and *Aphanomyces* root rot of peas under diverse environmental conditions, with *P. cepacia* (AMMD) being most effective.

Nelson et al. (1988) investigated a different potential antagonist

**Table 18** Compounds* promoting biological control activity of *T. koningii* and *T. harzianum*

| *T. koningii* | | *T. harzianum* | |
|---|---|---|---|
| Compound | Increase in biocontrol activity† (%) | Compound | Increase in biocontrol activity† (%) |
| Myristic acid | 48 | Arabitol | 44 |
| Linolenic acid | 48 | Starch | 44 |
| Caprylic acid | 36 | Inulin | 36 |
| *p*-hydroxybenzoic acid | 32 | Ribose | 32 |
| Maleic acid | 32 | Adonitol | 32 |
| Tannic acid | 32 | Malonic acid | 32 |
| Oleic acid | 32 | Mannitol | 28 |
| Palmitic acid | 28 | Laminarin | 20 |
| Chloroform | 28 | Xylan | 20 |
| Stearic acid | 28 | Sorbic acid | 24 |
| Tween 40 | 24 | Butyric acid | 20 |
| Lauric acid | 24 | | |
| Caproic acid | 24 | | |
| *p*-hydroxybenzaldehyde | 20 | | |
| Inulin | 20 | | |

* Includes only those compounds providing significant ($P = 0.05$) increases in activity over the *Trichoderma* seed treatments alone. Compounds were incorporated into a 2% methylcellulose (Methocel A4C) suspension (1 $mg/mL^{-1}$) containing $10^8$ to $10^9$ *Trichoderma* conidia $mL^{-1}$.

† Means represent % increase in activity calculated using the formula: % increase = [(TC) - C] - (T - M), where TC = % emergence from seeds treated with *Trichoderma* + compound; C = % emergence from seeds treated with compound only; T = % emergence from seeds treated with *Trichoderma* only; and M = % emergence from methylcellulose-treated seeds (control).

*Source:* From Nelson et al. (1988), with kind permission from Pergamon Press Ltd.

against pythium seed rot and preemergence damping-off of *Pisum sativum* seedlings. *Trichoderma koningii* and *T. harzianum* had been previously shown to exert biological control against *Pythium* spp., but various problems arose which reduced their efficacy. Nelson et al. tested several compounds for their abilities to promote biological control by these *Trichoderma* spp. Biological control activity of *T. koningii* was increased up to 48%, whereas activity of *T. harzianum* was increased up to 44% by adding some compounds in seed treatments (Table 18). Organic acids were most effective in promoting biocontrol by *T. koningii,* while polysaccharides and polyhydroxy alcohols were stimulatory to *T. harzianum.* The scientists found no relationship between the ability of the compounds to support in vitro growth and proliferation of *Trichoderma* strains in the spermosphere and increased biocontrol. The isolates of the microorganisms used were

*T. koningii* (T8), *T. harzianum* (T12), and *P. ultimum* (P4). Venus pea seeds were used in all the tests. It was concluded that addition of appropriate compounds with *T. koningii* or *T. harzianum* could give as good control as that provided by effective fungicides.

*Sclerotinia sclerotiorum* is a widely distributed pathogen of vegetable and field crops, and there is a great need of a biocontrol method for this pest, which is a serious pathogen of peas in the pea-growing areas of Himachal Pradesh in India. Singh (1991) isolated twelve fungi as potential antagonists from *Sclerotinia*-infested soils and found seven of them to be inhibitory to *S. sclerotiorum* on agar plates. The seven were *Penicillium cyclopium, P. sheari, Paecilomyces lilacinus, Aspergillus niger, A. fumigatus, Trichothecium roseum,* and *Acremonium implicatum. Trichoderma harzianum* grew with *S. sclerotiorum* without showing any clear zone, but it overran *S. sclerotiorum.* No sclerotia were produced, and attempts to isolate the host (pathogen) from the area where it first grew vigorously were futile. Microscopic examination revealed that direct penetration of the hyphae of the pathogen by hyphae of *Trichoderma* and *T. roseum* completely inhibited sclerotial germination after 30 days of coincubation in soil. These two antagonists were promising, but *T. harzianum* gave best control in vitro. Application of a wheat-bran culture of *T. harzianum* under field conditions gave significant biocontrol of the gray mold of peas, plus a yield increase. A mycelial preparation of *T. harzianum* was more effective than a spore preparation.

Knudsen et al. (1991) reported that *Trichoderma harzianum* (strain Thz IDI), formulated as mycelial fragments in alginate pellets with or without wheat bran, colonized sclerotia of *Sclerotinia sclerotiorum* in raw or steamed soil. Incidence of colonization was higher in steamed soil than in raw soil, higher at 25°C than at 15°C, and higher at –500 kPa than at –50 kPa, but it was not affected by wheat bran. Pellets containing *T. harzianum* were added to soil in a pea field at densities of $4 \times 10^2$ to $2 \times 10^4$ pellets/m$^2$, along with sclerotia of *S. sclerotiorum.* The researchers found that recovery of *Trichoderma* spp. from soil samples in the field was not significantly increased by adding the pellets. Addition of *T. harzianum* at high density ($1 \times 10^4$ or $2 \times 10^4$ pellets/m$^2$), however, did increase the proportion of sclerotia of the pathogen that was colonized by *T. harzianum.* This suggested that the biocontrol agent would have to be added periodically in the field in order to be effective.

Brown rot of stone fruit is the major cause of spoilage of postharvest peaches (*Prunus persica*). It is caused by *Monilinia fructicola,* which appears to be developing resistance to some of the common fungicides such as benomyl and related benzimidazoles (Gueldner et al., 1988). The result is that considerable work has gone into developing an effective biocontrol technique. Certain strains of *Bacillus subtilis,* such as strain B-3, have been shown to give relatively good control of the peach brown rot, and *B. subtilis* and related bacilli are known to produce several antibacterial and antifungal antibiotics. Gueldner et al. therefore isolated and identified the antibiotics produced by *B. subtilis* (B-3). They found that the cell-free culture liquor of *B. subtilis* (B-3) inhibited *Monilinia fructicola,* the peach brown rot fungus. They isolated several antifungal peptides of the Iturin family from the liquor and found that these peptides inhibited growth of *M. fructicola*.

Madrigal et al. (1991) isolated a compound with antibiotic activity toward *Monilinia laxa* from 10-day potato-dextrose stationary cultures of an isolate of *Epicoccum nigrum,* a component of the usual mycoflora of peach twigs previously found to antagonize *M. laxa*. By use of UV, IR, NMR, and MS, the researchers identified the antibiotic as flavipin. Flavipin was inhibitory to a wide range of phytopathogenic fungi and bacteria, including *M. laxa*. The $ED_{50}$ values for the germination of spores and germ-tube growth of *M. laxa* were 17.5 and 2.1 μg/mL, respectively. Crude filtrates of *E. nigrum* containing flavipin were relatively stable in a pH range of 6 to 8 with different $O_2$ tensions and with light for 54 days at room temperature. Application of the filtrates or propagules of *E. nigrum* to peach blossoms previously inoculated with *M. laxa* prevented blossom and twig infection.

De Cal et al. (1988) investigated the antagonistic action of isolate 909 of *Penicillium frequentans* against *Monilinia laxa* collected from a commercial apricot orchard in Almonacid de la Sierra, Spain. *P. frequentans* was cultured in PDB in darkness at 25°C. The filtrate was inhibitory to germination of spores and growth of the germ tubes of *M. laxa*. Two active substances were isolated and partially characterized; these were named antibiotics A and B. Both were active against 15 of 22 fungi against which they were tested, including *M. laxa*. The scientists suggested that the antibiotics were frequentin and palitantin. The antibiotics were toxic, in the same degree as benomyl, to *M. laxa* on detached peach twigs under laboratory conditions.

Under the same conditions, spores of *P. frequentans* inhibited *M. laxa* to a lesser degree, and it was postulated that this was probably because they did not have enough time to produce the antibiotics, which in culture conditions required about 10 days to start their production, reaching a peak at about 20 days.

In subsequent investigations, de Cal et al. (1990) tested six forms of *Penicillium frequentans* inocula for the biocontrol of peach-twig blight caused by *Monilinia laxa*. The forms used were conidia, conidia plus nutrients, conidia plus mycelium, conidia plus mycelium plus nutrients, wheat-bran cultures, and mycelial plugs. In field tests, some preparations containing nutrients gave significant reductions in severity of the disease (from 38 to 80%) over the control, comparable to that given by captan. Combinations of *P. frequentans* and captan gave a control level similar to or less than that given by the antagonist or chemical alone. In field experiments, *P. frequentans* applied in nutrients on shoots of peach retained higher populations than when the fungus was applied without nutrients. In general, those types of inocula giving rise to the highest populations of *P. frequentans* were generally the most effective in controlling shoot blight.

*M. laxa* is about eight times more sensitive to captan than to *P. frequentan,* and it appears that this difference could be used in integrated control. Nevertheless, attempts to use the two together gave no improvement in disease control over each alone. It is possible that alternating treatments with *P. frequentans* and captan could still have advantages over use of only one, by saving the amount of fungicide required and by reserving the fungicide for years when weather conditions do not favor the activity of the antagonist.

*Rhizopus* rot of *Prunus persica* is second only to *Monilinia fructicola* in causing postharvest losses of fruit (Wilson et al., 1987). There are several species of *Rhizopus* that can cause the fruit rot, but Wilson et al. experimented with *R. stolonifer,* which was isolated from soil and organic debris at the base of peach trees near Kearneysville, West Virginia. The researchers tested these species as potential antagonists against *R. stolonifer.* Thirteen isolates of bacteria inhibited growth and/or sporulation of the pathogen on agar plates.

The 13 bacteria were tested against artifically wounded peach fruits inoculated with *R. stolonifer.* Among the test bacteria, only one reduced fruit rot as effectively as dichloran did. The bacterium was identified as *Enterobacter cloacae* and was designated as strain SC843.

Infections by *R. stolonifer* were inhibited by *E. cloacae* in 70% of the fruits, and onset of rot in infected fruits was slowed by the antagonist. Decay was significantly reduced from day 2 through day 5, but most fruits treated with *E. cloacae* eventually developed *Rhizopus* lesions after eight days. Compared with a distilled-water control, the filtrate from cultures of *E. cloacae* accelerated colonization by *R. stolonifer.* Moreover, cells of *E. cloacae* resuspended in a comparable volume of deionized water were more effective in controlling *R. stolonifer* than were those applied in the culture medium. Thus *E. cloacae* apparently does not produce an antibiotic effective against *R. stolonifer.*

Concentrations of both the pathogen and antagonist affected *Rhizopus* root rot development. A 10- and 100-fold dilution of *E. cloacae* cells allowed more infection than did undiluted cells ($1.5 \times 10^{12}$ cfu). Only at a spore concentration of $10^3$ *R. stolonifer* spores/mL was the 100-fold dilution significantly different from the water control. Undiluted cells gave significantly better control than a water check at both $10^3$ and $10^4$ spores of *R. stolonifer* per milliliter. Less firm fruit were more susceptible to *R. stolonifer,* but control with *E. cloacae* was equally effective at all pressures tested.

Janisiewicz and Marchi (1992) screened numerous microorganisms as potential antagonists against blue mold of *Pyrus communis* (pear, cv. Anjou, Bosc, Bartlett, and Red Bartlett) caused by *Penicillium expansum* and gray mold of pear caused by *Botrytis cinerea.* One isolate that was promising for antagonistic activity was identified as *Pseudomonas syringae* pv. *lachrymans.* Wounded pears held in plastic mesh baskets were submerged for two minutes and agitated occasionally. The treatments were $10^4$ conidia of *B. cinerea* or *P. expansum* alone, or each mixed with 2.2, 3.2, or $5.4 \times 10^8$ cfu/mL of *P. syringae lachrymans.* The fruit were then placed on polystyrene fruit tray packs in 1-bu fruit boxes with a polystyrene liner. Fruit from each treatment were placed on separate trays and the trays were placed in boxes. There were four trays per box, one containing a pathogen control, the other three containing a treatment with three different concentrations of the antagonist. One set of fruit was stored at 18°C and 95% relative humidity for five days; another set stored at 1°C and 95% R.H. for 30 days.

Cultivar, wound type, antagonist concentration, and temperature significantly affected the severity of both gray mold and blue mold. On fruit inoculated only with *B. cinerea* and stored at 18°C, the severity of the disease was similar for Bosc and Anjou pears but less for Red Bartlett and least for Bartlett. Disease severity at 1°C occurred in the same pattern by cultivars. The incidence and severity of disease was higher on fruit with nail wounds than on fruit with cut wounds. Control of both diseases as high as 100% was attained in many tests with the addition of L-59-66 to a final concentration of $5.4 \times 10^8$ cfu/mL in the inocula. Disease control was best on Anjou pears, where frequently no rot developed, and worst on Bosc. Antagonist populations in wound sites increased from 6.86 to 9.51 log cfu/mL per site during storage at 1°C for 30 days. It was concluded that, using a fruit dip treatment, gray and blue mold of pears can be controlled by the application of the antagonist *P. syringae lachrymans* (L-59-66) under the conditions given.

*Erwinia amylovora* causes fire blight of all species of the Pomoideae, but it is especially destructive to apple and pear trees (Vanneste et al., 1992). *E. herbicola* is a closely related bacterium that is not pathogenic, and certain strains isolated from shoots, leaves, or blossoms of apple trees have been shown to inhibit *E. amylovora* in immature pear fruits. Moreover, several strains of *E. herbicola* isolated from fire-blight lesions were reported to produce an antibiotic inhibitory to *E. amylovora*.

Vanneste et al. therefore investigated the role of the antibiotic produced by *E. herbicola* (Eh252) in the biological control of fire blight in pears. They isolated Tn5-induced mutants of Eh252 that did not produce the antibiotic nor inhibit *E. amylovora* on agar and compared their abilities to inhibit *E. amylovora* in immature pear fruits with that of the wild-type strain. They found that strain Eh252 produced an antibiotic on a minimal medium that inhibited the growth of *E. amylovora*. This antibiotic was inactivated by histidine but not by iron, was sensitive to proteolytic enzymes, and showed a narrow range of host activities. It was inhibitory to 62 of the 67 *Erwinia* strains against which it was tested but not against many strains of *Pseudomonas, Agrobacterum,* and *Klebsiella*. The two mutants of Eh252, 10:12 and 17:12, had single Tn5 insertions, and although they grew in immature pear fruits at a rate similar to that of Eh252, neither mutant strain suppressed fire blight as well as Eh252 did. The Tn5-containing fragment isolated from strain 10:12 was used to mutagenize Eh252 by marker exchange. Derivatives that acquired the Tn5-containing fragment lost the ability to inhibit *E. amylovora* in immature pear fruits.

These results indicate that antibiotic production is necessary for the biological control of *E. amylovora* by *E. herbicola* strain Eh252. It was concluded that some other mechanism is also involved in the process.

Five isolates of *Bacillus subtilis* (a, b, c, d, e) and three virulent streptomycin-sensitive isolates of *Erwinia amylovora* (SI, SII, SIII) were used by El-Goorani and Hassanein (1991) in a biocontrol investigation against fire blight in pear (*Pyrus serotina* x *P. communis*). All tested isolates of *B. subtilis* strongly inhibited growth of the three isolates of *E. amylovora* in vitro. The strains of *B. subtilis* and the culture filtrates had no effect, however, when tested on immature green pear fruits in the laboratory or in a commercial pear orchard under natural conditions. Streptomycin completely suppressed the development of necrotic symptoms and ooze for as long as five days. According to the results of other work done with different strains of *B. subtilis* against other pathogens and the work of Vanneste et al. (1992) in tests of *E. herbicola* Eh252 against *E. amylovora,* the fact that the filtrates had no effects against *E. amylovora* suggests that no effective antibiotics were produced by the strains of *B. subtilis* used in these experiments.

Kwok et al. (1987) selected eight species of bacterial antagonists—including *Bacillus cereus, Enterobacter cloacae, Flavobacterium lividum, Pseudomonas fluorescens* biovar III, *P. putida, P. stutzeri,* and *Xanthomonas maltophilia*—from 562 isolates from *Raphanus sativus* (radish) roots, cucumber roots, and soil inoculum pieces of *Rhizoctonia solani,* the causative pathogen of *Rhizoctonia* damping-off of radish and other vegetable and crop plants. All the bacteria listed above were suppressive to *Rhizoctonia* damping-off of radish seedlings in container media amended with composted hardwood tree bark. Combinations of some strains of bacterial antagonists—*E. cloacae* 313, *Flavobacterium balustrinum* 299, *P. fluorescens* biovar VA1, *P. putida* 371, and *P. stutzeri* 280—with *Trichoderma hamatum* 382 were consistently more effective than was *T. hamatum* 382 alone. Some bacterial isolates, such as *P. fluorescens biovar* VA1 and A498, were not effective antagonists unless they were combined with *T. hamatum.* Spontaneous rifampicin-resistant mutants of *F. balustrinum, P. putida,* and *X. maltophilia* colonized cucumber roots and a conducive medium to high population levels in irrigation pots. Population levels of the antagonists were highest in the autoclaved container medium and lowest in the suppressive medium already colonized by a mesophilic microflora.

Of thirteen bacterial antagonists, only four were found to produce antifungal antibiotics against *Rhizoctonia solani* and three were found to produce antibiotics against *Trichoderma hamatum*. Overall, it was concluded that suppression of *Rhizoctonia* damping-off in the naturally suppressive container medium may be due to a variety of antagonists. The Ohio State University has a patent (#4642131) for the production of composts that are predictably suppressive to *Rhizoctonia* damping-off through introduction of specific antagonists.

Hassanein and El-Goorani (1991) isolated five strains of *Bacillus subtilis* from soil samples collected from the experimental farm of the Sabahia Research Station at Alexandria, Egypt. In vitro tests of these five strains were made against six isolates (I, II, III, IV, V, VI) of *Agrobacterium tumefaciens* (biovar I) obtained from the stock culture collection of the Department of Plant Pathology, Alexandria University. All five isolates of *B. subtilis* strongly inhibited all six strains of *A. tumefaciens* on all media used.

Inoculation of wounded *Ricinus communis* (castor bean) plants 30 minutes before, or simultaneous with inoculation with *A. tumefaciens* resulted in excellent control of the crown gall symptoms on the host plants. On the other hand, application of *B. subtilis* 30 minutes after inoculation of the castor bean plants with *A. tumefaciens* did not appreciably suppress the disease.

Treatment of castor bean plants with *B. subtilis* alone did not cause any injury or growth-retarding side effects to castor bean plants. *B. subtilis* may be an alternative to *Agrobacterium radiobacter* for the biocontrol of crown gall because of the discovery of *A. tumefaciens* with resistance to *A. radiobacter.* It was concluded that the ability of *B. subtilis* to produce antibiotics, to form endospores that are tolerant to heat and desiccation, and to stimulate plant growth makes it very suitable as a biocontrol candidate.

*Botrydis cinerea* causes many diseases of flowers, fruits, and other soft plant parts, and it is particularly bad in causing *Botrydis* blight of *Rosa* spp. (rose) (Redmond et al., 1987). Approximately 72 microorganisms (56 bacterial isolates and 16 yeast isolates) were isolated from rose petals and tested for antagonism against *B. cineria* on Golden Wave rose petals. Preliminary tests identified four microorganisms able to reduce the number of lesions caused by *B. cinerea* on rose. Biocontrol was demonstrated for *Exophiala jeanselmei, Crytococcus albidus, Erwinia* sp., and a coryneform bacterium by applying them to cut roses one day before inoculation with a suspension of 1000 conidia of *B. cineria*/mL.

The most effective antagonist tested was the yeast *Exophiala jeanselmei,* which reduced the number of lesions by 63%. This was similar to suppression due to the fungicide iprodione (74%), which was not significantly different. Thus *E. jeanselmei* appears to be a good biocontrol microorganism for *Botrydis* blight of roses.

Hajlaoui et al. (1992) sprayed Ruiredro rose leaves, naturally infected with *Sphaerotheca pannosa* var. *rosae,* the causative agent of rose powdery mildew, with $1 \times 10^6$ conidia/mm$^2$ of the known antagonist, *Sporothrix flocculosa.* The researchers' main objective was to locate the potential substrate for chitinase in healthy as well as in damaged hyphae of *S.p. rosae* and to gain a better insight into the antagonistic activity of *S. flocculosa* against the rose-powdery-mildew pathogen. They used a wheat germ agglutinin-ovomucoid gold complex as a specific probe for localizing chitin distribution in walls of the powdery mildew fungus and in the antagonistic fungus. The same probe was used to determine whether chitinolytic activity was involved in the antagonistic action of *S. flocculosa.* Within 12 hours of inoculation, the fungal host started to suffer some damage. Twenty-four hours after inoculation, close contact between the two fungi was sometimes associated with the penetration of the host fungus by the antagonist, with considerable changes in the cytoplasm but no observable alteration in the chitin-labeling distribution over the cell walls. By that time, *S. flocculosa* had caused complete plasmolysis of the pathogen hyphae, which were reduced to just walls. Close contact between the interacting fungi did not appear to be necessary for the antagonistic activity to be induced. The scientists inferred from their results that antibiosis, rather than chitinolytic activity, was responsible for the antagonistic activity.

The bacterial ring-rot disease of *Solanum tuberosum* (potato), caused by *Clavibacter michiganensis* subsp. *sepedonicus,* has been a threat to the potato industry since its introduction. De la Cruz et al. (1992) isolated several fluorescent pseudomonads as potential antagonists against the bacterial ring rot. The antagonists, designated as IS-1, IS-2, and IS-3, were isolated from potato underground stems, with roots collected from southern Idaho. IS-1 and IS-2 were identified as *Pseudomonas aureofaciens* and IS-3 as *P. fluorescens* biovar III. These three strains were tested for in vitro antibiosis against 30 strains of

*C. michiganensis* subsp. *sepedonicus* from several geographical locations. All three antagonists caused growth inhibition of each of the 30 strains of the pathogen. IS-1 was the most inhibitory, however, consistently producing inhibition zones greater than 50 mm in diameter. IS-2 was the least inhibitory to most strains of the pathogen, whereas IS-3 was moderately inhibitory in most cases.

Each of the three antagonists significantly reduced the *C. michiganensis sepedonicus* population levels in potato roots in greenhouse trials, but IS-3 had this effect only at the higher treatment level. Nevertheless, only IS-3 at the higher treatment level significantly reduced the percentage of plants infected with the pathogen. Different combinations of antagonist strains were never more effective than single-strain treatment in reducing ring-rot symptoms, infection, and root population levels of the pathogen. None of the antagonist treatments had a significant effect on root weights of the potato plants. Antibiotic production may have played a significant role in disease suppression, since there was a strong relationship between the level of in vitro antibiosis and in vivo suppression of ring rot. It is known that certain strains of *P. aureofaciens* produce phenazine-1-carboxylic acid (both in vitro and in vivo) and hydroxylated phenazine antibiotics, as well as pyrrolnitrin (de la Cruz et al., 1992). Further tests are necessary, however, with mutant strains of one of the antagonists that is negative for antibiotic production against the pathogen.

Prior greenhouse experiments with a nonantagonistic, avirulent strain of *Pseudomonas solanacearum* (B82) demonstrated that this strain suppressed infection of potato (*Solanum tuberosum*) with bacterial wilt, which is caused by virulent strains of *P. solanacearum* (McLaughlin et al., 1990). McLaughlin and co-workers therefore decided to investigate the ability of strain B82 to control bacterial wilt of potatoes in the field.

In field tests, seed pieces of potato cultivars Pungo and/or Atlantic (susceptible to bacterial wilt) and Ontario (tolerant to the wilt) were treated by immersion in a suspension of *P. solanacearum* (B82) prior to planting in naturally infested soil. Atlantic and Ontario cultivars were tested in 1984, whereas all three cultivars were tested in 1985. Relatively low populations of *P. solanacearum* (2 to 4 log cfu/g soil) were present in the soil before planting. There was approximately a 50% reduction, compared with controls, in incidence of bacterial wilt in B82-treated plants of cultivar Atlantic. With cultivar Ontario, however, disease incidence was very low and differences between treatments were not significant. In 1985, the incidence of bacterial wilt was low in all plots, but treatment with strain B82 reduced incidence of tuber brown rot in all cultivars; however, differences between treatments were not statistically significant.

Ciampi-Panno et al. (1989) investigated a different antagonistic strain (nonvirulent) of *Pseudomonas solanacearum* (strain BC8) to control potato wilt. This strain was originally isolated from potato and tested under in vitro and growth-chamber conditions. For seed tuber treatments used as controls and adult plant inoculations, a Chilean isolate of *P. solanacearum* (race 3) was used.

Seed-potato tubers treated with BC8 and coated with $CaCO_3$ and BC8 formulated with an amendment caused strong inhibition of the virulent *P. solanacearum* both in in vitro assays and under growth-chamber conditions. Field experiments with Corahila potato tubers and the same antagonist and pathogen indicated that treatments that included the antagonist BC8 gave the lowest amount of wilted plants and the fewest tubers latently infected. Using the system with BC8, about 80% of the tubers assayed from plants growing in the naturally infested soil were colonized by the isolate BC8 and were free of the pathogen, *P. solancearum*. Moreover, it was established that the pathogen was still present in the soil after two years of nonpotato cropping and that latent infections play an important role in the dispersal of the pathogen. Delivering the antagonist in an amendment was more efficient in the biological control of *P. solanacearum* than just coating the seed-potato tubers with the antagonist. Strain BC8 proved to be a relatively effective antagonist against potato wilt, but a better delivery method needs to be developed.

*Rhizoctonia* canker occurs in all potato-growing areas of the world but is considered to be the most important potato disease in North Carolina. Escande and Echandi (1991) decided to see if the binucleate (BNR) *Rhizoctonia* species might exert biocontrol on *Rhizoctonia* canker of potato. Certified seed tubers of Atlantic potatoes were immersed in 0.5% NaOCl for 5 minutes, rinsed in running water, selected to eliminate those with *Rhizoctonia* sclerotia, and cut into 10-g one-eyed seed pieces. Fifteen isolates of BNR were selected for testing as biocontrol agents against the pathogen *Rhizoctonia solani* (AG-3) isolated from a potato in Pamlico County, North Carolina. The isolates were tested first for pathogenicity to potato; only one fit this category, and it was eliminated from further tests. The rest were tested for antagonism of *R. solani* (Rs-8-87). Eight BNR strains significantly reduced both disease severity and disease incidence in greenhouse tests in soil.

In a field naturally infested with *R. solani* (AG-3), only PCNB suppressed stem canker to a level that was significantly less than that in the control treatment. However, BNR isolates JF-3S4-3, BN-160, 232-CG, BN29, and R-72 suppressed stem canker to levels that were not significantly different from those in the PCNB and Tops 2.5 D treatments. Protection of potato plants from *Rhizoctonia* stolon canker with BNR was also attained in a field naturally infested with *R. solani* (AG-3). JF-3S4-3 alone and JF-3S4-3 combined with 232-CG provided protection of potato from stolon canker similar to that provided by the fungicide PCNB. Various treatments had similar suppressive effects on both stem and stolon canker. Cultivars Atlantic, Kennebec, Irish Cobbler, Norchip, Russet Burbank, and Superior were equally protected from *Rhizoctonia* canker with 232-CG in both naturally and artificially infested fields; the interactions between cultivar and biocontrol treatments was not significant.

Consistency of protection is a major concern of researchers studying biological control, according to Escande and Echandi (1991). In their studies, BNR isolate 232-CG applied to soil in colonized oat kernels protected potato plants consistently against *Rhizoctonia* canker in four greenhouse tests, in three field experiments in soil artificially infested with *R. solani* AG-3, and in eleven experiments in fields naturally infested with *R. solani* AG-3.

Sclerotia of *Rhizoctonia solani* grown in soils with a pH below 6.5 are frequently colonized by *Verticillium biguttatum* (Jager and Velvis, 1988). To determine whether parasitism of the sclerotia of *R. solani* by *V. biguttatum* might be determined by temperature and relative humidity, experiments were conducted in which sclerotia on tubers were inoculated with the parasite and kept at different temperatures and relative humidities in the laboratory and under actual storage conditions on farms. At R.H. values below 96.7%, no *V. biguttatum* was visibly present, but the sclerotia were overgrown by *Penicillium* spp. Isolates of *Penicillium* spp. growing on sclerotia were tested and proved only slow killers of sclerotia. At an R.H. of 98.4% and higher, growth and sporulation of *V. biguttatum* became prominent, and other mycoparasites such as *Gliocladium roseum* and *G. nigrovirens* were observed. *Penicillium* spp. were still present, but at a much lower level than under drier conditions. The percentage of sclerotia with visible growth and sporulation of *V. biguttatum* increased from 12% at 98.4% R.H. to 60% at 100% R.H. At an R.H. of 99.7% and higher, all sclerotia were dead after one month. Jager and Velvis concluded that the failure of germination of the sclerotia was probably due principally to the activity of *V. biguttatum* and that for successful commercial application, 99% should be considered as the minimum R.H. At 14°C, the killing rate of the sclerotia was too slow, but was satisfactory at 17 and 20°C.

Sclerotia on freshly harvested potato tubers can be killed in a period of six to eight weeks, provided that a direct contact between sclerotia and conidia of *V. biguttatum* is obtained; that the temperature during storage is at least 15°C, but preferably closer to 20°C during the early weeks; and that the R.H. of the air between the tubers is at least 99%. Seed tubers of potatoes are certified as export quality only if the infection with *R. solani,* visible as sclerotia on the tubers, is assessed at below a specified incidence.

Potato early-dying disease occurs in all potato-producing areas of the United States, and *Verticillium dahliae* is often the major pathogen involved in this disease (Keinath et al., 1991). This pathogen can persist up to 10 years in the absence of potato plants. A bioassay method was developed to evaluate fungi for biocontrol of *V. dahliae* by inactivating microsclerotia of the pathogen before infection of the potato plants occurs. *Gliocladium roseum* was the most effective antagonist among 11 fungi tested. No viable microsclerotia were recovered from soil amended with *G. roseum* at a rate of 1.0% (w/w). The most effective isolate, *G. roseum* 632, significantly reduced the viability of microsclerotia by 36% or more when grown on a vermiculite-bran mixture and added to soil at a rate of 0.01% (w/w). Three isolates of *G. roseum* (632, W14, and Std) were effective in three different soils at –10 and –100 kPa soil matric potential. *G. roseum* could be applied to the soil before or at planting, or it could be applied to potato residues at the end of the season to reduce the formation of microsclerotia of *V. dahliae* in infested potato stems. Until *G. roseum* is tested in the field, the percent of biocontrol of potato early-dying disease obtainable cannot be known.

Spink and Rowe (1989) investigated a different fungus, *Talaromyces flavus* as a biological control agent against *Verticillium dahliae* on *Solanum rostratum*. Soil in field plots at Wooster, Ohio, was amended with various concentrations of the microsclerotial inoculum of *V. dahliae* and various rates of Pyrax-based pellets containing ascospores of *T. flavus*. Potato seed pieces free of *V. dahliae* were planted in the test soil, and plants that developed were inspected over a period of time. There were significant increases in disease (early dying) resulting from increasing levels of microsclerotia of *V. dahliae;* however, increasing rates of addition of pellets containing *T. flavus* had no effect on disease development. Tuber yields of plants grown in soil treated with *T. flavus* pellets were not affected by increasing levels of *V. dahliae*. Pellets of *T. flavus* had little effect on potato yields in the absence of *V. dahliae*. Populations of *T. flavus* recovered from nonrhizosphere soils at harvest were not affected in any consistent manner by initial levels of *V. dahliae* or by increasing rates of application of Pyrax-based pellets of *T. flavus*. Moreover, percent recovery of *T. flavus* from potato root sections at harvest was low, ranging from 0 to 1.4%.

Pyrax-based pellets of ascospores of *T. flavus* were compared with wheat-bran-based pellets of *T. flavus* ascospores and an untreated control during the next year. Compared with Pyrax-based pellets, wheat bran-based pellets gave a 6- to 10-fold increase in percent recovery of *T. flavus* from potato root sections, and a 1000-fold increase in percent recovery of the antagonist from nonrhizosphere soil at harvest. Despite the improved establishment of *T. flavus* with wheat-bran-based pellets, there were still no significant differences in percent recovery of *V. dahliae* from roots or soil at harvest, or in tuber yield between wheat-bran-based pellets and Pyrax-based pellets of *T. flavus*. According to the reported results, *T. flavus* (isolate Tf1), used throughout the tests, does not look promising as an effective antagonist against the early-dying disease of potato caused by *V. dahliae*.

Zaspel (1992) isolated several potential bacterial antagonists from the rhizosphere of a field used only for wheat (*Triticum aestivum*) culture and tested them against take-all of wheat caused by *Gaeumannomyces graminis* var. *tritici* (isolate 173). The bacterial isolates were identified as *Pseudomonas* spp., *Bacillus, Streptomyces,* and *Agrobacterium*. After initial tests, *P. putida* (P27), *B. subtilis, S. achromogenes,* and *A. radiobacter* were selected for further field tests.

In a few field tests with Miras wheat, *P. putida* (P27) and/or *A. radiobacter* reduced the disease index or increased the yield of the wheat significantly. In several other trials, these antagonists and

*B. subtilis* increased wheat yield or decreased the take-all disease index, but not at a statistically significant level. In most tests, it appeared that seed preparations of the antagonists were more effective than peat preparations, although once again not at a statistically significant level.

Maplestone and Campbell (1989) theorized that antagonists with the best ability to colonize the root system of the host plant would probably be the most effective antagonists. Therefore they selected two bacteria reported to be antagonistic to take-all of wheat and investigated their abilities to colonize both infected (with take-all) and noninfected root systems of wheat under gnotobiotic conditions. The bacteria selected were *Bacillus cereus* var. *mycoides* strain 31 and *B. pumilus* strain 87, both in the culture collection of the Department of Botany at Bristol University. For studies under more natural conditions, mutants of *B. cereus* var. *mycoides* resistant to rifampicin and nalidixic acid and of *B. pumilus* resistant to rifampicin only were selected.

The motile strain, *B. pumilus,* colonized the wheat root system rapidly and reached a population of $10^6$ to $10^8$ cfu/g dry root. Colonization was enhanced when roots were simultaneously inoculated with *Gaeumannomyces graminis* var. *tritici* and was impeded when the rooting medium contained clay. The nonmotile strain, *B. cereus* var. *mycoides,* spread more slowly through the root system and reached a population of $10^6$ cfu/g dry root only on the oldest parts of the root system. Colonization was inhibited by *G. graminis tritici,* but not by *B. pumilus,* and was stimulated by clay. Use of the mutant strains of *B. pumilus* and *B. cereus mycoides* indicated that the spread of the antagonists down roots in nonsterile soil was very similar to that in gnotobiotic systems.

*Fusarium culmorum* is a soil-borne pathogen that causes seedling and head blight and foot and root rot of wheat. Kempf and Wolf (1989) isolated 15 bacteria and 54 actinomycetes from roots of several species of Gramineae in different locations in Germany and tested them for antibiotic production against *F. culmorum* (isolate 102 from Landespflanzenschutzamt Rheinland-Pfalz, Mainz, Germany). Fifteen isolates produced clear inhibition zones on agar plates inoculated also with *F. culmorum.* This indicated that all of these antagonists produced antibiotics against the pathogen. One antibiotic-producing bacterium, designated as B247, was identified as *Erwinia herbicola,* and the antibiotic produced by *E. herbicola* B247 was identified as herbicolin A. A transposon Tn5 mutant of *E. herbicola* B247 was generated by conjugative transfer of the plasmid pSUP5011 from *Escherichia coli* S17–1. One mutant lacking in vitro antibiosis was used during the investigation. This mutant, *E. herbicola* Tn247, had a single Tn5 insertion. Antagonists were tested in seed treatments for their ability to suppress *F. culmorum* in greenhouse pot tests. Seeds of winter wheat (*Triticum aestivum* cv. Okapi), surface-disinfected in 1% NaOCl for 3 minutes, were shaken for 15 minutes in a suspension of the antagonist bacteria with 1% (w/v) methylcellulose 400 and then dried on filter paper.

Nonsterile sand peat (1300 to 1400 g) was blended with 3 g of wheat straw colonized by *F. culmorum* to give a pathogen concentration of $10^4$ to $10^5$ cfu/g. The wheat seeds coated with the antagonists were planted in the prepared medium containing *F. culmorum*. Fifteen antagonists applied to the seeds resulted in more than 40% suppression of seedling blight on wheat caused by *F. culmorum*. A highly significant correlation occurred between the size of the inhibition zone in vitro and disease suppression in vivo.

*E. herbicola* B247 effectively controlled *F. culmorum* in pot tests, and there was only a minor reduction in shoot length of wheat seedlings grown from treated seeds, compared with a marked stunting of shoots from nontreated seeds. Furthermore, treated plants had significantly greater shoot lengths than nontreated plants at all inoculation levels tested. There was about a 90% suppression of biomass of *F. culmorum* at all inoculum concentrations. The antibiosis-negative Tn5 mutant (Tn247) gave 38% disease suppression, compared with 54% suppression for the wild type. This indicated that only part of the disease suppression could be attributed to antibiosis.

Spraying a suspension of *E. herbicola* B247 on wheat leaves before inoculation with urediniospores of the rust *Puccinia recondita* f. sp. *tritici* suppressed the wheat rust by 76%, whereas the Tn5 mutant was not effective. Almost complete protection against this rust was attained by application of the cultural filtrate from the wild type. This indicated that the mechanism of control of *P. recondita tritici* by *E. herbicola* B247 was antibiosis. It is interesting that *E. herbicola* gave good biological control of two very different diseases of wheat—and, at least in part, through two different mechanisms.

In a search for alternative biocontrol antagonists against take-all of wheat, Kirk and Deacon (1987) isolated *Phialophora graminicola, Periconia macrospinosa,* and *Microdochium bolleyi* from wheat crops and turf grass near Edinburgh, Scotland. They also isolated strain Ggt of *Gaeumannomyces graminis* var. *tritici* from wheat. This strain was highly virulent, whereas the other fungi produced no disease symptoms on wheat roots in greenhouse tests. For most experiments the inocula consisted of cornmeal-sand (3% w/w) at 60% saturation, colonized by the fungi for 28 days at 25°C. The soil used in tests was a clay loam from beneath a sequence of potato crops near Edinburgh. It was air-dried, passed through a 2.4-mm sieve, and mixed with sand (1:2 w/w).

*Microdochium bolleyi* significantly reduced infection of Maris Huntsman wheat by *G. graminis tritici* (Ggt) when inocula were dispersed in soil at ratios of 10:1 (Mb:Ggt) or more. Spread of take-all lesions up roots from an inoculum layer was reduced when *M. bolleyi* was inoculated just below the crown. On the other hand, *P. macrospinosa* did not control take-all, even at an inoculum ratio of 100:1.

In the absence of *M. bolleyi, Phialophora graminicola,* a known biocontrol agent of take-all of wheat, became established on nearly all roots, often producing several runner hyphae on or in the cortex of the root. At inoculum ratios of 1:1 or 2:1, *M. bolleyi* did not affect growth of *P. graminicola,* but it significantly reduced growth of *P. graminicola* at inoculum ratios of 5:1 and 10:1. On the other hand, *P. graminicola* reduced infection of roots by *M. bolleyi* at an inoculum ratio of 1:1. Kirk and Deacon (1987) concluded that *M. bolleyi* has several advantages over weak parasites such as *P. graminicola* for commercial control of take-all: It sporulates readily in culture, it grows readily on stem base tissues (and thus might protect the crown and subcrown internode from colonization by *G. graminis tritici*), it is a normal and common inhabitant of the root and basal stem tissues of cereals in the field, and it controls several other cereal pathogens. Thus it has a broader spectrum of activity.

Take-all of wheat (*Triticum aestivum*) caused by *Gaeumannomyces graminis* var. *tritici* has been so widespread in the world and so destructive to wheat yields that many scientists have searched for numerous strains of fluorescent pseudomonads and other organisms that could possibly exert biocontrol of this disease (Brisbane and Rovira, 1988; Weller et al., 1988; Thomashow and Weller, 1988, 1990; Brisbane et al., 1989; Thomashow et al., 1990; Bull et al., 1991; Mazzola and Cook, 1991).

Brisbane and Rovira (1988) selected *Bacillus* sp. (ABG 4000) from the Plant Research Institute, Burnley, Victoria, and four fluorescent pseudomonads (strains 2-79, Victorian Crops Research Institute, Horsham, Victoria; PGPR, Plant Pathology Department, University of Wisconsin; NRRLB-15132, U.S.D.A., Pullman, Washington; and A37 isolated from a South Australian soil) as potential antagonists against *G. graminis* var. *tritici* (strain 500 from the University of Western Australia). Initial tests with Condor wheat in large tubes of sand demonstrated that *Bacillus* sp. (ABG 4000) did not significantly affect the weight of tops of the wheat seedlings inoculated also with *G. graminis tritici,* whereas the two pseudomonads tested (2-79 and PGPR) gave a significant increase in leaf weight. The numbers ($\log_{10}$) of bacteria added to each tube were 8.86 cfu/mL for 2-79, 8.81 cfu/mL for PGPR, and 7.04 cfu/mL for ABG 4000. At harvest, however, the numbers of ABG 4000 on the roots were significantly less than those of the pseudomonads.

Subsequent experiments were run to determine the mechanism of action of the pseudomonads. Five levels of FeNa EDTA, ranging from 0 to 1000 $\mu M$, were added to the plant-nutrient solution. There was no effect of the added iron on the inhibition of the take-all pathogen by the effective pseudomonads. It was concluded therefore by Brisbane and Rovira that the siderophores contributed very little if any to the inhibition of the take-all fungus by the pseudomonads.

In a different series of tests, a yellow compound was isolated from the culture filtrates of the four pseudomonads and found to have absorbance peaks of 365 and 252 nm. This compound was added at concentrations of 133 to 0.06 μg/mL to a series of potato-glucose broths buffered between pH 4.5 and 7.2 and inoculated with the take-all pathogen. The compound was nontoxic at pH 7.2, even at 133 μg/mL, whereas at pH 5.2, even 1.6 μg/mL completely prevented growth of *G. g. tritici.* Subsequent tests caused Brisbane and Rovira to conclude that the yellow compound was phenazine-1-carboxylic acid. They also concluded that at least part of the mechanism of action of the effective pseudomonads was production of this antibiotic, particularly at low soil pHs.

Brisbane et al. (1989) extended the experiments of Brisbane and Rovira to determine the effects of 9 commercial fungicides, 3 strains of *Pseudomonas fluorescens* (2-79, $K_2$-79, A-37), and 2 antibiotics (phenazine-1-carboxylic acid and a siderophore) against 22 test pathogens frequently found on diseased cereal roots. In agar plate tests, phenazine-1-carboxylic acid inhibited 17 of 22 test pathogens, but the siderophore was not active against any of the fungi. Of the 9 fungicides, bitertanol inhibited the growth of 20 of the 22 test pathogens, whereas the least active, metalaxyl, inhibited only 5 of the 22 pathogens. *P. fluorescens* (2-79) was added to three soils naturally infested with the take-all fungus, some treated with bitertanol or metalaxyl and some left untreated; wheat plants were then grown in the soil for three weeks. Inspection of the wheat roots indicated that *P. fluorescens* (2-79) reduced the numbers of *Rhizopus* spp. isolated and in one soil improved wheat germination. In one soil bitertanol increased the shoot weight of the wheat and reduced the numbers of the take-all fungus and *Fusarium* spp. isolated.

Weller et al. (1988) investigated the ability of several pseudomonads, from suppressive and nonsuppressive soils, to antagonize *G. graminis* var. *tritici* in vitro and control the take-all disease on wheat. They also investigated the mechanisms of action. This was one of a series of projects carried out by Weller and various colleagues. A soil suppressive to take-all was collected from a field near Moses Lake, Washington, that had been cropped to wheat for 22 yr. Nonsuppressive soil was collected from fields near Mt. Vernon, Washington, that had been cropped to pea-wheat rotation or to alfalfa. These soils were amended with fumigated virgin silt loam to minimize differences in physical and chemical factors. The mixtures were amended with fragmented oat-kernel inoculum 1.0% (w/w) of *G. g. tritici* and placed in pots; five seeds of Fielder wheat were then planted in each pot. For *controls*, part of the soil was amended with autoclaved fragmented oat-kernel inoculum of *G. g. tritici* before potting.

Fluorescent pseudomonads were isolated from roots of wheat plants in the different soils, and these pseudomonads were tested against *G. g. tritici* on petri plates, some with high iron and some with low. Inhibition on PDA medium (high iron) was assumed to be due mainly to the production of antibiotics, whereas inhibition on King's medium B (KMB) agar (low iron)—if inhibition was reduced by addition of $FeCl_3$ to the agar—was assumed to be due to the production of a siderophore. Rifampin-resistant strains were selected from *P. fluorescens* Rla-80 and R7z-80 and designated Rla-80R and R7z-80R. These resistant strains were treated with N-methyl-N'-nitro-N-nitroguanidine (NTG), and mutants were selected that were nonfluorescent, or reduced in ability to inhibit *G. g. tritici,* but that maintained growth rates similar to their respective parental strains.

Fluorescent pseudomonads from the suppressive soil were more inhibitory to *G. g. tritici* on agar than were strains from the conducive soils. In the tube assay with wheat, most strains from the suppressive soil were more suppressive to take-all on wheat than were the strains from the conducive soil. *P. fluorescens* 2-79$RN_{10}$, Rla-80R, and R7z-80R were from roots of wheat grown on suppressive soils from this study or a previous one. They were highly inhibitory to *G. g. tritici* on both PDA and KMB agar. The ability of these strains to inhibit *G. g. tritici* was not altered on PDA by addition of $FeCl_3$; on KMB agar, all three strains produced fluorescent pigment, presumed to be a siderophore since pigment production was suppressed by the addition of 10 to 50 $\mu M$ $FeCl_3$ to the medium. Addition of 10 to 100 $\mu M$ of $Fecl_3$ to AMB completely eliminated inhibition by Rla-80R and almost completely eliminated inhibition by R7z-80R. On KMB agar, antibiotic production by Rla-80R was not stimulated by $FeCl_3$. The mechanism of action of these two strains was probably iron shortage caused by the fluorescent siderophore. With strain 2-79$RN_{10}$, the zone of inhibition against *G. g. tritici* on KMB was reduced by 10 $\mu M$ $FeCl_3$, but with higher concentrations of iron, the inhibitory zone increased in size even though no fluorescent pigment was evident, which suggested that when siderophore production was inhibited by iron, more antibiotic was produced.

Mutants of *P. fluorescens* strains 2-79$RN_{10}$, Rla-80R, and R7z-80R, which lost or had a reduced ability to produce antibiotics and/or siderophores, also lost some of their ability to suppress take-all, but still colonized wheat roots to the same amount as their parent strains. The overall results confirmed previous conclusions that fluorescent pseudomonads do play a role in take-all decline and that suppression of take-all by at least some fluorescent pseudomonads is due in part to the production of antibiotics and/or siderophores.

Thomashow and Weller (1988) continued the investigation of Weller et al. (1988) concerning the role of a phenazine antibiotic produced by *P. fluorescens* 2-79 and its rifamsin-resistant derivative 2-79$Rn_{10}$ in the biological control of *G. g. tritici.* Culture techniques were generally the same as previously described. The suicide plasmid pGS9 in *E. coli* WA803 was used to generate Tn5 insertions in strain 2-79$RN_{10}$ and to finally develop several mutants defective in phenazine antibiotic production (Phz$^-$). Six prototrophic Tn5 mutants unable to produce detectable phenazine were isolated from seven independent matings, resulting in more than 6000 transconjugants. These mutant strains failed to inhibit growth of *G. g. tritici* in vitro, but reconstituted strains and the original strain suppressed the take-all of wheat (Table 19 and Fig. 12). Antibiotic synthesis, fungal inhibition in vitro, and suppression of take-all on wheat were all restored in two mutants complemented with cloned DNA from a 2-79 genomic library. These mutants contained Tn5 insertions in adjacent EcoRI fragments in the 2-79 genome; the restriction maps of the region flanking the insertions and the complementary DNA were colinear. These results indicated that sequences required for phenazine production were present in the cloned DNA and supported the importance of the phenazine antibiotic in disease suppression in the rhizosphere.

The improvement in efficacy of fluorescent pseudomonads as biocontrol agents depends in part on understanding and making use of the mechanisms involved in the interactions between the pathogens, pseudomonads and their plant hosts (Thomashow and Weller, 1990). In their investigation, Thomashow and Weller found that phenazine-carboxylic acid was produced in the rhizosphere of wheat plants previously inoculated with *P. fluorescens* 2-79. They found no antibiotic in roots and soil colonized by a Phz$^-$ mutant. The significant take-all control occurred in the presence of as little as 25 to 30 ng of the antibiotic per gram of root. Strains of mutants that produced only fluorescent siderophores and no phenazine-1-carboxylic acid conferred little protection against take-all. On the other hand, mutants producing anthranilic acid (Aff$^+$) were somewhat suppressive to take-all, but much less so than antibiotic-producing strains. It appears that iron-regulated antibiotics (nonphenazine antibiotics) may be produced more frequently than previously thought and may be responsible for suppressive effects that were previously attributed to fluorescent siderophores.

Thomashow et al. (1990) reported that no antibiotic was recovered from roots of seedlings grown from seeds treated with nonphenazine-producing mutants of *P. fluorescens* 2-79 and *P. aureofaciens* 30–84 or from seeds left untreated. They also found that 27 to 43 ng of phenazine-1-carboxylic acid per gram of roots was recovered from roots colonized by strain 2-79, whether or not *G. g. tritici* was present. In steamed and natural soils, roots from which the phenazine antibiotic was recovered had significantly less disease than roots with no antibi-

**Table 19** Effect of *P. fluorescens* 2-79RN$_{10}$ phenazine antibiotic on inhibition of *G. graminis* var. *tritici* and suppression of take-all*

| Strain | Phenazine production† | Relative inhibition‡ | Root disease§ in experment: | | | |
|---|---|---|---|---|---|---|
| | | | 1 | 2 | 3 | 4 |
| 2-79RN$_{10}$ | + | 1.0 | 4.89 d | 3.66 d | 3.80 c | 5.95 b |
| 2-79-10147 | ± | 0 | 5.11 d | 4.20 c | | |
| 2-79-2510 | – | 0 | | 4.20 c | | |
| 2-79-21455 | – | 0 | 5.71 c | 4.41 bc | | |
| 2-79-99 | – | 0 | 5.94 bc | 4.29 bc | | |
| 2-79-611 | – | 0 | 6.25 b | 4.54 bc | | |
| 2-79-B46 | – | 0 | 6.11 bc | 4.56 b | 5.99 b | |
| 2-79-782 | – | 0 | 6.15 bc | 4.28 bc | 5.88 b | 6.39 a |
| 2-79-B46R | + | 0.89 | | | 3.75 c | |
| 2-79-782R | + | 0.93 | | | | 5.75 b |
| Control (no bacteria) | | | 7.19 a | 4.91 a | 6.58 a | 6.44 a |

* Bacteria were tested in vitro for inhibition of *G. graminis* var. *tritici* or applied at $1 \times 10^8$ to $2 \times 10^8$ cfu/seed and tested for suppression of take-all, as described in "Materials and Methods."

† Presence of the antibiotic was detected by UV absorbance on KM-PDA medium, as described in the text. Symbols represent easily detectable (+), barely detectable (±), or not detectable (–) UV absorbance in the medium surrounding regions of heavy bacterial growth.

‡ Inhibition of fungal growth was measured on KM-PDA, as described in the text. The values are the means from three plates, with the bacteria spotted twice per plate.

§ Disease was rated on a 0 to 8 scale; 0 indicates no disease and 8 indicates that the plant was nearly or completely dead. After a significant *F*-test, least-significant-difference analysis was performed. Means in the same column followed by the same letter are not significantly different by least-significant-difference analysis ($P = 0.05$).

*Source:* From Thomashow and Weller (1988), with permission of Amer. Soc. for Microbiology.

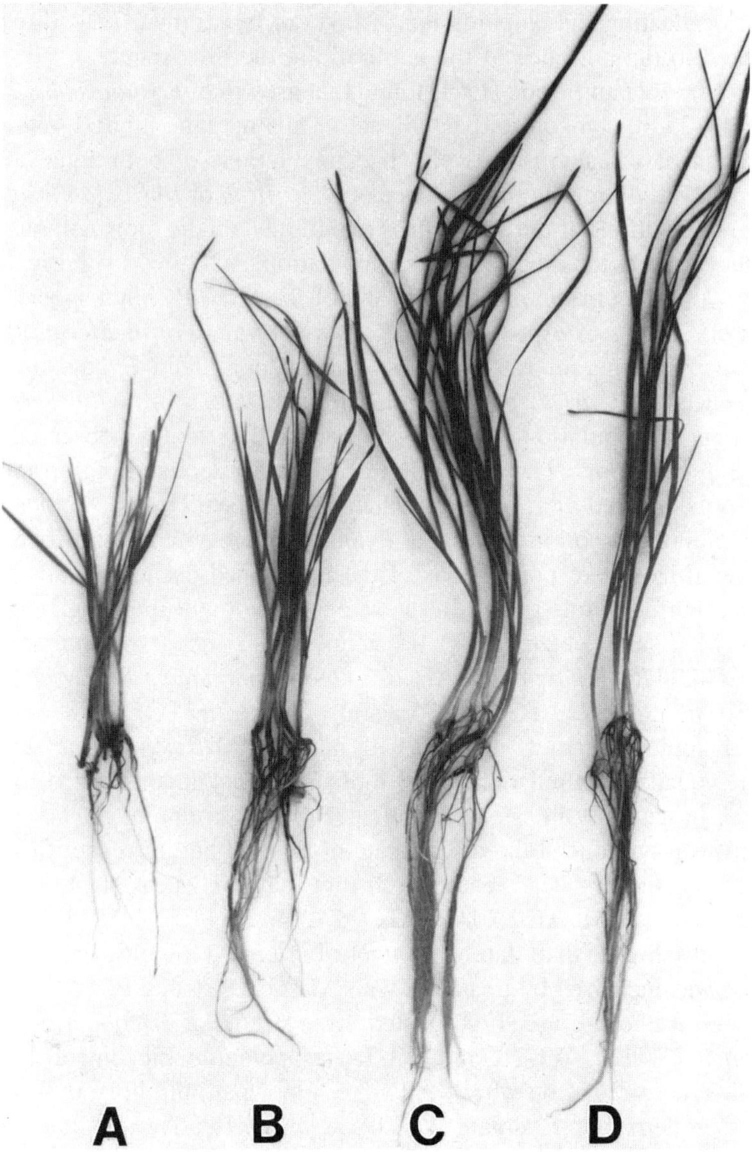


FIGURE 12 Suppression of take-all by phenazine-producing and non-phenazine-producing bacterial strains. Seedlings were grown from (A) nontreated seed or (B) seed treated with 2-79-B46, (C) seed treated with 2-79-B46R, and (D) seed treated with 2-79RN$_{10}$. See Table 19 for phenazine-producing ability of bacterial strains used. *From Thomashow and Weller (1988), with permission of Amer. Soc. for Microbiology.*

otic, indicating that suppression of take-all by strain 2-79 is directly related to the presence of the antibiotic in the rhizosphere.

Mazzola and Cook (1991) found that growth of *P. fluorescens* 2-79 and Q72a-80 (biocontrol bacteria active against take-all and pythium root rot of wheat, respectively) had populations of both equal to or significantly larger in the presence of *G. g. tritici* or *Rhizoctonia solani* than populations of the antagonists maintained on the roots without the pathogens. On the other hand, the population of 2-79 was significantly lower on roots in the presence of any of the three *Pythium* species (*P. irregulare, P. aristosporum,* and *P. ultimum* var. *sporangiiferum)* that cause pythium root rot of wheat than on noninfected roots. In the presence of *P. aristosporum* or *P. ultimum* var. *sporangiiferum,* the decline in population of Q72a-80 was similar to that observed on noninfected roots. The population of Q72a-80 declined more rapidly on roots infected by *P. irregulare* than on uninfected roots. Application of metalaxyl to soil infested with *Pythium* spp. resulted in significantly larger rhizosphere populations of the introduced pseudomonads over time than on plants grown in the same soil without the fungicide.

Maas and Kotze (1987) tested in vitro *Trichoderma harzianum* ATCC 20476, *T. harzianum* ATCC 24274, *T. harzianum* PREM 47942, and *T. polysporum* ATCC 20475 against *G. g. tritici* PREM 48869 isolated from wheat roots infected with take-all. All *Trichoderma* strains were equally antagonistic to growth of *G. g. tritici* under in vitro conditions. In naturally infested soil, all species and strains of *Trichoderma* significantly reduced the disease rate of take-all and also significantly increased the dry head mass. Fresh root mass of wheat plants treated with *T. harzianum* ATCC 20476, ATCC 24274, and PREM 47942 was significantly higher than that of the control after two months. Fresh shoot mass was increased by treatment with ATCC 20476 and PREM 47942.

Ghisalberti et al. (1990) tested three strains of *Trichoderma harzianum* (WU70, WU71, and WU73) as potential biocontrol agents against *G. g. tritici* on wheat. All were somewhat inhibitory in in vitro tests on petri plates. Isolate WU71 was most effective in suppressing take-all of wheat and was found to produce two pyrone antibiotics and other undetermined analogues. It was suggested that the success of WU71 was related to the pyrones it produces and that the abilities of strains WU70 and WU73 to reduce take-all may be due to other mechanisms, since they did not produce pyrones.

Another potential antagonist of take-all of wheat was isolated from soil in Western Australia. It was a strain of *Trichoderma koningii* (Simon, 1989). The soil used in the tests was collected from an experimental farm at Kapunda, South Australia. It was classified as a red-brown earth. The inocula of the pathogen and antagonist were mixed through the air-dried soil at rates of 30 propagules of the pathogen and 120 propagules of *T. koningii* per 250 g air-dried soil. *T. koningii* reduced stelar discoloration of wheat roots by 95% where soil was inoculated with the pathogen and antagonist two weeks before sowing, compared with 52% where soil was inoculated just before sowing the wheat seeds. *T. koningii* caused a reduction in length of seminal roots, but take-all reduced the lengths of the seminal roots by a greater amount. *T. koningii* increased shoot growth following inoculation two weeks before sowing. In the presence of the pathogen, *T. koningii* increased shoot and root growth in soil inoculated two weeks before and in soil inoculated just prior to planting.

*T. koningii* reduced take-all of wheat grown in pots containing natural field soil, but the reduction was greater when the soil was inoculated with the pathogen and antagonist two weeks prior to planting the wheat seeds. Simon's results supported the hypothesis that there is a link between the suppression of the saprophytic growth *G. g. tritici* and suppression of the disease caused by the pathogen. This suggests that the biological control of take-all may be increased by applying the biocontrol agent to the soil some time before seeding.

According to Weller and Cook (1986), wheat in eastern Washington and adjacent Idaho often yields only 75 to 80% (and sometimes only 50 to 60%) of its potential, even when diseases caused by well-known soil-borne pathogens are not problems. Treatment by methods that eliminate or significantly reduce the *Pythium* population of the soil generally eliminate the growth-limiting factor. Weller and Cook selected all the fluorescent pseudomonads from a collection stored at the U.S.D.A.-ARS Root Disease and Biological Control Research Unit at Washington State University and tested all of these for their ability to increase the growth of wheat. Of 64 fluorescent pseudomonad strains selected, 17 resulted in greater seedling heights than those produced with nontreated seeds. Four strains selected on the basis of their performance in the tube bioassay produced significant increases in stand, plant height, number of heads, or grain yield—or in all of these parameters—when tested as seed treatments on winter wheat planted in a field plot naturally infested with *Pythium* spp. *Pseu-*

*domonas fluorescens* biovar 1 (strain Q72a-80) improved growth the most and increased grain yield by 26% compared with nontreated wheat. Addition of the fungicide metalaxyl as a soil drench gave an increased seedling height equivalent to that obtained with addition of strain A72a-80. Both the bacterial and chemical treatments eliminated symptoms of pythium root rot; thus Weller and Cook suggested that the growth response was probably due to protection against *Pythium* spp.

*Pseudomonas cepacia* R55 and R85 and *P. putida* R104 are fluorescent pseudomonads that are known plant-growth-promoting rhizobacteria for winter wheat; therefore de Freitas and Germida (1991) investigated them as potential antagonists against *Fusarium solani* and *Rhizoctonia solani* on winter wheat. The pathogens selected for the investigation were *R. solani* strains AG-1, AG-2-1, and A3 and *Fusarium solani* from the Agriculture Canada Research Station, Saskatoon, Saskatchewan. Cultivar Norstar was the wheat used in the tests.

All three pseudomonad inoculants significantly increased biomass of wheat in soil infested with *R. solani* AG-1. Inoculants increased the root dry weight by 92 to 128% and shoot dry weight by 28 to 48%. The effects of the pseudomonads were less consistent in soil infested with *R. solani* AG-2-1. Inoculation with *P. cepacia* R55 significantly increased plant biomass, whereas *P. putida* R104 increased shoot and total plant weight. In no case did *P. cepacia* R85 cause significant increases in plant biomass. *R. solani* AG-3 had no effects on wheat development; nevertheless, *P. cepacia* R85 significantly increased wheat-shoot biomass in the *R. solani* AG-3-infested soil. Pseudomonad inoculants had no effects on winter wheat growth in the *F. solani*–infested soil. All the pseudomonad strains produced siderophores when cultured in a low-iron medium. The scientists inferred that their overall results suggested an antibiotic activity of the three fluorescent pseudomonads toward the pathogens. They felt that the fluorescent siderophores act in an antibiotic capacity, a feeling I share.

According to Levy et al. (1989), the use of biocontrol agents to suppress infections by foliar pathogens on cereals has been limited. They therefore decided to test a fluorescent *Pseudomonas* sp. (strain LEC-1), isolated from soil, against infection of leaves of Shafir wheat by *Septoria tritici* (ISR 398) or *Puccinia recondita* (ISR1010). When *Pseudomonas* sp. LEC-1 was applied to Shafir wheat seedlings three hours prior to inoculation with the pathogens, LEC-1 suppressed *S. tritici* by 88% and *P. recondita* by 98%.

Previous work suggested that LEC-1 produced one or more antibiotics, so Levy and his colleagues isolated and identified two antifungal compounds, 1-hydroxyphenazine (phOH) and chlororaphin. The former compound differentially inhibited growth in mycelial weight and colony diameter of *Fusarium oxysporum* f. sp. *vasinfectum, Pythium aphanidermatum, Sclerotium rolfsii,* and *Sclerotinia sclerotiorum.* The greatest reduction in both mycelial weight (92%) and colony diameter (65%) occurred with *P. aphanidermatum.* Statistically significant growth suppression of the other fungi ranged from 16 to 53%. The compound 1-hydroxyphenazine did not affect mycelial weight of *S. rolfsii,* but it did significantly suppress increase in colony diameter. Chlororaphin was tested only against growth of *Septoria tritici* on a malt-agar medium and was found to be inhibitory to that pathogen. Growth of *S. tritici* in liquid medium was reduced by phOH at 5 to 20 mg/L and was prevented at 40 mg/L.

The number of *P. recondita* pustules/cm$^2$ was reduced by 35 to 75% with phOH concentrations of 20 to 160 mg/L. Treatment with $10^7$ cfu/mL of LEC-1 cells reduced the number of pustules per cubic meter by 98%. Increasing phOH concentrations progressively reduced the pycnidial coverage by *S. tritici.* LEC-1 cells, however, were more effective in reducing symptoms than any of the tested concentrations of the antibiotic it produced. It certainly appeared that LEC-1 was an effective suppressor of both *S. tritici* and *P. condita.*

Levy et al. (1992) followed up their previous work with tests of a different pseudomonad, *Pseudomonas fluorescens* (PFM2), against *Septoria tritici* (ST83A) on wheat. It was found in in vitro tests that *P. fluorescens* (PFM2) inhibited growth of the four pathogenic fungi it was tested against—*F. oxysporum, S. rofsii, Pyrenophora tritici-repentis,* and *Pythium irregulare*—and five test bacteria—*A. tumefaciens, E. c. carotovora, P. solanacearum, X. c. phaseolus,* and *Clavibacter m. michiganense.* It was also antagonistic to *Septoria tritici.*

Three inhibitory compounds were extracted from a liquid glycerol-peptone-phosphate medium in which PFM2 had grown for four weeks. The chief antibiotic was identified as 2,4-diacetylphloroglucinol (DPG). The other two were produced only in much smaller amounts and were not identified. The DPG was very inhibitory to all the test fungi and all but one of the test bacteria. It was pointed out above that *S. tritici* was inhibited strongly by both PFM2 and the antibiotic, which strongly suggests that the antibiotic plays an important role in the antagonism of *P. fluorescens* PFM2 against the pathogen *S. tritici.*

Growing bulbs for bulb and flower production is very important in the Netherlands, and *Tulipa* spp. (tulips) have a prominent position, with 600 million flowers being produced each year (Westeijn, 1990). Root and bulb rot caused by *Pythium* spp. commonly occurs. *Pythium ultimum* is probably the most frequent species encountered, but *P. intermedium, P. irregulare,* and *P. dissotocum* are also associated with this disease. A good biocontrol microorganism could be very valuable; thus Westeijn isolated 26 fluorescent pseudomonad isolates from rhizosphere soil suspensions from diseased tulip roots. The pathogen selected for the tests was *P. ultimum* P17 isolated from *Pythium*-diseased tulip roots, and the tulip cultivar Paul Richter, prepared as ice tulip, was used as host. (Ice tulips are tulips whose bulbs, after development of flower and leaf initials and after the required period at low temperatures, are frozen in moist peat dust at –2°C until use.)

The root rot–suppressing capabilities of the selected pseudomonads were assessed in pot experiments, using pots or PVC tubes. Natural sandy soil was obtained from the bulb-growing field near Lisse, the Netherlands. The soil was mixed with the *Pythium* soil-cornmeal culture. Depending on the experiment, the *Pseudomonas* suspensions were also mixed thoroughly through the soil to a concentration of $10^8$ cells/g dry soil before planting the bulbs, or the bulbs were dipped in a bacterial suspension of $2 \times 10^9$ /mL, with or without 1% methylcellulose, immediately before planting. The disease severity was rated on a scale of 1 (healthy) to 5 (all roots severely affected). Based on preliminary tests, it was concluded that root fresh weight and shoot length were the most suitable parameters for measuring disease (root weight) and shoot length (effects of antagonists).

One pseudomonad, strain E11.3, gave consistent root-rot suppression. Beneficial effects were obtained after introduction of the bacteria, either by mixing them through the soil or by dipping the bulbs in a bacterial suspension immediately before planting. Application of bacteria in methylcellulose reduced disease but was of no practical value because methylcellulose itself increased disease. Moreover, disease was more severe in steamed soil than in natural soil. Because of the addition of isolate E11.3, the number of saleable flowers was increased from 29 to 79% in natural soil.

Shi and Brasier (1986) investigated the possibility of developing a biocontrol for Dutch elm disease on *Ulmus* caused by *Ceratocystis* (or *Ophiostoma*) *ulmi*. They selected four isolates of *Pseudomonas syringae* (M27m, MSU33, M323m, MSU83), six isolates of *P. fluorescens,* one isolate of *Enterobacter* sp., and eleven isolates of *Pseudomonas* spp. [BN, B321(b), BSI, B321(S2), B321(S3), B321(S4), B321(f), A321(r), C406–5, C406–6, and C406–7] isolated from soil at Rowledge, Surrey. All these strains were tested against *C. ulmi* H340 on agar plates.

None of the six isolates of *P. fluorescens* nor the isolate of *Enterobacter* showed any evidence of antagonistic action against *C. ulmi.* Three of the isolates of *P. syringae* showed positive antagonistic action, as did eight of the eleven isolates of *Pseudomonas* spp. Isolates selected for injection into elm were *P. syringae* (M27m, MSU33, and M323m). Additionally, soil isolates BN, BSI(s3), and C406–5 and two nonactive *P. fluorescens* isolates 1115 B and 1204 were selected for injection for comparison.

In May 1982, a block of 107 three-year-old clonal English elm (*U. procera*) were injected with the selected antagonistic and two nonantagonistic pseudomonads. All the trees were inoculated with *C. ulmi* H340 four weeks later and assessed for symptoms thereafter at two- or three-week intervals. A therapeutic injection of Commelin elm (*Ulmus* x *hollandica*), which had been injected with several aggressive isolates of *C. ulmi* in 1981, was made by using the antagonists M27m, BN, B321(S3), and C406-5 in May 1982. In the series of preventive bacterial injections in English elm, no reduction in disease levels, by comparison with noninjected controls, was obtained. In the curative injections of Commelin elm, only a slight reduction of disease recurrence in treated trees was observed. The injected antagonistic bacteria demonstrated limited upward distribution in the trees but may have become well distributed in the roots. It was inferred that bacterial injection does not offer potential for preventive control of Dutch elm disease in English elm, nor as an alternative to fungicide injection.

Many problems concerning the biological control of Dutch elm disease caused by *Ceratocystis* (*Ophiostoma*) *ulmi* depend on a knowledge of the distribution of the introduced antagonists within the elm (*Ulmus*) tree and on the development of the population over a period of time (Scheffer et al., 1989a). To this point in time the antagonists have been various strains of *Pseudomonas.* Scheffer et al. tested three approaches to identifying the *Pseudomonas fluorescens* strain WCS374 used in their investigations: (1) immunoagglutination of representative reisolates with an antiserum against the *P. fluorescens* isolate in use; (2) chemotaxonomy, using lipopolysaccharide pattern, cell-envelope protein pattern, or DNA-restriction fragment pattern; and (3) use of a metabolic marker (Lac ZY) to label bacteria with a transposon (Tn903) or a plasmid construct (pMON5003), coding for ß-galactosidase and lactose permease. All techniques proved reliable in identifications, but there were problems with methods 2 and 3. Method 2 was very time-consuming, and method 3 was not satisfactory because populations of the transposon-labeled bacteria in elms declined much faster than populations of the unlabeled wild type. The plasmid carrying the metabolic marker disappeared from the bacterial population over time. The immunoagglutination test proved to be specific and fast and was therefore employed for routine tests.

In experiments concerning the duration, population development, and movement of the introduced strains of *Pseudomonas,* several significant findings were made. Introduced populations in elm twigs (cv. Commelin, Vegeta, Belgica, 390) declined rapidly upon inoculation but apparently became relatively stable within three months, up to the end of the second growing season, when the experiment was terminated. The much lower antagonist population higher in the twigs, according to Scheffer et al., probably reflects the inability of *Pseudomonas* cells to escape from the xylem vessels into which they were introduced. In an experiment with 47 different elm clones, no, or hardly any, antagonists were isolated from 88% of the samples of the new year's shoots. This too demonstrates that the antagonists apparently cannot escape from the vessels in which they were introduced.

Scheffer et al. (1989b) continued their study of the location, persistence, and ecology of *Pseudomonas* injected into *Ulmus* (elm) trees to help control Dutch elm disease. Two aggressive strains of *Ceratocystis* (*Ophiostoma*) *ulmi* (H6 and H106), three strains of *Pseudomonas fluorescens* (WCS361, WCS374, and WCS374RJS101, adapted to low water potential), two strains of *P. putida* (WCS085, WCS358), and one strain of *P. syringae* ($M27^{+}$) were used in this investigation.

Gouge pistols were used to inoculate trees with antagonists at a height of 1.2 to 1.4 m every 5 cm of circumference, 1 mL of a shake culture per inoculation. In one experiment twigs of young elms of clones *Ulmus* x *hollandica* (Belgica and 390) were inoculated near their base with a 3-mm-wide chisel using 20 μL of a shake culture of

WCS374 or WCS374RJS122 at each of three points. Controls received a modified King B medium only. The trees (or twigs) were inoculated later with a 1:1 mixture of conidial suspensions of *C. ulmi* strains H6 and H106, with $5 \times 10^6$ conidia/mL. Stems were inoculated with 50 μL at opposite sides of the stem.

Reisolation of *Pseudomonas* from the elm twigs or trees was carried out at various times and positions. For immunofluorescence microscopy, small sections were taken from stems, and then thin sections were cut from them for observation. Observation by this method indicated that the antagonistic bacteria occurred only along the walls of the xylem vessels. The lumen of the vessels was low in nutrients, the pH was above the critical value for *Pseudomonas* spp., and at times the water potential was very low for bacteria. The isolate selected for growth at low water potential established itself better within elm than did the parent strain.

Spatial distribution studies indicated again that the antagonists did not escape from the vessels in which they were introduced. However, bacteria did regularly occur in the new annual ring in the second growing season, and Scheffer et al. postulated that the wound tissue or the root system probably served as an inoculum source. They suggested, therefore, that the suppressive effect against Dutch elm disease may last for two seasons and possibly longer if the root system is colonized by the antagonist.

Scheffer (1989) reported on field trials on the biological control of Dutch elm disease conducted throughout the Netherlands from 1982 through 1986. In one type of trial, only natural infections were monitored with or without injections of *Pseudomonas* antagonists. In the second type, trees were artificially injected with isolates H6 and H106 of *Ceratocystis* (*Ophiostoma*) *ulmi.* In the large-scale field trials, the results were based on the expected annual losses to Dutch elm disease of about 2%. Dutch elm disease incidence was 22 to 45% lower in the trees treated with a *Pseudomonas* isolate in the year of treatment and the year after.

The advantage of artificial infections with *C. ulmi* was a reproducible development of symptoms and the possibility of maintaining diseased trees, at least until the first signs of elm bark beetle breeding. Different bacterial treatments indicated that more injections per tree may result in better suppressive effects than increasing the number of bacteria inoculated at the usual number of positions around the trunk.

The host elm was very important in determining the results of bacterial treatment. In the case of Commelin elm, one bacterial treatment reduced the Dutch elm disease in trees inoculated with *C. ulmi* by 10 to 85% by the end of the second year, as compared with controls. This was also true at times with Belgica elms, but in Vegeta elms symptoms were only less severe and in *Ulmus carpinifolia* (field elms) some slight effect was observed in one experiment but none in two others. The same antagonistic *Pseudomonas* strains as were used by Scheffer et al. (1989b) were used in these widespread field experiments.

Fungicide resistance is becoming more common among plant-pathogenic fungi with the introduction of systemic fungicides. It was for this reason that Ravi and Anilkumar (1991). undertook the discovery and development of suitable methodology for the biocontrol of fungicide-resistant strains of *Colletotrichum truncatum,* a serious pathogen of *Vigna unguiculata* (cowpea). A virulent culture of *C. truncatum* $Ct_1$ from cowpea, two of its carbendazim-resistant UV mutants ($Ct_2$ and $Ct_3$), and three thiophanate resistant mutants ($Ct_4$, $Ct_5$ and $Ct_6$) were selected to use in the screening of three species of cowpea leaf resident fungi—*Phoma* sp., *Pestalotiopsis* sp., and *Penicillium* sp.—against conidial germination of the *C. truncatum* isolates. In the tests, a given concentration of conidia of a pathogen isolate was mixed with an equal volume of the same concentration of one of the potential antagonists, and the extent of conidial germination of the pathogen was assessed by a slide germination technique after eight hours.

*Penicillium* sp. inhibited conidial germination of the wild sensitive isolate $Ct_1$ by 6.9%. The fungicide-resistant strains were relatively more sensitive to *Penicillium* sp., with conidial germination being reduced by 7.65 to 27.04%. *Phoma* sp. was more inhibitory to conidial germination of *C. truncatum* isolates, ranging from 32.28 to 63.59%, with all fungicide-resistant strains, except $Ct_4$, being more sensitive to *Phoma* sp. *Pestalotiopsis* sp. was generally the least inhibitory to the conidial germination of the strains of the pathogen. *Phoma* sp. showed considerable promise as a biocontrol agent against *C. truncatum* on cowpea, if suitable management practices are developed.

Another serious disease of *Vigna unguiculata* is the web blight disease caused by an aerial ecotype of *Rhizoctonia solani* (Latunde-Dada, 1991). This disease may result in total crop loss in the warm and humid conditions of the lowland tropics of Africa. Latunde-Dada investigated the possibility of controlling that disease in field conditions through spray applications of spore suspensions of the potential antagonist *Trichoderma koningii.* The strain of *R. solani* (IMI334840) used in this study was isolated originally from a diseased cowpea plant showing symptoms of the web blight disease. The strain of *T. koningii* (IMI334839) used was isolated from the soils of the teaching and research farm of Orgun State University in Nigeria.

Seeds of cowpea cultivar IT82E-60 were sown in field plots. Plants were inoculated with *Rhizoctonia* at four weeks of age and protection treatments with *Trichoderma,* where administered, commenced one week after inoculation with *R. solani.* Some plots were sprayed once weekly and some twice weekly with *Trichoderma.* All plants were sprayed weekly with the insecticide Agrothion, from the onset of flowering at five weeks after planting until the eighth week.

*Trichoderma koningii* IMI334839 restricted the spread of web blight in all plots on which it was used, but the disease suppression and yield levels were significantly greater in plots sprayed twice weekly. The levels of disease suppression and yields compared favorably with those obtained from inoculated cowpea plants protected with a fungicide mixture. It appears that *T. koningii* IMI334839 has a high potential for the biocontrol of web blight of cowpea plants under field conditions.

Many of the pathogens of cowpea are seed-transmitted, but no research was previously done concerning seed microflora on seed-borne infection by *Xanthomonas campestris* pv. *vignicola* and on seed germination of cowpea. Jindal and Thind (1990) isolated five species of bacteria and seventeen species of fungi and soaked 200 cowpea seeds (Pusa Barsati and Pusa Dofasli) in a suspension of each bacterial or fungal culture containing between $10^5$ and $10^9$ cfu or spores/mL. Some seeds were also rolled in viscous cell-spore suspensions. The seeds in both cases were spread over moistened blotter sheets, after which the sheets were rolled. Seed germination was checked after seven days at 27 ± 1°C. Soaking or rolling cowpea seeds in cell suspensions of *Acinetobacter calcoaceticus, Pseudomonas acidovorans,* and spore suspensions of *Aspergillus flavus, Fusarium equiseti, F. semitectum, Rhizopus oryzae,* and *Trichothecium roseum* resulted in a significant reduction in seed germination.

When the five bacterial and seventeen fungal species were tested against the pathogen *X. c. vignicola* in vitro, *Acinetobacter calcoaceticus, Bacillus subtilis, Flavobacterium* sp., *Pseudomonas acidovorans, Aspergillus flavus, Epicoccum nigrum, Fusarium equiseti,*

*Penicillium citrinum, P. oxalicum, Rhizopus oryzae,* and *Trichothecium roseum* were found to be inhibitory. Several organisms, including *Erwinia herbicola* and *P. oxalicum,* were as effective as hot water (50°C for 30 minutes), solar heat, streptocycline plus captan, and streptocycline plus agallol for eradication of *X. campestris* pv. *vignicola* from cowpea seeds. Thus, *E. herbicola* and *P. oxalicum* appear to have great potential for the biocontrol of *X. c. vignicola* on cowpea.

*Eutypa lata* is the fungal pathogen that causes dieback of *Vitis vinifera* (grapevine), and Ferreira et al. (1991) found an isolate of *Bacillus subtilis* associated with the *E. lata* when collecting the pathogen from a grapevine at the Viticultural and Oenological Research Institute at Stellenbosch, South Africa. In vitro tests indicated that the *Bacillus* isolate caused 91.4% inhibition of mycelial growth of *E. lata* on potato-dextrose agar and 100% inhibition of ascospore germination. Thin-layer chromatography of a crude antibiotic extract of the *B. subtilis* isolate gave five bands, two of which inhibited mycelial growth of *E. lata.* Thus at least two antibiotics were present. A concentration of 0.8 mg/mL or greater of the chief antibiotic completely inhibited germination of ascospores of *E. lata.* It was inferred therefore that the chief mechanism of action of the *B. subtilis* isolate was by antibiosis. Spraying a suspension of the isolate on pruning wounds of Riesling grapevine before inoculation with ascopores of *E. lata* significantly reduced infection compared with the unsprayed inoculated controls. Ferreira et al. concluded that use of the *B. subtilis* isolate as a biocontrol agent against *E. lata* may be an economical way to suppress the dieback disease.

Corn anthracnose caused by the fungus *Colletotrichum graminicola* is one of the chief diseases of *Zea mays* (corn or maize) in the areas where it is grown. DaSilva and Pascholati (1992) decided to study the possible use of commercial yeast (*Saccharomyces cerevisiae*) as a biocontrol agent of corn anthracnose. Suspensions of *S. cerevisiae* cells and filtrates of the suspensions reduced the development of *C. graminicola* as well as the expression of anthracnose on leaves of corn when they were treated previously or concomitantly with the yeast preparations. The suppression of the development of the pathogen and of the disease on the corn leaves indicated that the presence of the yeast cells was not necessary to protect the leaves. In vitro tests demonstrated that *S. cerevisiae* cells exerted an antagonistic activity against *C. graminicola* due to antibiosis. It is noteworthy that the active compounds in the yeast preparations and their filtrates were thermolabile.

*Fusarium graminearum* is a well-known pathogen that causes stem and ear rot of corn in many countries, and two populations of this pathogen have been designated (van Wyk et al., 1988). Members of Group 1 are usually associated with diseases like the crown rot of wheat, whereas Group 2 is usually associated with diseases of the aerial parts of plants, such as stalk and ear rot of corn. Van Wyk et al. investigated the potential of *Fusarium moniliforme* as a biocontrol agent against *F. graminearum* and the stem and ear rot of maize. Nine strains of *F. moniliforme* (MRC826, 1069, 4315, 4318, 4323, 4341, 4346, 4348, and 4493) isolated from corn grains, two strains of *F. graminearum* Group 1 (MRC4517 and 4518) isolated from wheat crowns, and three strains of *F. graminearum* Group 2 (MRC4514, 4535, and 4536) isolated from ear rot of maize were used in the investigation. Two South African white corn cultivars (AX305W and A471W) were also selected for the tests.

Uncontaminated corn germlings were inoculated with strains of *F. moniliforme* (one at a time) by placing them on five-day cultures of the isolates in petri dishes for 10 hours. Controls consisted of corn germlings placed on sterile PDA for 10 hours. A mixture of maize meal and sand (autoclaved) was inoculated with the different *F. graminearum* isolates and later added to steam-treated soil, which was placed in pots in the greenhouse. Corn plants, inoculated and uninoculated with isolates of *F. moniliforme,* were subsequently planted in the pots of soil, some inoculated with *F. graminearum* and some not. The plants were harvested either 24 or 30 days after planting, the roots were kept for observations, and the tops were weighed after drying.

Weights of corn seedlings preinoculated with some isolates of *F. moniliforme* and exposed to some isolates of *F. graminearum* in soil were significantly higher than those of controls. This was particularly evident in corn seedlings exposed to an aggressive isolate of *F. moniliforme* Group 1. Although all eight isolates of *F. moniliforme* protected maize seedlings against infection by *F. graminearum* Group 1 MRC4517, none of them protected against all isolates of *F. graminearum.* Three isolates of *F. moniliforme* (MRC826, 1069, and 4341), however, provided significant protection against three of the isolates of *F. graminearum.* In all instances, cultures of *F. graminearum* Group 1 and Group 2 were reisolated from the necrotic tissue of plants subjected to the inoculum of *F. graminearum* but not preinoculated with *F. moniliforme.* The possibilities of using *F. moniliforme* in the biological control of corn diseases caused by *F. graminearum* are thus very evident.

Caryopses of 10 wild populations of *Tripsacum dactyloides* (eastern gamagrass) were screened for the presence of fungal-inhibiting bacteria (Anderson and Liberta, 1986). Three populations were found to harbor bacteria inhibitory to 10 of 11 fungi used in tests. The bacteria were identified as pseudomonads (*Pseudomonas* spp.). Because eastern gamagrass is a distant relative of maize and can be hybridized with maize, Anderson and Liberta decided to test the bacteria against some maize fungal pathogens to determine if they could protect germinating corn seeds in soil against certain diseases. They found that the three bacterial strains inhibited growth of the three pathogens of corn seeds or seedlings, *Diplodia maydis, Fusarium roseum,* and *F. moniliforme.* These were inhibited by the bacteria on agar plates, and a zone of fungal inhibition 1 to 2 cm wide occurred around surface-sterilized maize kernels that were treated with the bacteria and placed in culture dishes containing 2% maize-meal agar. In cold-soil germination tests, there were no significant differences in germination of scarified maize kernels that were treated with gum arabic (sticking agent) and one of the bacterial strains (98%) or the commercial fungicide captan (96%). Both treatments gave significantly higher germinations than scarification (50%) or scarification plus gum arabic alone (58%). Apparently, the pseudomonads isolated may be effective in suppressing pathogens capable of invading damaged corn grains and inhibiting germination and destroying seed embryos.

Cantone and Dunkle (1990) reported that resistance to *Helminthosporium carbonum* race 1 in maize was induced by prior inoculation of maize leaves with *H. carbonum* race 2. Moreover, the induced resistance was associated with the production of a compound in the host leaf that reversibly inhibited conidial germination and germ-tube elongation. The inhibitory diffusate from the host leaves was also very inhibitory to two phytopathogenic bacteria, *Corynebacterium michiganense* and *Erwinia amylovora.* The diffusates also inhibited lesion formation by race 1; lesions were strikingly absent in tissue inoculated with race 1 conidia in the presence of the diffusate. In fact, in the presence of wash fluid from susceptible leaves inoculated with race 2, less than 3% of race 1 conidia germinated, and these did not develop appressoria or penetrate the leaf. It therefore appears that the inhibitor plays a role in induced resistance to *H. carbonum* race 1 in corn leaves.

Lumsden and Locke (1989) screened over 50 fungal and bacterial isolates for their efficacy against damping-off of *Zinnia elegans* (zinnia) caused by *Rhizoctonia solani* and *Pythium ultimum.* The potential antagonists included six isolates of *Gliocladium* spp., seventeen of *Trichoderma* spp., eighteen of *Talaromyces flavus,* five of miscellaneous fungi, and eleven of bacteria. The isolates of the bacteria *T. flavus* and *Trichoderma* sp. were not effective. Only isolates of *Gliocladium* were effective in improving seedling stands of *Zinnia.* Isolate G20 of *Gliocladium virens* was selected for subsequent tests because of its ability to control diseases caused by both *R. solani* and *P. ultimum* (Fig. 13). Populations of *G. virens* G20 were consistently high after one-week of incubation ($10^6$ to $10^7$ cfu/g substrate).

Control of *P. ultimum* was effective when inoculation of the pathogen occurred at the time of planting the zinnia seed. Control of *R. solani,* however, required contact of *G. virens* with the inoculum of *R. solani.* Sodium alginate formulations of isolate G20 of *G. virens* maintained a high population density in a dry formulation when stored for two months at 4 and 20°C, but not at 30°C. Storage of an alginate formulation in air-dried soilless mix, however, was not successful. Lumsden and Locke concluded that alginate formulations of *G. virens* G20 added to a soilless mix before planting zinnia seed showed promise as a control of damping-off in the greenhouse production of plants.

In subsequent experiments, Lumsden et al. (1992) demonstrated that *G. virens* G20 produced the antibiotic gliotoxin in a soilless medium when the antagonist was introduced as a prill pellet. Moreover, the gliotoxin was produced throughout the medium away from the pellet, with the concentration decreasing with distance from the pellet. Aqueous extracts of the soilless medium amended with *G. virens* and added to flats planted with zinnia were as effective for control of damping-off as was the intact *G. virens*–amended medium. In other tests, gliotoxin was detected in composted mineral soil (0.36 $\mu g/cm^3$), clay soil (0.20 $\mu g/cm^3$), and sandy soil (0.02 $\mu g/cm^3$), all with natural biota but amended with 0.1% alginate prill containing *G. virens* G20. At a rate of 0.4% (w/v) amendment of the alginate prill, the amount of gliotoxin was quadrupled when compared with the 0.1 rate.

Gliotoxin was detectable over a wide range of incubation temperature (15 to 30°C) in a soilless medium amended with prill of G20,

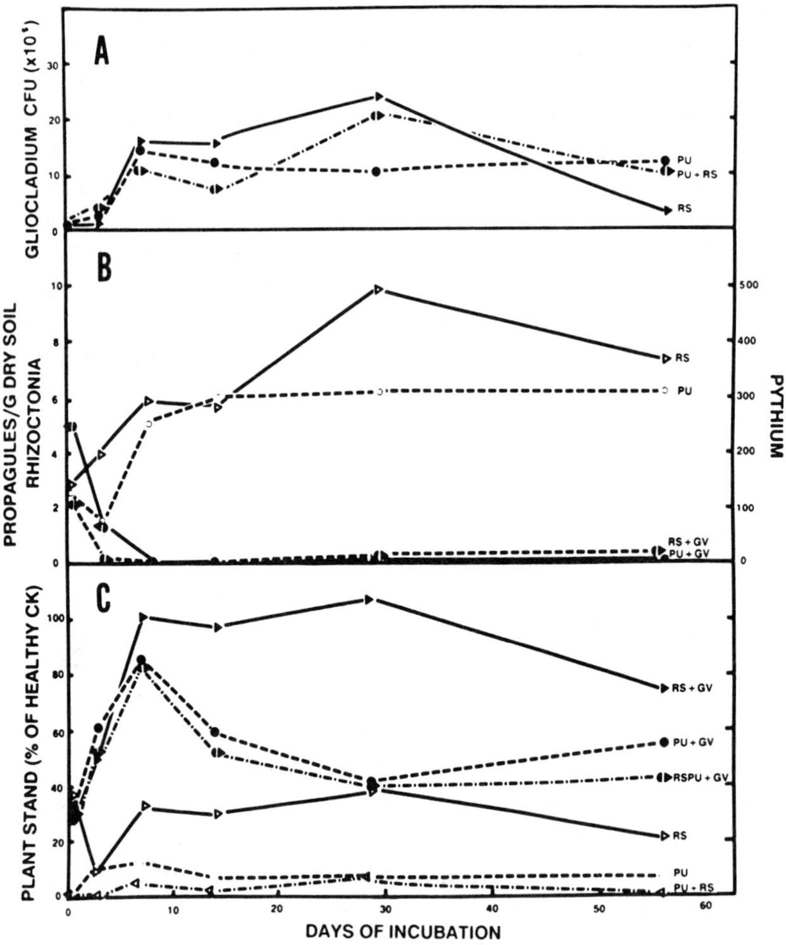


FIGURE 13 (A) Colony-forming units (cfu) of *Gliocladium virens* (GV) in a bran culture incubated in soilless mix with *Pythium ultimum* (PU) and *Rhizoctonia solani* (RS) alone and in combination. (B) Effects of the bran culture of *G. virens* on the number of propagules of *P. ultimum* and *R. solani* per gram of soilless mix. (C) Effects of the bran culture of *G. virens* on damping-off of *Zinnia elegans* in soilless mix caused by *P. ultimum* and *R. solani* alone and in combination (RSPU) during a two-month incubation. Each pathogen pair with and without *G. virens* was tested for significant differences at zero time and at intervals. *Adapted from Lumsden and Locke (1989), with permission of Amer. Phytopathological Soc.*

but the highest amount was produced at 30°C. It is obvious that gliotoxin can be produced by G20 soils and soilless media and is correlated with disease suppression of *P. ultimum* and *R. solani* in nonsterile growing media.

**Conclusion** Some readers have no doubt been wondering why I have devoted two chapters to the biological control of plant diseases when my overall topic of discussion is concerned with advances in allelopathy. In that connection, it should be recalled from Chap. 1 that Molisch, who coined the term *allelopathy,* defined it as chemical interactions between plants, including the microorganisms traditionally placed in the plant kingdom. These include, of course, algae, bacteria, and fungi. Thus production of chemicals by any of these microorganisms that get out of the cells and affect other microorganisms falls directly under the definition of allelopathy. It should always be remembered that the chemical interactions involved in allelopathy can be between plants and plants, plants and microorganisms, and microorganisms and microorganisms.

I recognize, of course, that plant pathologists do not generally use the term *allelopathy* in their writings. Nevertheless, much of their work is concerned with allelopathic interactions. Even a quick perusal of Chaps. 6 and 7 of this book will indicate that chemical interactions have been demonstrated in a relatively high percentage of suppressive actions by antagonists against plant pathogens. The chemicals involved do not have to fit someone's narrow definition of antibiotics to be allelochemicals. Siderophores are sometimes involved in important interactions between antagonists and pathogens and are therefore important allelochemicals; though many pathologists do not consider them as antibiotics, some pathologists do. In allelopathic interactions between plants, one of the mechanisms of action is the production of allelochemicals that affect the uptake of certain minerals by the affected plants (Rice, 1984, Chap. 13). Further, in the many cases where the mechanism of action of the demonstrated antagonists has not yet been elucidated, it may later be shown to be due to allelochemicals.

Another important point to keep in mind is that more than one phenomenon may be involved in interactions between organisms, including microorganisms. For example, competition probably plays a role in interactions between plants and between microorganisms, even where allelopathy or antibiosis has also been shown to play a role. One mechanism does not necessarily exclude the operation of another.

I feel so strongly that the entire phenomenon of biological control of plant diseases is crucial to sustainable agriculture and to the survival of humankind, that I do not hesitate to discuss a particular procedure just because the mechanism of control has not yet been shown or because the most likely mechanism has been shown to be something other than a chemical interaction. Even 10 years ago, there was very little direct evidence to support the important role of antibiosis in the biological control of diseases, yet recent developments in equipment, techniques, and genetic engineering have caused great strides to be made. I have no doubt that many more developments will be made in the next 10 years.

# CHAPTER 8

# ALLELOPATHY IN FORESTRY

IN THIS CHAPTER, allelopathic plants are discussed in alphabetical order of their scientific names except where several species have been discussed in the same scientific paper, in which case all species discussed in that paper are dealt with according to the alphabetical order of the first species considered. When several papers were published concerning the same allelopathic species, they are discussed in chronological order by date of publication.

## I. ALLELOPATHIC EFFECTS OF WOODY PLANTS

Jobidon (1986) was interested in the reforestation of abandoned fields in eastern Quebec, Canada, and was therefore concerned with the allelopathic potential of several forest trees against some common old-field weeds. In this connection he tested fresh-leaf and leaf-litter leachates of *Abies balsamea* (balsam fir), *Picea mariana* (black spruce), *Pinus divaricata* (jack pine), *P. resinosa* (Norway pine), and *Thuja occidentalis* (white cedar) against seed germination and seedling growth of *Phleum pratense* (timothy), *Poa pratensis* (Kentucky bluegrass), *Agropyron repens* (couch grass), and *Epilobium angustifolium* (fireweed).

Leachate solutions were made by soaking 10 g of intact leaves or leaf litter in 100 mL of distilled water for 24 hours at 20°C. The solutions were filtered and termed 10% solutions. Two and five percent solutions were made from the original. Germination of all weed species was inhibited in proportion to the dilutions used, but fresh-leaf leachates of *A. balsamea* and *P. resinosa* were most inhibitory to germination, and Kentucky bluegrass was usually the least affected of the weeds. Most of the leachates inhibited both height growth and root elongation of the three grasses; black spruce was not only the most inhibitory, it also caused severe root necroses, which affected normal root development. Overall, the litter leachates had effects similar to the fresh-leaf leachates in most tests. For some reason, effects of the fresh-leaf and litter leachates on root and shoot growth of fireweed were not mentioned. Nevertheless, the overall data suggested quite clearly that the fresh leaves and leaf litter of the common conifer trees used in reforestation in Quebec could be useful in controlling weeds during reforestation. The data also suggested that throughfall or leaf fall from the living trees could help control weeds after establishment.

Reigosa-Roger (1987) did an impressively thorough study of the allelopathic potential of *Acacia dealbata*. This tree, originally from Australia and Tasmania, is common now in Europe. Widely used as an ornamental tree in Spain, it has also become widespread in the wild, where it is a pest. *A. dealbata* was very dominant in the study area, where some of the herbaceous understory plants were *Lolium perenne* (perennial ryegrass), *Dactylis glomerata* (orchard grass), *Trifolium repens* (white clover), and *Trifolium pratense* (red clover). These were selected as target species of the potential allelopathic effects of *A. dealbata*. Tests were made with natural throughfall and artificial throughfall against seed germination and seedling growth of the target species. In both cases highly significant inhibitory effects occurred against seed germination and seedling growth compared with the effects of rainfall collected between the *A. dealbata* trees or distilled water in the tests with artificial throughfall.

Extracts of various parts of *A. dealbata* were generally inhibitory to seed germination and seedling growth of test plants. Flowers were strongly allelopathic. Soil under the trees and volatiles from the trees were also often allelopathic to test species. The various plant parts of *A. dealbata* had pronounced changes in allelochemicals throughout the year, as indicated by variations in effects of the various parts on test plants. This was also true of the soil under the *A. dealbata* trees. There were sometimes stimulatory effects on germination or seedling growth, and at other times and in other tests there were no significant effects.

The allelochemicals present in various parts of *A. dealbata* and the soil were identified as follows: (1) in the flowers, *p*-coumaric, *p*-hydroxybenzoic, protocatechuic, gentisic, and caffeic acids plus traces of vanillic, ferulic, and sinapic acids; (2) in the leaves, *p*-coumaric, *p*-hydroxybenzoic, gentisic, ferulic, protocatechuic, vanillic, caffeic, and sinapic acids; (3) in the soil, *p*-hydroxybenzoic and protocatechuic acids, with traces of gentisic, ferulic, and *p*-coumaric acids. It was clearly demonstrated that the flowers liberated volatile compounds that had allelopathic effects, but it was concluded that solution in rainfall was the chief pathway of liberation of allelochemicals that had the most important allelopathic and ecological effects. Reigosa-Roger concluded that there was a very definite allelopathic effect of *A. dealbata* on neighboring plants in the grove or forest.

Foresters have long been aware of the interference to establishment and growth of timber species that occurs when other plant species are present in abundance, as occurs following fire or logging. Tinnin and Kirkpatrick (1985) decided to test the effects that aqueous leachates of several shrubs and a fern have on seedlings of *Pseudotsuga menziezii* (Douglas fir), a very important timber species in Oregon and elsewhere. Leaves of the potentially allelopathic species were collected shortly after leaf fall, thoroughly air-dried, and stored in paper bags. Leachates were made by shaking 1 g of leaf material in 20 mL of distilled water for 30 minutes and filtering out the leaf material. The filtrate was then tested against seed germination and seedling growth of cucumber.

Five species were selected on the basis of preliminary tests: *Arctostaphylos patula, Arbutus menziesii, Castanopsis chrysophylla, Ceanothus velutinus,* and *Umbellularia californica*. Leachates were made this time by soaking 1 g of leaf material in 100 mL of water, filtering and then testing the leachates against pregerminated seeds of Douglas fir. The various leachates reduced growth of Douglas fir seedlings from about 44% of control growth for *U. california* to about 98% of control growth for *A. menziesii.* When leaves were added to an artificial soil mix, all shrubs significantly reduced root growth of Douglas fir and three of them significantly reduced shoot growth. When field soil was used in a similar experiment, only *U. californica* significantly reduced root growth of Douglas fir and none of the shrubs significantly reduced shoot growth. The severity of the effects of filtrates and artificial soil mixes was so great, however, that it was concluded that allelopathy is of potential importance under field conditions in reforestation.

The growing popularity of planting ground cover under trees suggested to Shoup and Whitcomb (1981) that interactions between certain trees and ground covers should be investigated. They selected *Acer saccharinum* (silver maple) and *Populus deltoides* (cottonwood) as the tree species and *Sasa pigmaea* (dwarf bamboo), *Liriope muscari* (liriope, or monkey grass), and *Hedera helix* (English ivy) as the ground cover plants. They used a connecting pot technique in which four root tips extended out of plastic bags in which each tree seedling (or sapling) was planted. In each case, each of the four root tips extended into the drain hole of a separate 3.8-L plastic jug. Ground cover was planted in each of three of the jugs; the fourth jug was left free of plants as a control for tree growth. Additional small containers (jugs) were planted to ground cover plants without tree roots to serve as controls for the ground cover plants.

Roots of cottonwood trees reduced English ivy top and root weight by 44 and 60%, respectively, and liriope top and root weight by 38%. On the other hand, dwarf bamboo top weight was not affected, but root weight was reduced 20% by the presence of cottonwood roots. Roots of silver maple had no effect on top and root weight of English ivy or liriope; however, liriope tuber production increased 28% when the silver maple roots were present. On the other hand, silver maple did reduce dwarf bamboo top and root weight by 0 and 43%, respectively, and number of rhizomes by 50%.

The ground cover had interesting effects on the trees. Cottonwood root development was reduced 32% by English ivy, 19% by liriope, and 24% by dwarf bamboo, whereas silver maple roots were reduced 64, 49, and 0% by English ivy, liriope, and dwarf bamboo, respectively. Unfortunately, the experiments were not designed to differentiate between allelopathic and competitive effects. Certainly, some of the effects (e.g., increase in rhizomes of dwarf bamboo by silver maple) were clearly allelopathic; in fact, it is likely that most of the effects were due to allelochemicals and not to competition. As I have noted before, it is not difficult to set up experiments that eliminate competition, but it is impossible to set up competition experiments that eliminate allelopathy.

Heisey and Delwiche (1983) surveyed 55 northern California plant species for water-soluble and volatile inhibitors of plant growth. The species were mostly woody, but several herbaceous plants were included in the survey. Target species used in the survey were *Bromus mollis* and *Hordeum vulgare.* Fresh shoots or leaves (most tests) were soaked in distilled water for 24 to 25 hours in a ratio of 1:10, 1:25, or 1:50 (g/mL). The plant material was filtered out and the pH of each filtrate was set at 7.0. Test seeds were moistened with the leachate and placed on water-moistened filter paper for tests of germination and root growth. Volatiles were assayed by placing plant material near water-moistened seeds in containers without touching the seeds; the containers were then sealed until germination percentages and radicle lengths were measured.

Sixty-nine percent of the 55 plants significantly inhibited, and none significantly stimulated, radicle growth of barley when tested with 1:25 or 1:50 (g tissue:mL water) leachates, whereas 38% inhibited and 15% stimulated *B. mollis.* Leachates of species in the Compositae and Labiatae, as well as *Aesculus californica, Ailanthus altissima, Brassica nigra, Ceanothus integerrimus, Lupinus arboreus, Nicotiana glauca, Ribes cereum,* and *Scrophularia californica* were the most inhibitory. Only *Artemisia tridentata, Heteromeles arbutifolia, Salvia sonomensis,* and *Trichostema lanceolatum* produced strongly inhibitory volatiles. The last species listed was the most toxic tested, both its leachates and volatiles being very active. *Chamaebatia, Salvia, Ceanothus, Baccharis,* and *Trichostema* leachates were also tested against six other important northern California plants; seed germination and radicle growth were inhibited by all of the potentially allelopathic plants. Several of these species certainly merit further research to determine if they are truly allelopathic.

*Ailanthus altissima,* which was introduced into the United States from Asia in the 1700s, has a remarkable ability to rapidly occupy and dominate various forest habitats. Heisey (1990a,b) followed up on work he and others had done on the allelopathic potential of this species. His papers are an excellent model of an investigation of the allelopathic properties of a species. Aqueous extracts were prepared by leaching 1 g of *Ailanthus* tissue in 100 mL of deionized water for 24 hours at 4 ± 2°C, followed by filtration through filter paper. These solutions were considered to have a concentration of 10 g tissue/L and were diluted to produce other concentrations. Allelochemical activity was quantified with bioassays using *Lepidium sativum* (garden cress).

Heisey found that inhibitory activity was highest in bark (particularly in the roots), intermediate in leaflets, and lowest in wood. Crude extracts of root bark and leaflets corresponding to 34 and 119 mg,

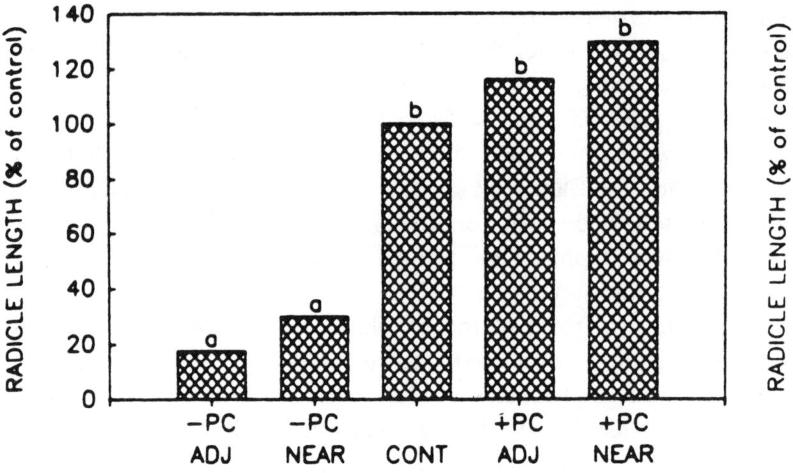


FIGURE 14 Radicle growth of garden cress seeds placed adjacent to (within 1 mm) or near (5 mm from) *Ailanthus* seeds having pericarp present (+PC) or absent (–PC). Control lacked *Ailanthus* seeds. Bars topped with same letter do not differ significantly ($P > 0.05$) in Duncan's multiple-range test. *Adapted from Heisey (1990a), with permission of Bot. Soc. of America.*

respectively, of water-extractable material per liter caused 50% inhibition of cress radicle growth. Tree-of-heaven seeds also contain one or more inhibitors, which are bound within the seed by the pericarp but diffuse out when the pericarp is removed. Under these conditions, the inhibitors even inhibit radicle growth of garden cress seeds placed near, but not touching, the *Ailanthus* seeds (Fig. 14). Since the samaras had been periodically leached by rain throughout the winter and still inhibited cress, the pericarp is apparently very effective in retaining the active compounds.

Plant response to the root bark extract, from lowest to the highest sensitivity, was as follows: velvetleaf < corn < pea < barnyard grass < foxtail < garden cress < redroot pigweed. One or more compounds in the methanol extract of the root bark had powerful herbicidal activity. Postemergence effects were so drastic that even the lowest doses tested caused nearly complete mortality of all species, except velvetleaf, within seven days or less. Even velvetleaf was killed at the higher doses used. The results suggested that the allelochemical(s) from *Ailanthus* may have potential for development as a natural-product herbicide.

Heisey (1990b) reported primarily on his experiments concerning allelopathic effects of *Ailanthus* in soil. He reported that powdered root bark and leaflets were very inhibitory to garden cress in soil in petri dishes, and pieces of root bark mixed with soil at 0.5, 1, or 2 g/pot reduced cress biomass in greenhouse tests. Two grams per pot reduced cress shoot biomass to 5% of the control receiving no *Ailanthus* tissue. *Ailanthus* roots in contact with soil reduced growth of cress nearby and fine roots appeared to release the toxin more rapidly and in greater amount. Both fine and coarse roots inhibited cress growth after six days, but fine roots remained more inhibitory. This pattern persisted for the duration of the test (13 days), with inhibition becoming greater as the days went on. This experiment and all others where plant material was incorporated in soil demonstrated that toxins did get out of plant material and cause growth effects, which definitely demonstrated allelopathy since the same phenomenon happens regularly in nature.

Lawrence et al. (1991) tested water leachates of leaves of *Acer saccharum, Platanus occidentalis, Liriodendron tulipifera, Ailanthus altissima,* and *Andropogon virginicus* and stem leachates of *Ailanthus* against germination and radicle growth of *Lactuca sativa.* Only *A. altissima* leaf and stem leachates were inhibitory to seed germination and seedling growth. Compounds inhibitory to the growth of neighboring plants were found in significant concentrations in the leaves and stems of young *Ailanthus* ramets, and the surrounding soil also contained appreciable concentrations of similarly acting toxins. Lawrence et al. found that neighboring plants had either incorporated active portions of inhibitory compounds or had responded to *Ailanthus* by producing growth-inhibiting compounds. They suspected this because, under greenhouse conditions, individuals of neighboring plant species not previously exposed to *Ailanthus* were more susceptible to the *Ailanthus* toxin than were plants previously exposed. Moreover, seeds produced by unexposed populations were also more susceptible to *Ailanthus* toxins than were seeds produced on previously exposed populations. The researchers inferred, therefore, that since the progeny of the two populations displayed a differential response to *Ailanthus* toxin, the phenotypic difference between the two populations may have a heritable basis. Thus it appears that allelopathic compounds may be directly responsible for detectable changes in the species composition of plant communities or the genetic composition of associated plant populations.

Datta et al. (1985) collected leaves of 11 woody species of plants often cultivated along roadsides near Calcutta, India, to determine what potential effects they might have on herbaceous species in their neighborhoods, including the cultivated species. The species studied were *Anthocephalus cadamba, Bauhinia purpurea, Bombax ceiba, Cestrum nocturnum, Mangifera indica, Mimusops elengi, Polyalthia longifolia, Pterospermum acerifolium, Samanea saman, Swietenia mahagoni,* and *Trema orientalis.* Leachates were made by soaking 20 g of leaves of each species in 100 mL of water. Each leachate was filtered and adjusted to 100 mL with water. Dilutions and tests were run from the stock solutions against germination, root growth, and hypocotyl growth of *Brassica campestris* var. *dichotoma.* In all seasons, leaf leachates of *C. nocturnum* and *S. saman* markedly inhibited both germination and seedling growth of mustard. Maximum inhibitions were encountered with leachates of *A. cadamba* and *P. longifolia* leaves during the summer and monsoon seasons, respectively. The phytotoxicity of *B. purpurea* and *B. ceiba* leachates seemed to be greatest during the winter and least during the monsoon. The inhibitory activity of *T. orientalis* was maximum in winter and minimum in summer, whereas leachates of the additional species did not present any particular trend in inhibition. Apparently, the trees involved could cause rather pronounced effects on herbaceous plants, but those effects changed markedly with seasons.

Goel and Sareen (1986) also investigated the allelopathic effects of three trees on the understory vegetation near New Delhi, India. The species were *Anthocephalus cadamba, Cassia siamea,* and *Thevetia neriifolia,* of which *A. cadamba* was also investigated by Datta et al. (1985). There was a pronounced reduction in herbaceous plant biomass under all test trees in January compared with that in September. Oven dry weights of all species under *Cassia* were significantly less than under *Thevetia* and *Anthocephalus,* and *Malvastrum tricuspidatum* was not present under *Cassia* in either season. *Malvastrum* was therefore selected as the test species to determine the allelopathic potentials of the three species. Effects of leaf litter of the tree species were determined by placing 0.2, 0.4, 0.6, 0.8, 1.0, 2.0, or 4.0 g of litter under filter paper in petri plates, with *M. tricuspidatum* seeds placed in distilled water on top of the filter paper. For soil tests, 2 or 4 g of soil were placed under the filter paper instead of leaf litter. Leaf leachates were made by placing 10 or 20 g of fresh leaves in 100 mL of distilled water for 24 hours, filtering, and making the filtrate up to 100 mL. These filtrates were also tested against seed germination and seedling growth of *M. tricuspidatum* on filter paper in petri dishes. Seed germination of *Malvastrum* was significantly inhibited by the leaf litter of *C. siamea,* but there were no significant differences due to leaf-litter concentration. The leaf litter of the other trees reduced germination, but differences from the control were not significant. Seed germination of *Malvastrum* was significantly reduced only by soil from near the roots of *Thevetia.* Leachates from *C. siamea* inhibited germination of *Malvastrum* seeds but other leachates did not.

The leachates of all the test trees reduced the radicle growth of *Malvastrum,* whereas plumule growth was increased in all cases. Seed germination was significantly reduced by soil collected under *Thevetia,* but soil collected under *Cassia* and *Anthocephalus* had no significant effect. Radicle growth of *Malvastrum* was reduced by soil from under *Thevetia* and *Anthocephalus,* whereas soil from under *Cassia* stimulated radicle growth. The radicle growth of *Malvastrum* was significantly reduced by leaf litter of all tree species, but the reduction was maximum with the litter of *Cassia.* The plumule growth was again increased by leaf litter of all tree species.

In summary, it appeared that all three tree species had allelopathic effects on understory species and that the patterns were different between different species, possibly because of leaf-litter effects.

Askham and Cornelius (1971) made extracts of *Atriplex polycarpa* (desert saltbush) by boiling 64 g of stems, twigs, and leaves in 250 mL of distilled water for 20 minutes, cooling, and filtering through cheesecloth. Three dilutions were made, equivalent to 32, 16, and 8 g of parts per 250 mL of distilled water. These extracts were tested against germination of four plant species: *Agropyron trichophorum* (trigo pubescent wheatgrass), *Medicago littoralis* (harbinger medic), *Ephedra californica* (California ephedra), and desert saltbush.

The extracts of desert saltbush reduced germination of all four test species, although the lowest concentration stimulated the germination of California ephedra seeds. Saponin was definitely present in the vegetative parts of *A. polycarpa,* as shown by the staining method with acetic anhydride and sulfuric acid mixture. The concentration of saponin in the leafy portions of the desert saltbush was 1.15% on a dry-weight basis. Saponin reduced seed germination of desert saltbush and harbinger medic most markedly. It appears that *A. polycarpa* thus has the potential to affect neighboring plants through allelopathic action.

Melkania et al. (1982) investigated the allelopathic potential of

*Artemisia vulgaris* and *Pinus roxburghii* against *Lepidium virginicum* (peppergrass) and *Lolium perenne* (perennial ryegrass). Leachates of the potential allelopathic plants were made by soaking 50 g of either young leaves or old leaves in 1 L of distilled water for 48 hours at 24 ± 2°C, and filtering through filter paper. Filtrates of both young and old leaves of both *A. vulgaris* and *P. roxburghii* were significantly inhibitory to seed germination of both *Lepidium* and *Lolium*.

Soil samples were collected from the top 0.5 cm of soil (litter removed) under *Artemisia* and *Pinus* and aqueous soil extracts were made. These extracts were found to be significantly inhibitory to both *Lepidium* and *Lolium*. The overall tests indicated that inhibitory allelochemicals were released into the soil from both *A. vulgaris* and *P. roxburghii*, by leaching, volatilization, or litter fall, or by some combination thereof, and were able to inhibit or reduce germination of test seeds.

*Azadirachta indica* (also *Melia azadirachta* or *M. indica*, neem tree) is native to the arid regions of India, where it grows 12 to 24 m high (Koul et al., 1990). It is presently grown in many Asian countries and in the tropical areas of the New World. Forty kilograms of fresh fruit yields nearly 24 kg of dry fruit (60%), which in turn gives 11.52 kg of pulp (48%), 1.1 kg of seed coats (4.5%), 6 kg of husk (25%), and 5.5 kg of kernel (23%). The kernel gives about 2.5 kg of neem oil (45%) and 3 kg of neem cake (55%). Neem timber is useful both as fuel and for making furniture. The fuel yield per 50-year-old tree has been reported to be 51 kg in arid regions of Rajasthan, India.

Various products of the neem tree have been shown to inhibit nitrification and thus conserve nitrogen in agriculture. Several other allelopathic actions have been demonstrated for the neem tree, including the inhibition of fungi by some of its products. Nearly 100 novel organic compounds have been isolated from the tree and their structures elucidated. These include nimbin, nimbinin, nimbidin, protolimonoids, and limonoids and tetranortriterpenoids, pentanortriterpenoids, hexanortriterpenoids, and nontriterpenoidal constituents. Among the various chemicals isolated to date, azaderachtin remains the most bioactive component.

Several years ago, a Brazilian shrub, *Baccharis megapotamica*, was found to contain baccharinoids, highly toxic macrocyclic trichothecenes; and another Brazilian shrub, *B. coridifolia*, was later reported to contain roridins A and E as well as smaller quantities of related toxins (Jarvis et al., 1988). After their first discovery, Jarvis and his colleagues postulated that fungal contaminants must have synthesized the baccharinoids, because the soil fungi, *Fusarium* and *Myrothecium,* were the only organisms known to that time to produce the trichothecene antibiotics. Much to their surprise, however, the researchers were able to confirm that the pistillate, Brazilian *B. coridifolia,* synthesized *de novo* a series of the highly phytotoxic trichothecene antibiotics—namely, roridins and related toxins. The staminate *B. coridifolia* cannot synthesize the compounds, but it and the pistillate plant are not sensitive to the toxic effects of the trichothecenes, whereas North American *Baccharis* species are sensitive to these toxins. The macrocyclic trichothecenes found in *B. coridifolia* are the same as those produced by the fungus *Myrothecium,* so Jarvis et al. suggested that the pistillate plant has acquired the toxin-producing genes from that fungus.

As part of a search to discover new naturally occurring bioactive substances, Ohigashi et al. (1989) investigated plants in the rain forest of Cameroon. One of the very interesting trees discovered was *Baillonella toxisperma* (Sapotaceae). The methanolic extracts of both the aerial parts and the roots markedly inhibited germination of cucumber and radish seed. The researchers subsequently found that leaves, stems, and roots of the tree contained plant-growth inhibitors. They selected dried stems to use in the isolation and identification of the toxin(s). By a combination of chromatographic techniques, the first inhibitor was isolated and identified as 3-hydroxy-uridine by $^1$HNMR. This allelochemical strongly inhibited both the hypocotyl and root growth of cucumber. It was also inhibitory against the hypocotyl and root growth of radish seedlings and against root growth of rice. When 3-hydroxy-uridine was sprayed on five weed species and corn in doses of 7.5 or 30 mg, it significantly inhibited growth of all the weeds—*Echinochloa crusgalli, Setaria viridis, Pharbitis purpurea, Abutilon avicennae,* and *Cassia tora*—at either dosage, but neither dosage affected growth of corn. The 30-mg dosage killed *Pharbitis* and *Abutilon.*

Ohigashi et al. were not able to recover 3-hydroxy-uridine from the soil under *B. toxisperma,* but they pointed out that 3-hydroxy-uridine could be readily transformed biologically or chemically in the soil to 3-hydroxy-uracil, which is less water-soluble and almost equally inhibitory to the tested seedlings and could thus aid in the allelopathic activity of *B. toxisperma.*

*Bambusa arundinacea* (bamboo) occurs in many parts of the world, often with many other genera and species of bamboos; it is raised in many districts of Tamilnadu, India, as a commercial crop. Eyini et al. (1989) observed that abscised leaves of the bamboo inhibited growth of weeds around them. They made extracts by placing 10 g of leaves in 100 mL of distilled water in a Waring blender. They concentrated the extract on a water bath and made dilutions of 1:5, 1:10, 1:15, and 1:20 with distilled water. They then tested the extracts against *Arachis hypogaea* (peanut or groundnut) and found that all concentrations decreased the leaf area, plant height, and total chlorophyll and protein content. All their measurements decreased with increases in extract concentrations. Six phenolic acids were identified in the fallen leaves of bamboo: chlorogenic, ferulic, coumaric, protocatechuic, vanillic, and caffeic acids. Since all these phenolic acids have been shown to be allelopathic compounds in other cases, all or some may act as the allelochemics responsible for the demonstrated allelopathic effects of *B. arundinacea.*

In order to obtain more information concerning interference interactions between forest species in the northern boreal forest regions of Europe, and particularly in Russia, Kolesnichenko and Andryushchenco (1979) selected *Betula verrucosa, Larix sibirica,* and *Pinus sylvestris* as donor species for tests of volatile compounds and water-soluble root exudates and *Picea excelsa* (Norway spruce) as the receptor species. The biochemical effects were measured by determining the effects of the donor species on the receptor using $^{32}P$ uptake in $K_2HPO_4$ labeled with $^{32}P$. They found that Siberian larch and Scotch pine had the greatest effects on Norway spruce but that white birch also had a significant effect.

Kolesnichenko and Andryuschenko also studied natural field mixtures of the same species to determine interactions under more typical conditions. They concluded that the results were amazingly similar and that similar kinds of biochemical interactions occur in the field and in the laboratory. It was clear that reforestation with Norway spruce and white birch would likely be more satisfactory than planting the spruce with the two test conifers under the given environmental conditions.

Well-drained sandy ridges of Florida and the southeastern coastal plain support two distinct types of vegetation—the sandhill being dominated by *Pinus palustris* (longleaf pine) and oaks, with a dense herbaceous cover of grasses and forbs, and the Florida scrub community being dominated by a closed canopy of *Pinus clausa* (sand pine), with a dense understory of evergreen oaks and nearly devoid of herbaceous cover (Fischer et al., 1988; Tanrisever et al., 1987). Fischer and Tanrisever and their colleagues postulated that allelochemicals released from plants of the Florida scrub community deter the invasion of fire-prone sandhill grasses. They thus identified numerous constituents of three endemic shrubs in the scrub community—*Calamintha ashei, Ceratiola ericoides,* and *Conradina canescens*—and tested the identified allelochemicals against germination and growth of lettuce and little bluestem (*Schizachyrium scoparium*), a native grass of the sandhill community.

In initial tests, fresh leaves were collected from the shrubs monthly from March through November, and filtrates were made by soaking 1 g of leaf material in 10 mL of water for 24 hours, then filtering and refrigerating at 8°C until tests were made. One to four of the leachates reduced seed germination and/or radicle growth of little bluestem during every month. The fourth leachate was made from leaf litter of *C. ericoides* each month.

Two triterpenes, ursolic acid and erythrodiol—as well as flavanones, catechins, proanthocyanidins, a chalcone, and dihydrochalcones—were subsequently identified from the hexane extract of *C. ericoides.* Additionally, the researchers identified ursolic acid and ceratiolin from leaf washes of this shrub. Ceratiolin had no significant activity but it decomposed spontaneously in water to hydrocinnamic acid, which was phytotoxic. Each of the flavanoids was dissolved in a saturated aqueous solution of an ursolic acid–erythrodiol mixture; most of these solutions inhibited both germination and radicle growth of lettuce, and three of them significantly inhibited radicle growth of little bluestem. Most of the catechins and epicatechins from *Ceratiola* significantly inhibited radicle growth of lettuce, one inhibited germination of little bluestem, and another significantly inhibited radicle growth of little bluestem.

The scientists identified copious amounts of ursolic acid and a series of monoterpenes from extracts of *Conradina canescens.* The major monoterpenes were 1,8-cineole, camphor, borneol, myrtenal, myrtenol, α-terpineol, and carvone. They tested camphor, myrtenal, borneol, and carvone against lettuce and little bluestem; all were found to be very inhibitory to lettuce but less so to little bluestem.

From *Calamintha ashei,* they identified ursolic acid, caryophyllene oxide, and the monoterpenes methofuran, epievodone, and calaminthone. A saturated aqueous solution of epievodone stimulated germination of little bluestem. However, in a saturated aqueous solution of ursolic acid (a natural mild detergent), epievodone strongly inhibited germination and growth of little bluestem. A mixture of epievodone, calaminthone, and caryophyllene oxide totally inhibited germination of little bluestem but had only minor effects on lettuce. It is obvious that pronounced effects could be brought about by the many aromatic compounds produced, particularly from the two mints, *Calamintha* and *Conradina*. The production of the natural surfactant ursolic acid by all three shrubs also appears to be very important in the allelopathic actions of the shrubs because it assists in transporting the relatively insoluble phytotoxic lipids into the soil, where they come in contact with target plants.

It certainly appears that the original hypothesis that certain plants of the Florida scrub community deter the invasion of fire-prone sandhill grasses by allelopathic action was strongly supported. The research is continuing. Tanrisever et al. (1988) and Williamson et al. (1989, 1992) have expanded on some of the studies previously reported concerning the allelochemicals produced by *Calamintha ashei, Ceratiola ericoides,* and *Conradina canescens* and their roles in the allelopathic potential of those shrubs.

Ballester and his colleagues previously reported on allelopathic effects of several heath species, mostly of the genus *Erica* (Rice, 1984). Ballester et al. (1982) discussed their investigations on the allelopathic effects of *Calluna vulgaris, Daboecia cantabrica,* and *Erica vagans.* Tops of these species were collected during the flowering stage at several sites in Gallicia, Spain, and were separated into leaves, stems, and flowers. In the initial test, 20 g of intact leaves, stems, or flowers were placed in 100 mL of distilled water for 24 hours in the dark at room temperature and filtered; 4 mL of the filtrate was then placed on filter paper in a petri plate with 50 red clover seeds. The seeds were then incubated for 72 hours at 24 to 25°C. Virtually all the leachates were significantly inhibitory to root and hypocotyl growth of red clover and also to seed germination. Leaf leachates of *C. vulgaris* and flower leachates of *D. cantabrica* were most inhibitory to germination and to root and hypocotyl growth. Conductivities of the extracts were so low that it was concluded that osmotic pressures played no part in the results.

In a second test, 5 g of intact leaves, stems, or flowers were placed in a petri dish and covered with a sheet of Whatman No. 1 filter paper. Five milliliters of distilled water was added and 50 red clover seeds were planted. All the plant parts significantly inhibited germination of red clover seeds, hypocotyl growth, and root growth.

In a third test, 4 mL of an aqueous leachate (equivalent to 0.8 g fresh weight) was concentrated and spotted on 3 × 57-cm Whatman No. 1 paper strips. These were developed in an isopropanol-ammonia-water mix (10:1:1), and each developed strip was cut into ten 3 × 3-cm pieces, moistened with 1.5 mL of distilled water, and used as seed beds for 15 red clover seeds. Inhibition of aqueous extracts remained after chromatographic separation and two well-defined inhibition zones were noted. The overall results of this project would not allow one to conclude that the three heath species involved do exert allelopathic action under field conditions, but they do suggest strongly that the very toxic, water-soluble compounds present play significant biological roles in nature.

Slepykh (1988) used a unique microbiological method to study the influence of air temperature and humidity on allelopathic dynamics in *Carpinus betulus* (white beech), *Fraxinus excelsior* (ash), and *Quercus robur* (oak) stands in the Beshtaugorskii Forest located in the Caucasus Mineral Springs District of Russia. The fundamental thesis of this study is that volatile, biologically active substances are released into the environment by terrestrial vegetation. The total amount of terpenes released by the earth's vegetation amounts to 175 million tons annually.

Slepykh's method consisted of placing petri plates containing 6% salt meat infusion agar and an average of 235 colonies per plate of *Staphylococcus aureus* 209 p on 1.3-m-high stands beneath the canopy of the stand to be investigated. Exposure took place between noon and 1 P.M. and exposure time was 10 minutes. To avoid the action of UV light and contamination, the open dishes were covered with sterile gauze. After exposure, the lids were replaced and sealed, and after 48 hours at 37°C, the number of staphylococcal colonies was counted.

The oak community began to exhibit pronounced phytoncidal activity at an air temperature above 16°C. The rate at which phytoncides were released diminished as air temperature rose, indicating the stressing effect of high temperatures on the plants. There was no extreme air temperature at which the phytoncidal activity of the white beech or oak communities dropped to zero. In the ash community, phytoncidal activity was maximal between 62 and 96%, but it continued at reduced rates below 62% to zero at 37%. The humidity levels at which white beech and oak communities exhibited phytoncidal activity covered specific individual ranges, but the parabolic form of the regression line remained. Thus very definite allelopathic effects were demonstrated in these three forest communities by a methodology that has not been utilized elsewhere and should definitely be tested in the same or a modified form.

Kuiters (1990) wrote an excellent review concerning the possible roles of phenolics from decomposing forest litter in plant-soil interactions. He pointed out initially that phenolics in litter are at least slightly water-soluble, are released by rainwater, and can be detected in small amounts in throughfall and stemflow. He emphasized, however, that their main source is their release from decomposing leaf litter deposited on the soil surface and from decomposing fine roots. Insofar as litter sources are involved, he stated that in European forests the highest amounts of phenolics are released from broad-leaved species, particularly from *Carpinus betulus, Corylus avellana,* and *Betula pendula.* Certainly, a careful review of this paper would be valuable to all persons concerned with ecology, allelopathy, forestry, or biological control of pests.

Suresh and Vinaya Rai (1988) reported that the biomass and its floristic diversity of ground cover at the Forestry Research Station, Mettupalayam, India, were considerably reduced under the trees as compared with open areas. Of the three tree species investigated, the maximum reduction of understory dry weight occurred under *Eucalyptus tereticornis* (87.2%), despite the fact that on the average more incident light (60.6%) was available under that tree. *Casuarina equisetifolia,* which had less light (53.5%), had more understory cover, with a reduction of 73.9%. *Leucaena leucocephala* was intermediate in understory biomass, but it had the smallest number of understory species (5), whereas *E. tereticornis* harbored 12 species and *C. equisetifolia* had 18 species. The chief species under *L. leucocephala* was that species itself. Light, as discussed above, did not appear to be a limiting factor in understory growth. Moreover, soil moisture was always as high or higher under the trees than between them and evaporation was always higher between the trees. N, P, and K concentrations were always higher under the trees than between them. It was concluded, therefore, that the paucity of understory vegetation was not due to competition for natural resources like light, water, and nutrients but occurred as a result of allelopathic effects caused by secondary metabolites leached from the tree leaves. This seems like a logical assumption, but the evidence was rather indirect and was based in part on published reports of others.

Establishment of natural *Abies concolor* (white fir) seedlings in fields with a high cover of *Ceanothus velutinus* is frequently spotty, and many established seedlings appear unhealthy for several years. Conard (1985) found no effects of *C. velutinus* on seed germination of white fir in several tests, but the presence of *C. velutinus* in pots with the white fir seedlings did affect survival after seven weeks, with the number of surviving seedlings being significantly lower than the number surviving in the same soil without *Ceanothus*. In a chromatography plate bioassay, radicle growth of germinated *Abies* seeds averaged $126.8 \pm 3.6$ mm for the control and $71.2 \pm 1.2$ mm for the control with filter paper strips. Foliage extracts of *C. velutinus* reduced radicle length of white fir seedlings to 45.4 mm versus 55.6 mm in the control. A foam-pad bioassay gave even more striking results than had the previous assay. It was concluded that *C. velutinus* may be expected to adversely affect the natural regeneration of white fir if similar effects occur in the field.

Lodhi and Johnson (1989) established a research area on the Maryville College campus in St. Louis County (Missouri) in a mixed bottomland forest in which the important tree species were *Celtis occidentalis* (hackberry), *Platanus occidentalis* (sycamore), *Quercus alba* (white oak), *Q. rubra* (red oak), and *Ulmus americana* (American elm), plus several minor species. Important species in the understory were *Eupatorium rugosum* (snakeroot), *Symphoricarpos orbiculatus* (coralberry), *Elymus virginicus* (wild rye), *Parthenocissus quinquefolia* (Virginia creeper), *Bromus tectorum* (downy brome), and several minor species. Ten randomly located quadrats were clipped in late May and late September under each of four tree species: American elm, sycamore, hackberry, and white oak. The May sampling was to estimate the exploitation of nutrient resources at the start of the growing season; the estimate in late September was considered to be the biomass accumulation of all species, which would reflect the utilization and incorporation of available nutrients and moisture. Soil moisture, pH, and nutrient analyses were made to see if the differences in the growth of understory vegetation under the four chief tree species corresponded to the differences in soil chemical and physical properties.

It was found that soil pH, Ca, Mg, Mn, Cu, $NH_4$ nitrogen, total N, and soil moisture were significantly different under all the dominant species; the pattern of soil-moisture levels under different tree species was not consistent from one sampling period to the next. The aboveground biomass of understory vegetation varied significantly under different tree species and was not correlated with variation in any of the soil properties. Moreover, maximum understory biomass gain (340%) from May to September occurred under oak trees, where soil moisture and most nutrients were the lowest. On the other hand, sycamore and hackberry had continuous release of allelochemics and had the smallest understory biomass gain (103%) during the growing season, even though soil under both species had more moisture and nutrients than the soils under white oak. It was therefore concluded that organic substances released in the immediate environment of dominant trees and their litter influenced soils and associated herb growth more than any other factor considered.

Matveev (1980) investigated the allelopathic potentials of five tree species—*Acer platanoides* (Norway maple), *Cotinus coggigria* (smoke tree), *Pinus sylvestris, Quercus robur,* and *Robinia pseudoacacia*—on about 30 understory herbaceous plants in a Russian steppe forest and on radish. Artificial-rain leaf leachates of black locust markedly inhibited seed germination of five of eleven herbaceous species, stem growth of three species, and root growth of five species. A similar type of leaf leachate of smoke tree inhibited seed germination of seven species, stem elongation of six species, and root growth of five species. In a field experiment with natural rain, English oak inhibited seed germination of ten of fifteen species and stem growth of eight species, whereas Scotch pine inhibited seed germination of nine of fifteen species and stem growth of nine species.

Under field conditions, volatile allelochemicals from Scotch pine inhibited seed germination of one of sixteen herbaceous species, stem growth of three species, and root growth of two species. On the other hand, the volatiles stimulated germination, stem growth, and root growth of several species. Volatiles from smoke tree inhibited seed germination, stem growth, and root growth in some species, yet stimulated these processes in other cases. Of course, as one would expect, there were no significant effects of the leaf leachates or volatiles on some species.

Root exudates and rhizosphere secretions (collected together in the field) of all tree species were inhibitory to most of the herbaceous species against which they were tested. The overall results demonstrated very strongly that all five tree species exhibited strong allelopathic effects against many of the herbaceous understory plants. Moreover, excellent techniques were developed, particularly for field conditions, which should be more widely used in allelopathic research.

Near Peshawar, Pakistan, relatively bare areas are frequent under and around *Datura innoxia* thickets, whereas several herbaceous species grow well under adjacent shrubs that cast a shade equal to that of *D. innoxia.* Hussain et al. (1979) grew *Datura innoxia* in mixed cultures with *Capsicum annuum, Lycopersicon esculentum, Pennisetum americanum,* and *Setaria italica* in alternate rows in field plots. Monocultures of each species were also grown in the same field. Five plants were randomly selected after eight weeks' growth from each type of stand, dried at 60°C for 72 hours, and weighed. *Datura* reduced height and dry weight of all test species in mixed cultures by more than 50%. *Datura* gained both in height and dry weight in mixed cultures with *Capsicum* and *Lycopersicon* but was retarded by *Pennisetum* and *Setaria.* This indicated that there were interference effects in several directions.

*Datura* was subsequently grown in combination with the same four plant species, using either root-mixed or root-separated treatments. The roots were separated by polyethylene partitions in the root-separated treatments. Competition for moisture and nutrients was avoided in all cases by frequent watering with Hoagland's nutrient solution. The dry weights of all test species decreased in the root-mixing condition, whereas the heights remained unaffected. On the other hand, *Datura* gained in dry weights in mixed cultures with all test species except with *Pennisetum.* This experiment certainly suggested that phytotoxins were being released by the roots of *Datura,* but other experiments were performed.

Extracts of all parts of *Datura* were made and tested against *Brassica campestris,* lettuce, *Pennisetum americanum,* and *Setaria italica.* Germination of most of the species was significantly inhibited by all parts of *Datura;* the radicle growth of all test species was also significantly inhibited. Root exudates, artificial-rain leachates of the top, and volatiles from *Datura* were also found to inhibit seed germination and seedling growth of test species. Moreover, soil collected under and around *Datura* was inhibitory to growth of the test plants. It was thus inferred that the bare areas under *D. innoxia* were due primarily to allelopathy.

*Delonix regia* was introduced to Taiwan several centuries ago and is now planted as an ornamental tree throughout southern Taiwan (Chou and Leu, 1992). There is a relatively thick mat of fallen flowers, leaves, seed pods, and twigs under the canopy, but few understory plants. Sampling of the understory plants under *D. regia* and the plants in a surrounding grassy control area indicated that 13 species of herbaceous plants abundant in the control were either absent or less abundant under the *Delonix* trees. Overall, there were 39 species of plants in the control and 29 species of understory plants.

Aqueous extracts of flowers, leaves, and twigs of *D. regia* were very inhibitory to the radicle growth of alfalfa, lettuce, and Chinese cabbage; measurements of the osmotic concentrations of the extracts demonstrated that none to very little of the inhibition could have been due to osmotic pressures of the extracts. Extracts of soil in the 0- to 5-cm level and of the litter from the soil surface did not affect radicle growth of lettuce, but extracts of soil collected from the 5- to 10-cm depth under *Delonix* did significantly inhibit such growth. A water extract of flowers of *D. regia* inhibited growth of two understory species, *Isachne nipponensis* and *Centella asiatica,* by more than 70%. The compounds responsible for the toxic action of the leaves, twigs, and flowers were identified as chlorogenic, 4-hydroxybenzoic, 3,4-dihydroxybenzoic, gallic, 3,4-dihydroxycinnamic, 3,5-dinitrobenzoic, and L-azetidine-2-carboxylic acids and 3,4-dihydroxybenzaldehyde. All were found to inhibit radicle growth of lettuce. It was concluded that the exclusion of many understory plants under *D. regia* was due largely to the allelopathic effects of the fallen flowers, leaves, and twigs of that tree.

Ahmed et al. (1987) decided to see if allelopathy plays a role in host specificity of *Cuscuta reflexa.* They collected shoots of *C. reflexa* from *Duranta plumeiri* hedges and shoots of *Ligustrum lucidum, D. plumeiri,* and *Zizyphus nummularia,* all confirmed hosts of *Cuscuta.* The researchers also collected shoots of *Euphorbia granulata* and *E. helioscopia;* leaves of *Datura innoxia;* shoots of *Artemisia scoparia;* berries of *Melia azedarach;* shoots of *Malvastrum coromandelianum;* and leaves, flower buds, and bark of *Eucalyptus terreticornis.*

Extracts were made by soaking 5 g of the parts collected in 100 mL of distilled water at 25°C and filtering out the plant parts. *C. reflexa* apical stem segments (60 mm long) were placed individually in vials containing 14 mL of an extract and incubated at 26°C in diffused light for 96 hours. Controls consisted of *C. reflexa* apical stem segments in distilled water. Extracts from shoots of *L. lucidum, D. plumeiri,* and *Z. nummularia,* berries of *M. azedarach,* and bark of *E. terreticornis* inhibited growth of *Cuscuta* segments, whereas *E. granulata, E. heliscopia, D. innoxia,* and *M. coromandelianum* leaves and flowers of *Eucalyptus* either stimulated stem growth of *Cuscuta* or had no effect. It is noteworthy that confirmed hosts retarded growth of *Cuscuta,* whereas nonhost species accelerated it. The original hypothesis was thus placed strongly in question.

Campbell et al. (1989) tested 11 common moist-tropical-forest species from the western Brazilian Amazon for allelopathy by determining the effects on seed germination and radicle growth of lettuce of soil collected underneath each species. Nine of the species showed some allelopathic action and five were strongly allelopathic. *Virola* sp. (Myristicaceae), *Rinorea racemosa* (Violaceae), *Protium* sp. (Burseraceae), *Mabea* sp. (Euphorbiaceae), and *Duroia hirsuta* (Rubiaceae) were very allelopathic, and *Micrandra* sp. (Euphorbiaceae), *Theobroma subincana* (Sterculiaceae), *Inga* sp. (Mimosaceae), and *Eschweilera coriacea* (Lecythidaceae) were mildly to intermediate in action. The fact that all tests were based on activity of soil under the trees definitely indicated that sufficient allelochemicals got out of the trees and into the soil to exert biological action against lettuce seeds and seedlings. This definitely demonstrated allelopathic activity on the part of nine of eleven tree species in the moist tropical forest under study, a surprisingly high percentage.

Wollenweber et al. (1992) reported that *Empetrum nigrum* leaves produce a considerable amount of lipophilic material that consists mostly of phenolic compounds. They identified two dioxygenated chalcones, three dioxygenated dihydrochalcones, and two dihydrophenanthrene derivatives. They also found that these compounds accumulated within the capitate glandular cells of trichomes on the upper and lower epidermis of the downward-rolled leaves. They suggested that their antimicrobial activity against viruses, bacteria, and pathogenic fungi is a particularly important function in *E. nigrum,* where the water vapor–saturated leaf cavity would provide most favorable growth conditions for microorganisms. Unfortunately, no evidence for this role was presented in their paper.

*Empetrum hermaphroditum* is an evergreen shrub forming extensive clones in postfire successions on acid mor humus soils in northern Scandinavia (Nilsson and Zackrisson, 1992). Regeneration of Scots pine (*Pinus sylvestris*) from seed trees and from planting operations on clear-cut areas dominated by *E. hermaphroditum* have often been unsuccessful. Green leaves of living *Empetrum* were collected in late autumn at Rovagern, Sweden, air dried and stored at –18°C until use. A stock 5% solution was made by soaking 50 g of dry leaves in 1 L of distilled water while slowly stirring for 48 hours. This leachate was filtered to give a clear, yellowish-brown solution with an osmotic potential of 43 mmol/kg and a pH of 4.2. Half-sib Scots pine seeds with a viability of 92% were used in most experiments.

Two milliliters of the 5% leaf leachate was placed in each 5-cm petri dish lined with a sheet of filter paper. Seven pregerminated Scots pine seeds (radicle of 1 mm) were exposed to the extract or to distilled water (control) for 7, 19, 25, 48, 72, or 240 hours, then rinsed in distilled water and transferred to petri dishes lined with filter paper and containing 2 mL of distilled water. These dishes were maintained in a growth chamber under 20-hour illumination daily at 20°C. The length of the radicle and, later, total seedling length (root plus hypocotyl) were measured daily. Growth of seedlings exposed to the 5% aqueous leachate of *Empetrum* was significantly inhibited, the amount of inhibition increasing with time of exposure. Although exposed seedlings grew, inhibition after exposure was persistent and growth continued to decline. When several leachate dilutions were used, mean dry weight of seedlings and elongation of roots decreased as extract concentration increased. In the 5% treatment, mortality reached 10% at day 22, 40% at day 31, and 100% by day 46. Seedlings exposed to *Empetrum* extract had discolored primary roots and strongly reduced side roots.

When 3 g of air-dried green leaves of *Empetrum* were added to each pot containing pregerminated Scots pine seed (3 g peat in control pots), seedlings grown with *Empetrum* had significantly higher mortality (58%) than did the control (13%). Surviving seedlings in the *Empetrum*-treatment had dry weights of shoots and roots and lengths of main roots about half those in the control. A much higher shoot/root ratio occurred in the *Empetrum*-treated seedlings. Moreover, rooting of pregerminated Scots pine in pots with *Empetrum* was inversely related to leaf weights in pots even after one day of treatment, and in the heaviest treatment most radicles were stunted and died before rooting occurred. Both green and brown *Empetrum* leaves strongly affected rooting of Scots pine seedlings, but green leaves were by far the most inhibitory.

Seed germination of aspen (*Populus tremula*) on different layers (A00, A1, A2) of *Empetrum* mor humus generally increased with depth in the soil profile. In field trials with a heavy growth of *E. hermaphroditum,* significantly more Scots pine seeds germinated and more seedlings became established in plots to which activated carbon was added than in untreated plots.

The evidence was very clear from Nilsson and Zackrisson's research that a very short exposure of germinating seeds to leaf extracts of *E. hermaphroditum* had a persistent inhibitory effect on radicle growth and seedling survival. Tests (not discussed here) demonstrated also that osmotic pressure and pH were not responsible for the action of the leachates. The researchers' evidence was clear that allelochemicals entered the soil and caused the soil to be as allelopathic as were the extracts. Moreover, if activated carbon was added to field plots where normal competition persisted, it removed the responsible allelochemicals and allowed normal germination and growth of Scots pine seedlings. These scientists did a beautiful job of demonstrating the allelopathic action of *E. hermaphroditum.*

Phosphorus is the nutrient-limiting phytoplankton growth in many freshwater habitats, and there is considerable evidence that sizeable amounts of inorganic phosphorus can be regenerated by the action of extracellular phosphatases. Serrano and Boon (1991) decided to investigate the effect of polyphenols extracted from *Eucalyptus camaldulensis* (river red gum) on phosphorus regeneration in some Australian aquatic habitats, specifically on activity of alkaline phosphatase. They found that aqueous extracts of wood and leachates of brown and green leaves of river red gum inhibited purified alkaline phosphatase activity from an animal source and from *Escherichia coli* and alkaline phosphatase activity from various Australian rivers and other freshwater bodies. The threshold for inhibition of purified enzymes and most native enzymes ranged from less than 2 to 10 mg/L of polyphenols, which is within the range of polyphenol concentrations commonly found in freshwater environments. Enzyme activity was increased after the extracts were treated with polyvinylpolypyrrolidone to remove polyphenols, thus indicating that the active compounds in the water extracts of plant parts of river red gum were the polyphenols.

Nishimura et al. (1982) isolated and identified two *p*-menthane-3, 8-diols from a volatile fraction of the acetone extract of fresh leaves of *Eucalyptus citriodora,* using IR, FI-MS, and $^1$HNMR. They tested a 200-ppm concentration of each against germination of 100 Wayahead lettuce seeds and found that the *cis* form reduced germination to 4%, whereas the *trans* form reduced germination to 73%. The controls usually germinated 100%.

*Eucalyptus* spp. have been planted widely in India, with government cooperation, for pulpwood, firewood, shade trees, ornamentals, and agroforestry (Kohli, 1987). Numerous complaints have been made against *Eucalyptus,* particularly concerning the lack of herbaceous growth under and near the trees. Kohli investigated the cause of the deficiency of herbaceous understory in relation to *E. globulus.* He prepared leachates of the leaves by soaking 100 g of fresh leaves in 200 mL of distilled water and filtering out the leaf material. Tests of the leachate against crop species indicated that it completely prevented seed germination of *Cicer arietinum* cv. C-235 and cv. PG-3, *Lens esculentum* cv. LG12, *Vigna unguiculata* cv. F0S1, and *Sorghum vulgare* and limited seed germination of *Vigna mungo* to 15% of the control, of *V. aureus* to 72% of the control, of *Avena sativa* to 28%, of *Vigna unguiculata* cv. HFC 42/1 to 75%, of *Trifolium alexandrinum* to 30%, and of *Medicago sativa* to 50%. Germination of *Cyamopsis tetragonaloba* and *Melilotus alba* was not affected. It was concluded that *E. globulus* produces water-soluble allelochemicals that can obviously exert negative effects on the regeneration of herbaceous vegetation.

May and Ash (1990) developed what they considered to be ecologically appropriate techniques for allelochemical extraction from trees and tested the leachates or extracts from different organs of several species of *Eucalyptus* against seeds of *Lolium perenne* and growth of *Lemna minor.* Some leachates or extracts were also tested on seedling growth of *Acacia saligna.* The tests of allelopathic activity or potential involved *E. globulus, E. maculata, E. macroryncha, E. rossii,* and *E. rubida.* Extraction techniques mimicked typical daily rainfall rates on quantities of foliage, leaf litter, and bark litter typically encountered in forests near Canberra, Australia. Root leachates were obtained hydroponically, stemflow was obtained following rainfall, soils were leached with water, and volatiles were studied in enclosed chambers as usual. Whole leaf litter, shed bark, and stemflow yielded suppressive leachates. Evaporative concentration of leachates in soils was demonstrated, and it increased the inhibitory effects. Decay was shown to reduce the allelopathic effects of leaf and bark litter leachates, but inhibitory chemicals still remained after five months. May and Ash concluded that allelopathy is likely to be a cause of understory suppression by various species of *Eucalyptus,* especially in drier areas.

Despite the statement by May and Ash (1990) that no one else had tested the allelopathic effects of throughfall and stemflow on any species, numerous persons and groups have done so. Previously in this chapter, I discussed a much more thorough project by Reigosa-Roger (1987) on allelopathic effects of throughfall and stemflow effects in *Acacia dealbata.* Moreover, Molina et al. (1991) investigated the allelopathic effects of natural leachates in the throughfall, stemflow, and soil percolates of *Eucalyptus globulus* on the seed germination and seedling growth of lettuce in Galicia, Spain. The site investigated was a 23-year-old *E. globulus* plantation near Santiago de Compostela. The density of the trees is 1289/ha, the average height is 28 m, and the average D.B.H. is 22.6 cm. The average annual precipitation is 1288 mm, of which only 137 mm fall during the three summer months. Leachates were collected daily during rainy spells in the vegetative period from February to July.

Foliar leachates were collected in 10 randomly placed PVC pluviometers 1 m deep and 20 cm in diameter, with a layer of glass wool at the bottom to prevent splashing and to retain organic debris that might otherwise contaminate the leachate. Stemflow was taken from 13 randomly selected trees by means of runnels encircling the tree trunk at least three times. Soil percolates were collected by means of 17 flat PVC lysimeters 22.5 cm wide, 50 cm long, and 15 cm deep, which were pushed horizontally into the soil without damage to its natural structure. All the leachates were collected over 24-hour periods in opaque plastic bottles, which, like the collectors, were washed with distilled water before each collection period. After the amount of each leachate was measured at the site, samples were rapidly returned to the laboratory for assay.

Leachates from decaying litter were obtained by collecting all the eucalyptus material that fell in one week onto randomly located 1-m squares. Thus all the litter came from natural abscission. After thorough mixing, thirty-five 30-g samples were placed in nylon litter bags of 2-mm mesh. On April 25, seven bags were placed at each of five randomly chosen sites in the area under investigation. One bag was removed from each site at 1, 15, 30, 60, and 90 days. Fifteen soil samples of the top 5 cm were taken at random using an auger 5 cm in diameter.

Bioassays of all leachates consisted of sowing 50 Great Lakes lettuce seeds on filter paper in petri plates and wetting them with 7.5 mL of the leachate being assayed. Controls were wetted with rainwater collected in gauges placed outside the plantation. The only date on which the biological activity of foliar leachates had a significant effect was May 30, after six days without rain. At that time, radicle growth of lettuce was significantly reduced. Germination was significantly depressed by stemflow leachates collected on March 14, and radicle growth was significantly reduced by samples collected on July 9. Again, these dates represented the ends of the longest exudate-accumulation periods. Soil percolates significantly affected radicle growth on April 18 and July 7 but did not affect germination. Topsoil did not affect germination but severely reduced radicle growth (up to 70% reduction) during many sampling periods. Leachates of decomposing litter were generally inhibitory to both germination and radicle growth of lettuce.

Molina et al. concluded that the most important mechanism by which *E. globulus* toxins are released into the soil is by decomposition of fallen litter and that litter leachates are more toxic than soil or soil percolates. Moreover, the researchers concluded that the toxic allelochemicals released by *E. globulus* may influence the composition and structure of the understory vegetation. It should be pointed out that Molina et al. did not consider the volatile allelochemicals that have been shown by many scientists to be very important in *Eucalyptus* spp.

Because of the numerous reports of allelopathy among various species of *Eucalyptus,* Ahmed et al. (1984) decided to investigate the allelopathic potential of *E. tereticornis,* which has been introduced into Pakistan and widely planted there. Fallen leaves, flower buds, and shed bark were collected from mature trees and dried. The samples were ground individually and 5 g of each was soaked individually in 100 mL of double distilled water and filtered. The filtrates were used in assays against seed germination and seedling growth of two cultivars of *Sorghum bicolor* (or *S. vulgare*), Dale and Wiay, *Phaseolus mungo, Brassica chinensis, B. campestris, Sisymbrium irio, Nigella sativa, Raphanus sativus,* and *Setaria italica* in filter-paper bioassays.

Aqueous extracts of the leaves of *E. tereticornis* significantly reduced germination of three of the nine test plants used and radicle growth of all nine. Aqueous extracts of the flower buds reduced germination of six species and radicle growth of all nine species; aqueous extracts of the bark did not significantly affect germination of any test species but significantly reduced radicle growth of all species. It is obvious, of course, that although allelopathy was not demonstrated in these experiments, a strong allelopathic potential *was* demonstrated.

Paul and Schütt (1987) observed damage to understory vegetation under diseased *Fagus sylvatica* (beech) trees in Germany and decided to test leachates of diseased and healthy beech leaves for their effects on seven understory plant species. The target species selected were *Agrostis tenuis, Digitalis purpurea, Mycelis muralis, Scrophularia nodosa, Verbascum densiflorum, Deschampsia flexuosa,* and *Stachys sylvatica.* In order to obtain leachates similar to those resulting from rainfall from leaves, the researchers placed the leaves in distilled water for 16 hours at 18 to 20°C, then filtered the material and used the filtrate in tests of germination and growth of the seven species listed above. Of course, some leachates were obtained from healthy leaves and some from diseased leaves. Seeds were disinfected with NaOCl or 30% $H_2O_2$ before being placed on an autoclaved agar medium with 5 mL of sterile filtrate representing 10, 20, 40, or 50%, depending on the ratios of leaf weight to weight of the water used in leaching the leaves. The researchers found that the leachates of beech leaves affected the germination and seedling development of the seven wild species under test; they also found that the leachates of diseased leaves inhibited germination and growth much more than did leachates of healthy leaves. Moreover, as is commonly true, radicle growth was generally more sensitive to toxins in the leachates than was seed germination or shoot growth.

Nicolai (1988) investigated the factors influencing decomposition of *Fagus sylvatica* (beech) in German forests dominated by that species. He found that the phenolic content of freshly fallen leaves of beech growing on nutrient-poor soils was higher than that of similar leaves found on nutrient-rich soils. This agrees, of course, with the findings of many other scientists (Rice, 1984) that stress factors of many kinds increase concentrations of phenolics and other allelochemicals in plants. Nicolai found that the phenolic content of beech leaves declined rather rapidly during the first winter on the ground. Nevertheless, after one year the phenolic content of beech leaves from trees growing on nutrient-poor soils was still twice as high as in similar leaves on nutrient-rich soils. He found also that the major decomposer was an isopod, *Oniscus asellus,* which preferred leaves that were low in phenolic content and high in mineral content.

Weissen and van Praag (1991) investigated the effects on root growth of beech, oak, and spruce of the holorganic moder humus layer under *Picea abies* (spruce) and *Fagus sylvatica* (beech) forests in Belgium. The mean litter thickness (holorganic layer) under beech is about 6 cm, which represents a dry weight of about 75 t/ha$^{-1}$. Under spruce, the holorganic layer is about 7 cm thick and contains 80 to 85 t/ha$^{-1}$ of dry matter. Field observations on rooting were combined with laboratory experiments to determine effects of the moder humus layer on root growth of seedlings. Field observations demonstrated root-growth inhibition of the target species by the moder humus material of the $Of_2$- and Oh-layers under beech and spruce. Growth-chamber experiments with spruce seedlings indicated that root growth on a natural $Of_2$-substrate was 10 times lower than on synthetic resin-quartz mixtures. Added protocatechuic acid inhibited root growth only in sterilized substrate. Leachates of $Of_2$-material inhibited root growth of spruce in a mineral substrate devoid of an absorption complex (pure quartz sand) but not in a resin-quartz mixture. It was concluded that the inhibiting substances were probably ionized molecules that can withstand biodegradation.

The project discussed above was continued by van Praag et al. (1991). Here the soil solutions were extracted from (1) soil of a beech stand about 120 years old, (2) brown acid soil with moder humus under a 90-year-old beech stand in the eastern Ardennes, and (3) brown acid soil with dysmoder humus under a 35-year-old spruce plantation at La Robinette in the eastern Ardennes. Soil-solution extractions were obtained by filter centrifugation of fresh material previously passed through a four-mesh sieve. Spruce seeds (Cedrogne 013A) were germinated until the white radicle appeared, and 10 such seed were then distributed in quartz sand (200 g in 14-cm-diameter petri dishes). The sand was moistened with 10 mL of either a soil solution adjusted to pH 4.0 or a control nutrient solution brought to a pH of 4.0. After 10 days of growth in the dark at 20°C, radicle length was measured. In another test, beech nuts of local origin from the Ardennes were kept up to five years at –10°C or lower and at 9% moisture content. After a three-month pregermination period at 4°C and 30% moisture content, the seeds started to germinate. They were then buried under fine sand in 1-L beakers, five in each beaker, moistened with 45 mL of either soil solution or the same control solution as in the previous experiment. Each beaker was covered with aluminum foil and kept for three weeks in an incubation chamber at 20°C. The length of the main root of each seedling was then measured.

Root growth of spruce was significantly reduced by the soil solutions from one-half dilution to concentration × 2; beech roots were inhibited considerably less. Numerous kinds of chemical tests were run on the various soil solutions, along with GC, GC-HPLC, HPLC, and MS-HPLC. Molecular gel filtration showed that the molecular weights of the inhibitory molecules varied over a wide range, from less than 100 to over 40,000 daltons. There were only negligible concentrations of simple aliphatic ($C_1$-$C_5$) and aromatic acids in the free state. Uronic acids represented only a small percentage of the carboxylic acids and had no inhibitory effects on root growth. It was concluded, by analogy with the results of others, that polycarboxylic acids in the soil solutions from the holorganic layers ($Of_2$-Oh) were the main cause of root-growth inhibition. A more general conclusion is that spruce and beech trees throughout Belgium exert allelopathic effects on their own reproduction through allelochemicals in the holorganic humus layers developed under them.

*Juglans nigra* (black walnut) produces a phytotoxic compound called juglone (5-hydroxy-1,4-naphthoquinone) that can cause plant mortality and effect changes in species composition, rates of succession, and growth rates (Ponder and Tadros, 1985). Mixed plantings of certain nitrogen-fixing plants, such as alder (*Alnus*) and Russian olive (*Eleagnus*), can substantially increase walnut growth. Little is known, however, of the long-term effects of black walnut, through its production and release of juglone, on the other species. Ponder and Tadros conducted an experiment on an upland site in southern Illinois in the Kaskaskia Experimental Forest. In the spring of 1969, nursery-grown autumn olive (*E. umbellata*) and European black alder (*A. glutinosa*) were interplanted alternately with black walnut seedlings in 29.6 × 38.4-m plots. Rows were arranged 3.7 m apart; within rows in pure walnut plots, walnut was spaced 4.9 m apart, and within rows in interplanted plots all seedlings were alternated 2.4 m apart. By the fall of 1982, height of black walnut seedlings ranged from 1.5 m in pure walnut plots to 5.5 m in mixed plots. Nitrogen-fixing seedlings averaged 5 m for autumn olive and 7 m for black alder. Soil samples were taken in mid-November of 1982 near five black walnut trees in each plot of walnut only, European alder and walnut, and autumn olive and walnut. Depth of sampling included 0 to 8, 8 to 16, 16 to 30, and 52 to 61 cm at 0.9 m from the trees.

Juglone concentrations declined with soil depth and distance from the walnut trees. Concentrations in autumn olive–walnut plots were statistically significantly lower than concentrations in European black alder–walnut, or pure walnut plots. The concentration of juglone in soil in European black alder–walnut plots was deemed sufficient to cause the onset of black alder mortality. Unfortunately, I have seen no later report so I do not know subsequent mortality rates.

*Kalmia angustifolia* is a native ericaceous shrub in Newfoundland; it is the most abundant forest undergrowth in the region, particularly in the more open stands of *Picea mariana* (black spruce) (Mallik, 1987). Plant and soil samples were collected from two areas on the Avalon Peninsula of Newfoundland, from a *Kalmia* stand in a forest opening and from an open heathland site dominated by *Kalmia*. In the laboratory, leachates were made by putting 10 g of leaves, litter, or roots in 500 mL of water. The leachates were shaken slowly for 2 hours, after which each was filtered through cheesecloth to remove the visible organic matter or soil.

All the leachates were tested against germination of black spruce seeds and growth of the seedlings at 20 ± 2°C. The leachates caused no statistically significant differences in seed germination of black spruce, but all of those from the forest opening significantly reduced root and shoot growth of black spruce seedlings. Only the soil leachate from the open-heath site significantly reduced root and shoot growth of the black spruce seedlings; the leaf and litter leachates had no significant effects. It is significant also that roots of black spruce seedlings treated with litter or soil leachate were not normal, lacking a root cap and root hairs. A stairstep sand culture experiment in the greenhouse indicated that water leachate passing through both *Kalmia* canopy and soil significantly reduced early growth of black spruce seedlings. Reductions of 23, 44, and 47% in height growth were observed in the first, second, and third month, respectively, after planting. The results definitely demonstrated that allelochemicals got out of the *Kalmia* and into the soil and brought about significant changes in black spruce seedlings. Allelopathy was therefore definitely demonstrated.

Mangrove detritus is composed chiefly of leaves and twigs that fall from trees onto the forest floor and into the water, where the material is degraded by bacteria and fungi. González-Farias and Mee (1988) investigated the effect of humuslike substances from mangrove leaves on the degradation rate of mangrove detritus. Senesced yellow leaves of *Laguncularia racemosa* (white mangrove) and water were collected in the Estero El Verde, a tropical coastal lagoon located 30 km north of Mazatlán on the Pacific coast of Mexico. The lagoon is formed by a 5-km-long shallow channel (0.5 to 1.0 m deep) parallel to the coast and surrounded by white mangrove trees. Humiclike substances, considered as the condensation polymers of polyphenols with amino acids and peptides, were obtained from aqueous leachates of the senesced white mangrove leaves, 20 g dry weight/L. The leachates were filtered through glass-fiber filters, evaporated to one-third the original volume, and added to an equal volume of methanol–ammonium hydroxide–water (3:2:5). The mixture was filtered again and evaporated to dryness under vacuum. Humic substances were obtained from lagoon water by passing filtered samples through Amberlite XAD-2 resin and subsequently eluting the column with a solution of methanol–ammonium hydroxide (2:1 v/v). The eluate was then evaporated to dryness. Tannins were isolated from senesced mangrove leaves by a standard procedure.

All the humic fractions were tested for their effects on the rate of degradation of 1.0 g (dry weight) of senesced *L. racemosa* leaves in 5 L of lagoon water, using the heterotrophic bacteria from the lagoon. Two tests were run with each leachate, one with 15 mg/L of the dried leachate and one with 80 mg/L. The mean weight loss after 90 days was determined for each leachate test and for the control. The mean weight loss of the control was 32.8%; for the group with 15 mg/L of humic substances, 26.2%; and for the group with 80 mg/L, 25.5%. Statistical analyses indicated that all groups spiked with humiclike substances had significant reductions in the rate of detrital degradation. Other tests were done with humiclike fractions separated chromatographically; all fractions gave strong inhibitory activity, indicating that they were responsible for the negative feedback between the input of detritus to the lagoon and its biodegradation. This represents, of course, another type of significant allelopathic activity in forestry.

Blaschke (1979) collected litter material from pure stands of *Larix decidua* (larch), *Picea abies* (Norway spruce), *Pinus sylvestris* (Scots pine), and *Pseudotsuga menziesii* (Douglas fir) located 30 km southwest of Munich, Germany. Water leachates were obtained by placing the litter of each species in distilled water at 24°C for 12 hours and filtering. One test used for bioactivity was the standard lettuce bioassay, and the other involved column chromatography followed by GC. Total phenols in the leachates were also determined.

The leachates from the needle litter had both growth-inhibitory and growth-promoting activity. Relatively large amounts of phenolics were leached from the litter of all four coniferous species after 24 hours. Eight phenolics were identified from the various litters: *p*-hydroxybenzoic acid, salicylic acid, cinnamic acid, gentisic acid, quinic acid, ferulic acid, caffeic acid, and DL-catechin. It was concluded that the phenolics produced in the litter of coniferous forests and subsequently introduced into forest soil and the rhizosphere may be of importance in regulating growth of herbaceous species and woody plant seedlings.

There are rarely ever any seedlings of *Larix leptolepis* (larch) on the forest floor of larch forests in Korea. Ko and Kil (1985) therefore collected roots, leaves, and stems of larch from forests near Iri, Korea. Aqueous leachates were made from each part and the extracts were tested against germination and seedling growth of about 11 weed and crop species. Virtually all leachates were inhibitory to germination and/or growth of the various species. Usually, leaf leachates were most inhibitory, roots were next, and stems were least inhibitory. Syringic and *p*-hydroxybenzoic acids were identified in the aqueous leachates of the larch leaves. Blaschke (1979) also identified *p*-hydroxybenzoic acid in leaves of *Larix decidua*. It was concluded that these and perhaps other phenolics are leached from larch leaves and litter by rain and accumulate in the soil to toxic proportions.

*Leucaena leucocephala* has been widely planted in Taiwan in recent years because of its economic value for producing nutritious forage, firewood, and timber. After a few years, the floors of the *Leucaena* plantations generally become relatively bare of understory plants. Chou and Kuo (1986) decided to investigate the reasons for the bare floors in the *Leucaena* plantations at Kaoshu in southern Taiwan.

In that area, six varieties of the Salvado-type of *L. leucocephala*—K8, K28, K29, K67, K72, and S1—and one variety of the native Hawaiian-type *Leucaena* were planted in different densities: 2500, 5000, 10,000, 20,000, and 40,000 plants per hectare. The scientists selected a study site with 10,000 plants per hectare. Fresh leaves and litter of *L. leucocephala* varieties were collected, air-dried in the field, and dried further in a laboratory hood. Soils were collected from the *Leucaena* and grassland control sites from the top 15 cm. A 1% aqueous leachate was made of the leaves, a 2.5% aqueous leachate was made of the litter, and soil extracts were made by shaking 150 g of each soil with 500 mL of distilled water for two hours, centrifuging at 2369 G and storing all at 4°C until use. Tests of the leachates and extracts were made against Great Lakes 366 lettuce, Taichung 65 rice, ryegrass, *Acacia confusa, Alnus formosana, Casuarina glauca, Pinus taiwanensis, Liquidambar formosana, Miscanthus floridulus, Mimosa pudica,* and *Ageratum conzoides.*

The leachates and extracts were generally inhibitory to radicle growth of test seedlings. K29 and K28 generally inhibited radicle growth of rice and lettuce the most. The original extract of the soil under *Leucaena* caused a 20% inhibition of ryegrass radicle growth, but when the soil extract was concentrated five times it caused a 45% inhibition of the radicle. The leachate of *Leucaena* seeds was very inhibitory to the radicle growth of lettuce. Leachates of leaves and decomposing *Leucaena* leaves in soil greatly inhibited seedling growth of several forest species. Ten phytotoxins were identified in leaf leachates, with mimosine generally being most concentrated and relatively toxic to rice and lettuce at 10 to 20 ppm and inhibitory to most test forest species at 50 ppm. It was concluded that the exclusion of understory plants in *Leucaena* plantations was due to the allelopathic effects of allelochemicals produced by *Leucaena.* Certainly, the evidence thoroughly supported the conclusion.

*Melicytus ramiflorus* (mahoe) is an abundant and widely dispersed small tree of lowland and montane New Zealand forests. Partridge and Wilson (1990) observed that large amounts of brown exudate washed out while they were germinating mahoe seed on a well-irrigated germination bed and that germination proceeded only after the exudate disappeared. They subsequently undertook to answer two questions: (1) Does the exudate play any role in the germination of mahoe seeds and (2) does the exudate affect other species? They collected several hundred fresh seeds from fruits in a forest at Kaituna Valley, New Zealand. These seeds were dried at 20°C. About 450 seeds were placed in 250 mL of distilled water; after 12 hours, some of the seeds were placed in freshwater to test for germination, and the rest were put in 250 mL of fresh distilled water. This process continued until the dark exudate was no longer released (after 10 extractions). The seeds were then dried.

The germination of both fresh and leached seed was determined by placing 25 seeds of each on filter paper with 5 mL of distilled water. Seeds were germinated in a lighted room with fluctuating temperature. No unleached seeds germinated, whereas after three weeks, the three leached replicates had 64, 74, and 78% germination. In the serial washing, germination of the mahoe seeds was low until after seven washings, even though the seeds were placed in freshwater to test for germination. None of them germinated in the original water used for leaching until after the sixth washing. At that time the washwater became light, and seed germination was significantly higher each time than after the first seven washings. When the same series of washings was tested against *Coprosma robusta, Fuchsia excorticata,* and *Kunzea ericoides,* none of the seeds germinated in the washwater until after the sixth washing, when *Fuchsia* germinated; the other two species did not have any germination until after the seventh washing. In another series, *Bromus unioloides* and *Spergula arvensis* reacted about like *Coprosma* and *Kunzea,* but *Sophora microphylla* germinated well in all the washwater. It was concluded that the exudate (not identified) acts mainly as an autoinhibitor and has little allelopathic activity. The results indicate to me, however, that the seeds, through the exudate from them, could have a pronounced allelopathic effect on seeds and other plants in the places where they fall to the ground.

*Myrica cerifera* (wax myrtle) is a widespread evergreen shrub or small tree found in Everglades National Park in south Florida. The leaves are covered above by minute dark glands and below by orange-colored glands (Dunevitz and Ewel, 1981). *Schinus terebinthifolius* (schinus or Florida holly) is an exotic shrub or tree in many areas of south Florida. In one area under study by Dunevitz and Ewel, *Myrica* has become dominant and the understory is open, even though there are large schinus plants in the area to serve as seed sources. Measurements indicated a significantly lower growth rate of schinus seedlings in the wax myrtle stand than in the other sites being monitored. The researchers hypothesized, therefore, that wax myrtle may produce allelochemicals that inhibit germination and/or growth of schinus.

A modified staircase experiment was designed to determine if wax myrtle might produce root exudates that inhibit growth of schinus. The scientists alternated pots of tendergreen beans (*Phaseolus vulgaris*) with wax myrtle or schinus on the steps, with bean at the top in the control series with schinus. The wax myrtle significantly reduced the numbers of leaves on schinus compared with the control and also significantly reduced the stem weight and total dry weight of schinus. In another experiment, a leaf extract was made by grinding 10 g of wax myrtle leaves in distilled water and filtering, after which the volume was made up to 120 mL with distilled water. The extract was then tested against germination of tendergreen bean seeds and was found to significantly reduce germination. Tests against germination of schinus seeds indicated a slight reduction also. It was concluded that allelopathy by wax myrtle reduces the vigor of schinus and may increase its susceptibility to competitors and pests. Overall, it appears to me that the experiments were not comprehensive enough to unequivocally infer that allelopathy was active against schinus. The researchers mentioned the possibility of volatiles being involved but then neglected to test that possibility, which seemed to be a good one.

Litter is recognized as a potential selective agent for germination and growth of tree seedlings and herbaceous woodland plants (Kuiters, 1989). Kuiters therefore investigated the effects of several commonly occurring phenolic compounds on germination and growth of four woodland species—*Scrophularia nodosa, Deschampsia flexuosa, Senecio sylvaticus,* and *Chamaenerion angustifoliam*—collected near Amsterdam, the Netherlands. The phenolic acids selected were *p*-hydroxybenzoic, vanillic, caffeic, salicylic, syringic, *p*-coumaric, and ferulic; they were tested in concentrations of 0, 0.01, 0.5, or 10 m*M*. In addition, an equimolar mixture of the seven phenolic acids, where the total phenolic acid concentration was 0, 0.01, 0.5, or 10 m*M,* was tested. Three variables were tested: total seed germination, 50% germination day, and radicle length after 30 days.

Germination was delayed rather than inhibited, which is a common allelopathic effect. Radicle elongation was strongly affected, with the lower concentrations being stimulatory and the higher concentrations being inhibitory to radicle growth. In a second experiment involving just *D. flexuosa* and *S. sylvaticus* and the phenolic acids ferulic and *p*-coumaric, primary root length, number and length of secondary roots, and dry weight were stimulated at 0.01 m*M* but were inhibited at 10 m*M* of both compounds. Kuiters suggested that, beside other factors, success in establishment of emerging seedlings in forests is strongly influenced by the chemical impacts of litter, which generally contains all or part of the phenolics under study. Moreover, the species composition of the herb layer in forests may be partly dependent on the litter type and not just on the abiotic conditions.

In some harvested forest areas of Newfoundland, unsatisfactory establishment of both natural and planted *Picea mariana* (black spruce) seedlings is a serious forestry problem. Mallik and Newton (1988) noted that there was a pronounced reduction in seedling growth when black spruce was grown in a forest soil in the greenhouse compared with growth in sand culture. Thus the objective of their study was to determine which, if any, of the naturally occurring seedbed substrates inhibit the initial growth and development of black spruce seedlings. The B28a-Grand Falls forest section of the boreal forest region occupying the plateau of north-central Newfoundland was selected for study. Within the area, two pure black spruce stands in each of four age classes (15, 30, 45, 60 years) on medium-quality podzolic sites were sampled. For each stand, replicate 10 black spruce seeds (95% viability) were planted on moist F and H organic matter and on Ae and Bf mineral soil contained in petri plates. Controls consisted of seeds sown on washed sand soaked in distilled water. Germination was recorded daily, and the length and fresh weight of the primary root and shoot were measured after 21 days. The roots were also examined microscopically if grown on the F and H organic matter.

The percentage germination of seeds was not affected by substrate, but the length and fresh weights of primary roots were significantly reduced when the seeds were grown on either F or H organic matter or on Ae horizon material. These inhibitory effects increased with age of stand. The Bf horizon showed a difference only in the 15-year-stand, where the root and shoot weights were significantly reduced. Additionally, in all affected seedlings in all stands, the roots were only blunt ends at the bottom of the shoots, which had no root hairs. The results clearly demonstrated allelopathic effects of the black spruce trees on its own reproduction through the effects of the toxins transferred from the litter into several substrate layers. It should be obvious that, in contrast to the claims of uninformed scientists, hundreds of investigations have demonstrated that plants have transferred phytotoxins into the substrate through the litter or otherwise, and these phytotoxins can affect growth of the same or other species. Allelopathy is a fact, not just a supposition!

Several scientists have reported allelopathic effects of *Pinus densiflora* (red pine), and Kil and Yim (1983) extended that research. They conducted a series of field and laboratory experiments to clarify the extent of the allelopathic effects of red pine on the understory plants. They selected 12 stands of red pine forest in the southwestern part of the Korean peninsula. They sampled the stands and collected fresh leaves, fallen leaves, roots, pine forest soil, and pine rain. They tested leachates and extracts of the collected materials against 15 species from within the forest, 6 species from marginal areas, and 13 species from outside the forest stands.

The leachates of fresh leaves inhibited seed germination the most (average rate, 78% of control), followed by leachates of fallen leaves (82%) and soil (97%). In seedling growth, the mean dry weight was 24% of the control in the leachates of fresh leaves, 53% in fallen leaves, and 59% in soil leachate. Eleven phenolic acids, plus benzoic acid, were identified by gas chromatography in the leachates of fresh and fallen leaves. The researchers found that pine toxic substances inhibited germination and seedling growth of low-frequency species more than of high-frequency species in the red pine forest and that these substances are released in descending concentration in fresh and fallen leaves, roots, pine forest soil, and pine pollen rain. The presence of phytotoxins in the soil of the pine forests definitely indicated that the toxins got out of the plant parts of the pine and into the soil and were able to inhibit germination and growth of many of the species on the forest floor and outside the forest. Kil and Yim concluded that the phenolic acids and benzoic acid identified obviously affect the germination and growth of understory species, and they suggested that benzoic acid was the most important allelochemical involved.

Kil (1987) extended the research of Kil and Yim (1983) on allelopathic effects of *Pinus densiflora* (red pine) on understory plants of red pine forests in Korea. He did additional tests on leachates and extracts of various parts of *P. densiflora* on understory species and determined the minimum active concentration of nine identified allelochemicals on the relative germination rates and relative growth rates of lettuce. He found that $5 \times 10^{-3}$ *M* was the minimum concentration of all nine compounds for inhibition of germination, but even $5 \times 10^{-5}$-*M* solutions of all the compounds reduced the relative elongation rates of the seedlings below about 70% of the control.

Kil (1989) compared the potential allelopathic effects of five species of the Pinaceae—*Pinus densiflora, P. thunbergii, P. rigida, Larix leptolepis,* and *Cedrus deodora.* Leachates (water) of fresh leaves, fallen leaves, and roots were made of all these species and were tested against several herbaceous species. All five species inhibited germination somewhat in some plant species, but the most severe inhibition in all cases was on dry-weight growth of species. Thirteen allelochemicals were identified from leaves of *P. thunbergii,* twelve from *P. densiflora* leaves, fourteen from leaves of *P. rigida,* and five from leaves of *C. deodora;* none were attempted from *L. leptolepis.* The overall inhibitory effects and other findings indicate that all five members of the Pinaceae tested have excellent allelopathic potential, although more tests need to be run on all species except *P. densiflora.*

Fox and Comerford (1990) analyzed many selected forest soils of the southeastern United States for low-molecular-weight organic acids. They compared concentrations in rhizosphere and nonrhizosphere soils and found the concentrations to be an order of magnitude greater in rhizosphere soils. Oxalic acid was found in all soils and was generally present in the highest concentration, ranging from 25 to 1000 $\mu M$. The concentration was usually greater in the Bh and Bt horizons than in the A horizon. High concentrations of formic acid were also identified in most soils, ranging from 5 to 174 $\mu M$ in soil solutions. Trace amounts of citric, acetic, malic, lactic, aconitic, and succinic acids were detected in some soil samples. Oxalic acid was the only low-molecular-weight organic acid found in the nonrhizosphere soil of *Pinus elliottii.* Fox and Comerford did not investigate the potential allelopathic effects of the identified organic acids, but acetic acid was identified as a very important allelochemical in wheat straw mulch in England (Chap. 1), and some of the other identified organic acids have been shown to be allelopathic in some specific cases.

The *Pinus koraiensis* (Korean pine) forest is the dominant forest formation in northeastern China, and Korean pine is the most valuable wood in the area (Tao et al., 1987). However, the pine is generally replaced by birch (*Betula*), aspen, and other deciduous trees after clear-cutting or fire, though seedlings of birch and aspen are rarely present under the canopy. A large number of plots were set, treated, sown, and periodically observed from 1963 to 1965 to study the effects of removing the litter layer and/or living ground cover on seed germination, and seedling survival of aspen, birch, spruce, larch, and linden. The scientists found an inverse correlation between litter layer and seed germination and seedling survival in all species except linden. These results were similar to those of several experiments previously described in this chapter for certain species of several of the same genera in Germany.

Kil et al. (1991) reported that few herbaceous plants grew in the understory of Korean pine in forests in Korea. Water extracts of *Pinus koraiensis* leaves, stems, and roots markedly inhibited seed germination and seedling growth of *Echinochloa crusgalli, Plantago asiatica, Achyranthes japonica, Oenothera odorata,* and lettuce. The researchers identified 16 phenolics plus benzoic acid in leaves of Korean pine, many of which were previously shown to inhibit seed germination and/or growth of several herbaceous plant species. They also identified 19 volatile compounds from Korean pine, many of which were previously shown to be inhibitory to germination and/or seedling growth of several herbaceous plants. They tested five concentrations of terpinene-4-ol, α-terpinene, α-pinene, caryophyllene, and ß-myrcene (1 to 10 μL/210 mL) on seed germination of *A. japonica.* They found that all except α-pinene and carophyllene were inhibitory to germination at one or more concentrations tested. They inferred from all the experiments that allelopathy due to Korean pine was responsible for the reduced understory beneath that tree species.

Samples of *Pinus monophylla* (single-leaf pinyon) litter and surface mineral soil below the litter were taken from two sites 113 km apart on the western edge of the Great Basin (Wilt et al., 1988). Litter samples taken 50 cm from the tree trunk comprised decomposing needles, twigs, cones, and bark. The litter and soil samples were analyzed for α-pinene, sabinene, camphene, 3-carene, myrcene, limonene, *p*-cymene, and γ-terpinene, all monoterpenes. Mineral soil samples were taken from the upper 2.5 cm of soils, immediately below the litter layer. The concentration of terpenes in μg/g air-dried litter sample was as follows: 340 μg/g ± 310 S.D., whereas in the mineral soil it was 6.6 μg/g ± 4.8. Thus the mineral soils had about 50 times less total terpenes than did the litter. These findings suggested that allelopathic effects would more likely occur in pinyon litter than in mineral soil. The findings also agreed with previous reports of decreased emergence of herbaceous species in pinyon litter but not in mineral soil. Of course, they also support previous suggestions that the litter is the chief source of terpenes in the mineral soil.

Wilt et al. (1993) followed up their previous research with a study of the fate of several monoterpenes in the natural environment. To do this, they determined total concentrations of monoterpenes in fresh, senescent, and decaying needles from 32 single-leaf pinyon pine trees growing at two locations in the western edge of the Great Basin. Total monoterpene content was highest in the fresh needles ($5.6 \pm 2.2$ mg/g) and intermediate in the senescent needles ($3.6 \pm 1.8$ mg/g). Decaying needles in a dark, decomposing layer of needle litter 5 to 20 cm from the surface were found to contain $0.12 \pm 0.06$ mg/g air-dried weight. The total monoterpene averages were based on analyses of nine monoterpenes in each case. The researchers suggested that senescent pinyon foliage in the uppermost layer of litter still retains relatively high concentrations of monoterpenes that are potentially available as phytotoxins.

Duchesne et al. (1988) investigated the origin of the antifungal compounds produced in the rhizosphere of *Pinus resinosa* seedlings inoculated with *Paxillus involutus.* Bioassay of the crude ethanol-soluble extracts from *Paxillus involutus* growing with *P. resinosa* seedlings, or on root exudates themselves, indicated that, compared with other substrates, root exudates stimulated the synthesis of antifungal compounds by *P. involutus.* Quantitatively, extracts of the growth medium of those seedlings inoculated with *P. involutus* had 4.7 times as much fungitoxic activity as did extracts from the growth medium of the uninoculated control seedlings. The results obtained from the cocultivation of *P. involutus* and *F. oxysporum* on root exudate were consistent with those from the bioassay of the extracts from the growth medium. *F. oxysporum* produced only one-fifth as many spores in cocultures as it did in pure culture on seedling root exudate. The increased resistance of *P. resinosa–P. involutus* associations to *Fusarium* root rot is apparently the result of the increased production of antifungal compounds by *P. involutus,* which was stimulated by *P. resinosa* root exudate. Here we have a plant (*P. resinosa*) producing allelochemicals that stimulated a microorganism (*P. involutus*) to produce other allelochemicals that are inhibitory to growth and reproduction of another microorganism. Each of these steps is an excellent example of allelopathy.

Rho and Kil (1986) tested aqueous extracts of the leaves and rain leachates of the leaves (obtained from *Pinus rigida* in the field) against seedling elongation and fresh and dry weight of the seedlings of nine species of herbaceous plants that occurred in or around the *P. rigida* forest in Korea. Both the leachates and extracts inhibited growth of the seedlings in most tests, but the amount of inhibition depended on the type of solution, the concentration, the time of extraction, and the seedling growth stage at which the extract or leachate was applied. Another test was done completely in the field in which 17 herbaceous species were planted in vermiculite under the crown of the *P. rigida* forest and allowed to grow there in the rain leachate falling from the tree leaves. Virtually all the species were inhibited in fresh- or dry-weight increase, although a very few were stimulated compared with a control away from the crown, which made use only of rainwater that did not pass through a tree crown. Nine phenolic acids and benzoic acid were identified in the aqueous extracts of *P. rigida* leaves, among which some no doubt represented effective allelochemicals.

Kil (1988) reported on additional research on the potential allelopathic effects of *Pinus rigida.* He found that root extracts and soil under *P. rigida,* which were not tested previously, were inhibitory to seed germination and seedling growth of many herbaceous species in the *P. rigida* forest. Moreover, additional allelochemicals were identified (14 in all), and more were tested against seed germination and seedling growth of several herbaceous species. All were found to be inhibitory to seed germination and seedling growth, at least at $5 \times 10^{-3}$-$M$ concentration. The previous report of the allelopathic potential of *P. rigida* was thus strongly supported.

Khosla et al. (1981) suspected from field observations that *Pinus roxburghii* (chir pine) had allelopathic potential. Therefore they powdered some of the bark and needles of chir pine and uniformly spread either 0.1, 0.2, 0.4, 0.8, 1, 2, 4, 8 or 10 g of one of the powders over 30 seeds of *Achyranthes aspera* or *Secale cereale* in three petri plates (15 cm). A known quantity of water was spread over all the petri plates, which were covered and let stand for 96 hours at room temperature, at which time percentage germination and seedling growth were determined. Most tests involving 0.2 g of powdered bark or powdered leaves significantly reduced germination and hypocotyl and radicle growth of *A. aspera* and germination and shoot and root growth of *S. cereale.* In fact, powdered leaves (2.0 to 10 g) killed the hypocotyls and radicles of *A. aspera* and 8 to 10 g killed the roots and shoots of *S. cereale.* The powdered bark was not as toxic as the leaves. Khosla et al. suggested that the allelochemicals released by rain leachates from *P. roxburghii* needles and bark and accumulated in the soil probably inhibit the growth of understory plants in forests where the tree is present. They might be correct, but they should definitely have tested the soil under the trees to see if it inhibited growth of the test species before suggesting that possibility.

Kil et al. (1989) observed that understory plants were sparse and dwarfed in size under *Pinus thunbergii* (black pine) forests in Korea. They found that seed germination, seedling elongation, and relative weights of seedling roots and shoots of six herbaceous species were reduced in many cases when treated with aqueous leachates of needles, stems, and roots of black pine and when planted in black pine forest soils. In some species, germination or growth was stimulated. In four of eight species, volatile allelochemicals from black pine needles inhibited germination significantly, and in six of the same eight species, seedling growth was significantly reduced by the volatiles. Twelve phenolic allelochemicals, plus benzoic acid, were identified in extracts of black pine needles. Eight phenolics and benzoic acid were tested against seedling growth of herbaceous species, and all the test allelochemicals significantly reduced growth of lettuce at concentrations of $5 \times 10^{-3}$ *M* or less. The overall results definitely indicated that water-soluble and volatile allelochemicals got out of black pine and into the soil under the black pine and significantly inhibited or, in a few cases, stimulated germination and/or seedling growth of herbaceous species from inside or outside the black pine forests. Thus allelopathy was definitely demonstrated for black pine.

I previously discussed at some length in this chapter the allelopathic effects of several shrubs in the Florida scrub community. Weidenhamer and Romeo (1989) worked on the same project but on a different shrub, *Polygonella myriophylla.* This species is endemic to the Florida scrub community, and striking bare zones surround mature *P. myriophylla* stands. Studies of the root distribution of *Polygonella* indicated that few of the roots extended into the bare zone. Therefore bioassays were conducted with soil collected biweekly for one year from beneath and around *Polygonella* shrubs to determine whether there is seasonal variation in the phytotoxicity of the soil under or near the shrubs. Samples of *Polygonella* soil and bare-zone and grassed-area soils were collected at each sampling period. Three 5-cm-diameter cores of the upper 5 cm of each soil type were collected at each of the four sampling sites randomly selected at sampling time. Prior to the bioassays, soil samples were air-dried, passed through a 2-mm sieve to remove roots and other large debris, and mixed thoroughly. Fifty seeds of bahiagrass (*Paspalum notatum*) or the native sandhill grass, little bluestem (*Schizachyrium scoparium*), were sown in 5 × 5-cm pots, each containing 200 g of soil. Pots were watered twice weekly to 12.5% moisture content. Germination percentage was determined at three weeks, after which plants were thinned to two of the largest seedlings per pot. This density was maintained by weekly thinnings until the plants were harvested and the number of leaves, length of the longest leaf, total shoot fresh weight per pot, and shoot dry weight per pot were determined.

Germination and growth of bahiagrass were reduced in *Polygonella* soil compared with adjacent grassed areas. Reductions of germination and growth were intermediate in the bare-zone soil. Average shoot dry weight of bahiagrass was more severely affected. For the year, growth (expressed as percent of grassed-area soil) was 48% in *Polygonella* soil and 81% in the bare-zone soil. The root mass of the bahiagrass was greatly reduced in the bare zone also, strongly suggesting a noncompetitive interaction between the *Polygonella* and the bahiagrass. Little bluestem germination was not affected by *Polygonella* soil, but average shoot weight for the year (expressed as percent of grassed-area soil) was 80% in *Polygonella* soil and 94% in bare-zone soil. In this study, variation in inhibition observed in soils collected at different sampling dates (as measured by the date and soil × date interaction terms of the analysis-of-variance model) was not significant for bahiagrass germination but was for bahiagrass dry weight. There was a significant effect of sampling date for little bluestem germination and growth only in the November through April samples. An investigation of the phytochemistry of *Polygonella* revealed the presence of high concentrations of several phenolic compounds in *Polygonella;* these data were to be published later. Overall, the results seem to provide strong evidence in support of the hypothesis that allelochemicals produced by *P. myriophylla* can reduce the germination and growth of other species.

*Prosopis glandulosa* (mesquite) is relatively widespread in Sind, Pakistan. Previous reports concerning possible allelopathic effects of this species caused Alam and Azmi (1989) to suspect that its leaves might be allelopathic to wheat if the leaves blew into a wheat field. They collected fresh leaves of mesquite, rinsed them with water, dried them, and ground them to pass a 20-mesh screen. The material was then stored in plastic bottles at room temperature until ready for assays. The researchers made three levels of extracts by soaking 0, 0.1, and 0.3% (w/v) of the powder in distilled water for 24 hours and then filtering out the powder. The extracts were tested against root and shoot growth of three cultivars of wheat (Sind-81, Sind-83, and Sarsabz). The growth of both shoots and roots decreased significantly with an increasing concentration of *Prosopis* leaf extract, and root growth was affected more than shoot growth. Wheat-growth reduction was 83% with Sarsabz, 65% with Sind-81, and 61% with Sind-83. The results indicated that some phytotoxin(s) leached out of the mesquite leaf residue in water and inhibited growth of the wheat seedlings. Allelopathy could not be inferred from the kinds of experiments performed because it was not shown that the allelochemicals moved out of the mesquite leaves under natural conditions and reduced root and stem growth of wheat.

Sankhla et al. (1965) investigated the toxicity of leaf and fruit extracts (water) of mesquite against several native plants plus cultivated plants at Jodhpur, India. They did this because they noticed that very little or no vegetation exists under or near mesquite trees. They made aqueous leachates by putting fruits or leaves in distilled water (20 g/100 mL) and filtering out the leaves and fruits. Both the leaf and fruit leachates markedly reduced the percentage germination of chrysanthemum, lettuce, marigold, *Merremia aegyptica, M. quinquefolia,* wheat, sesamum, and *Cicer*. The leachates also reduced seed germination in *Indigofera tinctoria* and *Tephrosia incana* rather strongly. The leaf extract markedly reduced germination of *M. aegyptica* until it was diluted 1:8-fold; the fruit extract markedly reduced germination of the same species until it was diluted 1:16-fold. But appropriate tests to clearly demonstrate allelopathy were not run by these scientists.

Sen and Chawan (1970) reported that aqueous extracts of *Prosopis juliflora* possess not only growth-inhibiting but also growth promoting effects on seed germination of *Euphorbia caducifolia*. The toxic effect was prominent in the concentrated aqueous extracts, while dilution of the extracts gave promoting effects. The best demonstration of growth enhancement was indicated by the aqueous stem extract, that of inhibition by the leaf extracts. Fruit extracts fell between the two. The inhibition of growth was more notable on the radicle than on the hypocotyl. The radicle appeared truncated and tapered in higher concentrations; growth of root hairs was suppressed, thus retarding absorption and further development. The extracts from dried material lost their growth-promoting effect after autoclaving. Although there were pronounced inhibiting and stimulating effects of the aqueous extracts, indicating that allelochemicals were dissolving in the water, no experiments were done to indicate that the allelochemicals got out of the various plant parts and into the soil under natural conditions and produced similar results. Thus Sen and Chawan could not infer that *Prosopis juliflora* was allelopathic.

Goel et al. (1989) compared the biological activity of water extracts of *Prosopis juliflora* and *P. cineraria* on plants of the understory. Field observations near New Delhi, India, indicated that the understory vegetation is mainly dominated by *Brachiaria ramosa, Dactyloctenium sindicum, Crotolaria medicagenia, Corchorum tridens, Indigofera linnaei,* and *Cassia occidentalis.* The growth was found to be better under *P. cineraria* than under *P. juliflora,* and bare areas were observed under *P. juliflora.* Pericarps, bark, and leaves were collected in the field, and aqueous extracts were made by soaking each under hot water for one hour, making 10 and 1% concentrations. Germination of radish was recorded daily, and radicle and plumule length were measured after 10 days, after which oven-dried weights were obtained. Field soils were tested for toxicity by collecting soil to a depth of 10 cm below the canopy of the tree, and the bioactivity was tested against radish and *I. linnaei.* Seed germination and radicle and plumule growth of radish were significantly inhibited by leaf extracts and 10% concentrations of the pericarp and bark extracts of *P. juliflora.* Leaf extract and 10% pericarp extracts of *P. cineraria* were toxic to radish. The litter of *P. juliflora* inhibited germination of *C. occidentalis* seeds, whereas the eight-week-old decaying litter did not inhibit growth or germination. The dry-weight yields of all species were significantly higher in all cases except when the litter was not allowed to decay previously. The litter of *P. cineraria* was not toxic to seed germination but was inhibitory to dry-weight yields before any decay. Field soil from under the canopy of the two species of *Prosopis* had about the same activity. Both 5- and 10-g quantities significantly stimulated the radicle length of radish but had no effect on *I. linnaei.* The allelochemicals in both species were found to be phenolic. Even though the results were not conclusive, the stimulatory effects of soil from under the canopies of the two species of *Prosopis* indicate that either small amounts of allelochemicals did get out into the soil or the allelochemicals that got into the soil were stimulatory in nature. In any event, still further work needs to be done on these species of *Prosopis* before definite inferences can be made.

In northeast Canada, *Rubus idaeus* (raspberry) appears to hinder artificial regeneration of black spruce (*Picea mariana*). Coté and Thibault (1988) decided to assess the allelopathic potential of raspberry foliar leachates on the growth of black spruce ectomycorrhizal fungi as interference factor in the *Picea-Rubus* relationship. Green foliage of *R. idaeus* was collected near Quebec City in May, July, and September to test the seasonal variation of toxicity. Then 50 g of air-dried intact material was soaked for 24 hours in 1000 mL of deionized water at room temperature. The leachates were then filtered through a series of Nalgene filters and finally through a 0.2-μm filter in order to sterilize them. Dilutions were made from the 5% stock solution to obtain 1, 0.2, and 0% aqueous solutions of the filtrate. The pH was adjusted to 5.5 and the leachate was added to a modified solid culture medium, leading to 2.5, 0.5, 0.1, and 0% concentrations of *R. idaeus* leachates. These media were put into petri plates along with plugs of different mycorrhizal fungi, and the effects on the growth of the mycorrhizal fungi were determined.

For the 2.5% treatment, *Paxillus involutus, Laccaria proxima, L. bicolor, Thelephora terrestris,* and *Cortinarius pseudonapus* grew only 6, 8, 32, 42, and 46% of their controls, whereas *Hebeloma cylindrosporum* and *Cenococcum geophilum* were stimulated. In a subsequent phase of the investigation, tree seedlings from a raspberry-invaded plantation showed a mycorrhizal infection rate over 75%, with fine roots mainly colonized by *C. geophilum.* Thus it appeared clear that allelopathic effects of *R. idaeus*–leaf leachates definitely affected the species of mycorrhizal fungi that infected the black spruce seedlings. It seems important, therefore, to select a well-adapted mycorrhiza when interference by raspberry is involved.

Parasorbic acid, a seed-germination inhibitor, was identified in 1859 from the juice of fruits of *Sorbus aucuparia* (mountain ash) (Oster et al., 1987). Since mountain ash seeds have a high degree of dormancy, Oster et al. decided to determine if other compounds with seed-germination activity occurred in the fruits or seeds of that species. Ripe berries of *S. aucuparia* were ground and 500 g were boiled in 2.5 L of water under reflux for 24 hours. The clear filtrate was adjusted to pH 8 with $NaHCO_3$, extracted with diethyl ether at that pH and subsequently extracted again at pH 1. Further fractionations were carried out with HPLC and identifications were made by GC and MS. The seeds (14.8 g) were boiled with 50-mL water under reflux for 15 hours, and the parasorbic acid was determined in the aqueous extract by HPLC.

The scientists determined values of 4 to 7 mg of parasorbic acid per gram fresh weight of fruit and 0.08 to 0.12 mg per gram fresh weight of seeds for various groups of berries. The main compound of the neutral fraction of fruit extracts was the lactone, parasorbic acid, which inhibited germination of seeds of *Amaranthus caudatus* and *Lepidium sativum* at a concentration equal to or greater than $5 \times 10^{-4}$ *M*. The acid fraction contained abscisic acid in the amount of 1.3 to 2.5 μg/g fresh weight of the fruits and isopropylmalic acid (1 to 1.5 μg/g fresh weight) as germination inhibitors. Whereas abscisic acid inhibited germination of *L. sativum* at concentrations of $5 \times 10^{7}$ *M* or greater, *A. caudatus* seeds were inhibited only at concentrations of abscisic acid equal to or greater than $10^{-5}$ *M*. The methyl ester of abscisic acid inhibited germination of the seeds of both species at concentrations equal to $5 \times 10^{-7}$ *M* or greater. Thus two new inhibitors were added to our known list of seed-germination inhibitors in fruits of *S. aucuparia*.

Oster et al. (1990) observed that seeds of garden cress seemed to be inhibited or slowed in germination in the immediate neighborhood of *Thuja occidentalis* (white cedar, or arbor vitae). They decided, therefore, to identify the volatile and nonvolatile compounds produced that affected germination of garden cress (*Lepidium sativum*) and foxtail (*Amaranthus caudatus*). In initial tests, they found that garden cress was increasingly inhibited in germination by volatiles from increasing amounts of white cedar leaves. In fact, about 140 mg of fresh leaves completely prevented germination of garden cress seeds. The volatile compounds were obtained by a vacuum method, by direct analysis of the content of secretory organs, and by solvent extraction of the leaves. All the bioactive compounds were identified by GC-MS as monoterpenes (19). The bioactivity of the terpenes was tested against garden cress and foxtail; all terpenes inhibited seed germination of one and/or both species at very low concentrations (all at less than 2 m*M*/L except for two). Three nonvolatile inhibitors isolated by hot water were abscisic acid (3 to 4 μg/g fresh leaves); 2-[2' acetyl-1'-isopropyl] cyclopropylacetic acid (Thujaketosäure); and 3-isopropyl-5-oxohex-2-enoic acid. Oster et al. clearly demonstrated allelopathy on the part of white cedar, because the volatile terpenes were collected initially from outside the plant tissues and were definitely shown to affect germination of seeds of garden cress and foxtail. This supported, of course, previous research on white cedar that demonstrated that it is allelopathic.

Natural regeneration of *Tsuga canadensis* (eastern hemlock), a commercial timber species, is usually poor, and a possible cause is allelopathy (Ward and McCormick, 1982). Aqueous extracts of eastern hemlock litter, foliage, roots, mycorrhizal roots, soil, root bark, and stem bark were made by soaking 250 g of each of the above materials (unmacerated) in 1 L of distilled water for 12 hours at 23°C. After soaking, the leachates were filtered through a 0.45-μm membrane filter, the pH was measured, and the filtrates were stored at 4 to 8°C. For bioassays, the litter filtrates were concentrated by evaporation into 2×, 4×, and 10× concentrates. The plants used for bioassays were eastern hemlock, northern red oak (*Quercus rubra*), and a commercial grass seed mixture containing *Poa pratensis, Festuca rubra,* and *Lolium perenne.*

Eastern hemlock litter extracts reduced seed germination of that species by 74% and killed all six-day-old seedlings of eastern hemlock. Root, mycorrhizal root, and litter extracts had no effects, however, on eastern hemlock seedlings over two weeks old. The litter extract of eastern hemlock reduced seed germination of northern red oak by 52% and mixed-grass seed germination by 10%. Growth of the northern red oak seedlings was not affected by the litter extracts. It was notable that root and mycorrhizal root extracts reduced the mixed-grass seed germination more than did the litter extracts. In conclusion, eastern hemlock is allelopathic to its own regeneration. Water-soluble allelochemicals leached from hemlock litter inhibited seed germination of eastern hemlock and killed very young seedlings (first six days) of that species. While the litter chemicals also inhibited the germination of northern red oak and grasses, the most dramatic effects were eastern hemlock–specific. Since older hemlock seedlings as well as northern red oak seedlings were unaffected by hemlock-litter extracts, hemlock allelopathy appears to specifically inhibit the initial stages of reproduction.

*Vitex negundo* dominates the coastal vegetation of Taiwan and is widely distributed in the southern parts of that country (Chou and Yao, 1983). Leaves of *V. negundo* were collected in the field and brought to the laboratory at the Academia Sinica for laboratory and greenhouse experiments. Aqueous extracts (1%) were made by soaking the leaves in water and filtering. Seeds of Great Lakes radish and *Lolium multiflorum* were initially bioassayed against the leaf extracts. Then cuttings of *Digitaria decumbens, Andropogon nodosus,* and *Mimosa pudica* were grown in pots in the greenhouse and watered with the 1% aqueous leaf extract of *Vitex*. Plants were harvested 30, 60, and 90 days after irrigation and their dry weights were determined. The leaf extracts of *V. negundo* were analyzed for the flavonoids as well as for the phytotoxic phenolics present. A field experiment was run in which quadrats were established in a grove of *Vitex* and in a surrounding pasture, both containing *D. decumbens* and *A. nodosus*. The *Vitex* plots received only rain drip that fell through the foliage of *Vitex,* whereas the pasture plots were watered by the natural rainfall.

In the field experiment, the rain drip passing through the *Vitex* foliage significantly retarded growth of *D. decumbens,* but stimulated growth of *A. nodosus* as compared with the rainfall control. In the greenhouse, the *D. decumbens* watered with the 1% water leachate of *Vitex* was significantly inhibited in growth, whereas growth of both *A. nodosus* and *M. pudica* was significantly stimulated. When the leaf leachates were passed through a polyamide column, some fractions inhibited lettuce and rice seedling growth and some stimulated growth. The responsible substances were identified as five phenolic acids and ten flavonoids. Only one of the flavonoids, 3'-hydroxy-vitexin, was structurally confirmed by MS. The rest of the flavonoids were present in very low concentrations. The correlations between the various experiments were excellent, so it seems clear that *V. negundo* definitely does have allelopathic effects, both inhibitory and stimulatory, on understory and surrounding plants.

## II. ALLELOPATHIC EFFECTS OF HERBACEOUS ANGIOSPERMS AND FERNS

It seems unfortunate that a relatively small amount of research has been done since 1983 on the allelopathic effects of herbaceous angiosperms and ferns in forestry. Nevertheless, it still seems desirable to discuss what *has* been done.

Drew (1988) decided to examine the hypothesis that herbaceous understory species of the northeastern United States interfere with the growth of *Prunus serotina* (black cherry) seedlings. He selected an area in the Heiberg Memorial Forest near Tully, New York, where the dominant forest type was a variant of the beech-birch-maple northern hardwood type, containing more black cherry and less yellow birch (*Betula alleghaniensis*) than usual. He thinned a 1-acre rectangular plot from a 141-$ft^2$ basal area of trees to a residual basal area of 54 $ft^2$. Sugar maple, red maple, black cherry, and American beech (*Fagus grandifolia*) were removed from the overstory, leaving the same species on the site but favoring black cherry, which accounted for 49% of the remaining overstory. This was done in January, and by September of the same year 90% of the area was dominated by two mutually exclusive ground-flora species, *Aster acuminatus* (whorled wood aster) and *Dennstaedtia punctilobula* (hayscented fern). These two plants rarely formed mixed stands but occurred separately as nearly pure species. Three mature black cherry trees were selected so that they were not closer than three survey or chain lengths. Each tree had areas beneath its crown encompassing populations of either whorled wood aster or hayscented fern. In each ground-flora community, two circular, contiguous 2-m-diameter plots were laid out, with centers permanently marked, for a total of 12 plots. Each plot was sampled immediately for numbers of woody and herbaceous species, and the height of shoot tips above ground was recorded. All black cherry seedlings were labeled with a split plastic ring. The same procedure was followed in September of 1982, 1983, and 1984, using different colored rings to identify the years of origin of the various black cherry seedlings. Following the original sampling in 1981, all fern and aster stems were removed from one of the two contiguous plots in each ground-flora type, leaving the remaining contiguous plot of each pair alone as a control. Thereafter, through September 1984, aboveground fern and aster parts were regularly removed, as soon as they appeared, from all the noncontrol plots.

Complete removal of the hayscented fern stimulated black cherry germination, height growth, and species diversity increase after two growing seasons. Removal of aster produced the same effect in the first growing season, but at the end of four years, black cherry growing beneath aster was twice as tall as black cherry of the same age growing beneath fern. Allelochemicals produced by the fern, competition for soil nitrogen, and the detrimental effects of reduced light intensity beneath the dense fern cover were suggested as the possible modes of interference by the hayscented fern on black cherry seedling growth. The experimental design did not allow a narrowing of the possibilities. It was previously demonstrated, however, by Horsley (see Rice, 1984) that hayscented fern, New York fern (*Thelypteris noveboracensis*), short husk grass (*Brachelytrum erectum*), and flat-topped aster (*Aster umbellatus*) were allelopathic to black cherry seedlings. Of course, other factors, along with allelopathy, were no doubt active in the interference.

Horsley and Marquis (1983) did a follow-up study on some of Horsley's previous research on interference by weeds on Allegheny hardwood reproduction, but they added additional research on interference by deer browsing. They reported again that weeds, particularly the ferns, caused significant interference with germination, survival, and growth of desirable species following a seed cut and a removal cut. Deer browsing had no direct effect on desirable species, since the species did not grow enough to emerge from the herbaceous cover. Deer browsing did affect growth of *Rubus idaeus, Betula alleghaniensis, B. lenta,* and *Prunus pennsylvanica,* which grew above the herbaceous cover. Following the shelterwood removal cut, fern and grass cover interfered with establishment of desirable reproduction. The ferns interfered more strongly than did the grass in this stand. Growth of black cherry seedlings was significantly reduced by both fern and grass cover, but ferns inhibited growth more than did grass. The experimental techniques were not designed to test the relative effects of competition and allelopathy, the latter having previously been demonstrated by Horsley. The important point is that interference (competition plus allelopathy) by the ferns and grass and browsing by the whitetail deer all affect the reproduction of the tree species in the forest.

Jobidon et al. (1989a) investigated the potential harmful and beneficial effects of *Avena sativa, Hordeum vulgare,* and *Triticum aestivum* straw residues on black spruce seedling growth and mycorrhizal status. This project was a forerunner to a proposal to use the same straw residues in weed control in *Picea mariana* (black spruce) forests in eastern Canada. The researchers found that the straw residues did not affect spruce seedling height growth over a two-month growth period, and the newly formed fine roots of both treated and control seedlings were mycorrhizal. Oat and wheat straw significantly enhanced foliar phosphorus content as compared with the control spruce seedlings. All straw treatments significantly depressed foliar manganese content, indicating that the residues could have a harmful effect on manganese uptake. The researchers recommended therefore that the status of manganese in planted black spruce should be monitored when using straw mulches for weed control during reforestation of black spruce forests.

Jobidon et al. (1989b) reported on additional research that further demonstrated that oat, barley, and wheat straw enhanced shoot and stem-diameter growth of *Picea mariana* seedlings but did not affect root infection by mycorrhizae. In this investigation, they found that treated black spruce showed a significantly higher foliar N content than did control seedlings without the mulch. There were no consistently significant differences otherwise in the nutrients analyzed. The researchers found that the number of $NH_4^+$ oxidizers were significantly decreased in all mulched plots in all study sites in June of the first year of mulching. All mulches, except the oat and mixed mulch, still reduced the number of $NH_4^+$ oxidizers significantly in August of the same year. All the individual phenolic acids identified in the straw mulches significantly reduced the numbers of $NH_4^+$ oxidizers in a concentration of $10^{-3}$ to $10^{-5}$ *M*, and two of them significantly reduced the numbers even at $10^{-6}$ *M*. The soil nitrate content was generally reduced in treated plots, and the ammonium content was generally increased, indicating along with the other data that nitrification was reduced by mulching, which lowered the loss of soil nitrogen and caused the N level in the black spruce leaves to increase.

Jobidon et al. (1989c) found that cold-water leachates of the three kinds of straw strongly inhibited red raspberry propagule growth. Shoot dry weights were 10, 44 and 68% of the control for oat, wheat, and barley straw leachates, respectively. The researchers tested the inhibitory effects of the mulches in the field by clear-cutting balsam fir–birch stands (5 ha each) located in three sites in eastern Quebec. In the fall of 1985, they made furrows in the stands and covered them with straw mulches—oat, barley, wheat, or a mixture of the three—plus a control using *Populus* wood shavings as a mulch. Vegetation analyses were made in the summers of 1986 and 1987. *Rubus idaeus* germination and growth were inhibited because of the straw treatments. In the summer of 1987, an overall reduction, 20%, in the total percent cover of all species was noted on the treated plots, compared with 61% on the control plots with the *Populus* wood-shaving mulch. The nitrogen content of the raspberry foliage from the treated plots was significantly lower than in seedlings from the control plots. The phytotoxic phenolics were previously identified in the mulch straws and were found to be toxic to some organisms. It was thus inferred that allelopathic effects of the straw mulches were responsible for the reduction in germination, frequency, and growth. The three mulches, singly or combined, were recommended for aiding in the partial control of raspberry in reforestation in eastern Quebec. Of course, it should be understood that the total interference effects of the mulches, not just the allelopathic effects, are involved in the raspberry control.

Hollis et al. (1982) collected foliage in late summer from species common to natural and prepared pine sites in the lower coastal plain flatwoods—*Andropogon virginicus* (broomsedge), *Eupatorium capillifolium* (dogfennel), *Hypericum fasciculatum* (St. John's wort), *Ilex glabra* (gallberry), *Lyonia lucida* (fetterbush), *Myrica cerifera* (wax myrtle), *Aristida stricta* (wiregrass), *Cliftonia monophylla* (black titi), and *Rhynchospora cephalantha* (beak rush). Most are herbaceous, but some are small shrubs. Foliage (300 g fresh weight) was loosely placed in a glass column and cold deionized water (1200 mL) was cycled through the column for 24 hours. The resulting leachate was filtered and tested against seed germination and radicle growth of *Pinus taeda* (loblolly pine) and *P. elliottii* (slash pine). Leaf leachate from six species interfered with either or both germination and radicle elongation of both pine species. In all cases, the fresh leachate was less inhibitory than was the leachate from frozen tissue. Even when the quantity of leachate applied was drastically reduced, dogfennel and fetterbush continued to show inhibition.

In greenhouse studies in soil, fetterbush mulch reduced the growth of both loblolly and slash pine, whereas dogfennel mulch stimulated growth of both pine species. Fetterbush significantly reduced the N content of both pines, but significantly increased the P concentration in slash pine and did not affect the P concentration of loblolly pine or the K content of either pine. Dogfennel significantly increased the N, P, and K concentrations in loblolly pine and the P content in slash pine but did not affect the N and K concentrations of slash pine. In field tests, first-season growth of seedlings planted within 0.5 m of residual debris or coppice of fetterbush was significantly less than the growth of control slash pine. The combination of experiments certainly demonstrated that at least *Lyonia lucida* and *Eupatorium capillifolium* have some allelopathic effects, along with their competitive effects, on slash and loblolly pine seedlings.

Todhunter and Beineke (1979) investigated differences in *Juglans nigra* (black walnut) growth in the Indiana-Purdue Black Walnut Tree Improvement Program in areas with *Festuca arundinacea* (tall fescue) and without. Fescue had a highly significant interference effect on height, sweep, diameter, and volume but did not influence survival. Moreover, black walnut growth was greatly reduced by fescue, compared with naturally occurring ground cover. It is obvious that tall fescue had a strong overall interference on black walnut trees, part of which was probably due to allelopathy, if we take into account all the prior research demonstrating that tall fescue has a strong allelopathic potential.

Unfortunately, most reported effects of tall fescue on black walnut trees have been labeled competition, even though with currently accepted techniques competition cannot be separated from allelopathy in any experiments and even though, as stated above, fescue has been shown to have allelopathic potential. For example, Walters and Gilmore (1976) showed that fescue leachates applied to *Liquidambar styraciflua* seedlings in a greenhouse stairstep apparatus reduced growth by up to 60%. Thus an allelopathic component of interference was definitely implicated. I should point out here that Weidenhamer et al. (1989) published a technique which they suggested held some promise of distinguishing between resource competition and allelopathic interference in plants. The technique needs much testing, however, before it can be known if it will work.

Rink and van Sambeek (1985) attempted to determine if genetic variation in resistance to interference is present in black walnut. A secondary objective was to identify the most significant environmental variables that contribute to interference. The researchers used 32 experimental chambers, which they divided into four blocks of eight chambers each. The same well-mixed soil was used in all chambers. Within blocks, each chamber was randomly assigned one of eight possible treatments. The eight treatments were all possible combinations of two moisture-stress regimes, two soil-fertility levels (fertilized or unfertilized), and two types of watering media (distilled water or tall fescue leachate).

Aside from fertilizer × moisture-stress interaction, survival of black walnut was not affected by any source of variation, including family. Survival was 90% in the fertilized, high-stress treatments as compared with 99% in all other treatments. Presence of tall fescue leachate under low moisture-stress conditions resulted in a 10% reduction in walnut seedling height and 14% reduction in dry weight. Although soil-moisture stress was the dominant variable in this study, response to moisture stress depended on the level of a corresponding variable. For example, fertilization affected seedling survival under high moisture stress but did not do so under low stress. The reduction in seedling height and dry weight in the presence of fescue leachate under moist conditions indicates that fescue leachate is phytotoxic to walnut seedlings. Interactions between moisture stress and family and between fescue leachate and moisture stress for both seedling height and dry weight suggested that selection for tolerance to moisture stress is possible, whereas selection for resistance to fescue leachate would be difficult. The only effect of fertilization was a 9% decrease in seedling survival under high moisture-stress conditions.

Rink and van Sambeek (1987) continued a similar line of research concerning the interference of tall fescue, but this time it was against four *Fraxinus americanus* (white ash) families. The experimental plan and techniques were similar to those involving black walnut. Analysis of three environmental variables affecting seedling growth of four half-sib white ash families was dominated by a three-way interaction between soil-moisture stress, fescue leachate, and family. Of these, soil-moisture stress contributed by far the most to the interaction and resulted in an average growth decline of 62%. Although fescue leachate appeared to be phytotoxic to ash seedlings, it had a lesser effect, resulting only in changes in height rankings among families. Seedling families that grew well under moisture stress also tended to be tallest in the presence of fescue leachate, but seedling families that were tallest under optimum conditions were not necessarily tallest when exposed to moisture stress. Such changes in family rank for height growth evident in this limited experiment suggest that substantial improvements in early selection for tolerance to drought and allelopathy are possible. Again, tall fescue was involved in interference against white ash through a combination of competition and allelopathy. Appropriate experimental techniques probably would indicate a somewhat similar result in most so-called competition experiments.

Ferns have been identified as a source of allelopathic agents

(Rice, 1984), and sensitive fern (*Onoclea sensibilis*) is known to reduce survival of *Quercus rubra* (northern red oak) seedlings more than shrubs and grass sod do (Hanson and Dixon, 1987). Hanson and Dixon investigated the effects of inoculation with an ectomycorrhizal fungus (*Suillus luteus*) on the survival and growth of northern red oak seedlings exposed to interrupted fern-frond leachates. Plants were grown under two light regimes to determine the effect of irradiance on the relationship between leachate additions and fungal inoculation.

Fern-frond leachates significantly reduced seedling survival. At 66 days after planting, leaves of *Onoclea sensibilis* leachate–treated oak seedlings developed interveinal necrosis and rapidly wilted, in some cases turning brown. Dead leaves remained attached to the stem, indicating that an abscision layer had not formed. Inoculation of seedlings with *S. luteus* enhanced the survival of the deionized-water controls. Moreover, there was an insignificant trend toward greater mortality of fern leachate–treated plants grown under full as compared with partial glasshouse light. Tissue analyses of pooled leaves and lateral roots from each treatment showed no trends associated with any of the major elements (P, K, Ca, Mg). However, oak seedlings treated with fern leachate showed elevated chromium concentrations over the deionized-water controls that were negatively correlated with seedling survival. Lateral-root chromium concentration was two orders of magnitude higher than that of the leaves. Seedlings that had been inoculated and treated with fern leachate had intermediate tissue levels of chromium that corresponded with their intermediate level of survival. The fern-leachate treatment had no effect on shoot-growth variables, but significantly reduced lateral and taproot dry weight under both full and partial glasshouse light. This study demonstrated that northern red oak seedlings inoculated with *S. luteus* had the capacity to withstand the deleterious effects of the fern leachate. The researchers hypothesized that the fern leachate chelated the existing soil chromium, making it available for uptake. It is possible that mycorrhizae may have ameliorated the uptake of chromium by sequestering the element. The results of these studies certainly confirmed the previous results of studies that implicated sensitive fern as being allelopathic to northern red oak seedlings.

While studying the survival and growth of *Pinus elliottii* (slash pine) on sandy dredge mine tailings in north Florida, Fisher and Adrian (1981) observed a strong effect of *Paspalum notatum* (Bahia grass) on the growth of slash pine. As the percentage of the ground covered with bahiagrass increased, the height of the three-year-old pine decreased markedly. The scientists decided to investigate the effect in more detail. Their first experiment was a complete factorial using high and low moisture; with and without supplemental nutrients (100 kg N, 45 kg P, 80 kg K/ha); and with mulch (peat moss or bahiagrass) and living bahiagrass. Six replications of dredge mine tailings were used in 1-kg plastic pots. For the living bahiagrass treatment, the grass was sown one month prior to planting a 1 + 0 slash pine seedling in each pot. For the mulch treatment, 25 g of air-dry material was applied to each pot after the pine seedling was planted. High-moisture pots were rewatered when 50% of the plant-available water by weight had been used. In the low-moisture pots, the plants were allowed to use 99% of the plant-available water by weight before the pots were watered again. After 20 weeks of growth, all treatments were harvested and dry weight per seedling was determined. The bahiagrass significantly reduced growth of the slash pine. It appeared to be a vigorous competitor for both water and nutrients, and the fertilizer stimulated the grass in both wet and dry treatments. The pine grew as well in the dry treatment, in the absence of grass or grass mulch and fertilizer, as it grew in the wet treatment. Although no tree seedlings survived the low-moisture-plus-fertilizer treatment, the grass grew well in that treatment.

In a second experiment, 10 newly germinated slash pine seedlings were planted into dredge mine tailings in 500-g pots. These were watered with a complete nutrient solution for two weeks. Half the pots were subsequently watered with leachate from pots containing living bahiagrass, and half were watered with leachate from pots containing slash pine. This was continued for 20 weeks, after which the plants were harvested and the root and shoot dry weights of the slash pine seedlings were measured. When slash pine was watered with bahiagrass leachates, its growth was significantly reduced. Root, shoot, and total dry weight of the seedlings were all significantly less than the corresponding values for seedlings watered with pine pot leachate. Moreover, in the first experiment, the pine plants mulched with bahiagrass were significantly reduced in dry weight in all treatments as compared with controls mulched with equivalent amounts of peat mulch. Additionally, pine seedlings growing in pots with living bahiagrass were lighter in weight than those growing in the mulch pots, which indicated a competitive effect over and above the effect due to allelopathy. It was clear therefore that bahiagrass interfered with growth of slash pine both through competition and allelopathy. This emphasizes again that it is unlikely that either competition or allelopathy ever occur alone, and adequate experiments should be conducted to test for both.

*Rudbeckia occidentalis* (western coneflower) has been suspected of being allelopathic, so Ferguson (1991) decided to determine the potential of that species for affecting seed germination and radicle elongation of two species of conifers and of lettuce under controlled conditions. Water extracts and volatiles from western coneflower stems, leaves, solid caudex, rotting caudex, and control were tested against Lilly Miller Great Lakes lettuce, *Pinus contorta* (lodgepole pine), and *Picea engelmannii* (Engelmann spruce). One hundred grams of fresh plant material was placed in 1500 mL of deionized water, oscillated for two hours, filtered to remove solids, and stored at 0 to 3°C until used in tests. The aqueous extracts were tested against seed germination and radicle elongation in petri plates. Volatilization tests were made in the usual fashion.

None of the aqueous extracts significantly affected seed germination. However, volatiles from the solid caudex significantly reduced germination of lodgepole pine (67.7%), and volatiles from the soft caudex significantly reduced germination of the same species (59.4%), versus 83.3% in the control. On the other hand, all the water extracts significantly reduced the radicle lengths of lodgepole pine and Engelmann spruce, but only the water extract of the soft caudex significantly affected length of the lettuce radicles, and they were stimulated. The only significant effect of the volatiles on radicle length was a significant reduction in length of lodgepole pine by volatiles from the soft caudex. Allelopathy by the volatiles was thus definitely demonstrated, but the possible allelopathic effects of aqueous leachates of the plant parts need more testing of a different type to demonstrate that the allelochemicals leach out of the plant under natural conditions and accumulate to the extent that they can exert biological effects.

Gilmore (1985) hypothesized that if substances are released from *Setaria faberii* (giant foxtail) that inhibit seed germination and radicle growth of *Pinus taeda* (loblolly pine), this might account at least in part for the erratic establishment of loblolly pine in an old field covered with foxtail. Water extracts of giant foxtail inhibited germination and radicle elongation of loblolly pine in petri dish tests. Most of the toxic effects came from extracts of dried foxtail tops, with lesser amounts from fresh tops and roots. Eight phenolic compounds were extracted from dry foxtail tops and were identified by GC-MS as *o*-hydroxyphenylacetic, vanillic, syringic, *p*-coumaric, ferulic, caffeic, and gentisic acids and scopoletin. Soils in which giant foxtail previously grew were extracted, and several compounds were identified in the extract, but none of the eight compounds previously named was identified. It is obvious from the results of various tests that allelopathy against loblolly pine was not established. The possibility of moderate to slight allelopathic potential was suggested, but appropriate tests were not carried out to confirm this possibility.

## III. ACTINOMYCETES AND REGENERATION IN CLEAR-CUT FOREST AREAS

Regeneration of *Pseudotsuga menziesii* (Douglas fir) after clear-cutting is successful in most regions of the Pacific Northwest. In some areas with well-developed indigenous forests, however, many successive replantings have failed (Friedman et al., 1989). In southwestern Oregon alone, poor regeneration has resulted in withdrawal of more than 68,000 ha of public forest land from the timber production base. Preliminary work in southwestern Oregon suggested that allelochemicals produced by soil actinomycetes contributed to regeneration failures. Like any other allelopathic organism, microorganisms may affect plants directly or indirectly through inhibition or stimulation of mycorrhizal associates or other beneficial organisms. Friedman et al. (1989) designed a study to determine whether soil-borne actinomycetes having phytotoxic or antifungal effects were present in a clear-cut area in which tree regeneration had failed and to evaluate the role of actinomycetes in the failure.

Soil samples were collected from Cedar Camp in the Siskiyou Mountains of southwestern Oregon at about 1700 m above sea level. Average annual precipitation there is about 1650 mm, with more than 50% as snow, and a drought period commonly occurs from July through September. Forests there are dominated by Douglas fir with some white fir (*Abies concolor*) and Shasta red fir (*Abies magnifica* var. *shastensis*). Soil is a sandy, skeletal, excessively drained, mixed Entic Cryumbrept formed from quartz diorite parent material. The study area was still free of trees 18 years after clear-cutting and after several replanting attempts. The control area was adjacent to undisturbed forest with a 3 to 5-cm organic layer over the inorganic soil. The total number of colony-forming units of actinomycetes per gram of dry soil was determined on tap-water agar (TWA) after 12 days and on sodium-albumen agar (SAA) after 6 days. The actinomycetes isolated were streaked on malt-yeast agar (MYA) plates. After a week, four rows of six seeds each were placed opposite the isolate. Because seeds of Douglas fir germinate slowly and are difficult to surface-sterilize, the fast-germinating seeds of *Anastatica hierochuntica* and lettuce were used in tests. This enabled germination to be recorded 12 or 24 hours after seed wetting, thus minimizing interference by associated seed-coat microflora. Isolates exhibiting phytotoxicity were further evaluated in a second series of tests that were replicated 10 times. Percentages of reduction in germination relative to the controls were determined from the later tests.

For evaluating the effects of actinomycetes on mycorrhizal fungi, two common Douglas fir ectomycorrhizal fungi, *Laccaria laccata* and *Hebeloma crustuliniforme,* were placed on MYA plates at a short distance from 10-day-old cultures of actinomycetes and on MYA plates without actinomycetes. Diameter growth of fungal colonies was measured after seven days of incubation, and growth of colonies plated with actinomycetes was compared with growth of colonies without actinomycetes.

More than twice as many actinomycete cfu formed from clear-cut soil as from forest soil. Of the 150 isolates from each area, 22.8% of those from the clear-cut and only 9.2% of those from the forest inhibited seed germination of *A. hierochuntica* and lettuce. Inhibitory isolates from both areas had similar effects on seeds of both plant species, reducing germination 12 to 19% below that of control seeds without actinomycetes. The population density of phytotoxic actinomycetes in clear-cut soil was nearly five times that in forest soil. Four percent of the actinomycetes from the clear-cut soil and 2.6% from the forest soil significantly reduced in vitro growth of the two common ectomycorrhizal fungi of Douglas fir, *L. laccata* and *H. crustuliniforme.* Two actinomycete isolates from the clear-cut area reduced fungal growth by 40 and 73%. Friedman et al. concluded that, along with other factors, phytotoxic and antifungal actinomycetes may suppress natural regeneration or establishment of planted seedlings, either directly or indirectly, through inhibition of seed germination or of mycorrhizal fungal growth.

Nursery inoculation of tree seedlings with ectomycorrhizal fungi is currently in use in some regions for reforestation to produce higher-quality outplanting stock with greater vigor while in the nursery and increased ability to survive and grow following planting in the field (Richter et al., 1989). The opportunity exists, therefore, to coinoculate tree seedlings, particularly conifer seedlings, with ectomycorrhizal fungi and mycorrhizae-associated bacteria or actinomycetes that mutually benefit tree growth. The purpose of the research of Richter et al. was to isolate mycorrhizoplane-associated actinomycetes from *Pinus resinosa* (red pine) seedlings and test their effects on three common mycorrhizal fungi.

Mycorrhizae were collected from field-planted red pine seedlings at the Toumey Tree Seedling Nursery at Watersmeet, Michigan, and incubated in a supplemented enrichment broth on a gyratory shaker. Actinomycetes were isolated from the enrichment broth using membrane filters, those with a pore size of 0.45 μm yielding the highest percent recovery. Over 20 actinomycete isolates were collected, and they all appeared to be streptomycetes, although no thorough identifications were attempted. The isolated actinomycetes were tested against three mycorrhizae common on nursery seedlings of red pine—*Laccaria bicolor, L. laccata,* and *Thelephora terrestris.* Interactions between the actinomycetes and mycorrhizal fungi were determined by growing colonies of the organisms adjacent to each other on Modified Melin-Norkrans agar medium in petri plates. The pH of the medium was 6.0. A single actinomycete colony was placed in the center of each plate, with one plug of each mycorrhizal fungus placed around the actinomycete colony on each plate. Plates were then incubated three days at 30°C, after which control and test plates were sealed with Parafilm and incubated at 18 to 22°C for four weeks, with weekly measurements.

Most actinomycetes exerted a range of effects on the three mycorrhizal fungi, inhibiting some while stimulating others. Several inhibited growth of all three mycorrhizal fungi, and some stimulated all the fungi. It was concluded that actinomycetes isolated from mycorrhizae show excellent potential for use as coinoculants with selected ectomycorrhizal fungi to optimize the microflora for developing seedlings. It seems obvious that conditions that would improve the establishment and growth of ectomycorrhizae on red pine and other tree seedlings could be very important to the establishment and growth of the tree seedlings, as I have discussed several places in this book. Inhibiting the growth of one or more mycorrhizae might also favor the development of certain naturally occurring mycorrhizae and thus favor the development of certain tree seedlings.

# AFTERWORD

EVEN A QUICK PERUSAL of this book should demonstrate to the reader that there have been tremendous strides in research in allelopathy since 1983, when most of the research discussed in these pages began. At that time, most researchers in the field of allelopathy were lamenting that we needed more information demonstrating that biologically active allelochemicals were getting out of the plants that produced them and into the surrounding environment, where they brought about observed results. Now much more elegant research techniques have been developed, and virtually all important investigations have demonstrated that biologically active allelochemicals volatilize into the atmosphere, leach out of living plant parts in rainfall or stemflow, leach out of plant residues, exude from roots, are transformed in the humus layer of soils, are released by the decay of plant residues, or are transformed in the soil either chemically or biologically to other phytotoxins.

It should be remembered at all times that the term *allelopathy* does not apply just to plants. It refers to chemical interactions (inhibitory or stimulatory) between plants, between plants and microorganisms, and between microorganisms (those traditionally placed in the plant kingdom—algae, fungi, and bacteria). We have seen few studies in the past concerning the role of allelopathy among microorganisms and between microorganisms and plants, but this oversight is now rapidly being corrected. (See, for example, Chaps. 3, 5, 6, and 7.) Microorganisms exert many inhibitory and stimulatory effects on other important microorganisms and on plants—and vice versa. Microorganisms can play many important roles in the biological control of weeds and plant diseases. It should be obvious from Chaps. 4, 5, 6, and 7 that we are already at the point that we could help greatly in the control of weeds and plant diseases by integrating biological control with chemical control, thus cutting down on the use of the synthetic chemicals that have caused so many serious environmental hazards. In numerous cases we could already substitute biological control for chemical control with no decrease in yield. The chief problems we face in the United States are resistance to change and political clout used to keep the U.S.D.A. from making needed changes.

It should be obvious from this book that rotating crops, as we used to do, is very important. It is clear that most crops that are planted year after year in the same field have a gradually decreasing yield, which has been clearly shown to be due to accumulation of phytotoxic chemicals. Rotation can also help prevent certain plant diseases from persisting in a field.

There is clear evidence that many important weeds interfere with certain crop plants and with other weeds, interference that is due to allelopathy as well as to competition. It should be remembered that experiments aimed only at demonstrating that plants compete for minerals and water are not valid experiments, since researchers cannot eliminate the possibility that the results are caused, at least in part, by allelopathy. On the other hand, it is very simple to set up experiments that completely eliminate competition and demonstrate allelopathy. Allelopathy and competition are both probably involved to some extent in all cases of interference, a fact that should always be recognized and stated. As seen in Sec. II of this chapter, several recent investigators in forestry have developed techniques in which they have been able to estimate the relative amount of interference due to allelopathy and the relative amount due to competition. This is what all investigators should strive for.

Many countries of the world are now exceeding the United States in innovative research in agriculture and forestry. This is definitely so in most aspects of research in allelopathy. (India now has developed an Indian Society of Allelopathy, the only such national society in the world.) The truly elegant research worldwide concerning allelopathic effects of woody plants in forestry is particularly encouraging. The structure, reproduction, and understory of many forests have clearly been demonstrated to be due to allelopathy, particularly in Germany,

Belgium, the Netherlands, Scandinavia, Switzerland, and to lesser extents in Russia, Australia, Canada, India, and the United States.

Despite the decline in research in this country, scientists in allelopathy can be proud of the advances in their field since 1983. I trust this rapid progress will continue, and I hope that this book will contribute to the movement.

# REFERENCES


Abbas, H. K., C. D. Boyette, R. E. Hoagland, and R. F. Vesonder. 1991. Bioherbicidal potential of *Fusarium moniliforme* and its phytotoxin, fumonisin. *Weed Sci.* 39:673–77.

Abdul-Rahman, A. A. and S. A. Habib. 1989. Allelopathic effect of alfalfa (*Medicago sativa*) on bladygrass (*Imperata cylindrica*). *J. Chem. Ecol.* 15:2289–2300.

Abu-Irmaileh, B. E., and J. R. Qasem. 1986. Aqueous extract effects of *Salvia syriaca* L. in various lines of four crops. *Dirasat* 13:147–70.

Achhireddy, N. R., and M. Singh. 1984. Allelopathic effects of lantana (*Lantana camara*) on milkweed vine (*Morrenia odorata*). *Weed Sci.* 32:757–61.

Achhireddy, N. R., M. Singh, L. L. Achhireddy, H. N. Nigg, and S. Nagy. 1985. Isolation and partial characterization of phytotoxic compounds from lantana (*Lantana camara* L.). *J. Chem. Ecol.* 11:979–88.

Adams, P. B., and D. R. Fravel. 1990. Economical biological control of sclerotinia lettuce drop by *Sporidesmium sclerotivorum. Phytopathology* 80:1120–24.

Ahmed, N., F. Hussain, and M. Akram. 1984. The allelopathic potential of *Eucalyptus tereticornis. Pak. J. Scient. Indust. Res.* 27:88–91.

Ahmed, N., F. Hussain, and M. Akram. 1987. Effect of aqueous extracts of some host and non-host species on *Cuscuta reflexa* Roxb. In *Modern Trends of Plant Science Research in Pakistan.* pp. 11–12 (I. Idahi and F. Hussain, eds.), Proc. Third Nat'l. Conf. of Plant Sciences, Dept. of Bot., Univ. of Peshawar, Peshawar.

Ahmed, S. A., M. Ito, and K. Ueki. 1982. Phytotoxic effect of waterhyacinth water extract and decayed residue. *Weed Res.* (Japan) 27:177–82.

Akhtar, N., H. H. Naqvi, and F. Hussain. 1978. Biochemical inhibition (allelopathy) exhibited by *Cenchrus ciliaris* Linn and *Chrysopogon aucheri* (Bioss) Stapf. *Pak. J. For.* 28:194–200.

Akiyama, M., and K. Nishigami. 1991. Allelopathic effects of *Solidago* plants on the growth of soil algae. *Phycology* 27 *Sup.*:5 (Abstr. #7).

Akram, M., and F. Hussain. 1987. The possible role of allelopathy exhibited by root extracts and exudates of chinese cabbage in hydroponics. *Pak. J. Scient. Indust. Res.* 30:918–21.

Alam, S. M., and A. R. Azmi. 1989. Influence of *Prosopis glandulosa* water extract on the seedling growth of wheat cultivars. *Pak. J. Scient. Indust. Res.* 32:708.

Alán, E., and U. Barrantes. 1988. Efecto alelopatico del madero negro (*Gliricidia sepium*) en la germinacion y crecimiento inicial de algunas malezas tropicales. *Turrialba* 38:271–78.

Ali, N. A., and R. M. Jackson. 1989. Stimulation of germination of spores of some ectomycorrhizal fungi by other micro-organisms. *Mycol. Res.* 93:182–86.

Alsaadawi, I. S., F. A. K. Sakeri, and S. M. Al-Dulaimy. 1990. Allelopathic inhibition of *Cynodon dactylon* (L.) Pers. and other plant species by *Euphorbia prostrata* L. *J. Chem. Ecol.* 16:2747–54.

Alström, S. 1987. Factors associated with detrimental effects of rhizobacteria on plant growth. *Plant & Soil* 102:3–9.

Ambika, S. R., and Jayachandra. 1980. Suppression of plantation crops by *Eupatorium* weed. *Curr. Sci.* 49:874–75.

Amritphale, D., and L. P. Mall. 1978. Allelopathic influence of *Saccharum spontaneum* L. on the growth of three varieties of wheat. *Sci. & Cult.* 44:28–30.

Anaya, A. L., L. Ramos, R. Cruz, J. G. Hernandez, and V. Nava. 1987. Perspectives on allelopathy in Mexican traditional agroecosystems: A case study in Tlaxcala. *J. Chem. Ecol.* 13:2083–2101.

Anaya, A. L., M. R. Calera, R. Mata, and R. Pereda-Miranda. 1990. Allelopathic potential of compounds isolated from *Ipomoea tricolor* Cav. (Convolvulaceae). *J. Chem. Ecol.* 16:2145–52.

Anaya, A. L., B. E. Hernandez-Bautista, M. Jimenez-Estrada, and L. Velasco-Ibarra. 1992. Phenylacetic acid as a phytotoxic compound of corn pollen. *J. Chem. Ecol.* 18:897–905.

Anbu, A. D., and S. B. Sullia. 1990. Antibiotic effect of *Rhizobium* sp. towards some soil fungi. *Acta Bot. Indica* 18:213–15.

Anderson, R. C., and A. E. Liberta. 1986. Occurrence of fungal-inhibiting *Pseudomonas* on caryopses of *Tripsacum dactyloides* L. and its implication for seed survival and agriculture application. *J. Appl. Bacteriol.* 61:195–99.

Anonymous. 1969. Natural weed killer. *Sci. Am.* 221:54.

Anuratha, C. S., and S. S. Gnanamanickam. 1990. Biological control of bacterial wilt caused by *Pseudomonas solanacearum* in India with antagonistic bacteria. *Plant & Soil* 124:109–16.

Aoki, M., K. Uehara, K. Koseki, K. Tsuji, M. Iijima, K. Ono, and T. Samejima. 1991. An antimicrobial substance produced by *Pseudomonas cepacia* B5 against the bacterial wilt disease pathogen, *Pseudomonas solanacearum*. *Agric. Biol. Chem.* 55:715–22.

Arnason, J. T., C. J. Bourque, C. Madhosingh, and W. Orr. 1986. Disruption of membrane functions in *Fusarium culmorum* by an acetylenic allelochemical. *Biochem. Syst. Ecol.* 14:569–74.

Ashrof, N., and D. N. Sen. 1980. Allelopathic influences of *Digera alterniflora* on cultivated crops in Indian arid zone. Indian J. Weed Sci. 12:69–74.

Askam, L. R., and D. R. Cornelius. 1971. Influence of desert saltbush saponin on germination. *J. Range Manage.* 24:439–42.

Asthana, A., N. N. Tripathi, and S. N. Dixit. 1986. Fungitoxic and phytotoxic studies with essential oil of *Ocimum adscendens. J. Phytopathol.* (Berlin) 117:152–59.

Aström, B. 1991. Role of bacterial cyanide production in differential reaction of plant cultivars to deleterious rhizosphere pseudomonads. *Plant & Soil* 133:93–100.

Aström, B., and B. Gerhardson. 1989. Wheat cultivar reactions to deleterious rhizosphere bacteria under gnotobiotic conditions. *Plant & Soil* 117:157–65.

Attafuah, A., and J. F. Bradbury. 1989. *Pseudomonas antimicrobica,* a new species strongly antagonistic to plant pathogens. *J. Appl. Bacteriol.* 67:567–73.

Auld, B. A., C. F. McRae, and M. M. Say. 1988. Possible control of *Xanthium spinosum* by a fungus. *Agric. Ecosyst. Environ.* 21:219–24.

Babczinski, P., M. Dorgerioh, A. Lobberding, H.-J. Santel, R. R. Schmidt, P. Schmitt, and C. Wunsche. 1991. Herbicidal activity and mode of action of vulgamycin. *Pestic. Sci.* 33:439–46.

Bakker, A. W., and B. Schippers. 1987. Microbial cyanide production in the rhizosphere in relation to potato yield reduction and *Pseudomonas* spp.-mediated plant growth-stimulation. *Soil Biol. Biochem.* 19:451–57.

Ballegaard, T. K., and E. Warncke. 1985. Observations on autotoxic effects on seed germination and seedling growth in *Cirsium palustre* from a spring area in Jutland, Denmark. *Holarctic Ecol.* 8:63–65.

Ballester, A., A. M. Vieitez, and E. Vieitez. 1982. Allelopathic potential of *Erica vagans, Calluna vulgaris,* and *Daboecia cantabrica. J. Chem. Ecol.* 8:851–57.

Bansal, R. P., and D. N. Sen. 1982. Effect of arid zone weed *Trichodesma sedgwickianum* on growth and yield of bajra and til crops. *Indian J. Bot.* 5:45–49.

Barnes, J. P., A. R. Putnam, and B. A. Burke. 1986. Allelopathic activity of rye (*Secale cereale* L.). In *The Science of Allelopathy.* pp. 271–286, (A. R. Putanam and C. S. Tang, eds.), John Wiley & Sons, New York.

Barnes, J. P., A. R. Putnam, B. A. Burke, and A. J. Aasen. 1987. Isolation and characterization of allelochemicals in rye herbage. *Phytochemistry* 26:1385–90.

Bar-Nun, N., and A. M. Mayer. 1990. Cucurbitacins protect cucumber tissue against infection by *Botrytis cinerea. Phytochemistry* 29:787–91.

Baumgartner, B., C. A. J. Erdelmeier, A. D. Wright, T. Rali, and O. Sticher. 1990. An antimicrobial alkaloid from *Ficus septica. Phytochemistry* 29:3327–30.

Beale, R. E., and D. Pitt. 1990. Biological and integrated control of *Fusarium* basal rot of *Narcissus* using *Minimedusa polyspora* and other micro-organisms. *Plant Pathol.* (London) 39:477–88.

Begum, I., and F. Hussain. 1980. Allelopathic effects of *Panicum antidotale* Retz. *Pak. J. Scient. Indust. Res.* 23:182–88.

Bell, A. A. 1977. Plant pathology as influenced by allelopathy. In *Report of the Research Planning Conference on the Role of Secondary Compounds in Plant Interactions (Allelopathy).* pp. 64–99 (C. G. McWhorter, A. C. Thompson, and E. W. Hauser, eds.), USDA, Agricultural Research Service, Tifton, Georgia.

Benallaoua, S., P. Nguyen Van, M. P. DeMeo, J. Coulon, G. Dumenil, et R. Bonaly. 1990. Recherches sur le mode d'action d'un antifongique non polyénique (désertomycine) produit par une souche de *Streptomyces spectabilis. Can. J. Microbiol.* 36:609–16.

Bennett, A. R., W. L. Bruckart, and N. Shishkoff. 1991. Effects of dew, plant age, and leaf position on the susceptibility of yellow starthistle to *Puccinia jaceae*. *Plant Dis.* 75:499–501.

Bernard, P., and D. Pesando. 1989. Antibacterial and antifungal activity of extracts from the rhizomes of the Mediterranean seagrass *Posidonia oceanica* (L.) Delile. *Bot. Mar.* 32:85–88.

Bhatia, R. K., and D. D. Chawan. 1976. Occurrence of phenolic substances in seed coat of *Cassia* species and their effect on early seedling growth. *Geobios* (Jodhpur) 3:214–16.

Bhattacharyya, D., S. Basu, J. P. Chattapadhyay, and S. K. Bose. 1985. Biocontrol of Macrophomina root-rot disease of jute by an antagonistic organism, *Aspergillus versicolor*. *Plant & Soil* 87:435–38.

Bhowmik, P. C., and J. D. Doll. 1983. Growth analysis of corn and soybean response to allelopathic effects of weed residues at various temperatures and photosynthetic photon flux densities. *J. Chem. Ecol.* 9:1263–80.

Bhowmik, P. C., and J. D. Doll. 1984. Allelopathic effects of annual weed residues on growth and nutrient uptake of corn and soybeans. *Agron. J.* 76:383–88.

Bin, L., G. R. Knudsen, and D. J. Eschen. 1991. Influence of an antagonistic strain of *Pseudomonas fluorescens* on growth and ability of *Trichoderma harzianum* to colonize sclerotia of *Sclerotinia sclerotiorum* in soil. *Phytopathology* 81:994–1000.

Biswas, E., and N. K. Chakraborti. 1984. Allelopathy in *Borreria articularis:* Inhibition of seed germination and seedling growth. *Sci. & Cul.* 50:263–64.

Biswas, E., and K. N. Ghosh. 1984. Allelopathic effects of *Chenopodium murale*—An aspect of weed control for non-associated species. *Sci. & Cul.* 50:99–102.

Blashke, H. 1979. Leaching of water-soluble organic substances from coniferous needle litter. *Soil Biol. Biochem.* 11:581–84.

Blum, U., T. M. Gerig, A. D. Worsham, L. D. Holappa, and L. D. King. 1992. Allelopathic activity in wheat-conventional and wheat-no-till-soils: Development of soil extract bioassays. *J. Chem. Ecol.* 18:2191–2221.

Blum, U., T. R. Wentworth, K. Klein, A. D. Worsham, L. D. King, T. M. Gerig, and S.-W. Lyu. 1991. Phenolic acid content of soils of wheat-no till, wheat-conventional till, and fallow-conventional till soybean cropping systems. *J. Chem. Ecol.* 17: 1045–68.

Bolton, H., Jr., L. F. Elliott, S. Gurusiddaiah, and J. K. Fredrickson. 1989. Characterization of a toxin produced by a rhizobacterial *Pseudomonas* sp. that inhibits wheat growth. *Plant & Soil* 114:279–87.

Bowen, G. D. 1961. The toxicity of legume seed diffusates toward Rhizobia and other bacteria. *Plant & Soil* 15:155–65.

Boyette, C. D., and H. L. Walker. 1985. Evaluation of *Fusarium lateritium* as a biological herbicide for controlling velvetleaf (*Abutilon theophrasti*) and prickly sida (*Sida spinosa*). *Weed Sci.* 34:106–9.

Boyette, C. D., G. J. Weidemann, D. O. Te Beest, and P. C. Quimby, Jr. 1991. Biological control of jimsonweed (*Datura stramonium*) with *Alternaria crassa. Weed Sci.* 39:678–81.

Bradow, J. M., and W. J. Connick, Jr. 1987. Allelochemicals from Palmer amaranth, *Amaranthus palmeri* S. Wats. *J. Chem. Ecol.* 13:185–202.

Bradow, J. M., and W. J. Connick, Jr. 1988a. Volatile methyl ketone seed-germination inhibitors from *Amaranthus palmeri* S. Wats. residues. *J. Chem. Ecol.* 14:1617–31.

Bradow, J. M., and W. J. Connick, Jr. 1988b. Seed-germination inhibition by volatile alcohols and other compounds associated with *Amaranthus palmeri* residues. *J. Chem. Ecol.* 14:1633–48.

Bradow, J. M., and W. J. Connick, Jr. 1990. Volatile seed germination inhibitors from plant residues. *J. Chem. Ecol.* 16:645–66.

Bradshaw-Smith, R. P., W. M. Whalley, and G. D. Craig. 1991. Interactions between *Pythium oligandrum* and the fungal footrot pathogens of peas. *Mycol. Res.* 95: 861–65.

Brian, P. W., J. M. Wright, J. Stubbs, and A. W. Way. 1951. Uptake of antibiotic metabolites of soil microorganisms by plants. *Nature* 167:347–49.

Brian, P. W., A. W. Dawkins, J. F. Grove, H. G. Hemming, D. Lowe, and G. L. F. Norris. 1961. Phytotoxic compounds produced by *Fusarium equiseti*. *J. Exp. Bot.* 12:1–12.

Brisbane, P. G., and A. D. Rovira. 1988. Mechanisms of inhibition of *Gaeumannomyces graminis* var. *tritici* by fluorescent pseudomonads. *Plant Pathol.* (London) 37:104–11.

Brisbane, P. G., J. R. Harris, and R. Moen. 1989. Inhibition of fungi from wheat roots by *Pseudomonas fluorescens* 2-79 and fungicides. *Soil Biol. Biochem.* 21:1019–25.

Brown, A. E., and J. T. G. Hamilton. 1992. Indole-3-ethanol produced by *Zygorrhynchus moelleri,* an indole-3-acetic acid analogue with antifungal activity. *Mycol. Res.* 96:71–74.

Brown, A. E., R. Finlay, and J. S. Ward. 1987. Antifungal compounds produced by *Epicoccum purpurascens* against soil-borne plant pathogenic fungi. *Soil Biol. Biochem.* 19:657–64.

Budge, S. P., and J. M. Whipps. 1991. Glasshouse trials of *Coniothyrium minitans* and *Trichoderma* species for the biological control of *Sclerotinia sclerotiorum* in celery and lettuce. *Plant Pathol.* (Oxford) 40:59–66.

Bull, C. T., D. M. Weller, and L. S. Thomashow. 1991. Relationship between root colonization and suppression of *Gaeumannomyces graminis* var. *tritici* by *Pseudomonas fluorescens* 2-79. *Phytopathology* 81:954–59.

Burr, T. J., and A. Caesar. 1984. Beneficial plant bacteria. *Critical Rev. Plant Sci.* 2:1–20.

Buta, J. G., and D. W. Spaulding. 1989. Allelochemicals in tall fescue-abscisic and phenolic acids. *J. Chem. Ecol.* 15:1629–36.

de Cal, A., E. M. Sagasta, and P. Melgarejo. 1988. Antifungal substances produced by *Penicillium frequentans* and their relationship to the biocontrol of *Monilinia laxa*. *Phytopathology* 78:888–93.

de Cal, A., E. M. Sagasta, and P. Melgarejo. 1990. Biological control of peach twig blight (*Monilinia laxa*) with *Penicillium frequentans*. *Plant Pathol.* 39:612–18.

Campbell, D. G., P. M. Richardson, and A. Rosas, Jr. 1989. Field screening for allelopathy in tropical forest trees, particularly *Duroia hirsuta,* in the Brazilian Amazon. *Biochem. Sys. Ecol.* 17:403–7.

Cantone, F. A., and L. D. Dunkle. 1990. Involvement of an inhibitory compound in induced resistance of maize to *Helminthosporium carbonum. Phytopathology* 80:1225–30.

Carballeira, A., and A. Cuervo. 1980. Seasonal variation in allelopathic potential of soils from *Erica australis* L. heathland. *Acta Oecologica* 1:345–53.

Carson, A. G. 1989. Effect of intercropping sorghum and groundnuts on density of *Striga hermonthica* in The Gambia. *Trop. Pest Manage.* 35:130–32.

Cast, K. G., J. K. McPherson, A. J. Pollard, E. G. Krenzer, and G. R. Waller. 1990. Allelochemicals in soil from no-tillage versus conventional-tillage wheat (*Triticum aestivum*) fields. *J. Chem. Ecol.* 16:2277–89.

Chandrasekaran, S., and T. Yoshida. 1973. Effect of organic acid transformations in submerged soils on growth of rice plant. *Soil Sci. Pl. Nutr.* 19:39–45.

Chanway, C. P., and F. B. Holl. 1991. Biomass increase and associative nitrogen fixation of mycorrhizal *Pinus contorta* seedlings inoculated with a plant growth promoting *Bacillus* strain. *Can. J. Bot.* 69:507–11.

Chanway, C. P., F. B. Holl, and R. Turkington. 1988a. Genotypic coadaptation in plant growth promotion of forage species by *Bacillus polymyxa. Plant & Soil* 106:281–84.

Chanway, C. P., L. M. Nelson, and F. B. Holl. 1988b. Cultivar-specific growth promotion of spring wheat (*Triticum aestivum* L.) by coexistent *Bacillus* species. *Can. J. Microbiol.* 34:925–29.

Chanway, C. P., F. B. Holl, and R. Turkington. 1990. Specificity of association between *Bacillus* isolates and genotypes of *Lolium perenne* and *Trifolium repens* from a grass-legume pasture. *Can. J. Bot.* 68:1126–30.

Chanway, C. P., R. A. Radley, and F. B. Holl. 1991. Inoculation of conifer seed with plant growth promoting *Bacillus* strains causes increased seedling emergence and biomass. *Soil Biol. Biochem.* 23:575–80.

Chapman, S. J., and J. M. Lynch. 1983. The relative role of microorganisms and their metabolites in the phytotoxicity of decomposing plant residues. *Plant & Soil* 74:457–59.

Chase, W. R., M. G. Nair, and A. R. Putnam. 1991a. 2,2'-Oxo-1,1'-azobenzene: Selective toxicity of rye (*Secale cereale* L.) allelochemicals to weed and crop species: II. *J. Chem. Ecol.* 17:9–19.

Chase, W. R., M. G. Nair, A. R. Putnam, and S. K. Mishra. 1991b. 2,2'-Oxo-1,1'-azobenzene: Microbial transformation of rye (*Secale cereale* L.) allelochemical in field soils by *Acinetobacter calcoaceticus:* III. *J. Chem. Ecol.* 17:1575–84.

Chaturvedi, R., A. Dikshit, and S. N. Dixit. 1987. *Adenocalymma allicea,* a new source of a natural fungitoxicant. *Trop. Agric.* 64:318–22.

Chaurasia, S. N. P., and D. Ram. 1990. Inhibition of mycelial growth and sclerotium formation in *Rhizoctonia solani* Kuhn and *Sclerotinia sclerotiorum* (Lib) de Bary by asafoetida. *Trop. Sci.* 30:15–19.

Cherrington, C. A., and L. F. Elliott. 1987. Incidence of inhibitory pseudomonads in the Pacific Northwest. *Plant & Soil* 101:159–65.

Chiang, M.-Y., C. G. van Dyke, and K. J. Leonard. 1989a. Evaluation of endemic foliar fungi for potential biological control of johnsongrass (*Sorghum halepense*): Screening and host range tests. *Plant Dis.* 73:459–64.

Chiang, M.-Y., C. G. van Dyke, and W. S. Chilton. 1989b. Four foliar pathogenic fungi for controlling seedling johnsongrass (*Sorghum halepense*). *Weed Sci.* 37:802–9.

Choesin, D. N., and R. E. J. Boerner. 1991. Allyl isothiocyanate release and the allelopathic potential of *Brassica napus* (Brassicaceae). *Am. J. Bot.* 78:1083–90.

Chou, C. H., and C. Yao. 1983. Phytochemical adaptation of coastal vegetation in Taiwan. I. Isolation, identification and biological activities of compounds in *Vitex negundo* L. *Bot. Bull. Acad. Sin.* 24:155–68.

Chou, C. H., and Y. L. Kuo. 1986. Allelopathic research of subtropical vegetation in Taiwan III. Allelopathic exclusion of understory by *Leucaena leucocephala* (Lam.) de Wit. *J. Chem. Ecol.* 12:1431–48.

Chou, C. H., and Y.-F. Lee. 1991. Allelopathic dominance of *Miscanthus transmorrisonensis* in an alpine grassland community in Taiwan. *J. Chem. Ecol.* 17:2267–81.

Chou, C. H., and L. L. Leu. 1992. Allelopathic substances and interactions of *Delonix regia* (Boj. Raf.) *J. Chem. Ecol.* 18:2285–2303.

Chou, C. H., M. L. Lee, and H. I. Oka. 1984. Possible allelopathic interaction between *Oryza perennis* and *Leersia hexandra. Bot. Bull. Acad. Sin.* 25:1–19.

Chou, C. H., F. J. Chang, and H. I. Oka. 1991. Allelopathic potentials of a wild rice, *Oryza perennis. Taiwania* 36:201–10.

Ciampi-Panno, L., C. Fernandez, P. Bustamante, N. Andrade, S. Ojeda, and A. Contreras. 1989. Biological control of bacterial wilt of potatoes caused by *Pseudomonas solanacearum. Am. Potato J.* 66:315–32.

Clark, C. A. 1989. Influence of volatiles from healthy and decaying sweet potato storage roots on sclerotial germination and hyphal growth of *Sclerotium rolfsii. Can. J. Bot.* 67:53–57.

Claydon, N., M. Allan, J. R. Hanson, and A. G. Avent. 1987. Antifungal alkyl pyrones of *Trichoderma harzianum. Trans. Br. Mycol. Soc.* 88:503–13.

Cochran, V. L., L. F. Elliott, and R. I. Papandick. 1977. The production of phytotoxins from surface crop residues. *Soil Sci. Soc. Am. J.* 41:903–8.

Cochrane, V. W. 1948. The role of plant residues in the etiology of root rot. *Phytopathology* 38:185–96.

Cole, M. D., P. D. Bridge, J. E. Dellar, L. E. Fellows, M. C. Cornish, and J. C. Anderson. 1991. Antifungal activity of neo-clerodane diterpenoids from *Scutellaria. Phytochemistry* 30:1125–27.

Conard, S. G. 1985. Inhibition of *Abies concolor* radicle growth by extracts of *Ceonothus velutinus. Madroño* 32:118–20.

Coté, J. F., and J. R. Thibault. 1988. Allelopathic potential of raspberry foliar leachates on growth of ectomycorrhizal fungi associated with black spruce. *Am. J. Bot.* 75:966–70.

Crookston, R. K., J. E. Kurle, P. J. Copeland, J. H. Ford, and W. E. Lueschen. 1991. Rotational cropping sequence affects yield of corn and soybean. *Agron. J.* 83: 108–13.

de la Cruz, A. R., A. R. Poplawsky, and M. V. Wiese. 1992. Biological suppression of potato ring rot by fluorescent pseudomonads. *Appl. Environ. Microbiol.* 58:1986–91.

Cruz-Ortega, R., A. L. Anaya, and L. Ramos. 1988. Effects of allelopathic compounds of corn pollen on respiration and cell division of watermelon. *J. Chem. Ecol.* 14:71–86.

Dahiya, J. S., D. L. Woods, and J. P. Tewari. 1988. Control of *Rhizoctonia solani,* causal agent of brown girdling root rot of rapeseed, by *Pseudomonas fluorescens. Bot. Bull. Acad. Sin.* 29:135–42.

Daigle, D. J., and P. J. Cotty. 1991. Factors that influence germination and mycoherbicidal activity of *Alternaria cassiae. Weed Technol.* 5:82–86.

Dandurand, L. M., and J. A. Menge. 1992. Influence of *Fusarium solani* on citrus root rot caused by *Phytophthora parasitica* and *Phytophthora citrophthora. Plant Soil* 144:13–21.

Daroesman, R. 1981. Vegetative elimination of alang-alang. *Bull. Indonesian Econ. Studies* XVII (March):83–107.

Das, T. M., and S. K. Pal. 1970. Effects of volatile substances of aromatic weeds on germination and subsequent growth of rice embryos. *Bull. Bot. Soc. Bengal* 24: 101–3.

DaSilva, S. R., and S. F. Pascholati. 1992. *Saccharomyces cerevisiae* protects maize plants, under greenhouse conditions, against *Colletotrichum graminicola. Z. Planzenkr. Pflanzenshutz* 99:159–67.

Datta, S. C., and S. D. Chakrabarti. 1982a. Allelopathy in *Clerodendrum viscosum:* inhibition of mustard (*Brassica*) germination and seedling growth. *Comp. Physiol. Ecol.* 7:1–7.

Datta, S. C., and S. D. Chakrabarti. 1982b. Allelopathic potential of *Clerodendrum viscosum* Vent. in relation to germination and seedling growth of weeds. *Flora* 172:89–95.

Datta, S. C., and K. N. Ghosh. 1982. Effect of pre-sowing treatment of mustard seeds with leaf and inflorescence extracts of *Chenopodium murale. Indian J. Weed Sci.* 14:1–6.

Datta, S. C., and S. P. Sinha-Roy. 1983. Characteristics of an inhibitory factor from *Croton bonplandianum* Baill. *Acta Agron. Acad. Sci. Hungaricae* 32:124–29.

Datta, S. C., and S. Dasmahapatra. 1984. Allelopathy in two legumes: Inhibition of mustard (*Brassica*) germination and seedling growth. *Comp. Physiol. Ecol.* 9: 285–89.

Datta, S. C., and K. N. Ghosh. 1987. Allelopathy in 2 species of *Chenopodium*—Inhibition of germination and seedling growth of certain weeds. *Acta Soc. Bot. Pol.* 56:257–70.

Datta, S. C., T. Das, and R. K. Bhakat. 1985. Inhibitors in leaves of road-side trees during various seasons. *Sci. & Culture* 51:313–15.

Deb, P. R. 1990. *In vitro* inhibitory activity of some rhizosphere fungi of soybean against *Sclerotium rolfsii* Sacc. growth. *Acta Bot. Indica* 18:159–62.

Deb, P. R, and B. K. Dutta. 1991. Studies on biological control of foot rot disease of soybean caused by *Sclerotium rolfsii* Sacc. *Z. Pflanzenkr. Pflanzenschutz* 98:539–46.

Dekker, J., and W. F. Meggitt. 1983. Interference between velvetleaf (*Abutilon theophrasti* Medic.) and soybean (*Glycine max* (L). Merr.) I. growth. *Weed Res.* 23:91–101.

Devi, T. V., R. Malar Vizhi, N. Sakthivel, and S. S. Gnanamanickam. 1989. Biological control of sheath-blight of rice in India with antagonistic bacteria. *Plant & Soil* 119:325–30.

Dewan, M. M., and K. Sivasithamparam. 1988a. A plant-growth-promoting sterile fungus from wheat and rye-grass roots with potential for suppressing take-all. *Trans. Br. Mycol. Soc.* 91:687–92.

Dewan, M. M., and K. Sivasithamparam. 1988b. Identity and frequency of occurrence of *Trichoderma* spp. in roots of wheat and rye-grass in Western Australia and their effect on root rot caused by *Gaeumannomyces gramini* var. *tritici. Plant Soil* 109:93–101.

Dewan, M. M., and K. Sivasithamparam. 1989. Growth promotion of rotation crop species by a sterile fungus from wheat and effect of soil temperature and water potential on its suppression of take-all. *Mycol. Res.* 93:156–60.

Dewan, M. M., and K. Sivasithamparam. 1991. Promotion of growth of wheat by a sterile red fungus in relation to plant density and soil fertility. *Plant & Soil* 135:306–8.

Dhyani, S. K. 1978. Allelopathic potential of *Eupatorium adenophorum* on seed germination of *Lantana camara* var. *aculeata. Indian J. For.* 1:311.

Didry, N., M. Pinkas, and L. Dubreuil. 1985. Antimicrobial activity of some naphthoquinones found in plants. *Ann. Pharm. Fr.* 44:73–78.

DiPietro, A., M. Gut-Rella, J. P. Pachlatko, and F. J. Schwinn. 1992. Role of antibiotics produced by *Chaetomium globosum* in biocontrol of *Pythium ultimum,* a causal agent of damping-off. *Phytopathology* 82:131–35.

Dirvi, A. D., and F. Hussain. 1979. Allelopathic effects of *Dichantium annulatum* (Forsk) Stapf on some cultivated plants. *Pak. J. Scient. Indust. Res.* 22:194–97.

Donnelly, D. M. X., and M. H. Sheridan. 1986. Anthraquinones from *Trichoderma polysporum. Phytochemistry* 25:2303–2304.

Downum, K. R., S. Villegas, E. Rodriquez, and D. J. Keil. 1989. Plant photosensitizers: A survey of their occurrence in arid and semiarid plants from North America. *J. Chem. Ecol.* 15:345–55.

Drew, A. P. 1988. Interference of black cherry by ground flora of the Allegheny uplands. *Can. J. For. Res.* 18:652–56.

Dube, S., A. Kumar, and S. C. Tripathi. 1990. Antifungal and insect repellent activity of essential oil of *Zanthoxylum alatum.* Ann. Bot. 65:457–59.

Dube, S., P. D. Upadhyay, and S. C. Tripathi. 1989. Antifungal, physicochemical, and insect-repelling activity of the essential oil of *Ocimum basilicum. Can. J. Bot.* 67:2085–87.

Dubey, P. S. 1973. Phytotoxicity of weeds to crops. I. Effect on germination. *Sci. & Cult.* 39:556–58.

Duchesne, L. C., R. L. Peterson, and B. E. Ellis. 1988. Pine root exudate stimulates the synthesis of antifungal compounds by the ectomycorrhizal fungus *Paxillus involutus. New Phytol.* 108:471–76.

Duchesne, L. C., R. L. Peterson, and B. E. Ellis. 1989a. The time course of disease suppression and antibiosis by the ectomycorrhizal fungus *Paxillus involutus. New Phytol.* 111:693–98.

Duchesne, L. C., B. E. Ellis, and R. L. Peterson. 1989b. Disease suppression by the ectomycorrhizal fungus *Paxillus involutus:* Contribution of oxalic acid. *Can. J. Bot.* 67:2726–30.

Dumas, M. T. 1992. Inhibition of *Armillaria* by bacteria isolated from soils of the boreal mixedwood forest of Ontario. *Eur. J. For. Pathol.* 22:11–18.

Dunevitz, V. and J. Ewel. 1981. Allelopathy of wax myrtle (*Myrica cerifera*) on *Schinus terebinthiofolius* (Florida holly). *Florida Scient.* 44:13–20.

Dunlop, R. W., A. Simon, K. Sivasithamparam, and E. L. Ghisalberti. 1989. An antibiotic from *Trichoderma koningii* active against soilborne plant pathogens. *J. Nat. Prod.* 52:67–74.

Edwards, M. E., E. M. Harris, F. H. Wagner, M. C. Cross, and G. S. Miller. 1988. Seed germination of American pokeweed (*Phytolacca americana*). I. Laboratory techniques and autotoxicity. *Am. J. Bot.* 75:1794–1802.

Einhellig, F. A., and J. A. Rasmussen. 1989. Prior cropping with grain sorghum inhibits weeds. *J. Chem. Ecol.* 15:951–60.

Einhellig, F. A., and I. F. Souza. 1992. Phytotoxicity of sorgoleone found in grain sorghum root exudates. *J. Chem. Ecol.* 18:1–11.

Einhellig, F. A., J. A. Rasmussen, A. M. Hejl, and I. F. Souza. 1993. Effects of root exudate sorgoleone on photosynthesis. *J. Chem. Ecol.* 19:369–75.

Eisikowitch, D., M. A. LaChance, P. G. Kevan, S. Willis, and D. L. Collins-Thompson. 1990. The effect of the natural assemblage of microorganisms and selected strains of the yeast *Metschnikowia reukaufii* in controlling the germination of pollen of the common milkweed *Asclepias syriaca. Can. J. Bot.* 68:1163–65.

Elakovich, S. D. 1989. Allelopathic aquatic plants for aquatic weed management. *Biol. Plant.* 31:479–86.

El-Ghareeb, R. M. 1991. Suppression of annuals by *Tribulus terrestris* in an abandoned field in the sandy desert of Kuwait. *J. Veg. Sci.* 2:147–54.

El-Goorani, M. A., and F. M Hassanein. 1991. The effect of *Bacillus subtilis* on *in vitro* growth and pathogenicity of *Erwinia amylovora. J. Phytopathology* 133:134–38.

Elliott, L. F., and J. M. Lynch. 1984. Pseudomonads as a factor in growth of winter wheat (*Triticum aestivum* L.). *Soil Biol. Biochem.* 16:69–71.

Elliott, L. F., and J. M. Lynch. 1985. Plant growth-inhibitory pseudomonads colonizing winter wheat (*Triticum aestivum* L.) roots. *Plant & Soil* 84:57–65.

Ells, J. E., and A. E. McSay. 1991. Allelopathic effects of alfalfa plant residues on emergence and growth of cucumber seedlings. *Hort. Sci.* 26:368–70.

El-Shanshoury, A. E.-R. R., M. A. Hassan, and B. A. Abdel-Ghaffar. 1989. Synergistic effect of vesicular-arbuscular-mycorrhizas and *Azotobacter chroococcum* on the growth and the nutrient contents of tomato plants. *Phyton* (Austria) 29:203–12.

Elsherif, M., and F. Grossmann. 1991. Investigations on biological control of some plant pathogenic fungi by fluorescent pseudomonads using different methods of application. *Z. Pflanzenkr. Pflanzenschutz* 98:236–49.

Enache, A. J., and R. D. Ilnicki. 1990. Weed control by subterranean clover (*Trifolium subterraneum*) used as a living mulch. *Weed Technol.* 4:534–38.

Endo, K., E. Kanno, and Y. Oshima. 1990. Structures of antifungal diarylheptenones, gingerenones A, B, C and isogingerenone B, isolated from the rhizomes of *Zingiber officinale. Phytochemistry* 29:797–99.

Escande, A. R., and E. Echandi. 1991. Protection of potato from *Rhizoctonia* canker with binucleate *Rhizoctonia* fungi. *Plant Pathol.* (Oxford) 40:197–202.

Eyini, M., M. Jayakumar, and S. Pannirselvam. 1989. Allelopathic effect of bamboo leaf extract on the seedling of groundnut. *Trop. Ecol.* 30:138–41.

Farquhar, M. L., and R. L. Peterson. 1990. Early effects of the ectomycorrhizal fungus

*Paxillus involutus* on the root rot organism *Fusarium* associated with *Pinus resinosa*. *Can. J. Bot.* 68:1589–96.

Farquhar, M. L., and R. L. Peterson. 1991. Later events in suppression of *Fusarium* root rot of red pine seedlings by the ectomycorrhizal fungus *Paxillus involutus*. *Can. J. Bot.* 69:1372–83.

Fay, P. K., and W. B. Duke. 1977. An assessment of allelopathic potential in *Avena* germplasm. *Weed Sci.* 25:224–28.

Ferguson, D. E. 1991. Allelopathic potential of western coneflower (*Rudbeckia occidentalis*). *Can. J. Bot.* 69:2806–8.

Ferreira, J. H. S., F. N. Matthee, and A. C. Thomas. 1991. Biological control of *Eutypa lata* on grapevine by an antagonistic strain of *Bacillus subtilis*. *Phytopathology* 81:283–87.

Figliola, S. S., N. D. Camper, and W. H. Ridings. 1988. Potential biological control agents for goosegrass (*Eleusine indica*). *Weed Sci.* 36:830–35.

Filippi, C., G. Bagnoli, M. Volterrani, and G. Picci. 1987. Antagonistic effects of soil bacteria on *Fusarium oxysporum* Schlecht f. sp. *dianthi* (Prill and Del.) Snyd. and Hans. III. Relation between protection against fusarium wilt in carnation and bacterial antagonists colonization on roots. *Plant & Soil* 98:161–67.

Fischer, N. H., N. Tanrisever, and G. B. Williamson. 1988. Allelopathy in the Florida scrub community as a model for natural herbicide actions. In *Biologically Active Natural Products: Potential Use in Agriculture*. pp. 232–249 (H. G. Cutler, ed.), ACS Symposium Series No. 380, Am. Chem. Soc., Wash., D. C.

Fischer, N. H., J. D. Weidenhamer, and J. M. Bradow. 1989a. Inhibition and promotion of germination by several sesquiterpenes. *J. Chem. Ecol.* 15:1785–93.

Fischer, N. H., J. D. Weidenhamer, and J. M. Bradow. 1989b. Dihydroparthenolide and other sesquiterpene lactones stimulate witchweed germination. *Phytochemistry* 28:2315–17.

Fischer, N. H., J. D. Weidenhamer, J. L. Riopel, L. Quijano, and M. A. Menelaou. 1990. Stimulation of witchweed germination by sesquiterpene lactones: A structure-activity study. *Phytochemistry* 29:2479–83.

Fisher, R. F., and F. Adrian. 1981. Bahiagrass impairs slash pine seedling growth. *Tree Planters' Notes* 32:19–21.

Forney, R. D., C. L. Foy, and D. D. Wolfe. 1985. Weed suppression in no-till alfalfa (*Medicago sativa*) by prior cropping of summer-annual grasses. *Weed Sci.* 33:490–97.

Fox, T. R., and N. B. Comerford. 1990. Low-molecular-weight organic acids in selected forest soils of the southeastern USA. *Soil Sci. Soc. Am. J.* 54:1139–44.

Fredrickson, J. K., and L. F. Elliott. 1985a. Colonization of winter wheat roots by inhibitory rhizobacteria. *Soil Sci. Soc. Am. J.* 49:1172–77.

Fredrickson, J. K., and L. F. Elliott. 1985b. Effects on winter wheat seedling growth by toxin producing rhizobacteria. *Plant & Soil* 83:399–409.

de Freitas, J. R., and J. J. Germida. 1991. *Pseudomonas cepacia* and *Pseudomonas putida* as winter wheat inoculants for biocontrol of *Rhizoctonia solani*. *Can. J. Microbiol.* 37:780–84.

French, R. C. 1990. Stimulation of germination of teliospores of *Puccinia punctiformis* by monyl, decyl, and dodecyl isothiocyanates and related volatile compounds. *J. Agr. Food Chem.* 38:1604–7.

French, R. C. 1992. Volatile chemical germination stimulators of rust and other fungal spores. *Mycologia* 84:277–88.

French, R. C., and M. D. Gallimore. 1971. Effect of some nonyl derivatives and related compounds on germination of uredospores. *J. Agr. Food Chem.* 19:912–15.

French, R. C., and M. D. Gallimore. 1972. Stimulation of germination of uredospores of stem rust of wheat in the pustule by n-nonanal and related compounds. *J. Agr. Food Chem.* 20:421–23.

French, R. C., A. W. Gale, C. L. Graham, and H. W. Rines. 1975. Factors affecting chemical stimulation of uredospore germination in pustules of crown rust of oats, common corn rust, stem rust of wheat, and leaf rust of wheat. *J. Agr. Food Chem.* 23:4–8.

Friedman, J., A. Hutchins, C. Y. Li, and D. A. Perry. 1989. Actinomycetes inducing phytotoxic or fungistatic activity in a Douglas-fir forest and in an adjacent area of repeated regeneration failure in southwestern Oregon. *Biol. Plantarum* 31:487–95.

Gagliardo, R. W., and W. S. Chilton. 1992. Soil transformation of 2(3H)-benzoxazolone of rye into phytotoxic 2-amino-3H-phenoxazin-3-one. *J. Chem. Ecol.* 18:1683–91.

Gaikwad, S. J., B. Sen, and S. U. Meshram. 1987. Effect of bottlegourd seedcoating with antagonists on seedlings, quantum of the pathogen inside the seedlings and population of the soil against *Fusarium oxysporum. Plant & Soil* 101:205–10.

Ganesan, P., and S. S. Gnanamanickam. 1987. Biological control of *Sclerotium rolfsii* Sacc. in peanut by inoculation with *Pseuodomonas fluorescens. Soil Biol. Biochem.* 19:35–38.

Gees, R., and M. D. Coffey. 1989. Evaluation of a strain of *Myrothecium roridum* as a potential biocontrol agent against *Phytophthora cinnamomi. Phytopathology* 79:1079–84.

Ghewande, M. P. 1989. Management of foliar diseases of groundnut (*Arachis hypogaea*) using plant extracts. *Indian J. Agric. Sci.* 59:133–34.

Ghewande, M. P. 1990. Biological control of groundnut (*Arachis hypogaea* L.) rust (*Puccinia arachidis* Speg.) in India. *Trop. Pest Manage.* 36:17–20.

Ghisalberti, E. L., M. J. Narbey, M. M. Dewan, and K. Sivasithamparam. 1990. Variability among strains of *Trichoderma harzianum* in their ability to reduce take-all and to produce pyrones. *Plant & Soil* 121:287–91.

Gibson, M. T., I. M. Welch, P. R. F. Barrett, and I. Ridge. 1990. Barley straw as an inhibitor of algal growth II: laboratory studies. *J. Appl. Phycol.* 2:241–48.

Gilmore, A. R. 1985. Allelopathic effects of giant foxtail on germination and radicle elongation of loblolly pine seed. *J. Chem. Ecol.* 11:583–92.

Goel, R. K., and J. S. Jhooty. 1987. Stimulation of germination in teliospores of *Urocystis agropyri* by volatiles from plant tissues. *Ann. Appl. Biol.* 111:295–300.

Goel, U., and T. S. Sareen. 1986. Allelopathic effect of trees on the understory vegetation. *Acta Bot. Indica* 14:162–66.

Goel, U., D. B. Saxena, and B. Kumar. 1989. Comparative study of allelopathy as exhibited by *Prosopis juliflora* Swartz and *Prosopis cineraria* (L.) Druce. *J. Chem. Ecol.* 15:591–600.

Goldberg, N. P., M. C. Hawes, and M. E. Stanghellini. 1989. Specific attraction to and infection of cotton root cap cells by zoospores of *Pythium dissotocum*. *Can. J. Bot.* 67:1760–67.

Golubev, V. I. 1989. Spectrum of action of killer toxins produced by *Rhodotorula glutinis* and its taxonomic significance. *Microbiology* 58:84–87. (English translation)

Gómez-Garibay, F., R. Reyes Chilpa, L. Quijano, J. S. Calderón Pardo, and T. Ríos Castillo. 1990. Methoxy furan auranols with fungistatic activity from *Lonchocarpus castilloi*. *Phytochemistry* 29:459–63.

González, A. G., T. Abad, I. A. Jiménez, A. G. Ravelo, J. G. Luis, Z. Aguiar, L. S. Andrés, M. Plasencia, J. R. Herrera, and L. Moujir. 1989. A first study of antibacterial activity of diterpenes from some *Salvia* species (Lamiaceae). *Biochem. Syst. Ecol.* 17:293–96.

González-Farias, F., and L. D. Mee. 1988. Effect of mangrove humic-like substances on biodegradation rate of detritus. *J. Exp. Mar. Biol. Ecol.* 119:1–13.

Goodman, D. M., and L. L. Burpee. 1991. Biological control of dollar spot disease of creeping bentgrass. *Phytopathology* 81:1438–46.

Gordon-Lennox, G., D. Walther, et D. Gindrat. 1987. Utilisation d'antagonistes pour l'enrobage des semences: efficacité et mode d'action contre les agents de la fonte des semis. *EPPO Bull.* 17:631–37.

Gören, N., J. Jakupovic, and S. Topal. 1990. Sesquiterpene lactones with antibacterial activity from *Tanacetum argyrophyllum* var. *argyrophyllum*. *Phytochemistry* 29: 1467–69.

Gorski, P. M., J. Miersch, and M. Ploszynski. 1991. Production and biological activity of saponins and canavanine in alfalfa seedlings. *J. Chem. Ecol.* 17:1135–43.

Govindasamy, V., and R. Balasubramanian. 1989. Biological control of groundnut rust, *Puccinia arachidis*, by *Trichoderma harzianum*. *Z. Pflanzenkr. Pflanzenschutz* 96:337–45.

Grant, N. T., E. Prusinkiewicz, R. M. D. Makowski, B. Holmstrom-Ruddick, and K. Mortensen. 1990. Effect of selected pesticides of survival of *Colletotrichum gloeosporioides* f. sp. *malvae*, a bioherbicide for round-leaved mallow (*Malva pusilla*). *Weed Technol.* 4:701–15.

Gubbels, G. H., and E. O. Kenaschuk. 1989. Agronomic performance of flax grown on canola, barley and flax stuble with and without tillage prior to seeding. *Can. J. Plant Sci.* 69:31–38.

Gueldner, R. C., C. C. Reilly, P. L. Pusey, C. E. Costello, R. F. Arrendale, R. H. Cox, D. S. Himmelsbach, F. G. Crumley, and H. G. Cutler. 1988. Isolation and identification of iturins as antifungal peptides in biological control of peach brown rot with *Bacillus subtilis*. *J. Agr. Food Chem.* 36:366–69.

Guenzi, W. D., W. D. Kehr, and T. M. McCalla. 1964. Water-soluble phytotoxic substances in alfalfa foliage; variation with variety, cutting, year and stage of growth. *Agron. J.* 56:499–500.

Gussin, E. J., and J. M. Lynch. 1981. Microbial fermentation of grass residues to organic acids as a factor in the establishment of new grass swards. *New Phytol.* 89:449–57.

Habeshaw, D. 1980. Endogenous growth and germination inhibitors and their role in grass survival and pasture management. *Grass Forage Sci.* 35:69–70.

Hadar, Y., and B. Gorodecki. 1991. Suppression of germination of sclerotia of *Sclerotium rolfsii* in compost. *Soil Biol. Biochem.* 23:303–6.

Hagedorn, C., W. D. Gould, and T. R. Bardinelli. 1989. Rhizobacteria of cotton and their repression of seedling disease pathogens. *Appl. Environ. Microbiol.* 55:2793–97.

Hagin, R. D. 1989. Isolation and identification of 5-hydroxyindole-3-acetic acid and 5-hydroxytryptophan, major allelopathic aglycons in quackgrass (*Agropyron repens* L. Beauv.). *J. Agr. Food Chem.* 37:1143–49.

Hajlaoui, M. R., N. Benhamou, and R. R. Belanger. 1992. Cytochemical study of the antagonistic activity of *Sporothrix flocculosa* on rose powdery mildew, *Sphaerotheca pannosa* var. *rosae. Phytopathology* 82:583–89.

Hall, K. C., S. J. Chapman, D. G. Christean, and M. B. Jackson. 1986. Abscisic acid in straw residues from autumn-sown wheat. *J. Sci. Food Agri.* 37:219–22.

Hall, M. H., and P. R. Henderlong. 1989. Alfalfa autotoxic fraction characterization and initial separation. *Crop Sci.* 29:425–28.

Hall, T. J., and W. E. E. Davis. 1990. Survival of *Bacillus subtilis* in silver and sugar maple seedlings over a two-year period. *Plant Dis.* 74:608–9.

Hamilton-Kemp, T. R., and R. A. Andersen. 1986. Volatile compounds from wheat plants: Isolation, identification, and origin. In *Biogeneration of Aromas.* pp. 193–200 (T. H. Parliment and R. Croteau, eds.), ACS Symposium Series 317, Washington, D.C.

Hamilton-Kemp, T. R., C. T. McCracken, Jr., J. H. Loughrin, R. A. Andersen, and D. F. Hildebrand. 1992. Effects of some natural volatile compounds on the pathogenic fungi *Alternaria alternata* and *Botrytis cinerea. J. Chem. Ecol.* 18:1083–91.

Handelsman, J., S. Raffel, E. H. Mester, L. Wunderlich, and C. R. Grau. 1990. Biological control of damping-off of alfalfa seedlings with *Bacillus cereus* UW85. *Appl. Environ. Microbiol.* 56:713–18.

Hanson, P. J., and R. K. Dixon. 1987. Allelopathic effects of interrupted fern on northern red oak seedlings: amerlioration by *Suillus luteus* L.: Fr. *Plant & Soil* 98:43–51.

ul Haq, I., and F. Hussain. 1979. Effect of root exudates of tobacco (*Nicotiana rustica*L.) on some cultivated plants. Pak. Tobacco 3:17–19.

Hardy, G. E. St. J, and K. Sivasithamparam. 1991. Suppression of *Phytophthora* root rot by a composted *Eucalyptus* bark mix. *Aust. J. Bot.* 39:153–59.

Harrison, H. F., Jr., and J. K. Peterson. 1986. Allelopathic effects of sweet potatoes (*Ipomoea batatas*) on yellow nutsedge (*Cyperus esculentus*) and alfalfa (*Medicago sativa*). *Weed Sci.* 34:623–27.

Harrison, L., D. B. Teplow, M. Rinaldi, and G. Strobel. 1991. Pseudomycins, a family of novel peptides from *Pseudomonas syringae* possessing broad-spectrum antifungal activity. *J. Gen. Microbiol.* 137:2857–65.

Hartung, A. C., M. G. Nair, and A. R. Putnam. 1990. Isolation and characterization of phytotoxic compounds from asparagus (*Asparagus officinalis* L.) roots. *J. Chem. Ecol.* 16:1707–18.

Hartung, A. C., A. R. Putnam, and C. T. Stephens. 1989. Inhibitory activity of asparagus root tissue and extracts on asparagus seedlings. *J. Am. Soc. Hort. Sci.* 114: 144–48.

Hasan, S. 1991. The biology and host specificity of the onion weed rust, *Puccinia barbeyi,* a potentially useful agent for biological control in Australia. *Ann. Appl. Biol.* 118:19–25. Abstr. in *Rev. Plant Pathol.* 70:663 (Abstr. No. 5228).

Hasan, S., and P. G. Ayres. 1990. The control of weeds through fungi: Principles and prospects. *New Phytol.* 115:201–22.

Hasan, S., and E. Aracil. 1991. Biology and effectiveness of *Uromyces heliotropii* Sred., a potential biological control agent of *Heliotropium europaeum* L. *New Phytol.* 118:559–63.

Hassanein, F. M., and M. A. El-Goorani. 1991. The effect of *Bacillus subtilis* on *in vitro* growth and pathogenicity of *Agrobacterium tumefaciens. J. Phytopathol.* 133:239–46.

Hazebroek, J. P., S. A. Garrison, and T. Gianfagna. 1989. Allelopathic substances in asparagus roots: Extraction, characterization and biological activity. *J. Am. Soc. Hort. Sci.* 114:152–58.

Hebbar, P., O. Berge, T. Heulin, and S. P. Singh. 1991. Bacterial antagonists of sunflower (*Helianthus annuus* L.) fungal pathogens. *Plant & Soil* 133:131–40.

Hegazy, A. K., K. S. Mansour, and N. F. Abdel-Hady. 1990. Allelopathic and autotoxic effects of *Anastatica hierochuntica* L. *J. Chem. Ecol.* 16:2183–93.

Hegde, R. S., and D. A. Miller. 1990. Allelopathy and autotoxicity in alfalfa: Characterization and effects of preceding crops and residue incorporation. *Crop Sci.* 30: 1255–59.

Hegde, R. S., and D. A. Miller. 1992. Scanning electron microscopy for studying root morphology and anatomy in alfalfa autotoxicity. *Agron. J.* 84:618–20.

Heisey, R. M. 1990a. Allelopathic and herbicidal effects of extracts from tree of heaven (*Ailanthus altissima*). *Am. J. Bot.* 77:662–70.

Heisey, R. M. 1990b. Evidence for allelopathy by tree-of-heaven *Ailanthus altissima. J. Chem. Ecol.* 16:2039–55.

Heisey, R. M., and C. C. Delwiche. 1983. A survey of California plants for water extractable and volatile inhibitors. *Bot. Gaz.* 144:382–90.

Heisey, R. M., and A. R. Putnam. 1986. Herbicidal effects of geldanamycin and nigericin, antibiotics from *Streptomyces hygroscopicus. J. Nat. Prod.* 49:859–65.

Hepperly, P. R., and M. Diaz. 1983. The allelopathic potential of pigeon peas in Puerto Rico. *J. Agric. Univ. P. R.* 67:453–63.

Hicks, S. K., C. W. Wendt, J. R. Gannaway, and R. B. Baker. 1989. Allelopathic effects of wheat straws on cotton germination, emergence, and yield. *Crop Sci.* 29:1057–61.

Höfte, M., J. Boelens, and W. Verstraete. 1991. Seed protection and promotion of seedling emergence by the plant growth beneficial *Pseudomonas* strains 7NSK2 and ANP15. *Soil Biol. Biochem.* 23:407–10.

Hogan, M. E., and G. D. Manners. 1990. Allelopathy of small everlasting (*Antennaria microphylla*): Phytotoxicity to leafy spurge (*Euphorbia esula*) in tissue culture. *J. Chem. Ecol.* 16:931–39.

Holl, F. B., and C. P. Chanway. 1992. Rhizosphere colonization and seedling growth promotion of lodgepole pine by *Bacillus polymyxa. Can. J. Microbiol.* 38:303–8.

Hollis, C. A., J. E. Smith, and R. F. Fisher. 1982. Allelopathic effects of common understory species on germination and growth of southern pines. *For. Sci.* 28:509–15.

Holm, R. E. 1972. Volatile metabolites controlling germination in buried weed seeds. *Plant Physiol.* 50:293–97.

Horsley, S. B., and D. A. Marquis. 1983. Interference by weeds and deer with Allegheny hardwood reproduction. *Can. J. For. Res.* 13:61–69.

Howell, C. R. 1991. Biological control of *Pythium* damping-off of cotton with seed-coating preparations of *Gliocladium virens. Phytopathology* 81:738–41.

Howie, W. J., and T. V. Suslow. 1991. Role of antibiotic biosynthesis in the inhibition of *Pythium ultimum* in the cotton spermosphere and rhizosphere by *Pseudomonas fluorescens. Mol. Plant-Microbe Interact.* 4:393–99.

Hradil, C. M., Y. F. Hallock, J. Clardy, D. S. Kenfield, and G. Strobel. 1989. Phytotoxins from *Alternaria cassiae. Phytochemistry* 28:73–75.

Hsiao, A. I., A. D. Worsham, and D. E. Moreland. 1981. Regulation of witchweed (*Striga asiatica*) seed conditioning and germination by dl-strigol. *Weed Sci.* 29: 101–4.

Hsiao, A. I., A. D. Worsham, and D. E. Moreland. 1983. Leaching and degradation of dl-strigol in soil. *Weed Sci.* 31:763–65.

Huang, J., A. R. Putnam, G. M. Werner, S. K. Mishra, and C. Whitenack. 1989. Herbicidal metabolites from a soil-dwelling fungus (*Scopulariopis brumptii*). *Weed Sci.* 37:123–28.

Huang, J. W., and E. G. Kuhlman. 1991a. Formulation of a soil amendment to control damping-off of slash pine seedlings. *Phytopathology* 81:163–70.

Huang, J. W., and E. G. Kuhlman. 1991b. Mechanisms inhibiting damping-off pathogens of slash pine seedlings with a formulated soil amendment. *Phytopathology* 81:171–77.

Huang, Y., B. L. Wild, and S. C. Morris. 1992. Post-harvest biological control of *Penicillium digitatum* decay on citrus fruit by *Bacillus pumilus. Ann. Appl. Biol.* 120:367–72.

Hubbell, S. P., D. F. Wiemer, and A. Adejare. 1983. An antifungal terpenoid defends a neotropical tree (*Hymenaea*) against attack by fungus-growing ants (*Atta*). *Oecologia* 60:321–27.

Hufford, C. D., S. Liu, and A. M. Clark. 1988. Antifungal activity of *Trillium grandiflorum* constituents. *J. Nat. Prod.* 51:94–98.

Hussain, F. 1980. Allelopathic effects of Pakistani weeds: *Euphorbia granulata* Forssk. *Oecologia* 45:267–69.

Hussain, F., and M. A. Gadoon. 1981. Allelopathic effects of *Sorghum vulgare* Pers. *Oecologia* 51:284–88.

Hussain, F., T. W. Khan, and A. Hussain. 1987. Allelopathic effects of *Cirsium arvense* (L). Scop. In *Modern Trends of Plant Science Research in Pakistan.* pp. 24–28 (I. Ilahi and F. Hussain, eds.), Proc. Third Nat'l. Conf. of Plant Scientists, Dept. of Bot., Univ. of Peshawar, Peshawar.

Hussain, F., B. Mubarak, I. ul Haq, and H. H. Naqvi. 1979. Allelopathic effects of *Datura innoxia* L. *Pak. J. Bot.* 11:141–53.

Hussain, F., H. H. Naqvi, and I. Ilahi. 1982. Interference exhibited by *Cenchrus ciliaris* and *Bothriochloa pertusa. Bull. Torrey Bot. Club* 109:513–23.

Hussain, F., M. I. Zaidi, and S. R. Chughtai. 1984. Allelopathic effects of Pakistani weeds: *Eragrostis poaeoides* P. Beauv. *Pak. J. Scient. Indust. Res.* 27:159–64.

Inam, B., F. Hussain, and F. Bano. 1987. Allelopathic effects of Pakistani weeds *Xanthium strumarium* L. *Pak. J. Scient. Indust. Res.* 30:530–33.

Inderjit and K. M. M. Dakshini. 1990. The nature of the interference potential of *Pluchea lanceolata* (DC) C. B. Clarke (Asteraceae). *Plant & Soil* 122:298–302.

Inderjit and K. M. M. Dakshini. 1991a. Investigations on some aspects of chemical ecology of cogongrass, *Imperata cylindrica* (L.) Beauv. *J. Chem. Ecol.* 17:343–52.

Inderjit and K. M. M. Dakshini. 1991b. Hesperetin 7-rutinoside (hesperidin) and taxifolin 3-arabinoside as germination and growth inhibitors in soils associated with the weed, *Pluchea lanceolata* (DC) C. B Clarke (Asteraceae). *J. Chem. Ecol.* 17:1585–91.

Inderjit and K. M. M. Dakshini. 1992a. Formononetin-7-O-glucoside (ononin), an additional growth inhibitor in soils associated with the weed, *Pluchea lanceolata* (DC) C. B. Clarke (Asteraceae). *J. Chem. Ecol.* 18:713–18.

Inderjit and K. M. M. Dakshini. 1992b. Interference potential of *Pluchea lanceolata* (Asteraceae): growth and physiological responses of asparagus bean, *Vigna unguiculata* var. *sesquipedalis. Am. J. Bot.* 79:977–81.

Inoue, M., H. Nishimura, H.-H. Li, and J. Mizutani. 1992. Allelochemicals from *Polygonum sachalinense* Fr. Schm. (Polygonaceae). *J. Chem. Ecol.* 18:1833–40.

Isenbeck, M., and F. A. Schulz. 1986. Biological control of fireblight (*Erwinia amylovora* (Burr) Winslow *et al.*) on ornamentals. 2. Investigation about the mode of action of the antagonistic bacteria. *J. Phytopathol.* 116:308–14.

Ito, M., T. Ichihashi, S. Hasebe, and K. Ueki. 1981a. Some observations on the root development and distribution in apple trees as affected by the different weed cover. *Weed Res.* (Japan) 26:24–29. (Japanese, English summary)

Ito, M., Y. Matsushita, Y. Umeki, and K. Ueki. 1981b. Determination of physiological effects of grass mulch in orchards: Bioassay of leachates from weed residues. *Weed Res.* (Japan) 26:221–27. (Japanese, English summary)

Jager, G., and H. Velvis. 1988. Inactivation of sclerotia of *Rhizoctonia solani* on potato tubers by *Verticillium biguttatum,* a soil-borne mycoparasite. *Neth. J. Plant Pathol.* 94:225–31.

Jain, R., M. Singh, and D. J. Dezman. 1989. Qualitative and quantitative characterization of phenolic compounds from lantana (*Lantana camara*) leaves. *Weed Sci.* 37:302–7.

James, K. L., P. A. Banks, and K. J. Karnok. 1988. Interference of soybean, *Glycine max,* cultivars with sicklepod, *Cassi obtusifolia. Weed Technol.* 2:404–9.

Janisiewicz, W. J. 1988. Biocontrol of postharvest diseases of apples with antagonist mixtures. *Phytopathology* 78:194–98.

Janisiewicz, W. J., and A. Marchi. 1992. Control of storage rots on various pear cultivars with a saprophytic strain of *Pseudomonas syringae. Plant Dis.* 76:555–60.

Janisiewicz, W. J., and J. Roitman. 1988. Biological control of blue mold and gray mold on apple and pear with *Pseudomonas cepacia. Phytopathology* 78:1697–1700.

Jarvis, B. B., J. O. Midiwo, G. A. Bean, M. B. Aboul-Nasr, and C. S. Barros. 1988. The mystery of trichothecene antibiotics in *Baccharis* species. *J. Nat. Prod.* 51:736–44.

Jensen, E. H., G. J. Hartman, F. Lundin, S. Knapp, and B. Brookerd. 1981. *Autotoxicity in alfalfa.* Nevada Agricultural Experiment Station Report R144.

Jha, P. K., and D. N. Sen. 1981. Allelopathic potential of *Echinops echinatus* Roxb. *Proc. 8th Asian-Pacific Weed Sci. Soc. Conf.* 2:197–202.

Jiménez, J. J., K. Schultz, A. L. Anaya, J. Hernández, and O. Espejo. 1983. Allelopathic potential of corn pollen. *J. Chem. Ecol.* 9:1011–25.

Jindal, K. K., and B. S. Thind. 1990. Microflora of cowpea seeds and its significance in the biological control of seedborne infection of *Xanthomonas campestris* pv. *vignicola*. *Seed Sci. Technol.* 18:393–403.

Jobidon, R. 1986. Allelopathic potential of coniferous species to old-field weeds in eastern Quebec. *For. Sci.* 32:112–18.

Jobidon, R. 1991. Potential use of bialaphos, a microbially produced phytotoxin, to control red raspberry in forest plantations and its effect on black spruce. *Can. J. For. Res.* 21:489–97.

Jobidon, R., J. R. Thibault, and J. A. Fortin. 1989a. Effect of straw residues on black spruce seedling growth and mineral nutrition, under greenhouse conditions. *Can. J. For. Res.* 19:1291–93.

Jobidon, R., J. R. Thibault, and J. A. Fortin. 1989b. Phytotoxic effect of barley, oat, and wheat-straw mulches in eastern Quebec forest plantations. 2. Effects on nitrification and black spruce (*Picea mariana*) seedling growth. *For. Ecol. Manage.* 29:295–310.

Jobidon, R., J. R. Thibault, and J. A. Fortin. 1989c. Phytotoxic effect of barley, oat, and wheat-straw mulches in eastern Quebec forest plantations. 1. Effects on red raspberry (*Rubus idaeus*). *For. Ecol. Manage.* 29:277–94.

Johnston, W. H. 1989. Consol lovegrass (*Eragrostis curvula* complex) controls spring burrgrass (*Cenchrus* spp.) in south-western New South Wales. *Aust. J. Exp. Agric.* 29:37–42.

Jones, R. W., W. T. Lanini, and J. G. Hancock. 1988. Plant growth response to the phytotoxin viridiol produced by the fungus *Gliocladium virens*. *Weed Sci.* 36:683–87.

de Jong, M. D., P. C. Scheepens, and J. C. Zadoks. 1990. Risk analysis for biological control: A Dutch case study in biocontrol of *Prunus serotina* by the fungus *Chondrostereum purpureum*. *Plant Dis.* 74:189–94.

de Jong, M. D., P. S. Wagenmakers, and J. Goudriaan. 1991. Modelling the escape of *Chondrostereum purpureum* spores from a larch forest with biological control of *Prunus serotina*. *Neth. J. Plant Pathol.* 97:55–61.

Joshi, S. 1990. An economic evaluation of control methods for *Parthenium hysterophorus* L. *Biol. Agric. Hort.* 6:285–91.

Joshi, S. 1991a. Interference effects of *Cassia uniflora* Mill on *Parthenium hysterophorus* L. *Plant & Soil* 132:213–18.

Joshi, S. 1991b. Biocontrol of *Parthenium hysterophorus* L. *Crop. Prot.* 10:429–31.

Joshi, S. 1991c. Biological control of *Parthenium hysterophorus* L. (Asteraceae) by *Cassia uniflora* Mill (Leguminosae), in Bangalore, India. *Trop. Pest Manage.* 37:182–84.

Joshi, S., and M. Mahadevappa. 1986. *Cassia sericea* S. to fight *Parthenium hysterophorus* Linn. *Curr. Sci.* 55:261–62.

Joye, G. F. 1990. Biocontrol of *Hydrilla verticillata* with the endemic fungus *Macrophomina phaseolina*. *Plant Dis.* 74:1035–36.

Kaiser, W. J., R. M. Hannan, and D. M. Weller. 1989. Biological control of seed rot and preemergence damping-off of chickpea with fluorescent pseudomonads. *Soil Biol. Biochem.* 21:269–73.

Kalantari, I. 1981. Stimulation of Corn Seedling Growth by Allelochemicals from Soybean Residue. Ph.D. dissertation, Iowa State Univ., Ames. (Diss. Abstr. Order No. DA8209137)

Kanchan, S. D., and Jayachandra. 1976. Parthenium weed problem and its chemical control. In *Parthenium-A Positive Danger.* pp. 6–10 (B. V. V. Rao, ed.), Univ. of Agricultural Sciences, Hebbal, India.

Kanchan, S. D., and Jayachandra. 1980. Allelopathic effects of *Parthenium hysterophorus* L. Part IV. Identification of inhibitors. *Plant & Soil* 55:67–75.

Karachi, M., and R. D. Pieper. 1987. Allelopathic effects of kochia on blue gram. *J. Range Manage.* 40:380–81.

Keel, C., U. Schnider, M. Maurhofer, C. Volsard, J. Laville, U. Burger, P. Wirthner, D. Haas, and G. Defago. 1992. Suppression of root diseases by *Pseudomonas fluorescens* CHAO; Importance of the bacterial secondary metabolite 2,4-diacetylphloroglucinol. *Mol. Plant-Microbe Interact.* 5:4–13.

Kehr, W. R., J. E. Watkins, and R. L. Ogden. 1983. Alfalfa establishment and production with continuous alfalfa and following soybeans. *Agron. J.* 75:435–38.

Keinath, A. P., D. R. Fravel, and G. C. Papavizas. 1991. Potential of *Gliocladium roseum* for biocontrol of *Verticillium dahliae. Phytopathology* 81:644–48.

Kempf, H. J., and G. Wolf. 1989. *Erwinia herbicola* as a biocontrol agent of *Fusarium culmorum* and *Puccinia recondita* f. sp. *tritici* on wheat. *Phytopathology* 79:990–94.

Kenfield, D., Y. Hallock, J. Clardy, and G. Strobel. 1989. Curvulin and O-methylcurvulinic acid: Phytotoxic metabolites of *Drechslera indica* which cause necroses on purslane and spiny amaranth. *Plant Sci.* 60:123–27.

Kennedy, B. S., M. T. Nielsen, R. F. Severson, V. A. Sisson, M. K. Stephenson, and D. M. Jackson. 1992. Leaf surface chemicals from *Nicotiana* affecting germination of *Peronospora tabacina* (Adam) sporangia. *J. Chem. Ecol.* 18:1467–79.

Kerwin, J. L., L. M. Johnson, H. C. Whisler, and A. R. Tuininga. 1992. Infection and morphogenesis of *Pythium marinum* in species of *Porphyra* and other red algae. *Can. J. Bot.* 70:1017–24.

Khanum, S., F. Hussain, and H. H. Naqvi. 1979. Allelopathic potentiality of *Chloris gayana* Kunth and *Panicum antidotale* Retz. *Pakistan J. For.* 29:245–49.

Khare, L. J. 1980. Phytotoxicity of the weed *Urgenia indica* Kunth on the seed germination of associated crops. *Indian J. Bot.* 3:87–91.

Khosla, S. N., K. Singh, and S. N. Sobti. 1981. Allelopathy of *Pinus roxburghii* Sargent. *Indian Perfumer* 25:100–4.

Kil, B.-S. 1987. Allelopathic effects of *Pinus densiflora* S. et Z. *J. Nat. Sci.* 6:27–33.

Kil, B.-S. 1988. Allelopathic effect of *Pinus rigida* Mill. *Korean J. Ecol.* 11:65–76.

Kil, B.-S. 1989. Allelopathic effects of five pine species in Korea. In *Phytochemical Ecology: Allelochemicals, Mycotoxins and Insect Pheromones and Allomones.* pp. 81–99 (C. H. Chou and G. R. Waller, eds.), Inst. of Bot., Academia Sinica Monogr. Ser. No. 9, Taipei, ROC.

Kil, B-S., and Y. J. Yim. 1983. Allelopathic effects of *Pinus densiflora* on undergrowth of red pine forest. *J. Chem. Ecol.* 9:1135–51.

Kil, B.-S., and K. W. Yun. 1992. Allelopathic effects of water extracts of *Artemisia princeps* var. *orientalis* on selected plant species. *J. Chem. Ecol.* 18:39–51.

Kil, B.-S., S.-H. Oh, and Y.-S. Kim. 1989. Effects of growth inhibitors from *Pinus thunbergii*. *Korean J. Ecol.* 12:21–35.

Kil, B.-S., D. Y. Kim, Y.-S. Kim, and S.-Y. Lee. 1991. Phytotoxic effects of naturally occurring chemicals from *Pinus koraiensis* on experimental species. *Korean J. Ecol.* 14:149–57.

Kim, K. K., D. R. Fravel, and G. C. Papavizas. 1988. Identification of a metabolite produced by *Talaromyces flavus* as glucose oxidase and its role in the biocontrol of *Verticillium dahliae*. *Phytopathology* 78:488–92.

Kim, Y.-S., and B.-S. Kil. 1987. Bioassay on susceptivity of selected species to phytotoxic substances from tomato plants. *Korean J. Bot.* 30:59–67.

Kimber, R. W. L. 1967. Phytotoxicity from plant residues. I. The influence of rotted wheat straw on seedling growth. *Aust. J. Agric. Res.* 18:361–74.

Kimber, R. W. L. 1973. Phytotoxicity from plant residues. II. The effect of time of rotting of straw from some grasses and legumes on the growth of wheat seedlings. *Plant & Soil* 38:347–61.

Kirk, J. J., and J. W. Deacon. 1987. Control of the take-all fungus by *Microdochium bolleyi,* and interactions involving *M. bolleyi, Phialophora graminicola* and *Periconia macrospinosa* on cereal roots. *Plant & Soil* 98:231–37.

Kishore, N., and R. S. Dwivedi. 1991. Fungitoxicity of the essential oil of *Tagetes erecta* L. against *Pythium aphanidermatum* Fitz. the damping-off pathogen. *Flavour Fragrance J.* 6:291–94.

Klecan, A. L., S. Hippe, and S. C. Somerville. 1990. Reduced growth of *Erysiphe graminis* f. sp. *hordei* induced by *Tilletiopsis pallescens. Phytopathology* 80:325–31.

Kleiman, R., R. D. Plattner, and D. Weisleder. 1988. Antigermination activity of phenylpropenoids from the genus *Pimpinella*. *J. Nat. Prod.* 51:249–56.

Kluge, R. L. 1991. Biological control of crofton weed, *Ageratina adenophora* (Asteraceae), in South Africa. *Agric. Ecosyst. Environ.* 37:187–91.

Knudsen, G. R., and H. W. Spurr, Jr. 1987. Field persistence and efficacy of five bacterial preparations for control of peanut leaf spot. *Plant Dis.* 71:442–45.

Knudsen, G. R., D. J. Eschen, L. M. Dandurand, and L. Bin. 1991. Potential for biocontrol of *Sclerotinia sclerotiorum* through colonization by *Trichoderma harzianum*. *Plant Dis.* 75:466–70.

Ko, B.-K., and B.-S. Kil. 1985. Allelopathic potentials of *Larix leptolepis* (S. et Z.) Gordon on germination and seedling growth of selected species. *Korean J. Ecol.* 8:15–19.

Kochhar, M., U. Blum, and R. A. Reinert. 1980. Effects of $O^3$ and (or) fescue on ladino clover: interactions. *Can. J. Bot.* 58:241–49.

Kochhar, M., R. A. Reinert, and U. Blum. 1982. Effects of fescue *Festuca arundinacea* and/or clover *Trifolium repens* debris and fescue leaf leachate on clover as modified by ozone and *Rhizoctonia solani*. *Environmental Pollution* (Series A) 28:255–64.

Kohli, R. K. 1987. *Eucalyptus*—an anti-social tree from social forestry. In *Social Forestry for Rural Development*. pp. 235–41. (P. K. Khosla and R. K. Kohli, eds.) Indian Soc. of Tree Specialists, Nauni, India.

Kohli, R. K., A. Kumari, and D. B. Saxena. 1985. Auto- and teletoxicity of *Parthenium hysterophorus* L. *Acta Univ. Agric. (Brne), Fac. Agron.* 33:253–64.

Kohli, R. K., P. Chaudhry, and A. Kumari. 1988. Impact of *Eucalyptus* on *Parthenium*-a weed. *Indian J. Range Manage.* 9:63–67.

Kolesnichenko, M. V., and G. S. Andryushchenko. 1979. Biochemical effect of some woody species on the Norway spruce. *Soviet J. Ecol.* 9:325–28.

Komai, K., and K. Ueki. 1975. Chemical properties and behaviour of polyphenolic substances in purple nutsedge (*Cyperus rotundus* L). *Weed Res.* (Japan) 20:66–71. (Japanese, English summary)

Komai, K., and K. Ueki. 1977. Physiological functions of polyphenolic substances in germination of purple nutsedge tubers. *Weed Res.* (Japan) 22:193–98.

Komai, K., and K. Ueki. 1980. Plant-growth inhibitors in purple nutsedge (*Cyperus rotundus* L.). *Weed Res.* (Japan) 25:42–47. (Japanese, English summary)

Komai, K., and C. S. Tang. 1989. Chemical constituents and inhibitory activities of essential oils from *Cyperus brevifolius* and *C. kyllingia*. *J. Chem. Ecol.* 15:2171–76.

Komai, K., J. Iwamura, and K. Ueki. 1977. Isolation, identification and physiological activities of sesquiterpenes in purple nutsedge tubers. *Weed Res.* (Japan) 22:14–18. (Japanese, English summary)

Komai, K., K. Osaki, and K. Ueki. 1978. Geographical variation of essential oils in tubers of purple nutsedge. *Weed Res.* (Japan) 23:160–64.

Komai, K., S. Sato, and K. Ueki. 1982. Antimicroorganisms activities of essential oil in purple nutsedge. *Mem. Fac. Agr. Kinki Univ.* 15:33–41. (Japanese, English summary)

Komai, K., J. Iwamura, and K. Ueki. 1983. Plant growth inhibitor in the seed of catchweed. *Weed Res.* (Japan) 28:205–9.

Komai, K., C. S. Tang, and R. K. Nishimoto. 1991. Chemotypes of *Cyperus rotundus* in Pacific rim and basin: Distribution and inhibitory activities of their essential oils. *J. Chem. Ecol.* 17:1–8.

Kope, H. H., and J. A. Fortin. 1989. Inhibition of phytopathogenic fungi *in vitro* by cell free culture media of ectomycorrhizal fungi. *New Phytol.* 113:57–63.

Kope, H. H., and J. A. Fortin. 1990. Antifungal activity in culture filtrates of the ectomycorrhizal fungus *Pisolithus tinctorius*. *Can. J. Bot.* 68:1254–59.

Kope, H. H., Y. S. Tsantrizos, J. A. Fortin, and K. K. Ogilvie. 1991. p-Hydroxybenzoylformic acid and (R)-(-)-p-hydroxymandelic acid, two antifungal compounds isolated from the liquid culture of the ectomycorrhizal fungus *Pisolithus arhizus*. *Can. J. Microbiol.* 37:258–64.

Koul, O, M. B. Isman, and C. M. Ketkar. 1990. Properties and uses of neem, *Azadirachta indica*. *Can. J. Bot.* 68:1–11.

Kremer, R. J., and L. K. Schulte. 1989. Influence of chemical treatment and *Fusarium theophrasti*. *Weed Technol.* 3:369–74.

Krishnamurthy, G. V. G., and G. H. Chandwani. 1975. Effects of various crops on the germination of *Orobanche* seed. *PANS* 21:64–66.

Kuiters, A. T. 1989. Effects of phenolic acids on germination and early growth of herbaceous woodland plants. *J. Chem. Ecol.* 15:467–79.

Kuiters, A. T. 1990. Role of phenolic substances from decomposing forest litter in plant-soil interactions. *Acta Bot. Neerl.* 39:329–48.

Kumar, A., and S. C. Tripathi. 1991. Evaluation of the leaf juice of some higher plants for their toxicity against soil borne pathogens. *Plant & Soil* 132:297–301.

Kumar, V., V. Karunaratne, M. R. Sanath, K. Meegalle, and J. K MacLeod. 1990. Two fungicidal phenylethanones from *Euodia lunu-ankenda* root bark. *Phytochemistry* 29:243–45.

Kumari, A., and R. K. Kohli. 1987. Autotoxicity of ragweed parthenium (*Parthenium hysterophorus*). *Weed Sci.* 35:629–32.

Kumari, A., R. K. Kohli, and D. B. Saxena. 1985. Allelopathic effects of *Parthenium hysterophorus* L. leachates and extracts on *Brassica campestris* L. *Ann. Biol.* 1: 189–96.

Kuti, J. O., B. B. Jarvis, N. Mokhtari-Rejali, and G. A. Bean. 1990. Allelochemical regulation of reproduction and seed germination of two Brazilian *Baccharis* species by phytotoxic trichothecenes. *J. Chem. Ecol.* 16:3441–53.

Kwok, O. C. H., P. C. Fahy, H. A. J. Hoitink, and G. A. Kuter. 1987. Interactions between bacteria and *Trichoderma hamatum* in suppression of *Rhizoctonia* damping-off in bark compost media. *Phytopathology* 77:1206–12.

LaBonte, D. R., and R. L. Darding. 1988. Noncompetitive effects of morning glory on the growth of soybeans. *Trans. Ill. Acad. Sci.* 81:39–44.

Lahiri, A. N., and B. C. Kharabanda. 1962. Germination studies of arid zone plants. II. Germination inhibitors in the spikelet glumes of *Lasiurus sindicus, Cenchrus ciliaris* and *Cenchrus setigerus. Ann. Arid Zone* 1:114–26.

Lakshmanan, P., R. Jeyarajan, and P. Vidhyasekaran. 1990. Leaf and stem blight of *Euphorbia geniculata* incited by *Bipolaris zeicola. Phytoparasitica* 18:353–55.

Lambert, J. D. H., G. Campbell, J. T. Arnason, and W. Majak. 1991. Herbicidal properties of alpha-terthienyl, a naturally occurring phototoxin. *Can. J. Plant Sci.* 71:215–18.

Lamprecht, S. C., W. F. O. Marasas, E. W. Sydenham, P. G. Thiel, P. S. Knox-Davies, and P. S. Van Wyk. 1989. Toxicity to plants and animals of an undescribed, neosolaniol monoacetate-producing *Fusarium* species from soil. *Plant & Soil* 114:75–83.

Lane, G. A., O. R. W. Sutherland, and R. A. Skipp. 1987. Isoflavonoids as insect feeding deterrents and antifungal components from root of *Lupinus angustifolius. J. Chem. Ecol.* 13:771–83.

Latunde-Dada, A. O. 1991. The use of *Trichoderma koningii* in the control of web blight disease caused by *Rhizoctonia solani* in the foliage of cowpea (*Vigna unguiculata*). *J. Phytopathol.* 133:247–54.

Lawrence, J. G., A. Colwell, and O. J. Sexton. 1991. The ecological impact of allelopathy in *Ailanthus altissima* (Simaroubaceae). *Am. J. Bot.* 78:948–58.

Lawton, M. B., and L. L. Burpee. 1990. Effect of rate and frequency of application of *Typhula phacorrhiza* on biological control of typhula blight of creeping bentgrass. *Phytopathology* 80:70–73.

Leather, G. R. 1983. Weed control using allelopathic crop plants. *J. Chem. Ecol.* 9:983–89.

Leather, G. R. 1987. Weed control using allelopathic sunflowers and herbicide. *Plant & Soil* 98:17–23.

Lehle, F. R., and A. R. Putnam. 1983. Allelopathic potential of sorghum (*Sorghum bicolor*): Isolation of seed germination inhibitors. *J. Chem. Ecol.* 9:1223–34.

Lemanceau, P., and C. Alabouvette. 1991. Biological control of fusarium diseases by fluorescent *Pseudomonas* and non-pathogenic *Fusarium. Crop Prot.* 10:279–86.

Lemanceau, P., A. H. M. Bakker, W. J. de Kogel, C. Alabouvette, and B. Schippers. 1992. Effect of pseudobactin 358 production by *Pseudomonas putida* WCS358 on suppression of fusarium wilt of carnations by nonpathogenic *Fusarium oxysporum* Fo47. *Appl. Environ. Microbiol.* 58:2978–82.

Lemos, M. L., C. P. Dopazo, A. E. Toranzo, and J. L. Barja. 1991. Competitive dominance of antibiotic-producing marine bacteria in mixed cultures. *J. Appl. Bact.* 71:228–32.

Lenné, J. M., and A. E. Brown. 1991. Factors influencing the germination of pathogenic and weakly pathogenic isolates of *Colletotrichum gloeosporioides* on leaf surfaces of *Stylosanthes guianensis. Mycol. Res.* 95:227–32.

Levin, H., R. Hazenfrantz, J. Friedman, and M. Perl. 1988. Partial purification and some properties of the antibacterial compounds from *Aloe vera. Phytotherapy Res.* 1:1–3.

Levitt, J., and J. V. Lovett. 1984a. *Datura stramonium* L.: Alkaloids and allelopathy. *Australian Weeds* 3:108–12.

Levitt, J., and J. V. Lovett. 1984b. Activity of allelochemicals of *Datura stramonium* L. (Thorn-apple) in contrasting soil types. *Plant & Soil* 79:181–89.

Levitt, J., J. V. Lovett, and P. R. Garlick. 1984. Effect of allelochemicals of *Datura stramonium* L. (Thorn-apple) on root tip ultrastructure of *Helianthus annuus* L. *New Phytol.* 97:213–18.

Levy, E., Z. Eyal, S. Carmely, Y. Kashman, and I. Chet. 1989. Suppression of *Septoria tritici* and *Puccinia recondita* of wheat by an antibiotic-producing fluorescent pseudomonad. *Plant Pathol.* 38:564–70.

Levy, E., F. J. Gough, K. D. Berlin, P. W. Guiana, and J. T. Smith. 1992. Inhibition of *Septoria tritici* and other phytopathogenic fungi and bacteria by *Pseudomonas fluorescens* and its antibiotics. *Plant Pathol.* 41:335–41.

Lewis, J. A., and G. C. Papavizas. 1987. Reduction of inoculum of *Rhizoctonia solani* in soil by germlings of *Trichoderma hamatum. Soil Biol. Biochem.* 19:195–201.

Lewis, J. A., and G. C. Papavizas. 1991. Biocontrol of cotton damping-off caused by *Rhizoctonia solani* in the field with formulations of *Trichoderma* spp., and *Gliocladium virens. Crop. Prot.* 10:396–402.

Lewis, J. A., T. H. Barksdale, and G. C. Papavizas. 1990. Greenhouse and field studies on the biological control of tomato fruit rot caused by *Rhizoctonia solani. Crop. Prot.* 9:8–14.

Li, H.-H., H. Nishimura, K. Hasegawa, and J. Mizutani. 1992. Allelopathy of *Sasa cernua.* J. *Chem. Ecol.* 18:1785–96.

Li, Y. 1988. A preliminary study on use of the medical herb peppermint in control of cotton fusarium wilt. *Sci. Agric. Sin.* 21:65–69.

Liao, C. H. 1989. Antagonism of *Pseudomonas putida* strain PP22 to phytopathogenic bacteria and its potential use as a biocontrol agent. *Plant Dis.* 73:223–26.

Liebl, R. A., and A. D. Worsham. 1983. Inhibition of pitted morning glory (*Ipomoea lacunosa* L.) and certain other weed species by phytotoxic components of wheat (*Triticum aestivum* L.) straw. *J. Chem. Ecol.* 9:1027–43.

Lindblad, P., K. Tadera, and F. Yagi. 1990. Occurrence of azoxyglycosides in *Macro-*

*zamia riedlei* and their effects on the free-living isolated *Nostoc* PCC73102. *Environ. Exp. Bot.* 30:429–34.

Liu, S. D. 1991. Biological control of adzuki-bean root rot disease caused by *Rhizoctonia solani.* Plant Prot. Bull. 33:63–71. (Chinese, English Summary)

Locken, L. J., and R. G. Kelsey. 1987. Cnicin concentrations in *Centaurea maculosa,* spotted knapweed. *Biochem. Sys. Ecol.* 15:313–20.

Lodhi, M. A. K., and F. L. Johnson. 1989. Forest understory biomass heterogeneity- Is "moisture complex" or associated litter the cause? *J. Chem. Ecol.* 15:429–37.

Lodhi, M. A. K., R. Bilal, and K. A. Malik. 1987. Allelopathy in agroecosystems: Wheat phytotoxicity and its possible roles in crop rotation. *J. Chem. Ecol.* 13: 1881–91.

Loper, J. E. 1988. Role of fluorescent siderophore production in biological control of *Pythium ultimum* by a *Pseudomonas fluorescens* strain. *Phytopathology* 78:166–72.

Lovett, J. V. 1982. Allelopathy and self defense in plants. *Australian Weeds* 2:33–36.

Lovett, J. V., and R. S. Jessop. 1982. Effects of residues of crop plants on germination and early growth of wheat. *Aust. J. Agric. Res.* 33:909–16.

Lovett, J. V., J. Levitt, A. M Duffield, and N. G. Smith. 1981. Allelopathic potential of *Datura stramonium* L. *Weed Res.* 21:165–70.

Lumsden, R. D., and J. C. Locke. 1989. Biological control of damping-off caused by *Pythium ultimum* and *Rhizoctonia solani* with *Gliocladium virens* in soilless mix. *Phytopathology* 79:361–66.

Lumsden, R. D., R. Garcia-E., J. A. Lewis, and G. A. Frías-T. 1987. Suppression of damping-off caused by *Pythium* spp. in soil from the indigenous Mexican chinampa agricultural system. *Soil Biol. Biochem.* 19:501–8.

Lumsden, R. D., J. C. Locke, S. T. Adkins, J. F. Walter, and C. J. Ridout. 1992. Isolation and localization of the antibiotic gliotoxin produced by *Gliocladium virens* from alginate prill in soil and soilless media. *Phytopathology* 82:230–35.

Luu, K. T., A. G. Matches, C. J. Nelson, E. J. Peters, and G. B. Garner. 1989. Characterization of inhibitory substances of tall fescue on birdsfoot trefoil. *Crop Sci.* 29:407–12.

Lynch, J. M. 1977. Phytotoxicity of acetic acid produced in the anaerobic decompositon of wheat straw. *J. Appl. Bacteriol.* 42:81–87.

Lynch, J. M. 1978a. Production and phytotoxicity of acetic acid in anaerobic soils containing plant residues. *Soil Biol. Biochem.* 10:131–35.

Lynch, J. M. 1978b. Microbial interactions around imbibed seeds. *Ann. Appl. Biol.* 89:165–67.

Lynch, J. M., and D. J. Penn. 1980. Damage to cereals caused by decaying weed residues. *J. Sci. Food & Agr.* 310:321–24.

Lynch, J. M., and L. F. Elliott. 1983. Minimizing the potential phytotoxicity of wheat straw by microbial degradation. *Soil Biol. Biochem.* 15:221–22.

Lynch, J. M., and S. J. Clark. 1984. Effects of microbial colonization of barley (*Hordeum vulgare* L.) roots on seedling growth. *J. Appl. Bacteriol.* 56:47–52.

Lynch, J. M., K. B. Gunn, and L. M. Panting. 1980a. On the concentration of acetic acid in straw and soil. *Plant & Soil* 56:93–98.

Lynch, J. M., K. C. Hall, H. A. Anderson, and A. Hepburn. 1980b. Organic acids from the anaerobic decomposition of *Agropyron repens* rhizomes. *Phytochemistry* 19:1846–47.

Lynch, J. M., R. D. Lumsden, P. T. Atkey, and M. A. Ously. 1991. Prospects for control of *Pythium* damping-off of lettuce with *Trichoderma, Gliocladium* and *Enterobacter* spp. *Biol. Fert. Soils* 12:95–99.

Maas, E. M. C., and J. M. Kotze. 1987. *Trichoderma harzianum* and *T. polysporum* as biocontrol agents of take-all of wheat in the greenhouse. *Phytophylactica* 19:365–67.

McChesney, J. D., A. M. Clark, and E. R. Silveira. 1991. Antimicrobial diterpenes of *Croton sonderianus,* I. hardwickic and 3,4-secotrachylobanoic acids. *J. Nat. Prod.* 54:1625–33.

McElgunn, J. D., and D. H. Heinrichs. 1970. Effects of root temperature and a suspected phytotoxic substance on growth of alfalfa. *Can. J. Plant Sci.* 50:307–11.

MacFarlane, M. J., D. Scott, and P. Jarvis. 1982. Allelopathic effects of white clover. I. Germination and chemical bioassay. *New Zealand J. Agric. Res.* 25:502–10.

McKee, N. D., and P. M. Robinson. 1988. Production of volatile inhibitors of germination and hyphal extension by *Geotrichum candidum. Trans. Br. Mycol. Soc.* 91:157–60.

McLaughlin, R. J., L. Sequeira, and D. P. Weingartner. 1990. Biocontrol of bacterial wilt of potato with an avirulent strain of *Pseudomonas solanacearum:* Interactions with root-knot nematodes. *Am. Potato J.* 67:93–107.

McQuilkin, M. P., J. M. Whipps, and R. C. Cooke. 1990. Control of damping-off in cress and sugar-beet by commercial seedcoating with *Pythium oligandrum. Plant Pathol.* (London) 39:425–62.

Madrigal, C., J. L. Tadeo, and P. Melgarejo. 1991. Relationship between flavipin production by *Epicoccum nigrum* and antagonism against *Monilinia laxa. Mycol. Res.* 95:1375–81.

Mallik, A. U. 1987. Allelopathic potential of *Kalmia angustifolia* to black spruce (*Picea mariana*). *For. Ecol., & Manage.* 20:43–51.

Mallik, A. U., and P. F. Newton. 1988. Inhibition of black spruce seedling growth by forest-floor substrates of central Newfoundland. *For. Ecol., & Manage.* 23:273–83.

Mallik, M. A. B., and K. Tesfai. 1987. Stimulation of *Bradyrhizobium japonicum* by allelochemicals from green plants. *Plant & Soil* 103:227–31.

Mallik, M. A. B., and K. Tesfai. 1990. Isolation of a factor stimulatory to *Bradyrhizobium japonicum* in broth culture. *Plant & Soil* 128:177–84.

Manners, G. D., and D. S. Galitz. 1986. Allelopathy of small everlasting (*Antennaria microphylla*): Identification of constituents phytotoxic to leafy spurge (*Euphorbia esula*). *Weed Sci.* 34:8–12.

Maplestone, P. A., and R. Campbell. 1989. Colonization of roots of wheat seedlings by bacilli proposed as biocontrol agents against take-all. *Soil Biol. Biochem.* 21:543–50.

Maplestone, P. A., J. M. Whipps, and J. M. Lynch. 1991. Effect of peat-bran inoculum of *Trichoderma* species on biological control of *Rhizoctonia solani* in lettuce. *Plant & Soil* 136:257–63.

Marston, A., F. Gafner, S. F. Dossaji, and K. Hostettmann. 1988. Fungicidal and molluscicidal saponins from *Dolichos kilimandscharicus. Phytochemistry* 27:1325–26.

Martin, F. N., and J. G. Hancock. 1987. The use of *Pythium oligandrum* for biological control of peremergence damping-off caused by *P. ultimum*. *Phytopathology* 77:1013–20.

Martin, V. L., E. L. McCoy, and W. A. Dick. 1990. Allelopathy of crop residues influences corn seed germination and early growth. *Agron. J.* 82:555–60.

Mason-Sedun, W., and R. S. Jessop. 1988. Differential phytotoxicity among species and cultivars of the genus *Brassica* to wheat. *Plant & Soil* 107:69–80.

Mason-Sedun, W., R. S. Jessop, and J. V. Lovett. 1986. Differential phytotoxicity among species and cultivars of the genus *Brassica* to wheat. I. Laboratory and field screening of species. *Plant & Soil* 93:3–16.

Masuda, T., A. Inazumi, Y. Yamada, W. G. Padolina, H. Kikuzaki, and N. Nakatani. 1991. Antimicrobial phenylpropanoids from *Piper sarmentosum*. *Phytochemistry* 30:3227–28.

Matveev, N. M. 1980. Role of intravitan secretions by trees in creating an allelopathic system in steppe forest stands. *Soviet J. Ecol.* 11:21–28.

Maurhofer, M., C. Keel, U. Schnider, C. Voisard, D. Haas, and G. Défago. 1992. Influence of enhanced antibiotic production in *Pseudomonas fluorescens* Strain CHAO on its disease suppressive capacity. *Phytopathology* 82:190–95.

May, F. E., and J. E. Ash. 1990. An assessment of the allelopathic potential of *Eucalyptus*. *Aust. J. Bot.* 38:245–54.

Mazzola, M., and R. J. Cook. 1991. Effects of fungal root pathogens on the population dynamics of biocontrol strains of fluorescent pseudomonads in the wheat rhizosphere. *Appl. Environ. Microbiol.* 57:2171–78.

Megharaj, M., A. P. Rao, K. Venkateswarlu, and A. S. Rao. 1987a. Toxicity of *Parthenium hysterophorus* to *Chlorella vulgaris* and *Synechococcus elongatus*. *Plant & Soil* 103:292–94.

Megharaj, M., A. P. Rao, K. Venkateswarlu, and A. S. Rao. 1987b. Influence of *Parthenium hysterophorus* L. on native soil algal flora. *Plant & Soil* 101:223–26.

Meissner, R., P. C. Nel, and E. A. Beyers. 1986. Allelopathic influence of *Tagetes*-and *Bidens*-infested soils on seedling growth of certain crop species. *S. Afr. J. Plant Soil* 3:176–80.

Meissner, R., P. C. Nel, and E. A. Beyers. 1989. Allelopathic effect of *Cynodon dactylon*-infested soil on early growth of certain crop species. *Appl. Plant Sci.* 3:125–26.

Meissner, R., P. C. Nel, and N. S. H. Smit. 1982. The residual effect of *Cyperus rotundus* on certain crop plants. *Agroplantae* 14:47–53.

Melkania, N. P., J. S. Singh, and K. K. S. Bisht. 1982. Allelopathic potential of *Artemisia vulgaris* L., and *Pinus roxburghii* Sargent: A bioassay study. *Proc. Indian Nat'l. Sci. Acad.* B 48:685–88.

Melouk, H. A., and C. N. Akem. 1987. Inhibition of growth of *Sclerotinia minor* and other pathogens by citrinin in the filtrate of *Penicillium citrinum*. *Mycopathologia* 100:91–96.

Menetrez, M. L., H. W. Spurr, Jr., D. A. Danehower, and D. R. Lawson. 1990. Influence of tobacco leaf surface chemicals on *Peronospora tabacina* Adam sporangia. *J. Chem. Ecol.* 16:1565–76.

Menges, R. M. 1987. Allelopathic effects of Palmer amaranth (*Amaranthus palmeri*) and other plant residues in soil. *Weed Sci.* 35:339–47.

Menges, R. M. 1988. Allelopathic effects of Palmer amaranth (*Amaranthus palmeri*) on seedling growth. *Weed Sci.* 36:325–28.

Mercer, K. L., D. S. Murray, and L. M. Verhalen. 1987. Interference of unicorn-plant (*Proboscidea louisianica*) with cotton (*Gossypium hirsutum*). *Weed Sci.* 35:807–12.

Merrill, G. B. 1989. Eupatoriochromene and encecalin, plant growth regulators from yellow starthistle (*Centaurea solstitialis* L.). *J. Chem. Ecol.* 15:2073–87.

Merriman, P. R., R. D. Price, J. F. Kollmorgen, T. Piggott, and E. H. Ridge. 1974. Effect of seed inoculation with *Bacillus subtilis* and *Streptomyces griseus* on growth of cereals and carrots. *Aust. J. Agric. Res.* 25:219–26.

Mersie, W., and M. Singh. 1987a. Allelopathic effect of lantana on some agronomic crops and weeds. *Plant & Soil* 98:25–30.

Mersie, W., and M. Singh. 1987b. Allelopathic effect of parthenium (*Parthenium hysterophorus* L.) extract and residue on some agronomic crops and weeds. *J. Chem. Ecol.* 13:1739–47.

Mersie, W., and M. Singh. 1988. Effects of phenolic acids and ragweed parthenium (*Parthenium hysterophorus*) extracts on tomato (*Lycopersicon esculentum*) growth and nutrient and chlorophyll content. *Weed Sci.* 36:278–81.

Mew, T. W., and A. M. Rosales. 1986. Bacterization of rice plants for control of sheath blight caused by *Rhizoctonia solani*. *Phytopathology* 76:1260–64.

Miedtke, U., and W. Kennel. 1990. *Athelia bombacina* and *Chaetomium globosum* as antagonists of the perfect stage of the apple scab pathogen (*Venturia inaequalis*) under field conditions. *Z. Pflanzenkr. Pflanzenshutz* 97:24–32.

Miersch, J., C. Jühlke, G. Sternkopf, and G.-J. Krauss. 1992. Metabolism and exudation of canavanine during development of alfalfa (*Medicago sativa* L. cv. Verbo). J. *Chem. Ecol.* 18:2117–29.

Miller, D. A. 1983. Allelopathic effects of alfalfa. *J. Chem. Ecol.* 9:1059–72.

Miller, H. G., M. Ikawa, and L. C. Peirce. 1991. Caffeic acid identified as an inhibitory compound in asparagus root filtrate. *Hort. Sci.* 26:1525–27.

Miller, R. V., E. J. Ford, N. J. Zidack, and D. C. Sands. 1989. A pyrimidine auxotroph of *Sclerotinia sclerotiorum* for use in biological weed control. *J. Gen. Microbiol.* 135:2085–91.

Miller, R. W., R. Kleiman, R. G. Powell, and A. R. Putnam. 1988. Germination and growth inhibitors of alfalfa. *J. Nat. Prod.* 51:328–30.

Mintz, A. S., D. K. Heiny, and G. J. Weidemann. 1992. Factors influencing the biocontrol of tumble pigweed (*Amaranthus albus*) with *Aposphaeria amaranthi*. *Plant Dis.* 76:267–69.

Misaghi, I. J., M. W. Olsen, J. M. Billotte, and R. M Sonoda. 1992. The importance of rhizobacterial mobility in biocontrol of bacterial wilt of tomato. *Soil Biol. Biochem.* 24:287–93.

Mishra, S. K., W. H. Taft, A. R. Putnam, and S. K. Ries. 1987. Plant growth regulatory metabolites from novel actinomycetes. *J. Plant Growth Regul.* 6:75–84.

Mishra, S. K., C. J. Whitenack, and A. R. Putnam. 1988. Herbicidal properties of metabolites from several genera of soil microorganisms. *Weed Sci.* 36:122–26.

Mishustin. E. N., and A. N. Naumova. 1955. Secretion of toxic substances by alfalfa and their effect on cotton and soil microflora. *Izvest. Akad. Nauk. SSSR, Ser. Biol.* 6:3–9. (Russian)

Misra, N., D. Mishra, and A. Rathore. 1989. Fungitoxic properties of the needle in *Pinus roxburghii* against aspergilli causing mycotoxicoses. *Pesticides* (Bombay) 23:23–25.

Mitchell, R. E. 1989. Biosynthesis of rhizobitoxine from L-aspartic acid and L-threo-hydroxythreonine by *Pseudomonas andropogonis. Phytochemistry* 28:1617–20.

Mitchell, R. E., and J. M. Coddington. 1991. Biosynthetic pathway to rhizobitoxine in *Pseudomonas andropogonis. Phytochemistry* 30:1809–14.

Molina, A., M. J. Reigosa, and A. Carballeira. 1991. Release of allelochemical agents from litter, throughfall, and topsoil in plantations of *Eucalyptus globulus* Labill in Spain. *J. Chem. Ecol.* 17:147–60.

Molisch, H. 1937. *Der Einflus einer Pflanze auf die andere-Allelopathie.* Gustave Fischer Verlag, Jena.

Mónaco, C., A. Perello, H. E. Alippi, and A. O. Pasquare. 1991. *Trichoderma* spp.: A biocontrol agent of *Fusarium* spp., and *Sclerotium rolfsii* by seed treatment. *Adv. Hort. Sci.* 1:92–95.

Montgomery, R. 1989. Personal communication with author.

Moore-Landecker, E. 1988. Response of *Pyronema domesticum* to volatiles from microbes, seeds, and natural substrata. *Can. J. Bot.* 66:194–98.

Moore-Landecker, E., and W. Shropshire, Jr. 1984. Effects of ultraviolet A radiation and inhibitory volatile substances on the discomycete, *Pyronema domesticum. Mycologia* 76:820–29.

Mori, A., N. Enoki, K. Shinozuka, C. Nishino, and M. Fukushima. 1987. Antifungal activity of fatty acids against *Pyricularia oryzae* related to antifungal constituents of *Miscanthus sinensis. Agric. Biol. Chem.* 51:3403–05.

Morris, M. J. 1989a. Host specificity studies of a leaf spot fungus, *Phaeoramularia* sp. for the biological control of crofton weed (*Ageratina adenophora*) in South Africa. *Phytophylactica* 21:281–83.

Morris, M. J. 1989b. A method for controlling *Hakea sericea* Schrad. seedlings using the fungus *Colletotrichum gloeosporioides* (Penz.) Sacc. *Weed Res.* 29:449–54.

Mortensen, K. 1986. Biological control of weeds with plant pathogens. *Can. J. Plant Pathol.* 8:229–31.

Moujir, L., A. M. Gutiérrez-Navarro, A. G. González, A. G. Ravelo, and J. G. Luis. 1990. The relationship between structure and antimicrobial activity in quinones from the Celastraceae. *Biochem. Syst. Ecol.* 18:25–28.

Mubarak, B., and F. Hussain. 1978. Biochemical inhibition exhibited by *Datura innoxia* seeds. *Pak. J. Bot.* 10:149–56.

Muller, C. H. 1969. Allelopathy as a factor in ecological process. *Vegetatio* 18:348–57.

Müller-Wilmes, U., and M. Zoschke. 1980. Allelopathie-eine mögliche Ursache fur Verträglichkeitsbeziehungen der Kulturpflanzen. *Angewandte Bot.* 54:109–23.

Murphy, S. D., and L. W. Aarssen. 1989. Pollen allelopathy among sympatric grassland species: *In vitro* evidence in *Phleum pratense* L. *New Phytol.* 112:295–305.

Murthy, M. S., and T. Zakharia. 1980. Allelopathic potentials of some common weeds of bajra crop. *Indian J. Exp. Biol.* 18:91–93.

Nair, M. G., C. J. Whitenack, and A. R. Putnam. 1990. 2,2'-Oxo-1,1'-azobenzene: A microbially transformed allelochemical from 2,3-benzoxazolinone: *I. J. Chem. Ecol.* 16:353–64.

Nakajima, H., T. Hamasaki, S. Maeta, Y. Kimura, and Y. Takeuchi. 1990. A plant growth regulator produced by the fungus, *Cochliobolus spicifer. Phytochemistry* 29:1739–43.

Nakashima, N., Z. Moromizato, and N. Matsuyama. 1991. *Chaetomium* spp. antagonistic microorganisms to phytopathogenic fungi. *J. Fac. Agri. Kyushu Univ.* 36:109–15.

Nelson, E. B. 1990. Exudate molecules initiating fungal responses to seeds and roots. *Plant & Soil* 129:61–73.

Nelson, E. B., and C. M. Craft. 1991. Introduction and establishment of strains of *Enterobacter cloacae* in golf course turf for the biological control of dollar spot. *Plant Dis.* 75:510–14.

Nelson, E. B., G. E. Harman, and G. T. Nash. 1988. Enhancement of *Trichoderma*-induced biological control of *Pythium* seed rot and pre-emergence damping-off of peas. *Soil Biol. Biochem.* 20:145–50.

Nelson, L. S. 1985. Isolating Potential Allelochemicals from Soybean-Soil Residues. M. S. thesis, Iowa State Univ., Ames.

Netzly, D. H., J. L. Riopel, G. Ejeta, and L. G. Butler. 1988. Germination stimulants of witchweed (*Striga asiatica*) from hydrophobic root exudate of sorghum (*Sorghum bicolor*). *Weed Sci.* 36:441–46.

Nicolai, V. 1988. Phenolic and mineral content of leaves influences decomposition in European forest ecosystems. *Oecolgia* (Berlin) 75:575–79.

Niggli, U., F. P. Weibel, and C. A. Potter. 1989. Weed control with organic mulches in apple orchards: Effects on yield, fruit quality, and dynamics of nitrogen in soil solution. *Gartenbauwissenschaft* 54:224–32.

Nilsson, M.-C., and O. Zackrisson. 1992. Inhibition of Scots pine seedling establishment by *Empetrum hermaphroditum. J. Chem. Ecol.* 18:1857–70.

Nishimura, H., K. Kaku, T. Nakamura, Y. Fukazawa, and J. Mizutani. 1982. Allelopathic substances (plus/minus)—p-methane-3,8-diols isolated from *Eucalyptus citriodora* Hook. *Agric. Biol. Chem.* 46:319–20.

Obiefuna, J. C. 1989. Biological weed control in plantains (musa AAB) with egusi melon (*Colocynthis citrullus* L.). *Biol. Agric. Hort.* 6:221–27.

Odamitten, G. T., and G. C. Clerk. 1988. Effect of metabolites of *Aspergillus niger* and *Trichoderma viride* on development and structure of radicle of cocoa (*Theobroma cacao*) seedlings. *Plant & Soil* 106:285–88.

Ohigashi, H., M. Koji, M. Sakaki, and K. Koshimizu. 1989. 3-Hydroxyuridine, an allelopathic factor of an African tree, *Baillonella toxisperma. Phytochemistry* 28: 1365–68.

Oleszek, W. 1987. Allelopathic effects of volatiles from some Cruciferae species on lettuce, barnyard grass and wheat growth. *Plant & Soil* 102:271–73.

Oleszek, W., and M. Jurzysta. 1987. The allelopathic potential of alfalfa root medicagenic acid glycosides and their fate in soil environments. *Plant & Soil* 98:67–80.

Orabi, K. Y., J. S. Mossa, and F. S. El-Feraly. 1991. Isolation and characterization of two antimicrobial agents from mace (*Myristica fragrans*). *J. Nat. Prod.* 54:856–59.

Ordentlich, A., Y. Elad, and I. Chet. 1987. Rhizosphere colonization by *Serratia marcescens* for the control of *Sclerotium rolfsii. Soil Biol. Biochem.* 19:747–51.

Ordentlich, A., Q. Migheli, and I. Chet. 1991. Biological control activity of 3 *Trichoderma* isolates against Fusarium wilt of cotton and muskmelon and lack of correlation with their lytic enzymes. *J. Phytopathology* 133:177–86.

Ordentlich, A., Z. Wiesman, H. E. Gottlieb, M. Cojocaru, and I. Chet. 1992. Inhibitory furanone produced by the biocontrol agent *Trichoderma harzianum. Phytochemistry* 31:485–86.

Oster, U., I. Blos, and W. Rüdiger. 1987. Natural inhibitors of germination and growth IV. Compounds from fruit and seeds of mountain ash (*Sorbus aucuparia*). *Z. Naturforsch.* C42:1179–84.

Oster, U., M. Spraul, and W. Rüdiger. 1990. Natural inhibitors of germination and growth, V. Possible allelopathic effects of compounds from *Thuja occidentalis. Z. Natuforsch. Sec. C.-J. Biosci.* 45:835–44.

Owens, L. D. 1973. Herbicidal potential of rhizobitoxine. *Weed Sci.* 21:63–66.

Owens, L. D., J. F. Thompson, and P. V. Fennessey. 1972. Dihydrorhizobitoxine, a new ether amino-acid from *Rhizobium japonicum. J. Chem. Soc.,* Chem. Commu. 1972, 715.

Padgett, M., and J. C. Morrison. 1990. Changes in grape berry exudates during fruit development and their effect on mycelial growth of *Botrytis cinerea. J. Am. Soc. Hort. Sci.* 115:269–73.

Panasiuk, O., D. D. Bills, and G. R. Leather. 1986. Allelopathic influence of *Sorghum bicolor* on weeds during germination and early development of seedlings. *J. Chem. Ecol.* 12:1533–43.

Pandya, S. M., V. R. Dave, and K. G. Vyas. 1984. Effect of *Celosia argentea* Linn. on root nodules and nitrogen contents of three legume crops. *Sci. & Cul.* 50:161–62.

Papavizas, G. C., and J. A. Lewis. 1989. Effect of *Gliocladium* and *Trichoderma* on damping-off and blight of snapbean caused by *Sclerotium rolfsii* in the greenhouse. *Plant Pathol.* 38:277–86.

Pardales, J. R., Jr., and A. G. Dingal. 1988. An allelopathic factor in taro residues. *Trop. Agric.* 65:21–24.

Park, C. S., T. C. Paulitz, and R. Baker. 1988. Biocontrol of fusarium wilt of cucumber resulting from interactions between *Pseudomonas putida* and nonpathogenic isolates of *Fusarium oxysporum. Phytopathology* 78:190–94.

Parke, J. L. 1990. Population dynamics of *Pseudomonas cepacia* in the pea spermosphere in relation to biocontrol of *Pythium. Phytopathology* 80:1307–11.

Parke, J. L., R. E. Rand, A. E. Joy, and E. B. King. 1991. Biological control of Pythium damping-off and Aphanomyces root rot of peas by application of *Pseudomonas cepacia* or *P. fluorescens* to seed. *Plant Dis.* 75:987–92.

Partridge, T. R., and M. D. Wilson. 1990. A germination inhibitor in the seeds of mahoe *(Melicytus ramiflorus). New Zealand J. Bot.* 28:475–78.

Paszkowski, W. L., and R. J. Kremer. 1988. Biological activity and tentative identification of flavonoid components in velvetleaf (*Abutilon theophrasti* Medik.) seed coats. *J. Chem. Ecol.* 14:1573–82.

Patil, T. M., and B. A. Hegde. 1988. Isolation and purification of a sesquiterpene lactone from the leaves of *Parthenum hysterophorus* L.—Its allelopathic and cytotoxic effects. *Curr. Sci.* 57:1178–80.

Patrick, Z. A., and L. W. Koch. 1963. The adverse influence of phytotoxic substances from decomposing plant residues on resistance of tobacco to black root rot. *Can. J. Bot.* 41:747–58.

Patrick, Z. A., T. A. Toussoun, and W. C. Snyder. 1963. Phytotoxic substances in arable soils associated with decomposition of plant residues. *Phytopathology* 53:152–61.

Paul, V. L., and P. Schütt. 1987. Auswaschung phytotoxischer Substanzen aus Blättern kranker und gesunder Buchen-Schäden an der Bodenflora. *European J. For. Pathol.* 17:356–61.

Paulitz, T. C. 1991. Effect of *Pseudomonas putida* on the stimulation of *Pythium ultimum* by seed volatiles of pea and soybean. *Phytopathology* 81:1282–87.

Paulitz, T. C., J. S. Ahmad, and R. Baker. 1990. Integration of *Pythium nunn* and *Trichoderma harzianum* isolate T-95 for the biological control of *Pythium* damping-off of cucumber. *Plant & Soil* 121:243–50.

Paulitz, T. C., and R. Baker. 1987. Biological control of pythium damping-off of cucumbers with *Pythium nunn;* Population dynamics and disease suppression. *Phytopathology* 77:335–40.

Paulitz, T. C., and R. G. Linderman. 1989. Interactions between fluorescent pseudomonads and VA mycorrhizal fungi. *New Phytol.* 113:37–45.

Paulitz, T. C., C. S. Park, and R. Baker. 1987. Biological control of fusarium wilt of cucumber with nonpathogenic isolates of *Fusarium oxysporum. Can. J. Microbiol.* 33:349–53.

Pearce, M. H. 1990. *In vitro* interactions between *Armillaria luteobubalina* and other wood decay fungi. *Mycol. Res.* 94:753–61.

Pedersen, C. T., G. R. Safir, J. O. Siqueira, and S. Parent. 1991. Effect of phenolic compounds on asparagus mycorrhiza. *Soil Biol. Biochem.* 23:491–94.

Pederson, M. W. 1965. Effect of alfalfa saponin on cotton seed germination. *Agron. J.* 57:516–17.

van Peer, R., A. J. van Kuik, H. Rattink, and B. Schippers. 1990. Control of *Fusarium* wilt in carnation grown on rockwood by *Pseudomonas* sp. strain WCS417r and by Fe-EDDHA. *Neth. J. Plant Pathol.* 96:119–32.

Peirce, L. C., and L. W. Colby. 1987. Interaction of asparagus root filtrate with *Fusarium oxysporum* f. sp. *asparagi. J. Am. Soc. Hort. Sci.* 112:35–40.

Penn, D. J., and J. M. Lynch. 1981. Effect of decaying couch grass (*Agropyron repens*) on the growth of barley (*Hordeum vulgare*). *J. Appl. Ecol.* 18:669–74.

Pérez, F. J. 1990. Allelopathic effects of hydroxamic acids from cereals on *Avena sativa* and *A. fatua. Phytochemistry* 29:773–76.

Pérez, F. J., and J. Ormeño-Nuñez. 1991a. Difference in hydroxamic acid content in roots and root exudates of wheat (*Triticum aestivum* L.) and rye (*Secale cereale* L.): Possible role in allelopathy. *J. Chem. Ecol.* 17:1037–43.

Pérez, F. J., and J. Ormeño-Nuñez. 1991b. Root exudates of wild oats: Allelopathic effect on spring wheat. *Phytochemistry* 30:2199–202.

Pesenti-Barili, B., E. Ferdani, M. Mosti, and F. Degli-Innocenti. 1991. Survival of *Agrobacterium radiobacter* K84 on various carriers for crown gall control. *Appl. Environ. Microbiol.* 57:2047–51.

Peters, E. J., and A. H. B. Mohammed Zam. 1981. Allelopathic effects of tall fescue (*Festuca arundinacea*) genotypes. *Agron. J.* 73:56–58.

Pfender, W. F. 1988. Suppression of ascocarp formation in *Pyrenophora tritici-repentis* by *Limonomyces roseipellis,* a basidiomycete from reduced-tillage wheat straw. *Phytopathology* 78:1254–58.

Politycka, B., D. Wójcik-Wojtkowiak, and T. Pudelski. 1984. Phenolic compounds as a cause of phytotoxicity in greenhouse substrates repeatedly used in cucumber growing. *Acta Hort.* 156:89–94.

Ponder, F., Jr., and S. H. Tadros. 1985. Juglone concentration in soil beneath black walnut interplanted with nitrogen-fixing species. *J. Chem. Ecol.* 11:937–42.

Porwal, M. K., and O. P. Gupta. 1986. Allelopathic influence of winter weeds on germination and growth of wheat. *Int. J. Trop. Agri.* 4:276–79.

van Praag, H. J., J. C. Motte, X. Monseur, J. Walravens, and F. Weissen. 1991. Root growth inhibition of the holorganic layer of moder humus under spruce (*Picea abies* Karst.) and beech (*Fagus sylvatica* L.). *Plant & Soil* 135:175–83.

Prasad, R., and B. R. R. Rao. 1981. Allelopathy in spring wheat mixtures (cultivars). *Experientia* 37:1078.

Pusey, P. L. 1989. Use of *Bacillus subtilis* and related organisms as biofungicides. *Pestic. Sci.* 27:133–40.

Putnam, A. R. 1988. Allelochemicals from plants as herbicides. *Weed Technol.* 2:510–18.

Putnam, A. R., and J. DeFrank. 1979. Use of cover crops to inhibit weeds. *Prox. IX Int. Cong. Plant Protection,* pp. 580–82.

Putnam, A. R., and J. DeFrank. 1983. Use of phytotoxic plant residues for selective weed control. *Crop Prot.* 2:173–81.

Putnam, A. R., and W. B. Duke. 1974. Biological suppression of weeds: Evidence for allelopathy in accessions of cucumber. *Science* 185:370–72.

Putnam, A. R., J. DeFrank, and J. P. Barnes. 1983. Exploitation of allelopathy for weed control in annual and perennial cropping systems. *J. Chem. Ecol.* 9:1001–10.

Qasem, J. R., and B. E. Abu-Irmaileh. 1985. Allelopathic effect of *Salvia syriaca* L. (Syrian sage) in wheat. *Weed Res.* 25:47–52.

Qasem, J. R., and T. A. Hill. 1989. Possible role of allelopathy in the competition between tomato, *Senecio vulgaris* L., and *Chenopodium album* L. *Weed Res.* 29:349–56.

Qureshi, I. H., S. Ahmed, and Z. Kapadia. 1989. Antimicrobial activity of *Salvia splendens. Pak. J. Scient. Indust. Res.* 32:597–99.

Qureshi, H. A., and F. Hussain. 1980. Allelopathic potential of Columbus grass (*Sorghum almum* (Piper) Parodi). *Pak. J. Scient. Indust. Res.* 23:189–95.

Qureshi, Z. M., F. Hussain, and S. Shaukat. 1987. Allelopathic effect of *Avena fatua* Linn. on some crops. In *Modern Trends of Plant Science Research in Pakistan.* pp. 53–58 (I. Ilahi and F. Hussain, eds.), Proc. Third Nat'l. Conf. of Plant Scientists, Dept. of Bot., Univ. of Peshawar, Peshawar.

Rafiq, M., M. A. Nasir, and M. A. R. Bhatti. 1984. Antifungal properties of certain common wild plants against different fungi. *Pak. J. Agric. Res.* 5:236–38.

Raimbault, B. A., T. J. Vyn, and M. Tollenaar. 1991. Corn response to rye cover crop, tillage methods, and planter options. *Agron. J.* 83:287–90.

Ram, D. 1989. Effect of Phenolic Compounds Released in Soil by Decomposing Medicinal Plants on Two Soil-Borne Plant Pathogens. Ph.D. dissertation, Banaras Hindu Univ., Varanasi, India (Abstr.).

Ram, D., and B. L. Jalali. 1992. The fungicidal efficacy of asafoetida (*Ferula foetida*

L.) on chickpea blight. In *Proceedings First National Symposium on Allelopathy in Agroecosystems.* pp. 194–95, (P. Tauro and S. S. Narwal, eds.), Indian Soc. of Allelopathy, Haryana Agricultural Univ., Hisar, India.

Ramachandra, G., and P. V. Monteiro. 1990. Preliminary studies on the nutrient composition of *Cassia sericea* Sw.—an unexploited legume seed. *J. Food Composition & Analysis* 3:81–87.

Rao, V. V., and Pandya, S. M. 1992. Allelopathic interference by *Triticum aestivum* on the growth of *Asphodelus tenuifolius.* In *Proceedings First National Symposium on Allelopathy in Agroecosystems,* pp. 41–44 (P. Tauro and S. S. Narwal, eds.), Indian Soc. of Allelopathy, Haryana Agricultural Univ., Hisar, India.

Ravi, K., and T. B. Anilkumar. 1991. Effect of cowpea phylloplane fungi on fungicide resistant strains of *Colletotrichum truncatum* (Schw.) Andrus & Moore. *Zentralbl. Mikrobiol.* 146:209–12.

Read, J. J., and E. H. Jensen. 1989. Phytotoxicity of water-soluble substances from alfalfa and barley soil extracts on four crop species. *J. Chem. Ecol.* 15:619–28.

Reddy, M. S., and Z. A. Patrick. 1992. Colonization of tobacco seedling roots by fluorescent pseudomonad suppressive to black root rot caused by *Thielaviopsis basicola. Crop Prot.* 11:148–54.

Redmond, J. C., J. J. Marois, and J. D. MacDonald. 1987. Biological control of *Botrytis cinerea* on roses with epiphytic microorganisms. *Plant Dis.* 71:799–802.

Reigosa-Roger, M. J. 1987. Estudio del Potencial Allelopatico de *Acacia dealbata* Link. Tese de Doutoramento, Universidad de Santiago. (Teses en Microficha Núm. 10, B-38729/88.)

Rho, B. J., and B.-S. Kil. 1986. Influence of phytotoxin from *Pinus rigida* on selected plants. *Korean J. Ecol.* 10:19–27. (Korean)

Rice, E. L. 1983. *Pest Control with Nature's Chemicals: Allelochemics and Pheromones in Gardening and Agriculture.* University of Oklahoma Press, Norman.

Rice, E. L. 1984. *Allelopathy.* 2d ed. Academic Press, Orlando, Florida.

Richter, D. L., T. R. Zuellig, S. T. Bagley, and J. N. Bruhn. 1989. Effects of red pine (*Pinus resinosa* Ait.) mycorrhizoplane-associated actinomycetes on in vitro growth of ectomycorrhizal fungi. *Plant & Soil* 115:109–16.

Riddle, G. E., L. L. Burpee, and G. J. Boland. 1991. Virulence of *Sclerotinia sclerotiorum* and *S. minor* on dandelion (*Taraxacum officinale*). *Weed Sci.* 39:109–18.

Riffle, M. S. 1988. Biological and Biochemical Interactions between Unicorn Plant (*Proboscidea louisianica*) and Cotton (*Gossypium hirsutum*). Ph.D. dissertation, Oklahoma State Univ., Stillwater.

Riffle, M. S., W. E. Thilstead, D. S. Murray, R. M. Ahring, and G. R. Waller. 1988. Germination and seed production of unicorn-plant (*Proboscidea louisianica*). *Weed Sci.* 36:787–91.

Rink, G., and J. W. van Sambeek. 1985. Variation among black walnut seedling families in resistance to competition and allelopathy. *Plant & Soil* 88:3–10.

Rink, G., and J. W. van Sambeek. 1987. Variation among four white ash families in response to competition and allelopathy. *For. Ecol. Manage.* 18:127–34.

Ristaino, J. B., K. B. Perry, and R. D. Lumsden. 1991. Effect of solarization and *Gliocladium virens* on sclerotia of *Sclerotium rolfsii* soil, microbiota, and incidence of southern blight of tomato. *Phytopathology* 81:1117–24.

Rizvi, S. J. H., D. Mukerji, and S. N. Mathur. 1980. A new report on a possible source of natural herbicide. *Indian J. Exp. Biol.* 18:777–78.

Rizvi, S. J. H., D. Mukerji, and S. N. Mathur. 1981. Selective phyto-toxicity of 1,3,7-trimethylxanthine between *Phaseolus mungo* and some weeds. *Agr. Biol. Chem.* 45:1255–56.

Rizvi, S. J. H., V. K. Singh, V. Rizvi, and G. R. Waller. 1988. Geraniol, an allelochemical of possible use in integrated pest management. *Plant Prot. Q.* 3:112–14.

Roberts, D. P., and R. D. Lumsden. 1990. Effect of extracellular metabolites from *Gliocladium virens* on germination of sporangia and mycelial growth of *Pythium ultimum. Phytopathology* 80:461–65.

Roder, W., S. S. Waller, and J. L. Stubbendieck. 1988. Allelopathic effects of sandbur leachate on switchgrass germination: Observations. *J. Range Manage.* 41:86–87.

Roiger, D. J., and S. N. Jeffers. 1991. Evaluation of *Trichoderma* spp. for biological control of Phytophthora crown and root rot of apple seedlings. *Phytopathology* 81:910–17.

Ruiz-Sifre, G. V., and S. K. Ries. 1983. Response of crops to sorghum residues. *J. Am. Soc. Hort. Sci.* 108:262–66.

Ryder, M. H., and D. A. Jones. 1991. Biological control of crown gall using *Agrobacterium* strains K84 and K1026. *Aust. J. Plant Physiol.* 18:571–79.

Rysheuvels, P., M. Laroche, and M. Verhoyen. 1984. Applicability study of biological control by *Agrobacterium radiobacter* var. *radiobacter* (K84 strain of Kerr) on *Agrobacterium radiobacter* var. *tumefaciens* in orchards. *Parasitica* (Gembloux) 40:183–96.

Rytter, J. L., F. L. Lukezic, R. Craig, and G. W. Moorman. 1989. Biological control of geranium rust by *Bacillus subtilis. Phytopathology* 79:367–70.

Sadhu, M. K. 1975. Nature of inhibitory substances in root exudates of rice seedlings. *Indian J. Exp. Biol.* 13:577–79.

Sadhu, M. K., and T. M. Das. 1971a. Root exudates of rice seedlings. The influence of one variety on another. *Plant & Soil* 34:541–46.

Sadhu, M. K., and T. M. Das. 1971b. Studies on root exudates of rice seedlings factors influencing the liberation of growth inhibiting substances. *Indian Agric.* 15:87–94.

Sajise, P. E., and J. S. Lales. 1975. Allelopathy in a mixture of cogon (*Imperata cylindrica*) and *Stylosanthes guyanensis. Kalikasan Philipp. J. Biol.* 4:155–64.

Saksena, N., and H. H. S. Tripathi. 1987. Effect of organic volatiles from *Saccharomyces* on the spore germination of fungi. *Acta Microbiol. Hungarica* 34:255–57.

Sakthivel, N., and T. W. Mew. 1991. Efficacy of bacteriocinogenic strains of *Xanthomonas oryzae* pv. *oryzae* on the incidence of bacterial blight disease of rice (*Oryza sativa* L.). *Can. J. Microbiol.* 37:764–68.

Salama, A-A. M., I. M. K. Ismail, M. I. A. Ali, and S. A-E. Ouf. 1988. Possible control of white rot disease of onions caused by *Sclerotium cepivorum* through soil amendment with *Eucalyptus rostrata* leaves. *Rev. Ecol. Biol. Sol.* 25:305–14.

Salmon, D. F., J. H. Helm, T. R. Duggan, and D. M. Lakeman. 1986. The influence of chaff extracts on the germination of spring triticale. *Agron. J.* 78:863–67.

Sands, D. C., E. J. Ford, and R. V. Miller. 1990. Genetic manipulation of broad host-range fungi for biological control of weeds. *Weed Technol.* 4:471–74.

Sankhla, N., D. Baxi, and U. N. Chatterji. 1965. Eco-physiological studies on arid zone plants. I. Phytotoxic effects of aqueous extracts of mesquite (*Prosopis juliflora* DC). *Curr. Sci.* 34:612–14.

Sarniguet, A., and P. Lucas. 1992. Evaluation of populations of fluorescent pseudomonads related to decline of take-all patch on turfgrass. *Plant & Soil* 145:11–15.

Savithiry, S., and S. S. Gnanamanickam. 1987. Bacterization of peanut with *Pseudomonas fluorescens* for biological control of *Rhizoctonia solani* and for enhanced yield. *Plant & Soil* 102:11–15.

Saxena, M. 1992. Allelopathic potential of terrestrial plants against the growth of aquatic weeds. In *Proceedings First National Symposium on Allelopathy in Agroecosystems.* pp. 147–48 (P. Tauro and S. S. Narwal, eds.). Indian Soc. of Allelopathy, Haryana Agricultural Univ., Hisar, India.

Scacchi, A., R. Bortolo, G. Cassani, G. Pirali, and E. Nielsen. 1992. Detection, characterization and phytotoxic activity of the nucleoside antibiotics, blasticidin S and 5-hydroxylmethyl-blasticidin S. *J. Plant Growth Regul.* 11:39–46.

Scheffer, R. J. 1989. *Pseudomonas* for biological control of Dutch elm disease. III. Field trials at various locations in the Netherlands. *Neth. J. Plant Pathol.* 95:305–18.

Scheffer, R. J., D. M. Elgersma, L. A. de Weger, and G. A. Strobel. 1989a. *Pseudomonas* for biological control of Dutch elm disease. I. Labeling, detection and identification of *Pseudomonas* isolates injected into elms: Comparison of various methods. *Neth. J. Plant Pathol.* 95:281–92.

Scheffer, R. J., D. M. Elgersma, and G. A. Strobel. 1989b. *Pseudomonas* for biological control of Dutch elm disease. II. Further studies on the localization, persistence and ecology of *Pseudomonas* isolates injected into elms. *Neth. J. Plant Pathol.* 95:293–304.

Schenk, S. U., and D. Werner. 1991. ß-(3-Isoxazolin-5-on-2-yL)-alanine from *Pisum:* Allelopathic properties and antimycotic bioassay. *Phytochemistry* 30:467–70.

Schumacher, W., D. Thill, and G. Lee. 1983. Allelopathic potential of wild oat (*Avena fatua*) on spring wheat. *J. Chem. Ecol.* 9:1235–46.

Selleck, G. W. 1972. The antibiotic effects of plants in laboratory and field. *Weed Sci.* 20:189–94.

Sen, D. N., and D. D. Chawan. 1970. Ecology of desert plants and observations on their seedlings. III. The influence of aqueous extracts of *Prosopis juliflora* DC. on *Euphorbia caducifolia* Haines. *Vegetatio* 21:277–98.

Serrano, L., and P. I. Boon. 1991. Effect of polyphenolic compounds on alkaline phosphatase activity: Its implication for phosphorus regeneration in Australian freshwaters. *Arch. Hydrobiol.* 123:1–19.

Sethi, A., and K. Mohnot. 1988. Allelopathic influence of leaf extract of *Trianthema portulacastrum* on germination and growth of moth bean, *Vigna aconitifolius. J. Curr. Biosci.* 5:61–63.

Shafer, S. R., and U. Blum. 1991. Influence of phenolic acids on microbial populations in the rhizosphere of cucumber. *J. Chem. Ecol.* 17:369–89.

Shanahan, P., D. J. O'Sullivan, P. Simpson, J. D. Glennon, and F. O'Gara. 1992. Isolation of 2,4-diacetyl-phloroglucinol from a fluorescent pseudomonad and investigation of physiological parameters influencing its production. *Appl. Environ. Microbiol.* 58:353–58.

Sharma, K. D., K. L. Sidana, and N. R. Singhvi. 1982. Allelopathic effect of *Peganum harmala* Linn. on *Pennisetum typhoideum* L. (Bajra). *Indian J. Bot.* 5:115–19.

Sharma, V., and G. S. Nathawat. 1987. Allelopathic effect of *Argemone mexicana* L. on species of *Triticum, Raphanus* and *Pennisetum. Curr. Sci.* 56:427–28.

Shi, J.-L., and C. M. Brasier. 1986. Experiments on the control of Dutch elm disease by injection of *Pseudomonas* species. *Eur. J. For. Pathol.* 16:280–92.

Shilling, D. G., R. A. Liebl, and A. D. Worsham. 1985. Rye and wheat mulch: The suppression of certain broadleaved weeds and the isolation and identification of phytotoxins. In *The Chemistry of Allelopathy.* pp. 243–71 (A. C. Thompson, ed.), ACS Symp. Ser. No. 268, Am. Chem. Soc., Washington, D. C.

Shilling, D. G., L. A. Jones, A. D. Worsham, C. E. Parker, and R. F. Wilson. 1986. Isolation and identification of some phytotoxic compounds from aqueous extracts of rye (*Secale cereale* L.). *J. Agr. Food Chem.* 34:633–38.

Shoup, S., and C. E. Whitcomb. 1981. Interaction between trees and ground covers. *J. Arboriculture* 7:186–87.

Silman, R. W., T. C. Nelsen, and R. J. Bothast. 1991. Comparison of culture methods for production of *Collelotrichum truncatum* spores for use as a mycoherbicide. *FEMS Microbiol. Lett.* 79:69–74.

Simon, A. 1989. Biological control of take-all of wheat by *Trichoderma koningii* under controlled environmental conditions. *Soil Biol. Biochem.* 21:323–26.

Simon, A., R. W. Dunlop, E. L. Ghisalberti, and K. Sivasithamparam. 1988. *Trichoderma koningii* produces a pyrone compound with antibiotic properties. *Soil Biol. Biochem.* 20:263–64.

Singh, D. 1991. Biocontrol of *Sclerotinia sclerotiorum* (Lib.) de Bary by *Trichoderma harzianum. Trop. Pest Manage.* 37:374–78.

Singh, M., R. V. Tamma, and H. N. Nigg. 1989. HPLC identification of allelopathic compounds from *Lantana camara. J. Chem. Ecol.* 15:81–89.

Singh, R. K., and R. S. Dwivedi. 1990. Fungicidal properties of neem and blue gum against *Sclerotium rolfsii* Sacc., a foot-rot pathogen of barley. *Acta Bot. Indica* 18:260–62.

Singh, U. P., V. B. Pandey, R. N. Singh, and R. D. N. Singh. 1988. Antifungal activity of some new flavones and flavone glycosides of *Echinops echinatus. Can. J. Bot.* 66:1901–03.

Singh, U. P., D. Ram, and V. P. Tewari. 1990. Induction of resistance in chickpea (*Cicer arietinum*) by *Aegle marmelos* leaves against *Sclerotinia sclerotiorum. Z. Pflanzenkr. Pflanzenschutz* 97:439–43.

Singh, U. P., V. B. Chauhan, K. G. Wagner, and A. Kumar. 1992. Effect of ajoene, a compound derived from garlic (*Allium sativum*), on *Phytophthora drechsleri* f. sp. *cajana. Mycologia* 84:105–8.

Sircar, S. M., and R. Chakravarty. 1961. The effect of growth-regulating substances of the root extract of water hyacinth (*Eichhornia speciosa* Kunth) on jute (*Corchorus capsularis* L.). *Curr. Sci.* 30:428–30.

Sivamani, E., and S. S. Gnanamanickam. 1988. Biological control of *Fusarium oxysporum* f. sp. *cubense* in banana by inoculation with *Pseudomonas fluorescens. Plant & Soil* 107:3–9.

Sivan, A., O. Ucko, and I. Chet. 1987. Biological control of fusarium crown rot of tomato by *Trichoderma harzianium* under field conditions. *Plant Dis.* 71:587–92.

Slepykh, V. V. 1988. Phytoncidal activity of forest phytocenoses. *Soviet J. Ecol.* 19:187–91.

Smeda, R. J., and A. R. Putnam. 1988. Cover crop suppression of weeds and influence on strawberry yields. *Hort. Sci.* 23:132–34.

Smith, A. E. 1989. The potential allelopathic characteristics of bitter sneezeweed (*Helenium amarum*). *Weed Sci.* 37:665–69.

Smith, J., A. Putnam, and M. Nair. 1990. *In vitro* control of fusarium diseases of *Asparagus officinalis* L. with a streptomyces or its polyene antibiotic, faeriefungin. *J. Agric. Food Chem.* 38:1729–33.

Sneh, B., M. Zeidan, M. Ichielevich-Auster, I. Barash, and Y. Koltin. 1986. Increased growth responses induced by a nonpathogenic *Rhizoctonia solani. Can. J. Bot.* 64:2372–78.

Sobiczewski, P., J. Karczewski, and S. Berczynski. 1991. Biological control of crown gall *Agrobacterium tumefaciens* in Poland. *Fruit Sci. Rep.* (Skierniewice) 18:125–32.

Sokolov, I. N., A. E. Chekhova, Y. T. Eliseev, G. I. Nilov, and L. R. Shcherbanovskii. 1972. Study of antimicrobial activity of certain naphthoquinones. *Prikl. Biokhim. Mikrobiol.* 8:261–63.

Solomon, M. J., and D. C. Bhandari. 1981. Allelopathic potential of *Gomphrena decumbens* on two rain-fed crops. *Geobios* (Jodhpur) 8:9–12.

Soni, S. R., and K. Mohnot. 1988. Presence of an autotoxic factor in fruit carp of *Echinops echinatus* Roxb. *J. Curr. Biosci.* 5:101–4.

Spencer, P. A., and G. H. N. Towers. 1988. Specificity of signal compounds detected by *Agrobacterium tumefaciens Phytochemistry* 27:2781–85.

Spencer, P. A., A. Tanaka, and G. H. N. Towers. 1990. An *Agrobacterium* signal compound from grapevine cultivars. *Phytochemistry* 29:3785–88.

Spink, D. S., and R. C. Rowe. 1989. Evaluation of *Talaromyces flavus* as a biological control agent against *Verticillium dahliae* in potato. *Plant Dis.* 73:330–36.

Spring, O., and A. Hager. 1982. Inhibition of elongation growth by two sesquiterpene lactones isolated from *Helianthus annuus* L. *Planta* 156:433–40.

Spring,O., U. Rodon, and F. A. Macias. 1992. Sesquiterpenes from noncapitate glandular trichomes of *Helianthus annuus. Phytochemistry* 31:1541–44.

Srivastava, A. K. 1969. Effect of seed extract of *Heliotropium eichwaldi* Steud. on seed germination of *Phaseolus aureus* Roxb. *Curr. Sci.* 38:440–41.

Srivastava, J. N., J. P. Shukla, and R. C. Srivastava. 1985. Effect of *Parthenium hysterophorus* Linn. extract on the seed germination and seedling growth of barley, pea and wheat. *Acta Bot. Indica* 13:194–97.

Srivastava, P. C., N. G. Totey, and O. Prakash. 1986. Effect of straw extract on water absorption and germination of wheat (*Triticum aestivum* L. variety RR-21) seeds. *Plant & Soil* 91:143–45.

Srivastava, P. P., and L. L. Das. 1974. Effect of certain aqueous plant extracts on the germination of *Cyperus rotundus* L. *Sci. and Culture* 40:318–19.

Stachel, S. E., E. Messens, M. van Montagu, and P. Zambryski. 1985. Identification of the signal molecules produced by wounded plant cells that activate T-DNA transfer in *Agrobacterium tumefaciens. Nature* 318:624–29.

Stephenson, R. J., and G. L. Posler. 1988. The influence of tall fescue on the germination, seedling growth and yield of birdsfoot trefoil. *Grass Forage Sci.* 43:273–78.

Sterling, T. M., and A. R. Putnam. 1987. Possible role of glandular trichome exudates in interference by velvetleaf (*Abutilon theophrasti*). *Weed Sci.* 35:308–14.

Sterling, T. M., R. L. Houtz, and A. R. Putnam. 1987. Phytotoxic exudates from velvetleaf (*Abutilon theophrasti*) glandular trichomes. *Am. J. Bot.* 74:543–50.

Stevens, G. A., Jr., and C. S. Tang. 1985. Inhibition of seedling growth of crop species by recirculating root exudates of *Bidens pilosa* L. *J. Chem. Ecol.* 11:1411–25.

Stierle, A. C., J. H. Cardellina II, and G. A. Strobel. 1989. Phytotoxins from *Alternaria alternata,* a pathogen of spotted knapweed. *J. Nat. Prod.* 52:42–47.

Stovall, M. E., and K. Clay. 1991. Fungitoxic effects of *Balansia cyperi. Mycologia* 83:288–95.

Sugawara, F., and G. A. Strobel. 1986. (-)-Dihydropyrenophorin, a novel and selective phytotoxin produced by *Drechslera avenae. Plant Sci.* 43:1–5.

Sugha, S. K. 1978. Allelopathic potential of superb lily (*Gloriosa superba* L.). *Sci. & Cult.* 44:461–62.

Sugha, S. K. 1979. Effect of weed extracts on wheat germination. *Sci. & Cult.* 45:65–66.

Suh, H.-W., D. L. Crawford, R. A. Korus, and K. Shetty. 1991. Production of antifungal metabolites by the ectomycorrhizal fungus *Pisolithus tinctorius* strain SMF. *J. Indust. Microbiol.* 8:29–36.

Sukumar, J., and A. Ramalingam. 1986. Antagonistic effects of phylloplane micro-organisms against *Cercospora moricola* Cooke. *Curr. Sci.* 55:1208–9.

Sun, W. H., Z.-W. Yu, and S.-W. Yu. 1988. Inhibitory effect of *Eichhornia crassipes* (Mart.) Solms on algae. *Acta Phytophysiol. Sin.* 14:294–300.

Sun, W.-H., Z.-W. Yu, and S.-W. Yu. 1989. The harness of an eutrophic water body by water-hyacinth. *Acta Scient. Circumstantiae.* 9:188–195. (Chinese, English summary)

Sun, W.-H, Z.-W Yu, G.F Tai, and S.-W. Yu. 1990. Sterilized culture of water hyacinth and its application in the study of allelopathic effect on algae. *Acta Phytophysiol. Sin.* 16:301–5. (Chinese, English summary)

Sundaramoorthy, S., and D. N. Sen. 1990. Allelopathic effects of *Tephrosia purpurea* Pers. I. On germination and growth of some arid zone crops. *J. Indian Bot. Soc.* 69:251–55.

Suresh, K. K., and R. S. Vinaya Rai. 1988. Allelopathic exclusion of understory by a few multipurpose trees. *Int. Tree Crops J.* 5:143–51.

Suseelamma, M., and R. R. Venkata Raju. 1992. Allelopathic effect of *Digera muricata* (L.) Mart. on the germination and early seedling growth of horsegram. In *Proceedings First National Symposium on Allelopathy in Agroecosystems.* pp. 89–91 (P. Tauro and S. S. Narwal, eds.), Haryana Agri. Univ., Hisar, India.

Suslow, T. V., and M. M. Schroth. 1982. Role of deleterious rhizobacteria as minor pathogens in reducing crop growth. *Phytopathology* 72:111–15.

Szewczuk, V., W. Kita, B. Jarosz, W. Truszkowska, and A. Siewinski. 1991. Growth inhibition of some phytopathogenic fungi by organic extracts from *Nigrospora oryzae* (Berkeley and Broome) Petch. *J. Basic Microbiol.* 31:69–73.

Tada, M., and K. Sakurai. 1991. Antimicrobial compound from *Cercidiphyllum japonicum. Phytochemistry* 30:1119–20.

Tahvonen, R., and M. L. Lahdenperä. 1988. Biological control of *Botrytis cinerea* and *Rhizoctonia solani* in lettuce by *Streptomyces* sp. *Ann. Agric. Fenn.* 27:107–16.

Tahvonen, R., and H. Avikainen. 1990. Effect of *Streptomyces* sp. on seed-borne foot rot diseases of wheat and barley. I. Pot experiments. *Ann. Agric. Fenn.* 29:187–94.

Tang, C. S., C. K. Wat, and G. H. N. Towers. 1987. Thiophenes and benzofurans in the undisturbed rhizosphere of *Tagetes patula* L. *Plant & Soil* 98:93–97.

Tanrisever, N., F. R. Fronczek, N. H. Fischer, and G. B. Williamson. 1987. Ceratiolin and other flavanoids from *Ceratiola ericoides. Phytochemistry* 26:175–79.

Tanrisever, N., N. H. Fischer, and G. B. Williamson. 1988. Menthofurans from *Calamintha ashei:* Effects on *Schizachyrium scoparium* and *Lactuca sativa. Phytochemistry* 27:2523–26.

Tao, D. L., Z. B. Xu, and X. Li. 1987. Effect of litter layer on natural regeneration of companion tree species in the Korean pine forest. *Environ. Exp. Bot.* 27:53–65.

Tapaswi, P. K., R. N. Banerjee, D. K. Bagchi, S. Adhikary, and R. L. Brahmachary. 1991. Interaction of two varieties of rice—a study of intervarietal interaction. *Indian Biologist* 23:8–9.

Tesar, M. B. 1984. Establishing alfalfa without autotoxicity. In *Report of the Twenty-ninth Alfalfa Improvement Conference,* p. 39, July 15–20, 1984, Lethbridge, Alberta, Canada.

Tewari, S. N., and M. Nayak. 1991. Activity of four plant leaf extracts against three fungal pathogens of rice. *Trop. Agric.* 68:373–75.

Thomashow, L. S., and D. M. Weller. 1988. Role of a phenazine antibiotic from *Pseudomonas fluorescens* in biological control of *Gaeumannomyces graminis* var. *tritici. J. Bacteriol.* 170:3499–3508.

Thomashow, L. S., and D. M. Weller. 1990. Role of antibiotics and siderophores in biocontrol of take-all disease of wheat. *Plant & Soil* 129:93–99.

Thomashow, L. S., D. M. Weller, R. F. Bonsall, and L. S. Pierson III. 1990. Production of the antibiotic phenazine-1-carboxylic acid by fluorescent *Pseudomonas* species in the rhizosphere of wheat. *Appl. Environ. Microbiol.* 56:908–12.

Thompson, D. P. 1989. Fungitoxic activity of essential oil components on food storage fungi. *Mycologia* 81:151–53.

Thompson, R. J., and R. G. Burns. 1989. Control of *Pythium ultimum* with antagonistic fungal metabolites incorporated into sugar beet seed pellets. *Soil Biol. Biochem.* 21:745–48.

Thorne, R. L. Z., G. R. Waller, J. K. McPherson, E. G. Krenzer, Jr., and C. C. Young. 1990. Autotoxic effects of old and new wheat straw in conventional-tillage and no-tillage wheat soil. *Bot. Bull. Acad. Sin.* 31:35–49.

Tinnin, R. O., and L. A. Kirkpatrick. 1985. The allelopathic influence of broadleaf trees and shrubs on seedlings of Douglas-fir. *For. Sci.* 31:945–52.

Todhunter, M. N., and W. F. Beineke. 1979. Effect of fescue on black walnut growth. *Tree Planters' Notes* 30:20–23.

Tomás-Barberán, F. A., J. D. Msonthi, and K. Hostettmann. 1988. Antifungal epicuticular methylated flavonoids from *Helichrysum nitens. Phytochemistry* 27: 753–55.

Tomás-Barberán, F., E. Iniesta-Sanmartín, F. Tomás-Lorente, and A. Rumberto. 1990. Antimicrobial phenolic compounds from three Spanish *Helichrysum* species. *Phytochemistry* 29:1093–95.

Tomás-Lorente, F., E. Iniesta-Sanmartín, F. A. Tomás-Barberán, W. Trowitzsch-

Kienast and V. Wray. 1989. Antifungal phloroglucinol derivatives and lipophilic flavonoids from *Helichrysum decumbens*. *Phytochemistry* 28:1613–15.

Toussoun, T. A., and Z. A. Patrick. 1963. Effect of phytotoxic substances from decomposing plant residues on root rot of bean. *Phytopathology* 53:265–70.

Toyota, M., and K. Hostettmann. 1990. Antifungal diterpenic esters from the mushroom *Beletinus cavipes*. *Phytochemistry* 29:1485–89.

Toyota, M., J. D. Msonthi, and K. Hostettmann. 1990. A molluscididal and antifungal triterpenoid saponin from the roots of *Clerodendrum wildii*. *Phytochemistry* 29: 2849–51.

Tripathi, R. S., R. S. Singh, and J. P. N. Rai. 1981. Allelopathic potential of *Eupatorium adenophorum*, a dominant ruderal weed of Meghalaya. *Proc. Indian Nat'l. Sci. Acad. Part B, Biol. Sci.* 47:458–65.

Tsantrizos, Y. S., H. H. Kope, J. A. Fortin, and K. K. Ogilvie. 1991. Antifungal antibiotics from *Pisolithus tinctorius*. *Phytochemistry* 30:1113–18.

Tschen, J. S.-M. 1991. Effect of antibiotic antagonists on control of basal stem rot of crysanthemum caused by *Rhizoctonia solani*. *Plant Prot. Bull.* 33:56–62.

Tsuzuki, E., A. Katsuki, S. Shida, and T. Nagatomo. 1977. On the growth inhibitors contained in buckwheat plants. 2. The effects of water and organic solvent extracts on the growth of rice seedlings. *Bull. Faculty Agri., Miyazaki Univ.* 24:41–46.

Tsuzuki, E., J. Watanabe, and S. Shida. 1978. Studies on allelopathy among higher plants I. Effects of residues and aqueous and various organic solvent extracts from the lawn weed pennywort (*Hydrocotyle sibthorpioides* Lam.) on the seed germination and growth of some vegetables and crop plants. *Bull. Fac. Agr., Miyazaki Univ.* 25:403–10. (Japanese, English summary)

Tsuzuki, E., Y. Yamamoto, and T. Shimizu. 1987. Fatty acids in buckwheat are growth inhibitors. *Ann. Bot.* 60:69–70.

Turchetti, T., and G. Maresi. 1991. Inoculation trials with hypovirulent strains of *Cryphonectria parasitica*. *Eur. J. For. Pathol.* 21:65–70.

Turco, R. F., M. Bischoff, D. P. Breakwell, and D. R. Griffith. 1990. Contribution of soil-borne bacteria to the rotation effect in corn. *Plant & Soil* 122:115–20.

Turner, J. T., and P. A. Backman. 1991. Factors relating to peanut yield increases after seed treatment with *Bacillus subtilis*. *Plant Dis.* 75:347–53.

Tuset, J. J., C. Hinarejos, and J. Garcia. 1990. *Phytophthora* foot rot control in citrus with *Myrothecium roridum*. *EPPO Bull.* 20:169–76.

Ueki, K., and M. Takahashi. 1984. Weed inhibition by *Lycoris radiata* Herb. *Proc. of Symposium on Allelopathy.* (1984 Int'l Chem. Cong. of Pacific Basin Societies, Dec. 16–21, Honolulu, Hawaii). (Abstract)

Umber, R. L. 1978. A Study of Allelopathy in *Helianthus rigidus*. M. S. thesis, Oklahoma State Univ., Stillwater.

Underwood, G. J. C., and J. H. Baker. 1991. The effect of various aquatic bacteria on the growth and senescence of duckweed (*Lemna minor*). *J. Appl. Bacteriol.* 70: 192–96.

Upadhyay, J. P., and A. N. Mukhopadhyay. 1986. Biological control of *Sclerotium rolfsii* by *Trichoderma harzianum* in sugarbeet. *Trop. Pest Manage.* 32:215–20.

Upadhyay, R. S., and B. Rai. 1987. Studies on antagonism between *Fusarium udum* Butler., and root region microflora of pigeon-pea. *Plant & Soil* 101:79–93.

Utkhede, R. S., and E. M. Smith. 1991. Biological and chemical treatments for control of *Phytophthora* crown and root rot caused by *Phytophthora cactorum* in a high density apple orchard. *Can. J. Plant Pathol.* 13:267–70.

Vail, S. L., O. D. Dailey, E. J. Blanchard, A. B. Pepperman, and J. L. Riopel. 1990. Terpenoid precursors of strigol as seed germination stimulants of broomrape (*Orobanche ramosa*) and witchweed (*Striga asiatica*). *J. Plant Growth Regul.* 9:77–83.

Vanneste, J. L., J. Yu, and S. V. Beer. 1992. Role of antibiotic production by *Erwinia herbicola* Eh252 in biological control of *Erwinia amylovora*. *J. Bacteriol.* 174: 2785–96.

Vierheilig, H., and J. A. Ocampo. 1990. Effect of isothiocyanates on germination of spores of *G. mosseae*. *Soil Biol. Biochem.* 22:1161–62.

Vrany, J., M. Rasochova, A. Fiker, and K. Dobias. 1990. Inoculation of potatoes with microorganisms under field conditions. I. Effect on plant growth, yields and physiological properties of tubers in potato and sugar-beet regions. *Folia Microbiol.* 35:326–35.

Wacker, T. L., G. R. Safir, and C. T. Stephens. 1990. Effects of ferulic acid on *Glomus fasciculatum* and associated effects on phosphorus uptake and growth of asparagus (*Asparagus officinalis* L.). *J. Chem. Ecol.* 16:901–9.

Wadhawani, C., and T. N. Bhardwaja. 1981. Effect of *Lantana camara* L. extract on fern spore germination. *Experientia* 37:245–46.

Walker, H. L., and W. J. Connick, Jr. 1984. Mycoherbicide-containing pellets. *Chem. Abstr.* 100:169956K.

Waller, G. R., E. G. Krenzer, Jr., J. K. McPherson, and S. R. McGown. 1987. Allelopathic compounds in soil from no tillage vs conventional tillage in wheat production. *Plant & Soil* 98:5–15.

Walters, D. T., and A. R. Gilmore. 1976. Allelopathic effects of fescue on the growth of sweetgum. *J. Chem. Ecol.* 2:469–79.

Walther, D., and D. Gindrat. 1987. Biological control of phoma and pythium damping-off of sugar beet with *Pythium oligandrum*. *J. Phytopathol.* 119:167–74.

Walther, D., and D. Gindrat. 1988. Biological control of damping-off of sugar-beet and cotton with *Chaetomium globosum* or a fluorescent *Pseudomonas* sp. *Can. J. Microbiol.* 34:631–37.

Wang, Y., M. Toyota, F. Krause, M. Hamburger, and K. Hostettmann. 1990. Polyacetylenes from *Artemisia borealis* and their biological activities. *Phytochemistry* 29:3101–5.

Ward, H. A., and L. H. McCormick. 1982. Eastern hemlock allelopathy. *For. Sci.* 28:681–86.

Watts, R., J. Dahiya, K. Chaudhary, and P. Tauro. 1988. Isolation and characterization of a new antifungal metabolite of *Trichoderma reesei*. *Plant & Soil* 107:81–84.

Weaks, T. E. 1988. Allelopathic interference as a factor influencing the periphyton community of a freshwater marsh. *Arch. Hydrobiol.* 111:369–82.

Webster, G. R., S. U. Khan, and A. W. Moore. 1967. Poor growth of alfalfa (*Medicago sativa* L.) on some Alberta soils. *Agron. J.* 59:37–41.

Weidenhamer, J. D., and J. T. Romeo. 1989. Allelopathic properties of *Polygonella myriophylla:* Field evidence and bioassays. *J. Chem. Ecol.* 15:1957–70.

Weidenhamer, J. D., D. C. Hartnett, and J. T. Romeo. 1989. Density-dependent phytotoxicity: Distinguishing resource competition and allelopathic interference in plants. *J. Appl. Eocl.* 26:613–24.

Weissen, F., and H. J. van Praag. 1991. Root growth inhibition effects of holorganic moder humus layer under spruce (*Picea abies* Karst.) and beech (*Fagus sylvatica* L.). *Plant & Soil* 135:167–74.

Welch, I. M., P. R. F. Barrett, M. T. Gibson, and I. Ridge. 1990. Barley straw as an inhibitor of algal growth I: Studies in the Chesterfield Canal. *J. Appl. Phycol.* 2:231–39.

Weller, D. M., and R. J. Cook. 1986. Increased growth of wheat by seed treatments with fluorescent pseudomonads, and implications of *Pythium* control. *Can. J. Plant Pathol.* 8:328–34.

Weller, D. M., W. J. Howie, and R. J. Cook. 1988. Relationship between *in vitro* inhibition of *Gaeumannomyces graminis* var. *tritici* and suppression of take-all of wheat by fluorescent pseudomonads. *Phytopathology* 78:1094–1100.

Weston, L. A., and A. R. Putnam. 1985. Inhibition of growth, nodulation, and nitrogen fixation of legumes by quackgrass (*Agropyron repens*). *Crop Sci.* 25:561–65.

Weston, L. A., and A. R. Putnam. 1986. Inhibition of legume seedling growth by residues and extracts of quackgrass (*Agropyron repens*). *Weed Sci.* 34:366–72.

Weston, L. A., R. Harmon, and S. Mueller. 1989. Allelopathic potential of sorghum-sudangrass hybrid (Sudex). *J. Chem. Ecol.* 15:1855–65.

Weststeijn, E. A. 1990. Fluorescent *Pseudomonas* isolate E11.3 as biocontrol agent for *Pythium* root rot in tulips. *Neth. J. Plant Pathol.* 96:261–72.

White, R. H., A. D. Worsham, and U. Blum. 1989. Allelopathic potential of legume debris and aqueous extracts. *Weed Sci.* 37:674–79.

Williamson, G. B., N. H. Fischer, D. R. Richardson, and A. de la Peña. 1989. Chemical inhibition of fire-prone grasses by the fire-sensitive shrub, *Conradina canescens. J. Chem. Ecol.* 15:1567–77.

Williamson, G. B., E. M. Obee, and J. D. Weidenhamer. 1992. Inhibition of *Schizachyrium scoparium* (Poaceae) by the allelochemical hydrocinnamic acid. *J. Chem. Ecol.* 18:2095–2105.

Wilson, C. L., and E. Chalutz. 1989. Postharvest biological control of *Penicillium* rots of citrus with antagonistic yeasts and bacteria. *Sci. Hort.* (Amsterdam) 40:105–12.

Wilson, C. L., J. D. Franklin, and P. L. Pusey. 1987. Biological control of rhizopus rot of peach with *Enterobacter cloacae. Phytopathology* 77:303–5.

Wilson, M., H. A. S. Epton, and D. C. Sigee. 1990. Biological control of fire blight of hawthorn (*Crataegus monogyna* ) with *Erwinia herbicola* under protected conditions. *Plant Pathol.* (London) 39:301–8.

Wilt, F. M., G. C. Miller, and R. L. Everett. 1988. Monoterpene concentrations in litter and soil of singleleaf pinyon woodlands of the western Great Basin. *Great Basin Nat.* 48:228–31.

Wilt, F. M., G. C. Miller, R. L. Everett, and M. Hackett. 1993. Monoterpene concentrations in fresh, senescent, and decaying foliage of single-leaf pinyon (*Pinus monophylla* Torr., and Frem.: Pinaceae) from the Western Great Basin. *J. Chem. Ecol.* 19:185–94.

Wójcik-Wojtkowiac, D., B. Politycka, M. Schneider, and J. Perkowski. 1990. Phenolic substances as allelopathic agents arising during the degradation of rye (*Secale cereale*) tissues. *Plant & Soil* 124:143–47.

Wokocha, R. C. 1988. Relationship between the population of viable sclerotia of *Sclerotium rolfsii* in soil to cropping sequence in the Nigerian savanna. *Plant & Soil* 106:146–48.

Wollenweber, E., M. Dörr, R. Stelzer, and F. J. Arriaga-Giner. 1992. Lipophilic phenolics from the leaves of *Empetrum nigrum*—chemical structures and exudate localization. *Bot. Acta* 105:300–5.

Worsham, A. D., and U. Blum. 1992. Allelopathic cover crops to reduce herbicide inputs in cropping systems. In *Proceedings of the First International Weed Control Congress.* pp. 577–579 (compiled by R. G. Richardson), Feb. 17–21, 1992, Monash Univ., Melbourne, Weed Science Soc. of Victoria, Melbourne.

van Wyk, P. S., D. J. Scholtz, and W. F. O. Marasas. 1988. Protection of maize seedlings by *Fusarium moniliforme* against infection by *Fusarium graminearum* in the soil. *Plant & Soil* 107:251–57.

Wyman-Simpson, C. L., G. R. Waller, M. Jurzsta, J. K. McPherson, and C. C. Young. 1991. Biological activity and chemical isolation of root saponins of six cultivars of alfalfa (*Medicago sativa* L.). *Plant & Soil* 135:83–94.

Yakle, G. A., and R. M. Cruse. 1983. Corn plant residue age and placement effects on early corn growth. *Can. J. Plant Sci.* 63:871–77.

Yakle, G. A., and R. M. Cruse. 1984. Effects of fresh and decomposing corn plant residue extracts on corn seedling development. *Soil Sci. Soc. Am. J.* 48:1143–46.

Yamamoto, Y., and K. Suzuki. 1990. Distribution and algal-lysing activity of fruiting myxobacteria in Lake Suwa. *J. Phycol.* 26:457–62.

Yamane, A., H. Nishimura, and J. Mizutani. 1992a. Allelopathy of yellow field cress (*Rorippa sylvestris*): Identification and characterization of phytotoxic constituents. *J. Chem. Ecol.* 18:683–91.

Yamane, A., J. Fujikura, H. Ogawa, and J. Mizutani. 1992b. Isothiocyanates as allelopathic compounds from *Rorippa indica* Hiern. (Cruciferae) roots. *J. Chem. Ecol.* 18:1941–54.

Yang, H. J. 1982. Autotoxicity of *Asparagus officinalis* L. *J. Am. Soc. Hort. Sci.* 107: 860–62.

Yang, S.-M., D. R. Johnson, and W. M. Dowler. 1990. Pathogenicity of *Alternaria angustiovoidea* on leafy spurge. *Plant Dis.* 74:601–4.

Yang, Y. S., and Y. Futsuhara. 1991. Inhibitory effects of volatile compounds released from rice callus on soybean callus growth: Allelopathic evidence observed *in vitro* cultures. *Plant Sci.* 77:103–10.

Young, C. C. 1984. Autointoxication in root exudates of *Asparagus officinalis* L. *Plant & Soil* 82:247–53.

Young, C. C., and D. P. Bartholomew. 1981. Allelopathy in a grass-legume association: I. Effects of *Hemarthria altissima* (Poir.) Stapf., and Hubb. root residues on the growth of *Desmodium intortum* (Mill.) Urb., and *Hemarthria altissima* in a tropical soil. *Crop Sci.* 21:770–74.

Young, C. C., and T. C. Chou. 1985. Autointoxication in residues of *Asparagus officinalis* L. *Plant & Soil* 85:385–93.

Young, C. C., and D. P. Bartholomew. 1987. Growth, nutrient uptake, and $N_2$ fixation of *Desmodium intortum* as influenced by root exudates of limpograsses (*Hemarthria altissima*). *Proc. Natl. Sci. Counc. B. ROC* 11:187–93.

Young, C. C., L. R. Zhu Thorne, and G. R. Waller. 1989. Phytotoxic potential of soils and wheat straw in rice rotation cropping systems of subtropical Taiwan. *Plant & Soil* 120:95–101.

Yun, K.-W., and B.-S. Kil. 1989. Phytotoxic effects on selected species by chemical substances of *Artemisia princeps* var. *orientalis. Korean J. Ecol.* 12:161–70.

Yun, K.-W., and B.-S. Kil. 1992. Assessment of allelopathic potential in *Artemisia princeps* var. *orientalis* residues. *J. Chem. Ecol.* 18:1933–40.

Zaspel, I. 1992. Influence of a treatment of seeds with bacterial antagonists against *Gaeumannomyces graminis* and the relation to yield and infestation by wheat. *Zentralbl. Mikrobiol.* 147:173–81. (German, English summary)

Zekhnov, A. M., Ya.O. Soom, and G. F. Nesterova. 1989. New test strains for detecting the antagonistic activity of yeast. *Mikrobiologiya* 58:807–11. (Russian)

Zhou, T., and R. D. Reeleder. 1990. Selection of strains of *Epicoccum purpurascens* for tolerance to fungicides and improved biocontrol of *Sclerotinia sclerotiorum. Can. J. Microbiol.* 36:754–59.

Zobel, A. M., and S. A. Brown. 1991. Psoralens in senescing leaves of *Ruta graveolens. J. Chem. Ecol.* 17:1801–10.

Zuberer, D. A., C. M. Kenerley, and M. J. Jeger. 1988. Populations of bacteria and actinomycetes associated with sclerotia of *Phymatotrichum omnivorum* buried in Houston black clay. *Plant & Soil* 112:69–76.

Zuberi, M. I., A. R. Biswas, G. B. Ghosh, P. C. Roy, and A. M. Choudhury. 1989. Allelopathic effects of diffusates and extracts of *Striga densiflora* on germination and early growth of rapeseed. *J. Indian Bot. Soc.* 68:317–19.

# INDEX

*Abies balsamea:* allelopathic to timothy, Kentucky bluegrass, couch grass, and fireweed, 317–18
*Abutilon theophrasti. See* Velvetleaf
*Acacia dealbata:* allelopathic effects on perennial ryegrass, orchard grass, white clover, and red clover, 318–19
*Acalypha indica:* inhibitory weed, 40
*Acer saccharinum* (silver maple): allelopathic to *Liriope muscari* and *Sasa pigmaea,* 320
*Acremonium breve* and *Pseudomonas* sp. in biocontrol of *Botrytis cinerea* and *Penicillium expansum* on apple fruits, 256–57
Actinomycetes: and bacteria from sunflower, antagonistic to *R. solani, Macrophomina phaseolina, Alternaria helianthi, Sclerotium rolfsii* pathogenic sunflower, 244; in clear-cut Douglas-fir area, allelopathic to lettuce, *Anastatica hierochuntica,* and ectomycorrhizal fungi *Laccaria laccata* and *Hebeloma crustuliniforme,* 376–77; collected from *Pinus resinosa,* allelopathy to three mycorrhizae, some inhibitory, some stimulatory, 377–78
Aerobic actinomycetes: culture filtrates bioactive to *Chlamydomonas* and some crop plants, 79–80
*Ageratum conyzoides:* inhibitory weed, 40
*Agrobacterium radiobacter:* biocontrol of take-all of wheat, 290
*Agrobacterium radiobacter* (K84): *A. tumefaciens* biotypes 1 and 2 usually sensitive to K84, 226; K84 not a good antagonist to *A. tumefaciens* biotype 3, 226–28
Agroecosystems in Tlaxcala, Mexico: 17
*Agropyron repens. See* Couch grass
*Agrostis stolonifera:* allelopathic potential, 21
*Ailanthus altissima:* allelopathy, 321–23
Ajoene (from garlic bulb): inhibited growth of many pathogenic fungi, 178–79
Alfalfa: allelochemicals, 8–10; allelopathic effects, 4–7, 131–32; and possible biocontrol of Cogon grass, 131–32
Allelopathic mulches: cover crops, 115; intercropping, 125; no-till farming, 115, 124; for weed control, 112–13, 117, 122, 124–25
Allelopathy, definition: 1
*Aloe:* allelopathy to *Bacillus* subtilis, 90; anthraquinone-type allelochemical, 90
*Alopecurus pratensis:* allelopathic potential, 21
Alpha-terthienyl: in Asteraceae, 155; as herbicide against several weeds, 155
*Alternaria alternata:* biocontrol fungus for weeds, 141
*Alternaria angustiovoidea:* in biocontrol of leafy spurge, 145–46
*Alternaria cassiae:* in biological control of sickle pod, 140–41, 154; four strong phytotoxins produced, 154
*Alternaria crassa:* in biocontrol of Jimsonweed, 143–44
*Alteromonas* (marine bacterium): some strains antibiotic producers, some not, 105
*Amaranthus palmeri:* allelopathy, 44–45; volatile inhibitors, 45
*Amaranthus retroflexus:* residue allelopathy, 39, 43–44
*Ambrosia artemisiifolia:* residue allelopathy, 39
*Anagallis arvensis, Chenopodium album, Colletotrichum falcatum, Cuscuta reflexa:* allelopathic to fungi, 90
*Anaphalis araneosa:* allelopathic weed, 59
*Anastatica hierochuntica:* allelopathic effects, 45
*Antennaria microphylla:* allelochemics produced, 46, 130; biocontrol of leafy spurge, 130
*Anthoxanthum odoratum:* allelopathic potential, 21
Anthraquinones: allelochemicals inhibitory to microorganisms, 91–92

*Aposphaeria amaranthi* (a fungal parasite of some *Amaranthus* spp.): in possible biocontrol of *Amaranthus* spp., 139
Aquatic weeds: biological control of, 126
*Argemone mexicana:* inhibitors produced by, 46–49
*Armillaria ostoyae:* antagonistic bacteria, 228–30
*Artemisia princeps. See* Wormwood
*Artemisia vulgaris:* allelopathy to *Lepidium virginicum* and perennial ryegrass, 326
Asafoetida (resin from *Ferula assa-foetida*): effects on two fungal pathogens, 182–83; inhibition of pathogen of chickpea blight, 183
Asparagus *(Asparagus officinalis):* allelopathic effects, 10–12; allelochemicals, 11, 12; stress of asparagus root leachate magnified by inoculation with *Fusarium oxysporum,* 172
*Aspergillus flavus:* 18 taxa allelopathic to, essential oils responsible, 101
*Aspergillus niger:* phytotoxic effects, 80; important antagonist against *Fusarium udum* (causes wilt of pigeon pea), 208. See also *Micromonospora globosa; Penicillium citrinum*
*Aspergillus versicolor:* good biocontrol agent of *Macrophomina phaseolina* on jute, 212–13
*Asphodelus fistulosus:* biological control with *Puccinia barfeyi,* 140
*Athelia bombacina:* strong biocontrol of apple scab caused by *Venturia inaequalis,* 259–60
*Atriplex polycarpa* (desert saltbush): allelopathy, 325
*Avena fatua. See* Wild oats
*Avena sativa. See* Oats
*Azotobacter chroococcum:* bioactivity against barley seeds, 80–81; with *Glomus fasciculatum,* enhanced root infection and stimulated root growth, 81

*Baccharis coridifolia* (Brazilian): allelopathic to American species of *Baccharis,* 48; trichothecene roridins and bacharinoids produced, 47
*Baccharis glutinosa:* produces no trichothecenes, 48
*Baccharis halimifolia:* produces no trichothecenes, 48
*Baccharis megapotamica* (Brazilian): allelopathic to American species of *Baccharis,* 326–27; trichothecene roridins and bacharinoids produced, 47
*Bacillus cereus* (UW85): in biocontrol of alfalfa damping-off caused by *Phytophthora megasperma* f. sp. *medicaginis,* 260
*Bacillus cereus* var. *mycoides* (strain 31, resistant to rifampicin and nalidixic acid): colonizes wheat root system, 291
*Bacillus polymyxa:* stimulated white clover, 82
*Bacillus pumilus* (87): colonized wheat root system, 291
*Bacillus pumilus* (B-PRCA-1): good in control of citrus fruit rot caused by *Penicillium digitatum,* 222, 224
*Bacillus* sp. *Pseudomonas fluorescens, P. putida:* as mobile organisms, gave biocontrol of tomato bacterial wilt caused by *Pseudomonas solanacearum,* but when nonmobile, did not, 254
*Bacillus* spp. (B33, B36): gave some biocontrol of banana, eggplant, and tomato wilt caused by *Pseudomonas solanacearum,* 262. See also *Pseudomonas fluorescens* (Pfcp)
*Bacillus subtilis:* bioactivity without disease control, 81; antagonist to *Verticillium dahliae* in *Acer,* 192. See also Bacillus subtilis *isolates; names of strains*
*Bacillus subtilis* isolates: isolates and *Enterobacter aerogenes* (B8) in biological control of *Phytophthora cactorum*–infected apple trees, 258–59; five isolates gave biocontrol of six isolates of *Agrobacterium tumefaciens* (biovar I), 284; antibiotics isolated in biocontrol of grapevine dieback caused by *Eutypa lata,* 310
*Bacillus subtilis* (B-3): good in biocontrol of apple rots caused by *Monilinia fructicola, Botrytis cinerea,* and *Glomerella cingulata* and for gray mold of grapes caused by *B. cinerea,* 257–58; produces antifungal peptides, 279
*Bacillus subtilis* (Bs-13, Bs-14): good in biocontrol of wilt of *Lagenaria siceraria* caused by *Fusarium oxysporum,* 249
*Bacillus subtilis* (M51): MZ51 strain protected carnation against wilt caused by *F. oxysporum* f. sp. *dianthi,* 218
*Bacillus subtilis* (13) and *B.* spp.: in biocontrol of garden-geranium rust caused by *Puccinia pelargonii-zonalis,* 269–70
Bacterial antagonists: of *Armillaria,* 228–30; of *Pythium ultimum,* 240–41; combinations vs. *Rhizoctonia solani,* 283–84
Bacteria with yellow pigment (rod shape): antagonistic to *Erwinia amylovora* on *Cotoneaster bullatus,* 213–14
*Bailonella toxisperma* (tree): allelochemicals identified, 327; allelopathic to cucumber, radish, and all weeds tested, 327
*Balansia cyperi:* inhibited all test fungal pathogens on solvent extract–infected *Cyperus rotundus* leaves, 225; mycelium extracts not inhibitory to test pathogens, 225
*Bambusa arundinacea:* allelopathic to peanut and several test weeds, 328
Barley: allelopathic effects, 12, 33–34
Barley straw, allelopathic effects: on algae, 98; on weeds in black spruce forest, 367–69
Bean root rot: due to *Fusarium solanii* f. sp. *phaseoli,* 171
Bermuda grass: inhibitory to crop growth, 54

Berseem clover: allelopathic effects, 14
*Betula verrucosa:* allelopathic to *Picea excelsa,* 328
*Bidens pilosa:* allelopathic effects, 48; roots produce phenylheptatriyne, 48
Binucleate *Rhizoctonia* (PCNB) and biocontrol of potato canker caused by *R. solani* (AG-3): 288
*Bipolaris halepense, Colletotrichum graminicola, Exserohilum turcicum, Gloeocercospora sorghi:* promising alone or in combination for biocontrol of Johnsongrass, 149–50
*Bipolaris setariae:* in biocontrol of grassy weeds, 145
Bitter sneezeweed: allelopathic weed, 61
*Boletinus cavipes:* an antifungal fungus, 106; five diterpenic esters identified, 106
*Borreria articularia:* allelopathic weed, 48
*Bothrichloa pertusa:* allelopathic effects, 20, 48
*Brassica:* allelopathic effects, 18, 19
*Brassica kaber:* seed allelopathy of, 38
*Brassica napus. See* Rape
Broad bean: allelopathy, 34
Buckwheat: allelopathic effects, 13

*Cajanus cajan. See* Pigeon pea
California shrubs allelopathic to *Bromus mollis* and *Hordeum vulgare:* 320–21
*Calluna vulgaris:* allelopathic to red clover, 330–31
*Cannabis sativa:* germination inhibitor, 49
*Carpinus betulus:* allelopathic to *Staphylococcus aureus,* 331–32
*Carthamus tinctorius:* 34
*Cassia auriculata:* allelopathic phenols and proanthocyanin produced, 49
*Cassia obtusifolia:* allelopathic weed, 50; variable effects on soybean cultivars, 50
*Cassia sophera* var. *purpurea:* allelopathic weed, 50; inhibitor(s) apparently phenolic, 50
*Cassia tora:* allelopathic weed, 49; phenols produced by, 49
*Cassia uniflora* (*C. sericea*): in biocontrol of some weeds, 133
*Ceanothus velutinus:* inhibited white fir roots, 333
*Celosia argentea:* allelopathic potential, 50
*Celtis occidentalis:* continuously releases allelochemicals in bottomland forest and low understory mass, 333–34
*Cenchrus ciliaris:* allelopathic effects, 19, 20, 50
*Cenchrus setigerus:* allelopathic effects, 50
*Centaurea maculosa:* allelopathic weed, 51; cnicin produced, 51
*Centaurea solstitialis:* allelopathic weed, 51; two chromenes produced, 51
*Centrosema pubescens:* in biocontrol of weeds, 130–31
*Ceratocysis ulmi:* antagonists of, 304–8
*Cercidiphyllum japonicum:* allelopathic to *Bacillus subtilis* and *Escherichia coli,* 92
*Chaetomium globosum* (Cg-1): antagonistic to damping-off of KWS Mono sugar beet and Acala cotton caused by *Pythium ultimum;* 199; antagonist to damping-off of sugar beets caused by *Phoma betae* and *Rhizoctonia solani,* 199–200; other strains effective antagonists also, 200–201
*Chaetomium globosum* (F6, NRRL 6296): good in biocontrol of apple scap caused by *Venturia inaequalis,* 259
*Chaetomium* spp.: several good in biocontrol of *Sclerotinia sclerotiorum,* 236
*Chenopodium album:* residue allelopathy, 39, 52
*Chenopodium ambrosiodes:* three active terpenes inhibitory to seed germination, 51–52
*Chenopodium murale:* inhibits mustard seed germination, 51–52; oxalic acid produced, 51
Chickpea, allelopathy, 34
*Chloris gayana:* allelopathic effects, 20
*Chondrostereum purpureum* (fungus): effective in biocontrol of black cherry, 148–49
*Chrysopogon aucheri:* allelopathic effects, 19, 20
*Cicer arietinum. See* Chickpea
*Cirsium arvense:* allelopathic phenolics produced, 52
*Cirsium palustre:* autoallelopathic, 52
Citral, citronellol, geraniol, as effective herbicidal terpenes: 153
*Cladosporium cladosporioides:* promising biocontrol agent of mulberry leaf-spot caused by *Cercospora moricola,* 261. See also *Curvularia lunata; Pseudomonas maltophila*
*Cladosporium cucumerinum:* fungal pathogen inhibited by several polyacetylenes from *Artemisia borealis,* 179
*Clerodendrum cucumerinum:* produces Mi-saponin A, 94; spore formation inhibited in *Cladosporium cucumerinum,* 94
*Clerodendrum viscosum:* allelopathic weed, 53; terpene produced in leaves is clerodin, 53
*Cochliobolus carbonum:* promising biocontrol fungus for *Euphorbia geniculata,* 146
*Cochliobolus spicifer* (D-5): produces spiciferin 1 and 2, plant growth stimulators or inhibitors, 84
Cogongrass: weed inhibitory to crop plants and fungi, 63
*Colletotrichum gloeosporioides* f. sp. *malvae:* good in biocontrol of *Malva pusilla,* 144; good in control of *Hakea sericea,* 146–47; not sensitive to synthetic herbicides, 144–45
*Colletotrichum graminocola:* for biocontrol of Johnsongrass, 149–50
*Colletotrichum orbiculare:* in biocontrol of *Xanthium spinosum,* 151
*Colletotrichum* spp.: cause disease of many plants, 138; *C. gloeosporioides* f. sp. *aeschynomene* sold as a mycoherbicide, 138

*Colocasia esculenta. See* Taro
Columbus grass: allelopathic potential of, 21
*Coniothyrium minitans:* as antagonist to *Sclerotinia sclerotiorum* on several crops, 194
*Coriolus versicolor:* good in control of the wood fungal pathogen *Armillaria luteobubalina,* 225–26
Corn: allelopathic effects, 14–17; monoculture vs. rotation of, 15, 16; pollen allelopathy, 17
*Cotinus coggigria* (smoke tree): allelopathy to understory, 334–35
Couch grass: allelochemicals, 40, 43; allelopathy, 40–43
Crabgrass: allelopathic effects, 46
Cress seeds: pelleting with antagonists successful, 204–5
Crimson clover: allelopathic effects, 13, 14
Crop rotation: and weed control, 113, 172–75
*Crotolaria pallida* var. *pallida:* allelopathic weed, 50; inhibitors apparently phenolic, 50
*Croton bonplandianum:* allelopathic, 53; phaseic and abscisic acids produced in leaf litter and soil, 54
*Croton sonderianus:* two acidic antimicrobial diterpenes produced, 94
Cruciferae root exudates: allelopathic to vesicular-arbuscular mycorrhizae, 91
*Cryphonectria parasitica:* hypovirulent strains of the parasite were antagonistic to the disease in chestnut inoculations, 209–10
*Cryptococcus albidus:* in biocontrol of *Botrytis cinerea* on rose, 284–85
Cucumber: allelochemicals and allelopathic effects, 19; breeding for allelopathic weed control, 111
*Cucumus sativus. See* Cucumber
*Cucurbitacins:* extracts of *Ecballium elaterium* prevented infection of cucumber fruits by *Botrytis cinerea,* 160
*Curvularia lunata:* in biocontrol of mulberry leaf-spot, 261
*Cuscuta reflexa* parasite: confirmed hosts inhibited growth of, 336–37
*Cyperus brevifolius:* allelopathic terpenes produced in roots and rhizomes, 55
*Cyperus kyllingia:* allelopathic terpenes produced, 55
*Cyperus rotundus. See* Purple nutsedge

*Daboecia cantabrica:* allelopathic to red clover, 330–31
*Datura innoxia:* allelopathic to seed germination and seedling growth, 55; inhibitory to most other species in a thicket in Pakistan, 335–36
*Datura stramonium. See* Jimsonweed
*Delonix regia:* allelopathic to most understory plants in Taiwan, 336
*Dennstaedtia punctiloba* (hayscented fern): possible allelopathic effect, 366; previously shown to be allelopathic to black cherry, 367
*Dicanthium annulatum:* allelopathic weed, 56
*Digera alternifolia:* allelopathic weed, 56–57
*Digera muricata:* allelopathic weed, 57
*Digitaria sanguinalis. See* Crabgrass
*Dolichos kilimandscharicus:* three saponins inhibited fungus spore formation, 95
Douglas fir: seed germination inhibited by actinomycetes, 80, 376; shrub species allelopathic to, 319; actinomycete population denser in clear-cut area, 80
*Drechslera avenae:* (-)-dihydropyrenophorin produced by, 152; as herbicide gives good control of wild oats and Johnsongrass, 151–52
*Drechslera indica:* curvulin and O-methylcurvilinic acid are herbicides produced, 153; effective pathogen of *Portulaca oleracea,* spiny amaranth, and rape, 153
*Drechslera oryzae* (leaf-spot of rice): essential oil of *Adenocalymma allicea* inhibited growth, 176

*Echinops echinatus:* allelopathic weed, 57; phenolics and terpenoids found in extracts, 57, 95; inhibits spore germination of a fungus, 95
*Eichhornia crassipes:* allelopathic to many algae, 96
*Eichhornia speciosa:* very stimulatory to jute, 57–58
*Empetrum hermaphroditum:* aqueous leaf leachates strongly inhibitory to Scots pine seedlings, 338–40
*Empetrum nigrum:* many phenolics in leaves may inhibit bacteria and pathogenic fungi and protect tree, 337–38
*Enterobacter cloacae:* (EcCt-501 and El): antagonistic to dollar-spot pathogen on *Agrostis-Poa* greens, 193
*Enterobacter cloaceae* (SC843): in biocontrol of *Rhizopus* rot of peach and other fruits, 280–81
*Epicoccum nigrum:* produced flavipin that prevented blossom and twig infection of peach, 279
*Epicoccum purpurascens* (M-20-A, fungicide-resistant strain): good in biocontrol of *Sclerotinia sclerotiorum* on bean, 237
*Epicoccum purpurascens* (I3): widespread red-pigmented fungus produces many antifungal compounds and is promising against *Phytophthora* and *Pythium,* 230
Epiphytic bacteria: 27 of 963 strains antagonistic to *Erwinia carotova* f. sp. *carotova,* 231; several promising in rot control, 231–32
*Eragrostis curvula* (Consol and 4660): in biological control of *Cenchrus longispinus* and *C. incertus,* 128
*Eragrostis poaeoides:* allelopathic potential, 58
*Erica australis:* allelopathic effects, 58; soil allelochemicals, 59

*Erica vagans:* allelopathic to red clover, 330–31
*Erwinia herbicola:* control of *Erwinia amylovora* on *Crataegus monogyna,* 224–225; control equivalent to several fungicides, 225
*Erwinia herbicola* (B247): in biocontrol of seedling blight and foot and root rot of wheat caused by *Fusarium culmorum,* 291; Tn5 antibiosis negative mutant gave less biocontrol, 292; in biocontrol of *Puccinia recondita* f. sp. *tritici* on wheat, 292
*Erwinia herbicola* (Eh252): in biocontrol of fire blight of Pomoideae caused by *E. amylovora,* 282–83
*Erwinia* sp.: in biocontrol of *Botrytris cinerea* on rose, 284–85
Essential oils, from several plants: effects on storage fungi, 183; inhibitory to *Pythium aphanidermatum,* 189–90
*Eucalyptus* bark compost and pine bark mix: suppressive to five *Phytophthora* spp. pathogens, 199
*Eucalyptus citridora:* the *cis*-form toxins produced inhibitory to lettuce seed germination, 340
*Eucalyptus globulus:* completely prevented seed germination of several crop plants and reduced the growth of others, 34
*Eucalyptus* sp.: in biocontrol of *Parthenium hysterophorus,* 134
*Eucalyptus* spp.: extracts or leachates of five species allelopathic to understory plants, 341–42; leachates of others prevented crop seed germination, 342–43
*Euodia lunu-ankenda:* allelopathic to *Cladosporium cladosporioides, 96*
*Eupatorium adenophorum:* allelopathic to plants and microorganisms, 59
*Eupatorium odoratum:* allelopathic weed, 59
*Eupatorium ripariam:* allelopathy, 59
*Euphorbia geniculata:* inhibitory weed, 40
*Euphorbia granulata:* allelopathic effects, 59
*Euphorbia prostrata:* allelopathic weed, 60
*Exophiala jeanselmei:* in biocontrol of *Botrytis cinerea* on roses, 284
*Exserohilum turcicum:* for biocontrol of Johnsongrass, 149–50

*Fagopyrum cymosum. See* Buckwheat
*Fagus sylvatica:* allelopathic to understory plants, 343; diseased leaves more allelopathic, 343; $Of_2$- and Oh-layers of moder humus very inhibitory to spruce seedlings, less so to beech, 344–45
Ferulic acid: produced in wheat residue, 119–20; biological control in wheat mulch, 120; decarboxylated to 4-hydroxy-3-methoxystyrene by a bacterium, better in weed control, 120
*Festuca arundinacea. See* Tall fescue
*Festuca arundinacea* (Kentucky 31). *See* Tall fescue
*Festuca rubra:* allelopathic potential of, 21
*Ficus septica:* allelopathic inhibitors to bacteria and fungi identified, 96–97
Flax: allelopathic effects, 19
Florida scrub community: allelochemicals, 328; allelopathic effects on sandhill plants, 329–30
Fluorescent bacteria: in biocontrol of sheath blight of rice caused by *Rhizoctonia* solani, 266–68
Fluorescent pseudomonad (hv37aR2): antibiotic oomycin A produced that controls *Pythium ultimum* on cotton, 241
Fluorescent *Pseudomonas:* antagonistic to *Gaeumannomyces graminis* var. *avenae,* 192; inhibitory to several other crop-plant pathogens, 208
Forest litter: phenolics produced, 332
*Fraxinus excelsior:* allelopathy to *Staphylococcus aureus,* 331–32
Freshwater bacterial species: stimulate plant growth, 89–90
Fungal pathogens: spores usually do not germinate at production site, 157; germination inhibitors produced, 157; many volatile compounds stimulate spore germination, 157–67
Fungal species antagonistic to *Puccinia arachidis:* 195–96
Fungi from rhizosphere of soybean: most antagonists to *Sclerotium rolfsii,* 239
*Fusarium heterosporum:* antagonist to dollarspot pathogen on *Agrostis* greens, 193
*Fusarium lateritium* conidia: good in weed control, 137
*Fusarium moniliforme* biocontrol: corn stem and ear rot caused by *Fusarium graminearum,* 311; Jimsonweed, 144
*Fusarium oxysporum:* nonpathogenic isolates and control of diseases caused by pathogenic *Fusarium* strains, 214; these antagonists plus *Pseudomonas putida* gave good control of cucumber wilt, 214–15
*Fusarium oxysporum* (Fo47) and *Pseudomonas* spp.: combined gave good biocontrol of tomato wilt caused by *Fusarium oxysporum* f. sp. *lycopersici* and tomato crown and root rot caused by *F. oxysporum* f. sp. *radicis lycopersici,* 249–51
*Fusarium solani:* gave some control of citrus root rot caused by *Phytophthora nicotianae* var. *parasitica* and *P. citrophthora,* 222
*Fusarium* spp.: culture filtrates toxic to pea, 78–79; important allelochemicals, 79, 84–85; isolates inhibited plant growth, 84

*Galinsoga ciliata:* as allelopathic weed, 59
*Galium aparine:* allelopathic weed, 60; asper-

uloside of a monoterpene lactone glucoside was the chief allelochemical, 60
*Geotrichium candidum:* sporostasis due to self-production of volatile signals, 170
*Gliocladium roseum:* reduced seed germination of barley, 80
*Gliocladium roseum* (632, 57d, W14): in biocontrol of potato early-dying disease caused by *Verticillium dahliae,* 289
*Gliocladium virens:* promising herbicidal activity correlated with viridiol production, 140; 20 strains produced different antibiotics inhibitory to *Pythium ultimum,* which causes damping-off of cotton, 239–40; 10 strains and *Trichoderma harzianum* (4 strains), in biocontrol of damping-off of snap beans caused by *Sclerotium rolfsii,* 271
*Gliocladium virens* (G20): produced gliotoxin and good biocontrol of plant diseases caused by *Pythium ultimum,* 237, 246; tested for biocontrol of *Rhizoctonia solani,* anastomosis group 2, type 1, 247; in biocontrol of zinnia damping-off caused by *Rhizoctonia solani* and *Pythium ultimum,* 313; gliotoxin produced, 313, 315
*Gliocladium virens* (Gl-3, Gl-17, Gl-21): in biocontrol of *Rhizoctonia solani* in cotton, tomatoes, snap beans, potatoes, beets, 242–44
*Gliocladium virens* (Gl-21): in biocontrol of *Sclerotium rolfsii* on tomato, 255–56
*Gliricidia sepium:* allelopathic weed, 61
*Gloeocercospora sorghi:* for biocontrol of Johnsongrass, 149–50
*Glycine max. See* Soybean
*Gomphrena decumbens:* as inhibitory weed, 60
Grape marc, composted: and germination of *Sclerotium rolfsii* sclerotia, 190–91
Griseofulvin: translocated through roots to leaves, 78

Hairy vetch: allelopathic effects, 13
*Hedera helix* (English ivy): allelopathic to *Populus deltoides* and *Acer saccharum,* 320
*Helenium amarum. See* Bitter sneezeweed
*Helianthus annuus. See* Sunflower
*Helianthus rigidus:* allelopathic weed, 62
*Helianthus tuberosus. See* Jerusalem artichoke
*Helichrysum decumbens* phloroglucinols: and growth of *Cladosporium herbarium,* 184
*Helichrysum* spp.: inhibitors found allelopathic to several fungi, 97–98
*Heliotropium eichwaldi:* allelopathic weed, 62
*Helminthosporium carbonum* (race 2): antagonism to *H. carbonum* (race 1), *Corynebacterium michiganense, Erwinia amylovora,* 312
*Hemarthria altissima. See* Limpograss
Herbicides from plants: for weed control, 114; rhizobitoxine from *Bradyrhizobium japonicum,* 114; caffeine from coffee, 114
*Holcus lanatus:* allelopathic potential, 21
*Hordeum vulgare. See* Barley
Humic substances from coastal lagoon: allelopathic to heterotrophic bacteria against breakdown of white mangrove leaves, 347–48
*Hydrocotyl sibthorpioides:* allelopathic weed, 62–63
*Hymenaea coubaril:* allelopathic to fungi, 98–99

*Imperata cylindrica. See* Cogongrass
Indian trees: leaf leachates inhibitory to herbs, 323–25; forest understory, allelopathy chiefly responsible for, 332–33
Interactions of antagonists: *Pseudomonas fluorescens* and *Trichoderma harzianum,* 238
*Ipomea batatas. See* Sweet potato
*Ipomoea lacunosa:* allelopathic weed, 63
*Ipomoea purpurea:* seed allelopathy, 38
*Ipomoea tricolor:* mixture of glycosides with jalapinolic acid as the aglycone in biological weed control, 126

Jerusalem artichoke: allelopathic effects, 19
Jimsonweed: allelopathic weed, 55–56; scopolamine and hyoscyamine allelochemicals in, 55–56
*Juglans nigra:* effects on $N_2$-fixing trees interplanted with walnut, 345–46

*Kalmia angustifolia* (ericaceous shrub): effects of leachates from open-heath sites on root and shoot growth of *Picea mariana,* 346
Kentucky bluegrass: allelopathic potential, 33
Kentucky 31 tall fescue: allelopathic potential, 22; abscisic acid production, 22
*Kochia scoparia:* allelopathic annual weed, 64

*Lantana camara:* allelopathic weed, 59, 64, 127; volatile allelochemicals in leaves, 64–65; allelochemicals identified, 66
*Larix decidua:* litter leachate allelopathic to lettuce, 348
*Larix leptolepis:* leachates allelopathic to several crops and weeds, 348; syringic and *p*-hydroxybenzoic acids identified, 348
*Larix sibirica:* allelopathic to *Picea excelsa,* 328
*Lasiurus sindicus:* allelopathic effects, 50
Leaf extracts of four plant species: inhibit three fungal pathogens of rice, 186
*Leersia hexandra:* allelopathic weed, 66
*Lens culinaris* (lentil): allelopathic effects, 33
Lettuce and spinach: allelopathy promotes invasion by low-grade pathogens, 171
*Leucaena leucocephala* leaf, litter, and soil leachates allelopathic to understory plants, 349
*Leucas cephalotes:* weed allelopathy, 40
*Limonomyces roseipellis:* occasionally isolated from ascocarps of *Pyrenophora tritici-repentis*

in wheat residue, 168–69; lowered spore production on wheat infected with *P. tritici-repentis,* 169
Limpograss: allelopathic effects, 23
*Linum usitatissimum. See* Flax
*Liriope muscari* (liriope): allelopathic to cottonwood and silver maple, 320
*Lolium multiflorum:* allelopathic effects, 46
*Lolium perenne:* allelopathic potential, 21
*Lonchocarpus castilloi* (tree): flavonoids inhibitory to fungal pathogen, 184
*Lupinus angustifolius* (lupin): allellopathy, 34, 99–100
*Lycopersicon esculentum. See* Tomato
*Lycoris radiata:* allelopathic weed, 66–67

*Macrophomina phaseolina:* in biocontrol of *Hydrilla verticillata,* 147–48
*Macrozamia riedlei:* azoxyglycosides inhibit nitrogenase in *Nostoc,* 100
Madero negro. See *Gliricidia sepium*
*Maytenus* and *Schaefferia* (Celastraceae): allelopathic to gram-positive bacteria, 100
*Medicago sativa. See* Alfalfa
*Melicytus ramiflorus:* leachate from seeds allelopathic to their germination, 349–50
*Metschnikowia reukaufii* (yeast): inhibits pollen germination of *Asclepias syriaca,* 85
*Microdochium bolleyi:* in biocontrol of take-all of wheat, 293
*Micromonospora globosa:* antagonist to *Fusarium udum,* 208
Milled pine-bark compost: inhibits several damping-off fungi in forest nurseries, 186–89
*Minimedusa polyspora* (128L): in biocontrol of basal rot of *Narcissus* caused by *Fusarium oxysporum* f. sp. *narcissi,* 263–64
*Miscanthus sinensis:* fatty acids' effects on rice blast fungus, 184–85
*Miscanthus transmorrisonensis:* allelopathic effects, 24; phenolic allelochemicals, 24
Moist-tropical-tree species: most slightly or strongly allelopathic to understory plants, 337
Mountain Pride tomato leaves: volatiles inhibitory to two fungal pathogens, 184
Mycorrhizae fungi: antagonistic to certain plant pathogenic fungi, 238–39
*Myrica cerifera* (wax myrtle): possible allelopathy to *Schinus terebinthifolius,* 350–51
*Myristica fragrans:* antimicrobial, 100
*Myrothecium roridum:* produced toxins potent against three *Phytophthora* fungus species, including chief cause of foot rot in Spain, 220
*Myrothecium roridum* (TW): antagonist of avocado root rot caused by *Phytophthora cinnamomi,* 270; disease reduction same as with the fungicide potassium phosphonate, 270–71
*Myxococcus* spp. (Lake Suiva, Japan): allelopathic effects, 106

Neem tree: allelochemicals and their uses, 326
*Nicotiana rustica* or *N. tabacum. See* Tobacco
*Nicotiana* spp.: allelochemicals from leaf surfaces effective against tobacco blue-mold pathogen, 185
*Nigrospora oryzae:* antibiotic producer, 234; controlled spore germination and mycelial growth in six species of *Fusarium* and *Botrytis cinerea,* 234
Nonfluorescent bacteria: biocontrol of sheath blight of rice caused by *R. solani* 268
Nonpathogenic *Rhizoctonia solani* as plant stimulators, 89

Oats: allelopathic effects, 15, 33–34; anaerobic vs. aerobic decomposition, 33; breeding for scopoletin production, 111; inhibitory effect in weed control, 367–69
*Ocimum sanctum:* volatile allelochemicals in leaves, 64–65
*Onoclea sensibilis:* leaf leachate allelopathic to *Quercus rubra,* 371–72
Organic acids (low molecular weight): in many forest soils in southeastern United States, 354
*Oryza perennis:* allelopathic native rice, 66–67; pronounced differences in biological activities of strains, 67
*Oryza sativa. See* Rice

*Panicum antidotale:* allelopathic effects, 20
*Parthenium hysterophorus:* allelopathic effects, 67–69, 100; important sesquiterpene lactones and phenolics, 68
*Paspalam notatum:* aqueous leachate allelopathic to *Pinus elliottii,* 372–74
Pathogenic microorganisms: for biological weed control, 113–14
*Paxillus involutus:* biocontrol of root pathogen *Fusarium oxysporum* f. sp. *pini* on pine, 272–74; stimulated synthesis of antifungal compounds by *Pinus resinosa,* 356
Peas: allelopathic effects, 24, 33, 34; chief allelochemical, ß-(3-isoxazolin-5-on-2 yl) alanine, 24–25
*Peganum harmala:* inhibit seed germination and seedling growth, 69
*Penicillium citrinum:* antagonism to *Fusarium udrum,* 208; important in biocontrol of *Sclerotinia minor, S. major, Sclerotium rolfsii, Rhizoctonia solani,* 235–36
*Penicillium claviforme* (IMI44744): in biocontrol of *Pythium ultimum,* 235
*Penicillium digitatum* and *P. italicum:* and citrus decay losses, 211; two bacterial species and two yeasts are promising control agents, 212
*Penicillium frequentans* (909): in biocontrol of *Monilinia laxa* on apricot, 279; toxins, 279–

80; in integrated control with fungicides, 279–80
*Penicillium griseofulvum:* griseofulvin produced in soil, 78
*Penicillium* sp.: antagonism to *Colletotrichum truncatum,* a cowpea pathogen, 308
*Pestalotiopsis* sp.: antagonism to *Colletotrichum truncatum,* a cowpea pathogen, 308
*Phaeoisariopsis personata* and *Puccinia arachidis* neem: and *Lawsonia inermis* leaf extracts gave fair control of these peanut pathogens, 179
*Phaeoramularia* sp. (leaf-spot fungus): gave partial biocontrol of crofton weed in Australia, 139
Phenazine carboxiamide, pyocyanin, pyrrolnitrin: antibiotics from *Pseudomonas fluorescens,* 207; effective against *Rhizoctonia solani* on rape, 207; pyocyanin inhibited other fungi associated with rape seedling disease complex, 207–8
Phenolic compounds: plant-produced, inhibitory to microorganisms, 94; stimulatory or inhibitory to four woodland species, 351–52
Phenylheptatriyne: allelopathic to *Fusarium culmorum,* 90–91
*Phleum pratense. See* Timothy
*Phoma* sp.: antagonistic to *Colletotrichum truncatum,* 308
*Phymatotrichum omnivorum:* causal pathogen of phymatotrichum root rot, 169; bacteria and fungi isolated from buried sclerotia of *P. omnivorum* inhibited growth of that pathogen, 169
*Phytolacca americana:* autoallelopathic to seed germination, 69
*Picea abies:* $Of_2$- and Oh-layers of moder humus allelopathic to spruce seedlings, 344; litter leachate allelopathic to lettuce, 348
*Picea mariana:* effects of leachates of *Kalmia angustifolia,* 346–47; allelopathic toxins transferred from litter to substrate, 352–53
Pigeon pea: allelopathic effects, 25; leaf terpenoids, 25; seed tannins, 29
Pinaceae, five species: 13 allelochemicals identified, 354; allelopathic to understory plants in Korean forests, 354
*Pumpinella* species: eight phenylpropenyl esters in seed extracts of six species, 69; epoxypropenyl groups caused considerable activity, while compounds with olefinic groups gave little activity, 69
*Pinus densiflora:* leachate and extracts allelopathic to understory plants, 353–54
*Pinus koraiensis:* litter layer allelopathic to understory plants in some Korean forests, 354–55
*Pinus monophylla:* allelochemicals in litter, 355; allelopathic effects on understory, 355–56
*Pinus resinosa:* allelopathic to timothy, Kentucky bluegrass, couch grass, and fireweed, 317–18
*Pinus rigida:* root extracts and soil allelopathic to understory plants, 357
*Pinus roxburghii:* leaves have 100% toxicity to *Aspergillus flavus* and *A. parasiticus,* 102; needles allelopathic to peppergrass and perennial ryegrass, 325–26; needles allelopathic to *Achyranthes aspera* and rye, 357–58
*Pinus sylvestris* (Scots pine): allelopathic to *Picea excelsa,* 328; allelopathic to understory in Russian forest, 334–35; litter leachate allelopathic to lettuce, 348
*Pinus taeda* and *P. elliottii:* aqueous leachates of grasses, rush, and shrubs allelopathic to, 369–70
*Pinus thunbergii:* aqueous leachates allelopathic to understory species, 358
*Piper sarmentosum:* allelopathic to *B. subtilis* and *E. coli,* 102; four phenylpropanoids identified in leaves, 102
*Piricularia grisea:* in biocontrol of several grassy weeds, 145
*Pisum sativum. See* Peas
*Plantago lanceolatus:* inhibitory weed, 40
Plant survey (up to 115 species): many allelopathic to *E. Scherichia coli* and *Saccharomyces cerevisae,* 99; some extracts inhibited all or part of four fungal pathogens, 182
*Platanus occidentalis:* continuous release of allelochemicals in bottomland forest and low understory mass, 333–34
*Pluchea lanceolata:* allelopathic action, 69–70; inhibitors identified, 70
*Poa pratensis. See* Kentucky bluegrass
*Poa trivialis:* allelopathic potential, 21
*Polygonella myriophylla:* allelopathic to *Paspalum notatum* and little bluestem, 359
*Polygonum longisetum:* allelopathic effects, 46
*Polygonum sachalinense:* inhibitory activity, 71; physcion and emodin most active, 71
*Populus deltoides* (cottonwood): allelopathic to *Hedera helix* and *Liriope muscari,* 320
Potential antagonists of *Rhizoctonia solani:* tested against basal stem rot in *Chrysanthemum* sp., 210; four plant species (six strains) with good control, 210
*Posidonia oceanica:* allelopathic to many microorganisms, 102
*Proboscidea louisianica:* allelopathic effects, 71; allelopathic compounds from essential oils, 71–72
*Prosopis cineraria:* allelopathic to about same degree as *P. juliflora,* 361–62
*Prosopis glandulosa* (mesquite): allelopathic to wheat and other plants, 360
Pseudomonad (E11.3): good biocontrol of tulip root rot caused by *Pythium* spp., 304

*Pseudomonas* (F113): antibiotic produced, 233; inhibits some plant fungal pathogens and *E. coli,* 233
*Pseudomonas* (RD:1): in biocontrol of black root rot of tobacco caused by *Thielaviopsis basicola,* 265
*Pseudomonas antimicrobica:* inhibitory to most bacteria and fungi against which it was tested, 230–31; good potential antagonist against pathogenic fungi and some bacteria, 231
*Pseudomonas aureofaciens* (IS-1, IS-2): in biocontrol of bacterial ring-rot disease of potato caused by *Clavibacter michiganensis* subsp. *sepedonicus,* 285–86
*Pseudomonas cepacia:* as pyrrolnitrin producer, good in biocontrol of gray and blue mold on apples and pears, 257
*Pseudomonas cepacia* (AMMD, AMMDRI): in biocontrol pythium dumping-off on peas, 274–75; *Aphanomyces* root rot caused by *A. euteiches* f. sp. *pisi,* 275
*Pseudomonas cepacia* (B5, B5-R): antibiotic 2-keto-D-gluconic acid, 264–65; biocontrol of wilt of tobacco caused by *Pseudomonas solanacearum,* 264–65
*Pseudomonas cepacia* (Pc742): antagonist to *Cercospora* leaf spot disease of peanut, 195
*Pseudomonas cepacia* (R55, R85): antagonists of *Rhizoctonia solani* and *Fusarium solani* in winter wheat, 302
*Pseudomonas fluorescens* biovar III (IS-3): in biocontrol of bacterial ring-rot of potato caused by *Clavibacter michiganensis* subsp. *sepedonicus,* 285–86
*Pseudomonas fluorescens* (CHAO): antagonist to several diseases and produces several antibiotics, 217–65; two strains produced more pyoluteorin and 2,4-diacetylphloroglucinol and better control of damping-off of cucumber, 217–18; in biocontrol of black root rot of tobacco caused by *T. basicola,* 265
*Pseudomonas fluorescens* (CHAO mutant engineered by Tn5 insertion): 2,4-diacetylphloroglucinol not produced, no biocontrol of black root rot of tobacco caused by *T. basicola,* 266
*Pseudomonas fluorescens* (Pfb, Pfbg, Pfco, Pfcp, Pfgn): Pfcp good in biocontrol of Panama wilt of banana caused by *Fusarium oxysporum* f. sp. *cubense,* 261–62; Pfcp good in biocontrol of wilt of banana, eggplant, and tomato caused by *Pseudomonas solanacearum,* 262
*Pseudomonas fluorescens* (pfc, pfp): antagonists to *Rhizoctonia solani* on peanuts, 196–97; biotype II strains effective against *Sclerotium rolfsii* on peanut, 197–98
*Pseudomonas fluorescens* (PFM2): 2,4-diacetylphloroglucinol produced, 303; antagonistic to *Septoria tritici* on leaves of wheat and to several other pathogenic fungi, 303
*Pseudomonas fluorescens* (PRA25): biocontrol and combined with fungicide against *Pythium* and *Aphanomyces* root rot of peas, 275–76
*Pseudomonas fluorescens* (Q29z-80, M8z-80): antagonistic to *Pythium ultimum,* with control equivalent to captan, metalaxyl, or *Penicillium oxalicum* treatment, 210–11
*Pseudomonas fluorescens* (Q72a-80): population on roots infected with pythium root rot of wheat, 300
*Pseudomonas fluorescens* rifampin resistant (R1a-80R, R7z-80R): antagonism low to take-all of wheat, 296–97
*Pseudomonas fluorescens* (2-79): phenazone antibiotic in wheat roots populated with this strain, 297; population in wheat roots infected with take-all, 300
*Pseudomonas fluorescens* (2-79, $K_2$-79, A-37): antagonism to fungal pathogens on wheat roots, compared with antibiotics, 294–95
*Pseudomonas fluorescens* (2-79-$RN_{10}$, R1a-80R, R7z-80R): from suppressive soils, 296; good biocontrol of take-all fungus of wheat, 296; low biocontrol with tn5 insertion to develop low antibiotic production from strain 2-79 $RN_{10}$, 297
*Pseudomonas* isolates: inhibited crop plants and microorganisms, 85–88
*Pseudomonas maltophila:* in biocontrol of leafspot of mulberry, 261
*Pseudomonas putida* (P27): in biocontrol of take-all of wheat, 290–91
*Pseudomonas putida* (R104): antagonist of *Fusarium solani* and *Rhizoctonia solani* in winter wheat, 302
*Pseudomonas putida* (WCS358): antibiotic pseudobactin produced increases control of fusarium wilts when combined with nonpathogenic *F. oxysporum* (Fo47), 220
*Pseudomonas solanacearum* (B82): avirulent, 286; good in biocontrol of potato wilt caused by *P. solanacearum,* 286–87
*Pseudomonas solanacearum* (BC8): avirulent, 287; good biocontrol of potato wilt and tuber infection caused by *P. solanacearum,* 287
*Pseudomonas* sp. (F113G22): engineered from strain F113 with Tn5-lac, 233; unable to produce 2-4-diacetylphloroglucinol or to control plant fungal diseases, 233
*Pseudomonas* sp. (LEC-1): in biocontrol of *Septoria tritici* or *Puccinia recondita* infection of wheat leaves, 302; two antifungal compounds, 303
*Pseudomonas* sp. (Ps-4): antagonist to *Pythium ultimum* on KWS Mono sugar beet and Acala cotton, 199; antagonist to *Phoma betae,* 200
*Pseudomonas* sp. (WCS417r): part of control of

fusarium wilt of carnation due to competition for iron, 218
*Pseudomonas* spp.: in biocontrol of three fungal pathogens of corn, 312
*Pseudomonas syringae* 174 (MSU16H): a transposon-generated mutant that produces a family of peptides (pseudomycins) and may have promise as antagonist, 231
*Pseudomonas syringae* pv. *lachrymans* (L-59-66): in biocontrol of blue and gray mold of pear, 281–82
*Pseudotsuga menziesii* (Douglas fir): litter leachate allelopathic to lettuce, 348
Psoralens: in fruits of Rutaceae, Umbelliferae, Leguminosae, 46
*Puccinia jaceae* (rust): potential for biocontrol of some weeds, 142–43
*Pulicaria wightiana:* weed allelopathy, 40
Purple nutsedge: allelopathic effects, 46, 54, 95; terpenes produced, 54; volatile allelochemicals from leaves, 64–65; essential oils produced, 95
*Pyronema domesticum:* apothecial formation prevented by own volatiles, 170; several bacteria prevent volatile production by *P. domesticum,* 170
*Pythium marinum:* common on red algae (particularly *Porphyra*), 167; readily enter sexual or asexual stages, 167; chemical signals required for zoospore encystment on red algae, 167–68
*Pythium nunn* (N3) and *P. ultimum* (N1): good in biocontrol of cucumber damping-off due to *P. ultimum,* 215–16; better control with *Trichoderma harzianum* (T-95), 216–17
*Pythium oligandrum:* antagonist against *P. ultimum* on sugar beets, 201, 203–4; antagonist to *Phoma betae,* 201–3; seed coating mechanically a success, 203–4
*Pythium oligandrum* (IMI133857): biocontrol of foot rot of peas caused by one or more fungi, 274

Quackgrass. *See* Couch grass
*Quercus robur:* allelopathy to *Staphylococcus aureus,* 331–32; allelopathy to understory in Russian forest; 335–36

Rape: allelopathy, 33, 34, 49; anaerobic vs. aerobic decay, 33
Rebel tall fescue: allelopathic potential of, 22
Red clover: allelopathic effects of, 7
Rhizobacteria: deleterious genera identified, 79; from cotton, biocontrol of *Pythium ultimum* on cotton seedlings, and fair control of *Rhizoctonia solani,* 241–42
Rhizobitoxine: effective herbicide, 153
*Rhizobium* sp. (from peanut plants): allelopathic to one or more of 19 rhizosphere fungi, 107–8; six fungi inhibited growth of one or more rhizobial strains, 108
Rice: allelopathic effects, root-exudate allelochemicals, 25–26
*Robinia pseudoacacia* (black locust): allelopathic to understory in Russian forest, 334–35
*Rorippa sylvestris:* allelopathic weed with several inhibitory compounds identified, 72
*Rubus idaeus:* aqueous leachate allelopathic to *Picea mariana,* 362; leaf leachate affected mycorrhizal fungus species on *P. mariana,* 362
*Rudbeckia occidentalis:* aqueous leachates allelopathic to lettuce, *Pinus contorta, Picea engelmannii,* 374
*Rumex japonicus:* allelopathic effects, 46
Rye: allelopathic effects, 26–29, 119; inhibitory chemicals, 27–29, 119; used in biological weed control, 119

*Saccharomyces cerevisiae:* volatile organic signals generally inhibited spore germination, 168; in biocontrol of anthracnose of corn caused by *Collectotrichum graminicola,* 310
Saccharomycetes: killer toxins, produced by numerous yeasts, apparently protein or glycoprotein in structure, 108–9
*Saccharum spontaneum:* allelopathic weed, 72–73
Safflower: allelopathy of, 34
*Salvia splendens:* effects of extracts on bacteria, 189
*Salvia* spp. (3): 20 diterpenes, all allelopathic to some of four bacteria, 103
*Salvia syriaca:* allelopathic weed, 73
*Sasa cernua:* allelopathic effects, 73; phenolics in soil, 73–74
*Sasa pigmaea* (dwarf bamboo): allelopathic to cottonwood and silver maple, 320
*Sclerotinia sclerotiorum:* in biocontrol of *Centaurea maculosa, Cirsium arvense,* and dandelion, 141, 150; mutants for biocontrol of weeds, 142; fungal growth inhibited by *Aegle marmelos,* 178
*Sclerotium cepivorum: Eucalyptus rostrata* leaf extract slowed growth of pathogen, 180
*Sclerotium rolfsii:* 7 plant species inhibited its growth, 178; *Eucalyptus* sp. distillate and neem oil gave good control of the pathogen, 179–80
*Scopulariopsis brumptii:* good biocontrol of 7 species, three herbicides produced, 154
*Scutellaria violaceae, S. woronowii:* neo-clerodone diterpenoids vs. fungal pathogens, 189
*Secale cereale. See* Rye
*Serratia marcescens:* good in biocontrol of damping-off of bean caused by *Sclerotium rolfsii,* 236

*Sesbania* leaves: increased acetic butyric and propionic acids in soils, 29
Sesquiterpene lactones (with lactone rings similar to strigol): stimulate witchweed seed germination, 136–37; several synthetic terpenoids stimulate *Striga* and *Orobanche* seeds, 137
*Setaria faberii* (giant foxtail): eight phenolics identified, 374–75; water extracts inhibited *Pinus taeda* growth, 374–75
*Setaria glauca. See* Yellow foxtail
Sicklepod. See *Cassia obtusifolia*
*Sisymbrium irio:* allelopathy, 44
*Solidago* sp.: allelopathic to algae, 103; *cis*-dehydromatricaria ester, menthol, camphene, chief allelochemicals, 103
*Sorbus aucuparia* (mountain ash): abscisic and parasorbic acids allelopathic to *Lepidium sativum* and *Amaranthus caudatas,* 363–64
Sorghum: allelopathic effects, 30, 34, 122; sorgoleone in root exudate, 31; weed control, 122, 124; four p-benzoquinones in root exudate, 135–36; dihydroquinone forms stimulate germination of witchweed seeds, 135–36
*Sorghum almum. See* Columbus grass
*Sorghum bicolor. See* Sorghum
*Sorghum bicolor* X *S. sudanense. See* Sudex
Sorghum Bird-a-Boo: for weed control, 117–19
Soybean: allelopathic effects, 15, 29
Spore germination, of ectomycorrhizal hymenomycetes: stimulated by compounds produced by many bacteria and fungi, 106–7; stimulators not yet identified, 107
*Sporisdesmium sclerotivorum:* model antagonist in field system to control *Sclerotinia minor,* cause of leaf drop of lettuce, 248–49
*Sporothrix flocculosa:* biocontrol of rose powdery mildew by antibiosis, 285
Spring triticale: autoinhibition of seed germination by, 32
*Stevia eupatoria:* allelopathic weed with volatile allelochemicals, 74
*Streptomyces griseoviridis:* promising biocontrol agent for *Botrytis cinerea* on lettuce, 247
*Streptomyces griseoviridis* (Mycostop): antagonistic to *Fusarium culmorum* and *Bipolaris sorokiniana* pathogenic on barley and wheat, 244–45
*Streptomyces griseus:* bioactive without disease control, 81
*Streptomyces griseus* var. *autotrophicus* (ATCC 53668): antifungal antibiotic produced, faeriefungin, good antagonist to *Fusarium oxysporium* f. sp. *asparagi,* 198
*Streptomyces spectabilis* (strain BT 352): allelopathic to several fungi, 108; desertomycin identified, 108
*Streptomyces* sp. (131): in biocontrol of basal rot of *Narcissus* caused by *Fusarium oxysporum* f. sp. *narcissi,* 263–64
*Streptomyces* sp. (SD-702): blasticidins and hydroxymethyl-blasticidins produced, 156; effective herbicides to certain dicots, 156
*Streptomyces* spp.: produce many important phytotoxins, 152; geldanamycin and nigericin very potent herbicides, 152
*Streptomyces viridochromogenes:* bialaphos produced, is an effective herbicide against some weeds, 155
*Striga densiflora:* allelopathic weed, 74
Strigol: in biocontrol of *Striga asiatica,* 135
*Stylosanthes guyanensis:* and biocontrol of weeds, 130–31
Subterranean clover: and biological weed control, 132–33
Sudex: allelopathic effects, 31, 120; phenolics produced, 31; weed control, 120
Sunflower: allelopathy, 34, 44, 97; sesquiterpene lactones in leaves and stems, 62, 97; rotation of oats and sunflower in weed control, 117; sunflower plus herbicide in integrated weed control, 117
Sweet potato: periderm allelopathic to purple nutsedge, 128–29; thirteen allelochemicals in periderm, 129
Synthetics plus biocontrol, often more effective than either alone: 276–78

*Tagetes minuta:* allelopathic, 48; roots produce alphaterthienyl, 48
*Tagetes patula:* allelopathic weed, 74: produces thiophenes and benzofurans, 74
*Talaromyces flavus:* allelochemical kills microsclerotia of *Verticillum dahliae,* 170–71; biocontrol of early-dying disease of potato caused by *V. dahliae,* 289–90
Tall fescue: allelopathic to ash, 371; allelopathic potential, 22, 129–30; interference against black walnut, 370; in weed control, 129–30
*Tanacetum argyrophyllum* var. *argyrophyllum:* allelopathic to numerous microorganisms, 103; important allelochemicals, 103
Taro: allelopathic effects, 31
*Tephrosia purpurea:* allelopathic weed, 74–75; volatiles and water leachates very inhibitory, 75
*Thielaviopsis basicola:* black root rot of tobacco promoted by its allelochemicals, 171
Thompson Seedless grape: berry exudates vs. *Botrytis cinerea,* 190; effective compounds, 190
*Tilletiopsis pallescens:* effective biocontrol agent for *Erysiphe graminis* f. sp. *hordei,* causative agent of powdery mildew on barley, 245
Timothy: pollen allelopathy, 23–24
Tobacco: allelopathic effects, 32, 185; inhibition of tobacco blue mold and brown spot pathogens, 185
Tomato: allelopathic effects, 32
*Trianthema portulacastrum:* apparently allelopathic, 75

*Tribulus terrestris:* inhibitory weed, 75
*Trichoderma hamatum:* good in biocontrol of take-all fungus (of wheat) on both wheat and ryegrass, 251–52
*Trichoderma harzianum* (ATCC20476, ATCC24274, PREM47942, and *T. polysporum* ATCC20475): all strains equally antagonistic to take-all of wheat in vitro, 300
*Trichoderma harzianum* (IMI-238493): effective antagonist to *Sclerotium rolfsii* on sugar beets, 206–7
*Trichoderma harzianum* (IMI275950): biocontrol of damping-off of lettuce caused by *Pythium ultimum,* 245
*Trichoderma harzianum* (IMI275950, IMI284726): two antibiotics produced were promising biocontrol agents for *R. solani,* causative agent of damping-off of lettuce, 247
*Trichoderma harzianum* (new strain): antibiotic 3-(2-hydroxypropyl)-4-(2-hexadienyl)-2(5H)-furanone produced, 234; good biocontrol possible against some fungal pathogens, 234; biocontrol of *Sclerotinia sclerotiorum,* 278
*Trichoderma harzianum* (092): in biocontrol of basal rot of *Narcissus* caused by *Fusarium oxysporum* f. sp. *narcissi,* 263–64
*Trichoderma hamatum* (094): in biocontrol of basal rot of *Narcissi* caused by *Fusarium oxysporum* f. sp. *narcissi,* 263–64
*Trichoderma hamatum* (TRI-4): effective antagonist to *Rhizoctonia solani* under various conditions, 205
*Trichoderma hamatum* (TRI-4, TM-23) and *Gliocladium virens* (Gl-21): in biocontrol of fruit rot of tomato caused by *R. solani,* 254–55
*Trichoderma hamatum* (TRI-4, Tm-23, 31–3): biocontrol of *Rhizoctonia solani* in cotton, tomatoes, snap beans, potatoes, beets, 242–44
*Trichoderma harzianum* (T-35): good biocontrol of *F. oxysporum* f. sp. *radicis lycopersici* on tomato, 252
*Trichoderma harzianum* (Thl: T35, T12, T95, Wt, WT6): autoclaved and unautoclaved, gave good biocontrol of *R. solani* causative fungus for damping-off of lettuce, 247–48; unautoclaved activity indicated role of antibiotics, 248
*Trichoderma harzianum* (WT): biological control of damping-off of lettuce caused by *P. ultimum,* 246
*Trichoderma harzianum* (WT-6-24, Th-23-R9, Th-58): biocontrol of *Rhizoctonia solani* from cotton, snap beans, potatoes, beets, 242–44
*Trichoderma harzianum* (WU71): produces two pyrone antibiotics, superb biocontrol of take-all of wheat, 300
*Trichoderma koningii:* good in biocontrol of take-all on both wheat and ryegrass, 252
*Trichoderma koningii* (IMI334839): good in biocontrol of web-blight of cowpea caused by *Rhizoctonia solani,* 309
*Trichoderma koningii* (not strain 7a): produced new antibiotic inhibitory to some plant fungal pathogens, 233
*Trichoderma koningii* (strain 7a, IMI No. 308475): antibiotic 6-n-pentyl-2-H-pyran-2-one produced, inhibitory to several fungal pathogens, 233
*Trichoderma koningii* (T12-R33-D25 mutant): in biocontrol of root rot of adzuki bean caused by *Rhizoctonia solani,* 271–72
*Trichoderma polysporum:* antibiotics identified, 272; biocontrol of *Fomes annosus,* a pathogenic fungus, 272
*Trichoderma reesii* (P-12, mutant of *T. reesei* QM9414): antibiotics produced, 234; antagonistic to 9 test fungal pathogens, 234
*Trichoderma* spp.: counts on wheat and *Lolium rigidum* in Australia, 251; biocontrol of *Sclerotium rolfsii, Fusarium solani, F. equiseti,* and *F. oxysporum* on tomato, 252, 254; changed germination percentage of tomato seeds, 252–54
*Trichoderma* spp. (T-68, Gh-2): overgrew *Fusarium* wilt of *Cucumis melo* and Pima cotton, S5, 215
*Trichoderma virens* (TW. 055): good biocontrol of phytophthora crown and root rot of apple trees, 259
*Trichoderma viride:* phytotoxic effects, 80; good in control of *Penicillium digitatum,* 224
*Trichoderma viride* (IMI29375): in biocontrol of damping-off of lettuce caused by *P. ultimum,* 246
*Trichoderma viride* (O89L): in biocontrol of basal rot of *Narcissus* caused by *Fusarium oxysporum* f. sp. *narcissi,* 263–64
*Trichoderma viride* (T-1-9, TS-1-R3): in biocontrol of *Rhizoctonia solani* from cotton, tomatoes, snap beans, potatoes, beets, 242–44
*Trichodesma sedgwickianum:* allelopathic effects, 75–76
*Tridax procumbens:* volatile allelochemicals in leaves, 64–65
*Trifolium alexandrinum. See* Berseem clover
*Trifolium incarnatum. See* Crimson clover
*Trifolium pratense. See* Red clover
*Trifolium repens. See* White clover
*Trifolium subterraneum. See* Subterranean clover
*Trillium grandiflorum:* allelopathic to *Candida albicans,* 104; two saponin glycosides identified, 104
*Triticum aestivum. See* Wheat
*Tsuga canadensis:* aqueous leachates allelopathic to trees and herbs in forest, 364
*Typhula phacorrhiza* (T016): antagonist to *Typhula ishikariensis* and *T. incarnata,* 193–94

*Urgenia indica:* allelopathic effects, 76
*Uromyces heliotropii* (rust): biocontrol of *Heliotropium europaeum,* 147

Velvetleaf: allelopathy by seeds, 38; residue allelopathy, 39; seed flavonoids, 39–40
*Verticillium biguttatum:* biocontrol of *Rhizoctonia solani* affected by temperature and humidity, 288–89
*Vicia faba. See* Broad bean
*Vicia villosa. See* Hairy vetch
*Vitex negundo:* leachates allelopathic to some grasses and shrubs, 364–65
Vulgamycin (enterocin): produced by *Streptomyces hygroscopicus* and *S. candidus,* herbicidal to all weeds tested, 156

Water extracts of plants: stimulatory to *Bradyrhizobium japonicum* growth, 92
Wheat: allelopathic effects of straw, 32–37, 119, 127–28; aerobic and anaerobic decomposition of straw, 33; mixtures of cultivars vs. pure stands, 33; important allelochemicals, 33–36, 127–28; tolerance of cotton cultivars to wheat straw, 37; wheat residue in biological weed control, 119, 127–28, 132, 155, 367–69; biocontrol of take-all may be improved by applying antagonist to soil before wheat planting, 301
White clover: allelopathic effects, 13; treatment with *Rhizoctonia solani* and $O_3$ caused more stress with addition of fescue leachate, 172
*Wilcoxina mikolae:* decreased shoot mass of lodgepole pine, 82; with *Bacillus polymyxa,* increased shoot mass, 82–83
Wild oats: allelopathic effects and phenolics produced, 47
Wormwood: allelopathic effects, 46–47; volatile allelochemicals, 47

*Xanthium strumarium:* allelopathic weed, 76; phenols in shoot extracts and rain leachates, 76
*Xanthomonas campestris* pv. *vignicola,* antagonists of, 309–10
*Xanthomonas oryzae* pv. *oryzae* (nonpathogenic): bacteriocin producers, 268; biocontrol of bacterial blight of rice caused by *X. oryzae* pv. *oryzae* (pathogen), 268–69
X *Triticosecale. See* Spring triticale

Yellow foxtail: allelopathic effects, 43

*Zanthoxylum alatum:* fruits allelopathic to *Aspergillus flavus* and *A. parasiticus,* 104; toxic oil isolated, 104
*Zea mays. See* Corn
*Zingiber officinale:* allelochemicals from rhizomes, 191; complete inhibition of *Pyricularia oryzae* hyphae, 191
*Zygorrhynchus moelleri:* antibiotic control of fungal pathogens, 235